MULTI-CARRIER SPREAD-SPECTRUM

Multi-Carrier Spread-Spectrum

Proceedings from the
5th International Workshop,
Oberpfaffenhofen, Germany,
September 14–16, 2005

Edited by

KHALED FAZEL

*Marconi Communications GmbH,
Backnang, Germany*

and

STEFAN KAISER

*DoCoMo Communications Laboratories Europe GmbH,
Germany*

 Springer

A C.I.P. Catalogue record for this book is available from the Library of Congress.

ISBN-10 1-4020-4435-6 (HB)
ISBN-13 978-1-4020-4435-9 (HB)
ISBN-10 1-4020-4437-2 (e-book)
ISBN-13 978-1-4020-4437-3 (e-book)

Published by Springer,
P.O. Box 17, 3300 AA Dordrecht, The Netherlands.

www.springer.com

Printed on acid-free paper

All Rights Reserved
© 2006 Springer
No part of this work may be reproduced, stored in a retrieval system, or transmitted
in any form or by any means, electronic, mechanical, photocopying, microfilming, recording
or otherwise, without written permission from the Publisher, with the exception
of any material supplied specifically for the purpose of being entered
and executed on a computer system, for exclusive use by the purchaser of the work.

Printed in the Netherlands.

TABLE OF CONTENTS

Editorial Introduction   xi

Acknowledgement   xvii

I   GENERAL ISSUES

Experiments on Space Diversity Effect in MIMO Channel Transmission with Maximum Data Rate of 1 Gbps in Downlink OFDM Radio Access   3
K. Higuchi, H. Taoka, N. Maeda, M. Sawahashi (NTT DoCoMo, Japan)

Reduced-complexity Maximum-Likelihood Detection in Multiple-antenna-aided Multicarrier Systems   21
J. Akhtman, L. Hanzo (University of Southampton, UK)

On Coding for OFDM and the Broadcast Channel   29
M. Bossert (University of Ulm, Germany)

Adaptivity in MC-CDMA Systems   45
I. Cosovic (German Aerospace Center (DLR), Germany), S. Kaiser (DoCoMo EuroLabs, Germany)

Parameter Estimation for Interference Cancellation in Power-controlled MC-CDMA Downlink Transmissions   55
M. Morelli, L. Sanguinetti, U. Mengali (University of Pisa, Italy)

Performance Evaluation and Parameter Optimization of MC-CDMA   67
M. Guenach, H. Steendam (Ghent University, Belgium)

A Novel Algorithm of Inter-subchannel Interference Cancellation for OFDM Systems   75
M.-X. Chang (National Cheng-Kung University, Taiwan)

Comparative Multicarrier System Performance in Jamming and Fading   83
D. W. Matolak, W. Xiong, I. Sen (Ohio University, USA)

CDMA in the Context of OFDM and SC/FDE - A Comparative Study   91
H. Witschnig (Philips, Austria), M. Huemer (University of Erlangen, Germany), R. Stuhlberger, A. Springer (University of Linz, Austria)

II   CELLULAR ASPECTS

Impact of the Intercell Interference in DL MC-CDMA Systems   101
X. G. Doukopoulos, R. Legouable (France Telecom R&D, France)

Quantifying the Impact of the MC-CDMA Physical Layer Algorithms
on the Downlink Capacity in a Multi-cellular Environment 109
A.-M. Mourad, A. Guéguen (Mitsubishi Electric ITE-TCL, France), R. Pyndiah
(Ecole Nationale Supérieure des Télécommunications de Bretagne, France)

Radio Resource Management for MC-CDMA over Correlated
Rayleigh Fading Channels 119
S. Plass, A. Dammann (German Aerospace Center (DLR), Germany)

Throughput of Heterogeneous Multi-Cell Multi-User MIMO-OFDM
Systems 127
B. Mielczarek, W. A. Krzymień (TRLabs/University of Alberta, Canada)

Real-link Performance of a SS-MC-MA High Frequency Radio Modem 137
H. Santana-Sosa (Universidad des Las Palmas de Gran Canaria, Spain),
I. Raos, S. Zazo-Bello (Universidad Politecnica de Madrid, Spain), I. A. Pérez-Álvarez,
J. López-Pérez (Universida des Las Palmas de Gran Canaria, Spain)

III   SPREADING AND DETECTION

Analysis of Iterative Successive Interference Cancellation in SC-CDMA
Systems 147
P. Weitkemper, V. Kühn, K.-D. Kammeyer (University of Bremen, Germany)

Bandwidth and Power Efficient Digital Transmission using Sets of Orthogonal
Spreading Codes 157
W. G. Teich, P. Kaim, J. Lindner (University of Ulm, Germany)

Simplified Realization of Pseudo-Orthogonal Carrier Interferometry OFDM by FFT
Algorithm 167
K. Anwar, M. Saito, T. Hara, M. Okada, H. Yamamoto (Nara Institute of Science
and Technology, Japan)

Exploiting Rotated Spreading Features for CDM-OFDMA 175
R. Raulefs, A. Dammann, S. Ayaz (German Aerospace Center (DLR), Germany)

Detection of Pre-coded OFDM by Recovering Symbols on Degraded
Carriers 183
S. Tamura, M. Fujii, M. Itami, K. Itoh (Tokyo University of Science, Japan)

Semi-blind Multiuser Detection in Zero-padded OFDM-CDMA Systems 191
R. Boloix-Tortosa, F. J. Payan-Somet, J. J. Murillo-Fuentes (Universidad de
Sevilla, Spain)

A Study on Adaptive Successive Detection using M Algorithm based on ML Criterion for Down-link MC-CDMA Systems   199
Y. Morishige, M. Fujii, M. Itami, K. Itoh (Tokyo University of Science, Japan)

IV CHANNEL ESTIMATION

Channel Estimation in the Presence of Timing Offsets for MC-CDMA Uplink Transmissions with Combined Equalization   211
T. Mazzoni, L. Sanguinetti, M. Morelli (University of Pisa, Italy)

Iterative Channel Estimation for MIMO MC-CDMA   221
S. Sand, R. Raulefs, A. Dammann (German Aerospace Center (DLR), Germany)

Performance Investigation of Improved Channel Estimation Exploiting Long Term Channel Properties   231
T. Weber, I. Maniatis, M. Meurer (Technical University of Kaiserslautern, Germany), W. Zirwas (Siemens, Germany)

Pilot Symbol-Aided Channel Estimation for MC-CDMA Systems   239
Y. Lee, Y. Kang, H. Park (Information and Communications University, Korea), M. Noh (LG Electronics, Korea)

Superimposed Pilot-based Channel Estimation for MIMO OFDM Code Division Multiplexing Uplink Systems   247
L. Cariou, J.-F. Hélard (Electronics and Telecommunications Institute of Rennes (IETR) INSA, France)

Pilot Time-Frequency Location Adjustment in OFDM Systems Based on the Channel Variability Parameters   257
F. Bader (Centre Tecnologic de Telecomunicacions de Catalunya (CTTC), Spain), R. Gonzalez (Universitat Politécnica de Catalunya, Spain)

V MIMO AND ADAPTIVITY

Reduced Feedback Closed-Loop Spatial Multiplexing for B3G Systems   267
R. Malik, T. P. Yew (Panasonic Singapore Laboratories, Singapore)

A MIMO-OFDM Transmission Scheme Employing Subcarrier Phase Hopping   275
S. Suyama, K. Tochihara, H. Suzuki, K. Fukawa (Tokyo Institute of Technology, Japan)

Multicarrier SDMA System with Reduced Intra-user Cross-correlations   283
N. H. Dawod, R. Hafez, I. Marsland (Carleton University, Canada)

Effect of Adaptive Modulation and Error Correction on Multi-Band OFDM-
MIMO System 291
M. Ohkawa (Next-generation Laser Communication Satellite Technology Research Center, Japan), R. Kohno (Yokohama National University, Japan)

Near Optimal Performance for High Data Rate MIMO MC-CDMA Scheme 303
P.-J. Bouvet, M. Hélard (France Telecom R&D, France)

STBC-TCM for MC-CDMA Systems with SOVA-based Decoding and Soft-
interference Cancellation 311
L. A. Paredes Hernández, M. García Otero (Universidad Politécnica de Madrid, Spain)

OFDMA with Subcarrier Sharing 319
S. Pfletschinger, F. Bader (Centre Tecnològic de Telecomunicacions de Catalunya (CTTC), Spain)

New Loading Algorithms for Adaptive SS-MC-MA Systems over Power Line
Channels: Comparisons with DMT 327
M. Crussière, J.-Y. Baudais, J.-F. Hélard (Electronics and Tele-communications Institute of Rennes (IETR) INSA, France)

An Investigation of Optimal Solution for Multiuser Sub-carrier Allocation in
OFDMA Systems 337
Y. Peng, S. Armour, A. Doufexi, J. McGeehan (University of Bristol, UK)

Optimal Solution to Adaptive Subcarrier-and-bit Allocation in Multiclass Multiuser
OFDM System 345
K. Zhou, Y. H. Chew, Y. Wu (National University of Singapore, Singapore)

Dynamic and Scalable Bandwidth Allocation for Beyond 3G CDMA
Systems 353
M.-H. Fong (Nortel Networks/University of Victoria, Canada), G. Wu, W. Tong, J. Li (Nortel Networks, Canada), T. A. Gulliver (University of Victoria, Canada), V. K. Bhargava (University of British Columbia, Canada)

Effective SINR Mapping for an MC-CDMA System 361
R. Elliott (TRLabs/University of Alberta, Canada), A. Arkhipov, R. Raulefs (German Aerospace Center (DLR), Germany), W.A. Krzymień (TRLabs / University of Alberta, Canada)

Combination of H-ARQ and Iterative Multi-user Detection for OFDMA-
CDM 371
A. Arkhipov, R. Raulefs, M. Schnell (German Aerospace Center (DLR), Germany)

Table of Contents

VI   SYSTEM PERFORMANCE AND IMPLEMENTATION ASPECTS

A Comparative Analysis of CDM-OFDMA and MC-CDMA Systems   385
W. Zhang, J. Lindner (University of Ulm, Germany)

Multi-User Transmit Power Control for Multi-carrier Modulation Systems in Quasi-synchronous Uplink Channel   393
M. Fujii, M. Itami, K. Itoh (Tokyo University of Science, Japan)

On the Compensation of IQ Imbalance in an SC/FDE System   401
C. Wicpalek (University of Linz, Austria), H. Witschnig (Philips, Austria), A. Springer (University of Linz, Austria)

On Implementation Aspects of Uplink MC-CDMA Multi-user Detection   409
A. Happonen (Nokia Technology Platforms, Finland), F. Bauer (Nokia Research Center, Germany), A. Burian (Nokia Research Center, Finland)

Analytical Performance of a Frequency Offset Multi-user Multi-array System   417
A. Renoult (Univ de Cergy-Pontoise/Thales Communications, France), I. Fijalkow (Univ de Cergy-Pontoise, France), M. Chenu-Tournier (Thales Communications, France)

Effects of Subcarrier Interleaving on LDPC Coded MC-CDMA Systems   425
Y. Lee, K. Kim, H. Park (Information and Communications University, Korea), D. Kwon (ETRI, Korea),

One-shot and Iterative Symbol Predistortion Techniques for PAPR Reduction in OFDM Systems   433
S. Sezginer, H. Sari (Ecole Supérieure d'Electricité (SUPELEC), France)

Iterative Correction and Decoding of OFDM Signals Affected by Clipping   443
W. Rave, P. Zillmann, G. Fettweis (Dresden University of Technology, Germany)

PAPR Reduction of MC-CDMA Signals by Selected Mapping with Interleavers   453
M. Saito, A. Okuda, M. Okada, H. Yamamoto (Nara Institute of Science and Technology, Japan)

Iterative Nonlinear Channel Compensation in MC-CDMA Systems   461
V. Lottici, F. Giannetti (University of Pisa, Italy)

Sidelobe Suppression in OFDM Systems   473
I. Cosovic (German Aerospace Center (DLR), Germany), V. Janardhanam (Munich University of Technology, Germany)

Blind Phase Noise Estimation in OFDM Systems by Sequential Monte Carlo
Method 483
E. Panayirci (Bilkent University, Turkey), H. A. Çirpan (Istanbul University, Turkey), M. Moeneclaey, N. Noels (Ghent University, Belgium)

Sensitivity Comparison of Multi-carrier and Spread Spectrum Systems to
Phase Noise 491
C. Garnier, M. Loosvelt, Y. Delignon (GET/INT/ENIC, France)

*Invited paper

EDITORIAL INTRODUCTION

Khaled Fazel
Radio System Design
Marconi Communications
D-71522 Backnang, Germany

Stefan Kaiser
DoCoMo Euro-Labs
Landsberger Strasse 312
D-80687 Munich, Germany

The field of multi-carrier and spread spectrum communications has became an important research topic with increasing number of research activities [1]. Especially in the last two years, beside deep system analysis of various multiple access schemes, new standardization activities in the framework of beyond 3G (B3G) concepts have been initiated. Multi-carrier transmission is considered to be a potential candidate to fulfil the requirements of the next generation system. The two important requirements of B3G/4G can be summarized as: i) much higher data rate for cellular mobile radio and ii) a unique physical layer specification for indoor/hot spot and outdoor/cellular applications, including fixed wireless access (FWA) schemes. The activities within the 3GPP and WiMAX fora are examples of such trends (see Fig. 1).

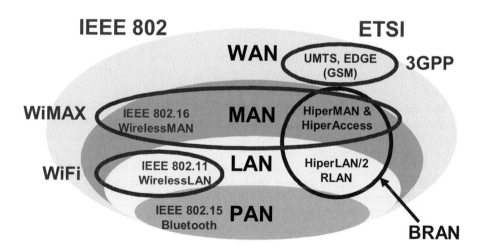

Figure 1 Beyond 3G: *Worldwide Standardization Activities*

The WiMAX (Worldwide Interoperability for Microwave Access [2]) vision is to provide broadband wireless access with its primary goal to promote IEEE 802.16a-e and ETSI-BRAN standards through interoperability testing and certification. In the first step the broadband access to the so-called last mile applications with fixed positioned terminals is envisaged. Further steps will cover portability and in long term even mobility. Note that OFDM/OFMA techniques with multiple antennas will play an important rule to cover the non line-of-sight reception conditions.

Meanwhile, in order to fulfil the long term strategy of 3GPP (third generation partnership project) with the targets [3]
- increased peak data rates (100/50 Mbps DL/UL),
- scalable and increased user throughput,
- improved spectrum efficiency (e.g., DL up to 5 bit/s/Hz),
- improved latency (user plan: below 10 ms),
- scaleable bandwidth (from 1.5 MHz up to 20 MHz),
- reduced CAPEX and OPEX,
- compatibility with earlier releases and with other systems and
- optimised for low mobile speed but supporting high mobile speed,

multi-carrier transmission is considered to be a potential candidate for the physical layer.

Adaptivity and trade-off between coverage, data rate and mobility with unique air interface architecture will be the key features for the success of next generation wireless systems. Fixed receivers with short coverage distance (pico cells) and ideal channel condition shall be able to receive the highest data rate, whereas subscribers with high mobility conditions and large coverage area (macro cells) shall be able to receive the necessary data rate to establish the required communication link.

Multi-carrier and spread-spectrum systems with their generic air interface and adaptive technologies will be considered as potential candidate to fulfill the above mentioned requirements of B3G/4G systems [1].

SCOPE OF THIS ISSUE

The aim of this issue, consisting of six parts, is to edit the ensemble of contributions presented during the three days of the 5th International Workshop on Multi-Carrier Spread-Spectrum (MC-SS 2005), held from September 14-16, 2005 in Oberpfaffenhofen, Germany.

The first part is devoted to *general issues* of MC-SS. First, Higuchi *et al.* give an overview on the 1 Gpbs transmission experiences at NTT DoCoMo with space diversity effects in MIMO channels. Their concept to achieve a tradeoff between data rate and coverage is based on using variable spreading in frequency and time domain.

Akthman and Hanzo analyze the performance of near-maximum-likelihood detection in multiple-antenna-aided multi-carrier systems that allow a high reduction of the system complexity. The idea of superposition of information as it was proposed by T. Cover for broadcast channel to offer a higher channel capacity is treated by Bossert with a new scope of application, which is the downlink of a cellular system. Here, the superposition of information is performed by using multi-level coding. The notion of adaptivity in MC-CDMA systems is studied by Cosovic and Kaiser. The estimation of the parameters needed for interference cancellation in a power-controlled MC-CDMA downlink transmission is analyzed by Morelli *et al.* Guenach and Steendam evaluate the performance and parameter optimization of MC-CDMA systems. An efficient low-complexity algorithm of OFDM inter-channel interference is proposed by Chang. A performance comparison of multi-carrier systems in the presence of jamming and fading is done by Matolak *et al.* Finally, a detailed comparison of CDMA in the context of OFDM and SC/FDE is performed by Witschnig *et al.*

The second part of this issue is devoted to *cellular aspects*. The impact of intercell interference in a downlink of MC-CDMA systems is presented by Doukopoulos and Legouable. Quantification of the impact of the physical layer based on the downlink capacity of a multi-cellular environment is performed by Mourad *et al.* Strategies for the management of the radio resource for an MC-CDMA system over correlated fading channels is presented by Plass and Dammann. The analysis of the throughput of heterogeneous multi-cell multi-user MIMO-OFDM systems is done by Mielczarek and Krzymien. Finally, a performance analysis of a real HF radio link based on spread-spectrum multi-carrier multiple access (SS-MC-MA) modem is done by Santana-Sosa *et al.*

The *spreading and detection* aspects are discussed in the third part of this issue. First, Weitkemper *et al.* analyze the performance of a successive iterative interference cancellation algorithm for a single carrier spread spectrum system. A bandwidth and power efficient digital transmission scheme using sets of orthogonal codes is presented by Teich *et al.* A simplified realization of a pseudo-orthogonal carrier interferometry OFDM by FFT algorithm is proposed by Anwar *et al.* The performance of a CDM-OFDMA system exploiting the uniqueness of rotated spreading is given by Raulefs *et al.* Tamura *et al.* analyze the detection of pre-coded OFDM by symbol recovering on degraded carriers. A semi-blind multiuser detection algorithm in a zero padded OFDM-CDMA scheme is proposed by Boloix-Tortosa *et al.* At the end of this section a study on adaptive successive detection algorithm using the M algorithm based on ML criterion for downlink MC-CDMA systems is proposed by Morishige *et al.*

The fourth part is devoted to *channel estimation*. Mazzoni *et al.* present an uplink channel estimation for MC-CDMA systems with timing offsets. An iterative channel estimation algorithm for MIMO MC-CDMA systems is proposed by Sand *et al.* The performance investigation of an improved channel estimation algorithm exploiting the long term channel properties is done by Weber *et al.* Space-frequency coding and signal processing for downlink MIMO MC-CDMA is analyzed by Lee *et al.* Cariou and Helard present a superimposed pilot-based channel estimation algorithm for

MIMO OFDM code division multiplexing uplink systems. Finally, Bader and Gonzales propose a location adjustment algorithm of pilot symbols in time and frequency in OFDM systems based on the channel variability parameters.

The fifth part assembles all issues related to ***MIMO and adaptivity*** techniques. A receiver architecture for closed-loop spatial multiplexing applied in a B3G system is presented by Malik and Yew. The performance of a MIMO-OFDM transmission scheme employing sub-carrier phase hopping is analyzed by Suyama *et al*. A multi-carrier SDMA system with reduced intra-user cross-correlations is studied by Dawod *et al*. In relation to adaptivity, the paper by Ohkawa and Kohno discusses the effect of adaptive modulation and forward error correction coding on multi-band OFDM-MIMO systems. Near optimal performance for high data rate MIMO MC-CDMA schemes is presented by Bouvet and Helard. The STBC-TCM for MC-CDMA systems with SOVA-based decoding and soft-interference cancellation method is proposed by Paredes Hernandez and Garcia Otero. An algorithm for an adaptive assignment of subcarriers and spreading codes for throughput maximization in a multi-user MC-SS system is presented by Pfletschinger and Bader. Crussiere *et al.* present a new loading algorithm for adaptive SS-MC-MA systems over power line channels. The system performance is compared with DMT. An investigation of an optimal solution for multiuser sub-carrier allocation in OFDMA systems is done by Peng *et al*. In a similar way an optimal solution to adapt sub-carrier-and-bit allocation in multi-class multiuser OFDM systems is presented by Zhou *et al.* A dynamic and scalable bandwidth allocation algorithm for beyond 3G CDMA systems is proposed by Fong *et al*. Elliott *et al.* analyze the performance of an effective SINR mapping for an MC-CDMA system. Finally, a combination of hybrid ARQ and iterative multi-user detection for OFDMA-CDM is presented by Arkhipov *et al.*

The last part of this book is devoted to ***system performance and implementation aspects***. A detailed comparative performance analysis of CDM-OFDMA and MC-CDMA systems is presented by Zhang and Lindner. The performance of a multi-user transmit power control algorithm for multi-carrier modulation systems in a quasi-synchronous uplink channel is analysed by Fujii *et al*. Wicpalek *et al.* present the compensation of IQ imbalance in a single carrier system with frequency domain equalization. The implementation aspects of an uplink MC-CDMA multi-user detector are presented by Happonen *et al.* The analytical performance of a frequency offset multi-user multi-array system is studied by Renoult *et al.* The effects of subcarrier interleaving on low density parity check coding is studied by Lee *et al*. Regarding combating the effect of non-linearity, Sezginer and Sari analyse one-shot and iterative symbol predistortion techniques for peak-to-average power reduction (PAPR) in OFDM systems. On a similar topic, Rave *et al.* present an iterative correction and decoding scheme for OFDM transmission systems affected by clipping. The reduction of PAPR in MC-CDMA signals by selected mapping with interleavers is presented by Saito *et al*. An iterative nonlinear channel compensation algorithm in MC-CDMA systems is analysed by Lottici and Giannetti. To gain more channel bandwidth in a crowded spectrum allocation, Cosovic and Janardhanam propose a technique for sidelobe suppression in an OFDM system. Blind phase noise estimation in OFDM systems by sequential Monte Carlo method is presented by Panayirci *et al.* A

Editorial Introduction

sensitivity comparison of multi-carrier and spread spectrum systems to phase noise is finally analyzed by Garnier *et al.*

In conclusion, we would like to thank all the authors who have contributed to this issue and all those in general who responded enthusiastically to the call. We hope that this edited book may serve to promote further research in this area and with that can contribute to the success of the next generation wireless technology.

REFERENCES

[1]   Fazel K., Kaiser S., "Multi-Carrier and Spread Spectrum Systems", John Wiley and Sons Ltd., Sept. 2003.
[2]   WiMAX Forum (www.wimax-forum.com).
[3]   3GPP, Technical Report TR 25.913, ETSI- Sophia Antipolis.

ACKNOWLEDGEMENT

The editors wish to express their sincere thanks for the support of the chairmen of the different sessions of the workshop, namely Prof. M. Bossert from University of Ulm, Prof. H. Cirpan from Istanbul University, Dr. U.-C. Fiebig from DLR, Prof. L. Hanzo from University of Southampton, Prof. K.-D. Kammeyer from University of Bremen, Prof. W. A. Krzymien from University of Alberta/TRLabs, Prof. J. Lindner from University of Ulm, Prof. M. Morelli from University of Pisa, and Prof. H. Steendam from University of Ghent. Many thanks to the invited speaker Prof. H. Berndt, CTO and Vice President of DoCoMo Euro-Labs, and to all authors for their contribution to the success of the workshop. Furthermore, many thanks to Mr. S. Plass, Ms. J. Uelner and Ms. Ch. Burger from DLR for their active support in the technical and local organization of the workshop.

This 5th International Workshop on Multi-Carrier Spread-Spectrum could not have become a success without the

- assistance of the *TPC members*

P. W. Baier (Germany)   K.-D. Kammeyer (Germany)   S. Pasupathy (Canada)
Y. Bar-Ness (USA)   W. Koch (Germany)   R. Prasad (Denmark)
K. Fazel (Germany)   W. A. Krzymien (Canada)   M. Renfors (Finland)
G. Fettweis (Germany)   J. Lindner (Germany)   H. Rohling (Germany)
G. B. Giannakis (USA)   U. Mengali (Italy)   H. Sari (France)
J. Hagenauer (Germany)   L. B. Milstein (USA)   M. Sawahashi (Japan)
S. Hara (Japan)   M. Moeneclaey (Belgium)   R. E. Ziemer (USA)
H. Imai (Japan)   W. Mohr (Germany)
S. Kaiser (Germany)   M. Nakagawa (Japan)

- technical and financial support of

   German Aerospace Center (DLR)
   DoCoMo Communications Laboratories Europe GmbH
   Marconi Communications GmbH

- technical support of

   IEEE Communication Society
   Information Technology Society (ITG) within VDE
   NEWCOM

Section I

GENERAL ISSUES

EXPERIMENTS ON SPACE DIVERSITY EFFECT IN MIMO CHANNEL TRANSMISSION WITH MAXIMUM DATA RATE OF 1 GBPS IN DOWNLINK OFDM RADIO ACCESS

Kenichi Higuchi, Hidekazu Taoka, Noriyuki Maeda, and Mamoru Sawahashi
IP Radio Network Development Department, NTT DoCoMo, Inc., 3-5, Hikari-no-oka, Yokosuka-shi, Kanagawa, 239-8536 JAPAN

Abstract: This paper presents experimental results on the space diversity effect in MIMO multiplexing/diversity with the target data rate up to 1 Gbps using OFDM radio access based on laboratory and field experiments including realistic impairments using the implemented MIMO transceivers with the maximum of four transmitter/receiver branches. The experimental results using multipath fading simulators show that at the frequency efficiency of less than approximately 2 bits/second/Hz, MIMO diversity using the space-time block code (STBC) increases the measured throughput compared to MIMO multiplexing owing to the high transmission space diversity effect. At a higher frequency efficiency than approximately 2 to 3 bits/second/Hz, however, MIMO multiplexing exhibits performance superior to that of MIMO diversity since the impairments using higher data modulation and a higher channel coding rate in MIMO diversity overcomes the space diversity effect. The results also show that the receiver space diversity effect is very effective in MIMO multiplexing for maximum likelihood detection employing QR-decomposition and the M-algorithm (QRM-MLD) signal detection. Finally, we show that the real-time throughput of 500 Mbps and 1 Gbps in a 100-MHz transmission bandwidth is achieved at the average received E_b/N_0 per receiver antenna of approximately 8.0 and 14.0 dB using 16QAM modulation and Turbo coding with the coding rate of 1/2 and 8/9, respectively, in 4-by-4 MIMO multiplexing in a real propagation environment.

Key words: VSF-Spread OFDM, MIMO multiplexing, MIMO diversity, QRM-MLD, Field experiments

For future mobile communication systems beyond the 3G system, an all packet-based highly-efficient radio access scheme is necessary with a short delay, i.e., low latency, and with high affinity to IP-based core networks. Among the requirements for radio access, the supportable data rate is the most directly related to providing customer services. In Recommendation ITU-R M.1645, the maximum data rate supported in the new mobile access scheme is defined as 100 Mbps and that in the new nomadic/local area wireless access scheme is greater than 1 Gbps. Although two different target data rates in the respective radio environments are defined in the recommendation, our concept is to support these two data rate requirements using the same radio access, i.e., the same air interface, by only changing the radio parameters. This concept is motivated by the desire to achieve seamless support for various radio environments using one radio access network deployment at low cost [1]. We previously presented field experimental results on the measured throughput of greater than 100 Mbps using Variable Spreading Factor (VSF)-Orthogonal Frequency and Code Division Multiplexing (OFCDM) transceivers employing a 100-MHz channel bandwidth in real propagation environments [2]. Furthermore, the effectiveness of key techniques relevant to packet access such as adaptive modulation and channel coding (AMC) and hybrid ARQ with packet combining for broadband radio access with a 100-MHz channel bandwidth was clarified. On the other hand, to achieve extremely high-speed data rates such as 1 Gbps, multiple-input multiple-output (MIMO) channel transmission is very beneficial because it takes advantage of space division multiplexing (SDM) using multiple transmitter and receiver branches [3, 4]. MIMO channel transmissions are categorized into MIMO multiplexing and MIMO diversity such as space time block coding (STBC) and space time trellis coding (STTC), e.g. [5, 6]. The former and latter schemes are very beneficial to enhancing the achievable data rate and providing the space diversity effect, respectively. We reported experimental results on real-time 1-Gbps high-speed packet transmission (corresponding frequency efficiency of 10 bits/second/Hz) employing the implemented 4-by-4 MIMO multiplexing transceiver in laboratory experiments using hardware fading simulators [7]. In MIMO channel transmissions, however, the optimum transmission scheme is different according to mainly the target data rate assuming the same number of transmission antenna branches.

This paper presents laboratory and field experimental results on the space diversity effect in MIMO channel transmissions using VSF-Spread OFDM radio access achieving the maximum data rate of 1 Gbps with a

100-MHz signal bandwidth. More specifically, in the paper, we investigate the optimum MIMO transmission scheme according to the target data rate of up to a maximum of 1 Gbps, i.e., frequency efficiency of 10 bits/second/ Hz, from the experimental approach including realistic impairments using the implemented MIMO transceivers with the maximum of four trans mitter/receiver antenna branches. The implemented transceiver can generate MIMO multiplexing or MIMO diversity (STBC based on the Alamouti algorithm [5]) transmitted signals and maximum likelihood detection (MLD) [8] based signal detection. The radio parameters such as the carrier frequency, channel bandwidth, and sub-carrier spacing of the implemented MIMO transceivers are identical to those of the transceivers with which we previously attained the throughput of greater than 100 Mbps in field experiments according to our proposed radio access concept. In the implemented MIMO transmitter/receiver, the data modulation is QPSK, 16QAM, or 64QAM (up to two transmitter branches), and Turbo coding with the rate of $R = 1/2 - 8/9$ is employed together with soft-decision decoding. The maximum information bit rate achieved by 4-by-4 MIMO multiplexing is greater than 1 Gbps associated with 16QAM data modulation and Turbo coding with $R = 8/9$. The rest of the paper is organized as follows. Section 2 describes the configuration of the implemented MIMO multiplexing/diversity transmitter and receiver. Then, laboratory experimental results focusing on the space diversity effect are given in Section 3. Finally, Section 4 presents field experimental results that show the real-time 1-Gbps packet transmission in a real propagation environment.

2. CONFIGURATION OF IMPLEMENTED MIMO TRANSCEIVER

2.1 Structure of overall MIMO channel transmitter and receiver

The implemented MIMO transmitter and receiver configurations are illustrated in Figs. 1(a) and 1(b), respectively, and they employ the major radio link parameters given in Table 1. The transmission bandwidth and the number of sub-carriers of the OFDM signal are 101.5 MHz and 768, respectively (thus, the sub-carrier separation is 131.836 kHz). The length of the packet frame is 0.5 msec, which contains 54 OFDM symbols as shown in Fig. 2. In the base station (BS) transmitter (we focus on the downlink) binary information data bits are channel-encoded followed by data modulation mapping according to the MIMO channel transmission scheme, i.e., either MIMO multiplexing or MIMO diversity. We employed Turbo coding with the coding rate of $R = 1/2$, $2/3$, or $8/9$ and with the constraint length of four bits. The employed data modulation schemes are QPSK, 16QAM, and 64QAM (only for the case where the number of transmitter antennas is less than $N_{Tx} = 2$). After data modulation mapping, symbol interleaving is performed to randomize the burst errors due to frequency-selective fading in the frequency domain over 768 sub-carriers. The contiguous four pilot symbols used for channel estimation are time-multiplexed within a packet frame. Then, each encoded data sequence is converted into an OFDM signal with 768 sub-carriers using the Inverse fast Fourier transform (IFFT) with 1024 points, and a cyclic prefix (CP) (1.674 µsec) is appended at the beginning of each OFDM symbol (7.585 µsec). The resultant sub-carrier spacing is 131.836 kHz. After conversion into baseband in-phase (I) and quadrature (Q) components by digital-to-analog (D/A) converters, quadrature-modulation is performed. Finally, the IF modulated signal is up-converted into the RF signal and amplified by the power amplifier where the center carrier frequency is 4.635 GHz.

At the mobile station (MS) receiver, the frequency down-converted IF signal is first linearly amplified by an automatic gain control (AGC) amplifier. The received spread signal is converted into baseband I and Q components by a quadrature detector. The I and Q signals are converted into digital format by the analog-to-digital (A/D) converters. The OFDM symbol timing is estimated by taking the maximum cross-correlation peak between the received baseband digital signal and a reference signal such as a pilot symbol (this symbol timing is updated every 0.5 msec). After the CP is removed, 768 parallel data sequences are de-multiplexed by FFT processing from the multi-carrier signal with 768 sub-carriers. The channel gain of each packet frame at each sub-carrier is estimated using a two-dimensional multi-slot and sub-carrier-averaging (MSCA) channel estimation filter employing an orthogonal pilot channel [9]. More specifically, the channel estimate of the k-th sub-carrier and the n-th frame is derived by coherently averaging pilot symbols with the appropriate weights over three sub-carriers and three frames, i.e., in total, 36 pilot symbols are averaged, with the k-th sub-carrier and the n-th frame as the center. This channel estimate is used commonly for 48 data symbols in one frame for each sub-carrier. Weights are determined such that the required average received signal energy per bit-to-noise power spectrum density ratio (E_b/N_0) is minimized based on the tradeoff relationship between the increase in the signal power and mitigation of the fluctuation of the channel gain due to fading as investigated in the subsequent section. By using the obtained channel estimates, the composite signals at each receiver antenna are separated using MLD-based signal detection to produce the log likelihood ratio (LLR) at each bit. Finally, the output LLR streams are parallel-to-serial-converted and soft-decision Turbo decoded by Max-Log-MAP decoding with eight iterations to recover the transmitted binary data.

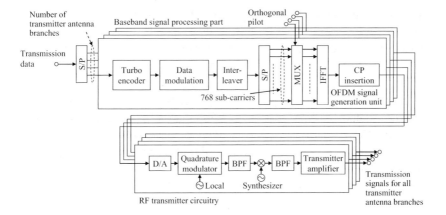

(a) Transmitter

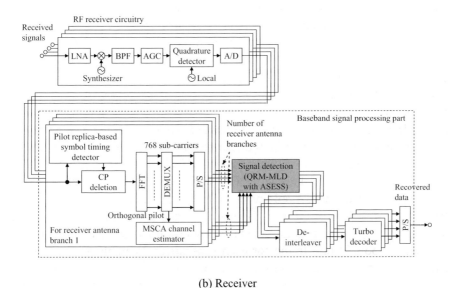

(b) Receiver

Figure 1. Structure of MIMO channel transmitter/receiver (MIMO multiplexing case)

Table 1. Major parameters in MIMO transmitter/receiver

Radio access	VSF-Spread OFDM
Bandwidth	101.5 MHz
Number of antenna branches	1 – 4 (Transmitter and receiver)
Number of sub-carriers	768 (131.836 kHz Sub-carrier separation)
OFDM symbol duration	7.585 msec + GI 1.674 msec (1024 + 226 samples)
Spreading factor	1
Data Modulation	QPSK, 16QAM, 64QAM
Channel coding / decoding	Turbo code ($R = 1/3 – 8/9$, $K = 4$) / Max-Log-MAP decoding
STBC in MIMO diversity	Alamouti's method
Signal separation in MIMO multiplexing	QRM-MLD with adaptive surviving symbol replica candidate selection method
OFDM symbol timing detection	Pilot symbol replica-based detection
Channel estimation	Orthogonal pilot channel-based MSCA channel estimation filter

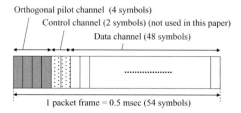

Figure 2. Frame structure

2.2 MIMO multiplexing

In MIMO multiplexing, the original information bit sequence is serial-to-parallel-converted according to the number of transmission antennas. After that, each information bit sequence is encoded by Turbo code independently at each antenna. Then, the encoded bits are data modulated by QPSK or 16QAM followed by symbol interleaving in the frequency domain. The achievable information bit rate (simply data rate hereafter) is increased in proportion to the number of transmission antennas, N_{Tx}. When $N_{Tx} = 4$ associated with 16 QAM modulation and the channel coding rate of $R = 8/9$, the peak data rate of 1.028 Gbps is achieved.

At the receiver, we apply the proposed adaptive selection of surviving symbol replica candidates (ASESS) that takes advantage of the maximum reliability [10] based on MLD using QR decomposition and the M-algorithm (QRM-MLD) [11] for the signal detection algorithm. Figure 3 illustrates the configuration of QRM-MLD using ASESS. From the estimated channel gain, we generate channel matrix $\hat{\mathbf{H}}_k(n)$ at the k-th sub-carrier of the n-th frame. Let $\mathbf{Q}_k(n)$ be the unitary matrix with the size of N_{Rx} x N_{Tx}. In QRM-MLD, by performing QR decomposition for the estimate of the channel matrix as $\hat{\mathbf{H}}_k(n) = \mathbf{Q}_k(n)\mathbf{R}_k(n)$, the obtained matrix, $\mathbf{R}_k(n)$, becomes an upper triangular matrix with the size of N_{Tx} x N_{Tx}.

$$\mathbf{R}_k(n) = \mathbf{Q}_k(n)^H \hat{\mathbf{H}}_k(n) = \begin{bmatrix} r_{1,1,k}(n) & r_{1,2,k}(n) & \cdots & r_{1,N_{TX},k}(n) \\ 0 & r_{2,2,k}(n) & \cdots & r_{2,N_{TX},k}(n) \\ \vdots & & & \\ 0 & \cdots & r_{N_{TX}-1,N_{TX}-1,k}(n) & r_{N_{TX}-1,N_{TX},k}(n) \\ 0 & \cdots & 0 & r_{N_{TX},N_{TX},k}(n) \end{bmatrix} \quad (1)$$

By multiplying $\mathbf{Q}_k(n)^H$ with the received signal in vector notation of the k-th sub-carrier, $\mathbf{Y}_k(n,b)$, the received signal in vector notation after nulling, i.e., orthogonalization, $\mathbf{Z}_k(n,b)$, is generated as

$$\mathbf{Z}_k(n,b) = \mathbf{Q}_k(n)^H \mathbf{Y}_k(n,b) = \mathbf{R}_k(n) \begin{bmatrix} d_{rank_k(N_{Tx}),k}(n,b) \\ d_{rank_k(N_{Tx}-1),k}(n,b) \\ \vdots \\ d_{rank_k(1),k}(n,b) \end{bmatrix} + \begin{bmatrix} n'_{1,k}(n,b) \\ n'_{2,k}(n,b) \\ \vdots \\ n'_{N_t,k}(n,b) \end{bmatrix}, \quad (2)$$

where $[n'_{1,k}(n,b) \ \ldots \ n'_{N_{Tx},k}(n,b)]^T = \mathbf{Q}_k(n)^H [n_{1,k}(n,b) \ \ldots \ n_{N_{Tx},k}(n,b)]^T$.

In QRM-MLD comprising stages corresponding to the number of transmitter branches, N_{Tx}, the symbol replicas with high reliability are successively selected stage-by-stage using $\mathbf{Z}_k(n,b)$ and $\mathbf{R}_k(n)$ based on the M-algorithm. In the ASESS method, we calculate the squared Euclidian distance only for the minimum number of the remaining symbol replica candidates at each stage by taking advantage of the reliability information of the symbol replica candidates newly-added at each stage [10].

We use likelihood function generation for bits when the surviving symbol replica candidates do not remain to the last stage in QRM-MLD as follows [12]. The space diversity order in MIMO multiplexing becomes N_{Rx} when the numbers of transmitter and receiver branches are N_{Tx} and N_{Rx}, since MLD-based signal detection is used. When $N_{Tx} < N_{Rx}$, the ($N_{Rx} - N_{Tx}$)-th order space diversity effect is gained in the orthogonalization process by multiplying the Hermitian transposition of the Q matrix to the received signal vector and then, the N_{Tx}-th space diversity effect is obtained in the MLD processing using the M-algorithm with the N_{Tx} stage.

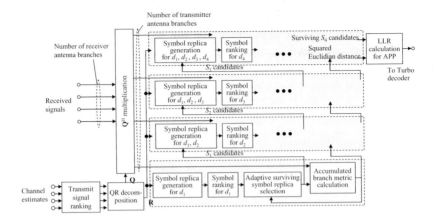

Figure 3. Structure of QRM-MLD with ASESS

2.3 MIMO diversity

In MIMO diversity with the maximum N_{Tx} value of two, the channel-encoded symbol sequence after symbol interleaving is space-time-block-coded by the Alamouti code [5] based on encoder matrix G,

$$\mathbf{G} = \begin{pmatrix} x_1 & x_2 \\ -x_2^* & x_1^* \end{pmatrix} \quad (3)$$

where $[x]^*$ denotes the complex conjugate of x. The symbol sequences after STBC, x_1 and x_2 from Antenna 1 and $-x_2^*$ and x_1^* from Antenna 2 are simultaneously transmitted by the contiguous OFDM symbols at the same sub-carrier. At the receiver, by multiplying the complex conjugate of the channel estimate at each transmission branch, the received data sequences are STBC-decoded. When N_{Rx} is two, the STBC-decoded symbol sequence, $\bar{x}_1$ and $\bar{x}_2$ are expressed as

$$\begin{cases} \bar{x}_1 = \dfrac{1}{|h_{11}|^2 + |h_{21}|^2 + |h_{12}|^2 + |h_{22}|^2} \left(r_{11} h_{11}^* + r_{12}^* h_{21} + r_{21} h_{12}^* + r_{22}^* h_{22} \right), \\ \bar{x}_2 = \dfrac{1}{|h_{11}|^2 + |h_{21}|^2 + |h_{12}|^2 + |h_{22}|^2} \left(r_{12} h_{11}^* - r_{11}^* h_{21} + r_{22} h_{12}^* - r_{21}^* h_{22} \right) \end{cases} \quad (4)$$

where $r_{q,1}$ and $r_{q,2}$ are two contiguous OFDM symbols at the same sub-carrier at receiver branch q ($1 \leq q \leq 2$), and $h_{p,q}$ denotes the channel impulse response between transmitter antenna branch p ($1 \leq p \leq 2$) and receiver gain, we generate channel matrix $\hat{\mathbf{H}}_k(n)$ at the k-th sub-carrier of the n-th frame. Let $\mathbf{Q}_k(n)$ be the unitary matrix with the size of $N_{Rx} \times N_{Tx}$. In QRM-MLD, by performing QR decomposition for the estimate of the channel matrix as $\hat{\mathbf{H}}_k(n) = \mathbf{Q}_k(n)\mathbf{R}_k(n)$, the obtained matrix, $\mathbf{R}_k(n)$, becomes an upper triangular matrix with the size of $N_{Tx} \times N_{Tx}$.

$$\mathbf{R}_k(n) = \mathbf{Q}_k(n)^H \hat{\mathbf{H}}_k(n) = \begin{bmatrix} r_{1,1,k}(n) & r_{1,2,k}(n) & \cdots & r_{1,N_{TX},k}(n) \\ 0 & r_{2,2,k}(n) & \cdots & r_{2,N_{TX},k}(n) \\ \vdots & & & \\ 0 & \cdots & r_{N_{TX}-1,N_{TX}-1,k}(n) & r_{N_{TX}-1,N_{TX},k}(n) \\ 0 & \cdots & 0 & r_{N_{TX},N_{TX},k}(n) \end{bmatrix} \quad (1)$$

By multiplying $\mathbf{Q}_k(n)^H$ with the received signal in vector notation of the k-th sub-carrier, $\mathbf{Y}_k(n,b)$, the received signal in vector notation after nulling, i.e., orthogonalization, $\mathbf{Z}_k(n,b)$, is generated as

$$\mathbf{Z}_k(n,b) = \mathbf{Q}_k(n)^H \mathbf{Y}_k(n,b) = \mathbf{R}_k(n) \begin{bmatrix} d_{rank_k(N_{Tx}),k}(n,b) \\ d_{rank_k(N_{Tx}-1),k}(n,b) \\ \vdots \\ d_{rank_k(1),k}(n,b) \end{bmatrix} + \begin{bmatrix} n'_{1,k}(n,b) \\ n'_{2,k}(n,b) \\ \vdots \\ n'_{N_t,k}(n,b) \end{bmatrix}, \quad (2)$$

where $[n'_{1,k}(n,b) \quad \cdots \quad n'_{N_{Tx},k}(n,b)]^T = \mathbf{Q}_k(n)^H [n_{1,k}(n,b) \quad \cdots \quad n_{N_{Tx},k}(n,b)]^T$.

In QRM-MLD comprising stages corresponding to the number of transmitter branches, N_{Tx}, the symbol replicas with high reliability are successively selected stage-by-stage using $\mathbf{Z}_k(n,b)$ and $\mathbf{R}_k(n)$ based on the M-algorithm. In the ASESS method, we calculate the squared Euclidian distance only for the minimum number of the remaining symbol replica candidates at each stage by taking advantage of the reliability information of the symbol replica candidates newly-added at each stage [10].

We use likelihood function generation for bits when the surviving symbol replica candidates do not remain to the last stage in QRM-MLD as follows [12]. The space diversity order in MIMO multiplexing becomes N_{Rx} when the numbers of transmitter and receiver branches are N_{Tx} and N_{Rx}, since MLD-based signal detection is used. When $N_{Tx} < N_{Rx}$, the ($N_{Rx} - N_{Tx}$)-th order space diversity effect is gained in the orthogonalization process by multiplying the Hermitian transposition of the Q matrix to the received signal vector and then, the N_{Tx}-th space diversity effect is obtained in the MLD processing using the M-algorithm with the N_{Tx} stage.

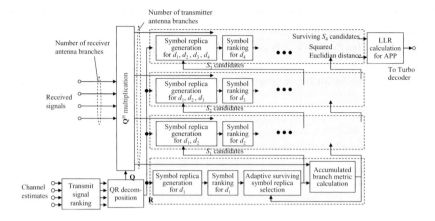

Figure 3. Structure of QRM-MLD with ASESS

2.3 MIMO diversity

In MIMO diversity with the maximum N_{Tx} value of two, the channel-encoded symbol sequence after symbol interleaving is space-time-block-coded by the Alamouti code [5] based on encoder matrix G,

$$\mathbf{G} = \begin{pmatrix} x_1 & x_2 \\ -x_2^* & x_1^* \end{pmatrix} \qquad (3)$$

where $[x]^*$ denotes the complex conjugate of x. The symbol sequences after STBC, x_1 and x_2 from Antenna 1 and $-x_2^*$ and x_1^* from Antenna 2 are simultaneously transmitted by the contiguous OFDM symbols at the same sub-carrier. At the receiver, by multiplying the complex conjugate of the channel estimate at each transmission branch, the received data sequences are STBC-decoded. When N_{Rx} is two, the STBC-decoded symbol sequence, $\bar{x}_1$ and $\bar{x}_2$ are expressed as

$$\begin{cases} \bar{x}_1 = \dfrac{1}{|h_{11}|^2 + |h_{21}|^2 + |h_{12}|^2 + |h_{22}|^2} \left(r_{11} h_{11}^* + r_{12}^* h_{21} + r_{21} h_{12}^* + r_{22}^* h_{22} \right), \\ \bar{x}_2 = \dfrac{1}{|h_{11}|^2 + |h_{21}|^2 + |h_{12}|^2 + |h_{22}|^2} \left(r_{12} h_{11}^* - r_{11}^* h_{21} + r_{22} h_{12}^* - r_{21}^* h_{22} \right) \end{cases} \qquad (4)$$

where $r_{q,1}$ and $r_{q,2}$ are two contiguous OFDM symbols at the same sub-carrier at receiver branch q ($1 \le q \le 2$), and $h_{p,q}$ denotes the channel impulse response between transmitter antenna branch p ($1 \le p \le 2$) and receiver antenna branch q using the MSCA channel estimation filter. As shown in Eq. (4), by combining the $2N_{Rx}$ received signals by MRC for the two transmitter branches and N_{Rx} receiver branches, the $2N_{Rx}$-th order space diversity effect is gained. In contrast, higher-level data modulation or a coding rate higher than that of MIMO multiplexing is necessary in MIMO diversity to achieve the same data rate.

3. LABORATORY EXPERIMENTS

3.1 Experiment setup

In the laboratory experiments, we employed 16 multipath fading simulators to simulate frequency-selective Rayleigh fading channels. A six-path channel is generated using multipath fading simulators where each path is independently Rayleigh-faded with the root mean squared (r.m.s.) delay spread, σ, of 0.26 μsec, and where the average signal power is decreased by 2 dB in descending order from the first path. Furthermore, the fading maximum Doppler frequency is set to 20 Hz, which corresponds to the moving speed of approximately 4.3 km/h at 4.6-GHz carrier frequency. We set the fading correlation between the transmitter and receiver antenna branches to zero.

3.2 Transmitter space diversity effect in the low frequency efficiency region

We first experimentally compare MIMO multiplexing and MIMO diversity from the viewpoint of the achievable data rate at the frequency efficiency of less than approximately 4 bits/second/Hz with the same transmission bandwidth. Table 2 (a) shows the achievable data rates using the modulation and coding schemes (MCSs) for MMO multiplexing/diversity modes. Figures 4(a) and 4(b) show the measured throughput comparison employing MIMO multiplexing and MIMO diversity with N_{Tx} and N_{Rx} of two for the data rate of 90 and 190 Mbps and 290 and 380 Mbps, respectively, as a function of the average received E_b/N_0 per receiver antenna. Computer simulation results with the same conditions are also given as dotted lines in the figure. Figure 4 shows that the loss in the required average received E_b/N_0 of the experimental results from the simulation results is suppressed to within 1 dB to achieve the same throughput even when 64QAM modulation is used in 2-by-2 MIMO transmissions. The figure also shows that when the data rate is 90 Mbps, i.e., corresponding frequency efficiency of 0.9 bits/sec/Hz, the measured throughput using MIMO diversity becomes higher than that of MIMO multiplexing regardless of the average received E_b/N_0 region. This is because the improvement due to the transmitter space diversity in MIMO diversity is larger than the degradation due to a reduction in the channel coding gain when the coding rate is twice for QPSK modulation. On the other hand, we find that the throughput employing MIMO multiplexing is better than MIMO diversity when the data rate is increased to 290 and 380 Mbps as shown in Fig. 4(b). In order to achieve a data rate higher than approximately 200 Mbps, i.e., 2 bits/second/Hz, high-level data modulation such as 64QAM associated with a high channel coding rate is necessary in MIMO diversity. Accordingly, the Euclidian distance among signal constellations becomes long and the channel coding gain is reduced. As a result, it is considered that the degradation exceeds the increased transmitter space diversity gain.

Table 2. Achievable data rates
(a) Various data rate for 2-antenna MIMO multiplexing / diversity with different MCSs
(b) Achievable data rate for N_{Tx}-antenna MIMO multiplexing

Data rate (Mbps)	MIMO multiplexing	MIMO diversity
90	QPSK, $R = 1/3$	QPSK, $R = 2/3$
190	QPSK, $R = 2/3$	16QAM, $R = 2/3$
290	16QAM, $R = 1/2$	64QAM, $R = 2/3$
380	16QAM, $R = 2/3$	64QAM, $R = 8/9$

N_{Tx}	Data rate (Mbps)	
	16QAM, $R = 1/2$	16QAM, $R = 8/9$
1	144	257
2	290	514
3	433	771
4	578	1028

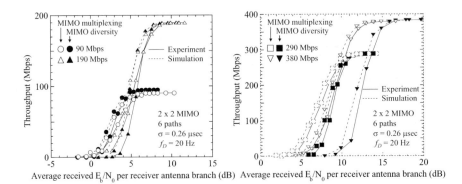

(a) Data rate of 90 and 190 Mbps (b) Data rate of 290 and 380 Mbps

Figure 4. Throughput comparisons between MIMO multiplexing and MIMO diversity

3.3 Receiver space diversity effect in MIMO multiplexing

We find that MIMO multiplexing is more promising than the MIMO diversity approach to achieve a data rate higher than approximately 200 Mbps (2 bits/sec/Hz). We investigate the space diversity effect at a receiver in MIMO multiplexing for a data rate greater than 200 Mbps up to 1 Gbps over a 100-MHz transmission bandwidth hereafter. Table 2(b) gives the achievable data rates employing the assumed MCSs for the respective N_{Tx} values. The maximum data rate of 1.028 Gbps is achieved when the 4-by-4 MIMO multiplexing is used together with 16QAM modulation and $R = 8/9$.

Figures 5(a) and 5(b) show the average packet error rate (PER) performance with 2-anntena MIMO multiplexing, i.e., $N_{Tx} = 2$, and $N_{Tx} = 3$, respectively, with the number of receiver antenna branches N_{Rx} as a parameter and as a function of the average received E_b/N_0 per antenna. The figures show that the space diversity effect is large when the N_{Rx} is increased from two to three, i.e., the required average received E_b/N_0 at the average PER of 10^{-2} is decreased by approximately 6.5 dB for $N_{Tx} = 2$, although a large frequency diversity effect over 100-MHz bandwidth and channel coding gains are attained. Moreover, we find that according to the increase in N_{Tx}, the improvement in the required average received E_b/N_0 by increasing N_{Rx} becomes large. This is because due to improving space diversity at a receiver, signal detection accuracy of the ASESS in the QRM-MLD is significantly improved. We show that the receiver space diversity is very beneficial to improve the achievable throughut through the improving the accuracy of signal detection in the QRM-MLD with ASESS.

Next, Fig. 6 shows the measured average PER performance parameterizing N_{Tx} from 1 to 4 for $N_{Rx} = 4$, as a function of the average received E_b/N_0 per receiver antenna with 16QAM and $R = 8/9$. From the figure, the loss in the required average received E_b/N_0 at the average PER of 10^{-2} of the experiments from the simulation results is suppressed to within 2 dB even in 4-by-4 MIMO multiplexing with $N_{Tx} = 4$, although the loss increases according to the increase in N_{Tx}. We can comprehend that by applying QRM-MLD with ASESS, four fold increase in the data rate such as 1.028 Gbps with $N_{Tx} = 4$ is achieved compared to the case with $N_{Tx} = 1$ (this situation corresponds to the achievement of 1.028 Gbps by four times the transmission bandwidth) allowing an increase in the required average received E_b/N_0 of approximately 4.5 dB.

Finally, Fig. 7(a) shows the measured throughput with $N_{Tx} = 2$ as a parameter of N_{Rx} and as a function of the average received E_b/N_0 per receiver antenna with 16QAM modulation and $R = 1/2$ and 8/9. Similarly, Fig. 7(b) shows the throughput with N_{Tx} = 4. Similar to the PER performance, the throughput is increased according to the increase in the N_{Rx} value owing to the increasing space diversity effect at the receiver. The throughput values of 500 Mbps and 1 Gbps are achieved at the average received E_b/N_0 of approximately 17.0 and 12.0 dB employing 2-by-2 and 4-by-4 MIMO multiplexing, respectively, along with 16QAM modulation and $R = 8/9$.

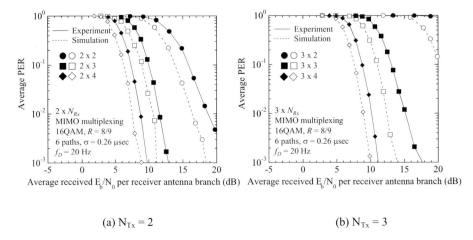

(a) $N_{Tx} = 2$ (b) $N_{Tx} = 3$

Figure 5. Average PER performance with N_{Rx} as a parameter

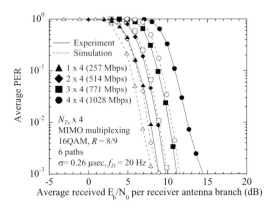

Figure 6. Average PER performance with N_{Tx} as a parameter

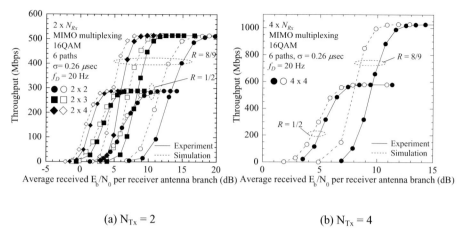

(a) $N_{Tx} = 2$ (b) $N_{Tx} = 4$

Figure 7. Throughput performance with N_{Rx} as a parameter

4. FIELD EXPERIMENTS

We also investigate high-speed packet transmission up to 1 Gbps using MIMO multiplexing in a real propagation channels based on field experiments. However, it should be noted that we consider that in the end, the adaptive selection of MIMO multiplexing or MIMO diversity is promising according to the required data rate and propagation channel conditions such as the fading correlation among antennas based on the results in Section 3. The field experiments in an actual multipath fading channel are conducted in the urban area of Yokosuka city nearby Tokyo without considering other-cell interference. Figure 8 shows the measurement course and the location of the BS. The heights of the BS and MS antennas are approximately 50 and 3.5 m from the ground, respectively. The maximum transmission power of the BS is 2.5W per antenna (total 10W). A measurement vehicle equipped with the MS is driven along the measurement course at the average speed of approximately 30 km/h, where the distance from the BS is approximately 300 m at Points B and C. Furthermore, most of the measurement course is under non-line-of-sight conditions.

Figures 9(a) and 9(b) show examples of the measured instantaneous power delay profiles for four transmission signals in 4-by-4 MIMO multiplexing at Points A and C in Fig. 8 respectively. The vertical axis is the received signal power relative to the background noise power. We see that many paths are observed, which are distributed over approximately 2 μsec. The average root mean squared delay spread value in the course is approximately 0.35 μsec. The figure shows that the measured power delay profiles are slightly different although the locations of the transmitter and receiver antennas are almost identical. This indicates that the propagation channels are different due to the different reflectors. Finally, Fig. 10 plots the measured throughput performance using MCSs such as QPSK with $R = 1/2$, 2/3, and 6/7, and 16QAM with $R = 1/2$, 3/4, and 8/9 as a function of the average received signal-to-interference plus noise power ratio (SINR) per receiver antenna in 4-by-4 MIMO multiplexing. The figure shows that according to the increase in the data modulation level and the channel coding rate, the measured throughput is increased. The figure also shows that the throughput values of 500 Mbps and 1 Gbps are achieved at the average received SINR of approximately 8 and 14 dB employing 16QAM modulation with $R = 1/2$ and 8/9, respectively, in 4-by-4 MIMO multiplexing. Therefore, these results show that the QRM-MLD signal detection with the ASESS is very beneficial to achieving 1-Gbps throughput at a low required received SINR to provide a high throughput with a wide coverage area and limited transmission power.

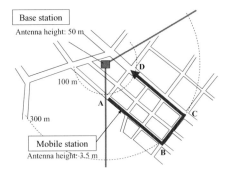

Figure 8. Measurement course in field experiments

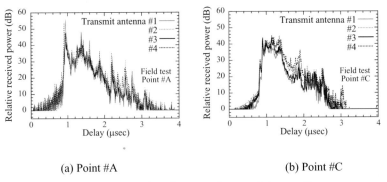

(a) Point #A　　　　　　　　　　(b) Point #C

Figure 9. Examples of measured power delay profile

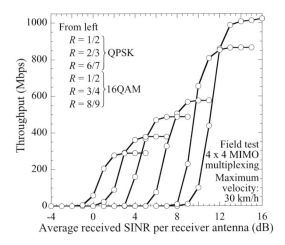

Figure 10. Throughput as a function of average received SINR per receiver antenna

5. CONCLUSION

This paper presented experimental results on the space diversity effect in MIMO multiplexing/diversity with the target data rate up to 1 Gbps using OFDM radio access based on laboratory and field experiments including realistic impairments using the implemented MIMO transceivers with the maximum of four transmitter/receiver branches. First, the experimental results using multipath fading simulators showed that in the frequency efficiency of less than approximately 2 bits/second/Hz, MIMO diversity using STBC can increase the measured throughput compared to MIMO multiplexing owing to the high transmission space diversity effect. At a higher frequency efficiency than approximately 2 to 3 bits/second/Hz, however, MIMO multiplexing exhibited performance superior to that of MIMO diversity since the impairments using higher data modulation and a higher channel coding rate in MIMO diversity overcomes the space diversity effect. It was also shown that the receiver space diversity effect was very effective in MIMO multiplexing for QRM-MLD-based signal detection using our proposed ASESS throughput-improving channel estimation accuracy method. Finally, we showed that real-time 500-Mbps and 1-Gbps throughput in a 100-MHz transmission bandwidth was achieved at the average received E_b/N_0 per receiver antenna of approximately 8.0 and 14.0 dB using 16QAM modulation and Turbo coding with the coding rate of 1/2 and 8/9, respectively, in 4-by-4 MIMO multiplexing, to confirm the possibility of 1-Gbps high-speed packet transmission in a real propagation environment.

REFERENCES

[1] H. Atarashi, S. Abeta, and M. Sawahashi, "Variable spreading factor orthogonal frequency and code division multiplexing (VSF-OFCDM) for broadband packet wireless access," IEICE Trans. Commun., vol. E86-B, no. 1, pp. 291-299, Jan. 2003.

[2] Y. Kishiyama, N. Maeda, K. Higuchi, H. Atarashi, and M. Sawahashi, "Transmission performance analysis of VSF-OFCDM broadband packet wireless access based on field experiments in 100-MHz forward link," in Proc. IEEE VTC2004-Fall, Sept. 2004.

[3] G. J. Foschini, Jr., "Layered space-time architecture for wireless communication in a fading environment when using multi-element antennas," Bell Labs Tech. J., pp. 41-59, Autumn 1996.

[4] R. D. Murch and K. B. Letaief, "Antenna systems for broadband wireless access," IEEE Commun. Mag., vol. 40, no. 4, pp. 76-83, April 2002.

[5] S.M. Alamouti, "Simple transmit diversity technique for wireless communications," IEEE J. Sel. Areas Commun., vol. 16, pp. 1451-1458, Oct. 1998.

[6] V. Tarokh, N. Seshadri, A.R. Calderbank, "Space-time codes for high data rate wireless communication: performance criterion and code construction," IEEE Transactions on Information Theory, vol. 44, no. 2, pp. 744-765, Mar. 1998.

[7] H. Kawai, N. Maeda, J. Kawamoto, K. Higuchi, and M. Sawahashi, "Experiments on real-time 1-Gbps packet transmission using QRM-MLD with ASESS in MIMO-OFDM broadband packet radio access," in Proc. IST Summit 2005 conference, Dresden, Germany, 19-23 June 2005.

[8] A. van Zelst, R. van Nee, and G.A. Awater, "Space division multiplexing (SDM) for OFDM systems," in Proc. IEEE VTC2000-Spring, pp. 1070-1074, May 2000.

[9] H. Kawai, K. Higuchi, N. Maeda, and M. Sawahashi, "Performance of QRM-MLD employing two-dimensional multi-slot and carrier-averaging channel estimation using orthogonal pilot channel for OFCDM MIMO multiplexing in multipath fading channel," in Proc. Wireless 2004, July 2004.

[10] K. Higuchi, H. Kawai, N. Maeda, and M. Sawahashi, "Adaptive selection of surviving symbol replica candidates based on maximum reliability in QRM-MLD for OFCDM MIMO multiplexing," in Proc. IEEE Globecom2004, Nov. 2004.

[11] K. J. Kim et al., "A QRD-M/Kalman filter-based detection and channel estimation algorithm for MIMO-OFDM systems," IEEE Trans. on Wireless Commun., vol. 4, no. 2, pp. 710-721, March, 2005.

[12] H. Kawai, K. Higuchi, N. Maeda, M. Sawahashi, T. Itoh, Y. Kakura, A. Ushirokawa, and H. Seki, "Likelihood function for QRM-MLD suitable for soft-decision turbo decoding and its performance for OFCDM MIMO multiplexing in multipath fading channel," IEICE Trans. Commun., vol. E88-B, no. 1, pp. 47-57, Jan. 2005.

REDUCED-COMPLEXITY MAXIMUM-LIKELIHOOD DETECTION IN MULTIPLE-ANTENNA-AIDED MULTICARRIER SYSTEMS

Jos Akhtman and Lajos Hanzo*
University of Southampton
Southampton, SO17 1BJ, UK
{yja02r,lh}@soton.ecs.ac.uk

Abstract In this contribution we explore a family of novel Optimized Hierarchy Reduced Search Algorithm (OHRSA)-aided space-time processing methods, which may be regarded as an advanced extension of the Complex Sphere Decoder (CSD) method. The algorithm proposed extends the potential application range of the CSD method, as well as reduces the associated computational complexity.

1. Introduction

Multi-Carrier (MC) modulation techniques [1, 2] have found their way into several wireless broadband communications standards as well as into local area networks. Owing to their advantageous properties they also constitute strong contenders for the next-generation cellular mobile communications standards.

The relevant information-theoretical analysis predicts [3] that substantial capacity gains are achievable in wireless communication systems employing a Multiple Input Multiple Output (MIMO) architecture using multiple antennas. Specifically, provided that the fading processes corresponding to different transmit-receive antenna pairs may be assumed to be independently Rayleigh distributed, the associated attainable

---

*Acknowledgements: The work reported in this paper has formed part of the Wireless Enabling Techniques work area of the Core 3 Research Programme of the Virtual Centre of Excellence in Mobile and Personal Communications, Mobile VCE, www.mobilevce.com, whose funding support, including that of EPSRC, is gratefully acknowledged. Fully detailed technical reports on this research are available to Industrial Members of Mobile VCE.

capacity was shown to linearly increase with the smaller of the numbers of the transmit and receive antennas. Additionally, the employment of a MIMO architecture allows for the efficient exploitation of the spatial diversity available in a wireless MIMO environment, thus an improvement of the system's transmission integrity, as well as a further increase in the system's capacity becomes possible.

Hence, in this contribution we propose a novel Optimized Hierarchy Reduced Search Algorithm (OHRSA)-aided space-time processing method, which may be regarded as an advanced extension of the Complex Sphere Decoder (CSD) method, portrayed in [4]. The algorithm proposed extends the potential application range of the CSD methods of [5] and [4], as well as reduces the associated computational complexity. Moreover, the OHRSA-aided SDM detector proposed combines the near-optimum performance of the ML SDM detector with the low-complexity of the linear MMSE SDM detector, which renders it an attractive design alternative for practical systems.

The rest of this paper is structured as follows. The system model as well as the principles of ML space-time detection are briefly outlined in Section 2.1. A novel recursive ML detection technique is derived in Section 2.2. The corresponding bitwise real-valued system model is derived in Section 2.3. Furthermore, the search optimization rules are described in Section 2.4. The achievable performance of the proposed technique is quantified using extensive computer simulations and the corresponding results are provided in Section 3, before offering our conclusions in Section 4.

2. Optimized Hierarchy Reduced Search Algorithm

2.1 Maximum Likelihood Detection

The subcarrier-related MIMO-OFDM system model considered is given by [1]

$$\mathbf{y} = \mathbf{Hs} + \mathbf{w}, \qquad (1)$$

where $\mathbf{y}, \mathbf{w}$ and $\mathbf{x}$ denote the n_r-dimensional received signal and AWGN sample vectors as well as the m_t-dimensional transmitted signal vector, respectively. Furthermore, $\mathbf{H}$ represents a $(n_\mathrm{r} \times m_\mathrm{t})$-dimensional matrix of subcarrier-related CTF coefficients. Note that for the sake of brevity we omit the OFDM subcarrier and symbol indices k and n. As outlined in [1], the ML SDM detector provides an m_t-antenna-based estimated signal vector candidate $\hat{\mathbf{s}}$, which maximizes the objective function defined as the conditional *a posteriori* probability $\mathsf{P}\{\check{\mathbf{s}}|\mathbf{y}, \mathbf{H}\}$ over the set $\mathcal{M}^{m_\mathrm{t}}$

of legitimate solutions. More explicitly, we have

$$\hat{\mathbf{s}} = \arg \max_{\check{\mathbf{s}} \in \mathcal{M}^{m_t}} P\{\check{\mathbf{s}}|\mathbf{y}, \mathbf{H}\}, \tag{2}$$

where $\mathcal{M}^{m_t}$ is the set of **all possible** m_t-dimensional candidate symbol vectors of the m_t-antenna-based transmitted signal vector $\mathbf{s}$.

Furthermore, it was shown in [1] that the probability maximization problem of Equation (2) is equivalent to the corresponding Euclidean distance minimization problem. Specifically, we have

$$\hat{\mathbf{s}} = \arg \min_{\check{\mathbf{s}} \in \mathcal{M}^{m_t}} \|\mathbf{y} - \mathbf{H}\check{\mathbf{s}}\|^2, \tag{3}$$

where the probability-based objective function of Equation (2) is substituted by the objective function determined by the Euclidean distance between the received signal vector $\mathbf{y}$ and the corresponding product of the channel matrix $\mathbf{H}$ with the *a priori* candidate of the transmitted signal vector $\check{\mathbf{s}} \in \mathcal{M}^{m_t}$.

2.2 Recursive ML Detection

Subsequently, our detection method relies on the observation, which may be summarized in the following lemma. For the sake of brevity in this contribution we omit the proof of Lemma 1.

LEMMA 1 *The ML solution of Equation (2) of a noisy linear problem described by Equation (1) is given by*

$$\hat{\mathbf{s}} = \arg \min_{\check{\mathbf{s}} \in \mathcal{M}^{m_t}} \left\{ \|\mathbf{U}(\check{\mathbf{s}} - \hat{\mathbf{x}})\|^2 \right\}, \tag{4}$$

where $\mathbf{U}$ *is an upper-triangular matrix having positive real-valued elements on the main diagonal and satisfying*

$$\mathbf{U}^H \mathbf{U} = (\mathbf{H}^H \mathbf{H} + \sigma_w^2 \mathbf{I}), \tag{5}$$

while

$$\hat{\mathbf{x}} = (\mathbf{H}^H \mathbf{H} + \sigma_w^2 \mathbf{I})^{-1} \mathbf{H}^H \mathbf{y} \tag{6}$$

is the unconstrained MMSE estimate of the transmitted signal vector $\mathbf{s}$.

Observe that Lemma 1 imposes no constraints on the dimensions, or rank of the matrix $\mathbf{H}$ of the linear system described by Equation (1). This property is particularly important, since it enables us to apply our proposed detection technique to the scenario of *over-loaded* systems, where the number of transmit antenna elements exceeds that of the receive antenna elements.

Using Lemma 1, in particular the fact that the matrix $\mathbf{U}$ is an upper-triangular matrix, we may introduce an objective function $J(\check{\mathbf{s}})$ described as follows

$$J(\check{\mathbf{s}}) = \|\mathbf{U}(\check{\mathbf{s}} - \hat{\mathbf{x}})\|^2 = (\check{\mathbf{s}} - \hat{\mathbf{x}})^H \mathbf{U}^H \mathbf{U}(\check{\mathbf{s}} - \hat{\mathbf{x}})$$

$$= \sum_{i=1}^{m_t} \left| \sum_{j=i}^{m_t} u_{ij}(\check{s}_j - \hat{x}_j) \right|^2 = \sum_{i=1}^{m_t} \phi_i(\check{\mathbf{s}}_i), \quad (7)$$

where $J(\check{\mathbf{s}})$ and $\phi_i(\check{\mathbf{s}}_i)$ are positive real-valued cost and sub-cost functions, respectively. Elaborating a little further, we have

$$\phi_i(\check{\mathbf{s}}_i) = \left| \sum_{j=i}^{m_t} u_{ij}(\check{s}_j - \hat{x}_j) \right|^2$$

$$= \left| u_{ii}(\check{s}_i - \hat{x}_i) + \underbrace{\sum_{j=i+1}^{m_t} u_{ij}(\check{s}_j - \hat{x}_j)}_{a_i} \right|^2. \quad (8)$$

Note that the term a_i is a complex-valued scalar, which is independent of the specific symbol value $\check{s}_i$ of the ith element of the *a priori* candidate signal vector $\check{\mathbf{s}}$.

Furthermore, let $J_i(\check{\mathbf{s}}_i)$ be a Cumulative Sub-Cost (CSC) function recursively defined as

$$J_{m_t}(\check{\mathbf{s}}_{m_t}) = \phi_{m_t}(\check{\mathbf{s}}_{m_t}) = |u_{m_t m_t}(\check{s}_{m_t} - \hat{x}_{m_t})|^2 \quad (9a)$$

$$J_i(\check{\mathbf{s}}_i) = J_{i+1}(\check{\mathbf{s}}_{i+1}) + \phi_i(\check{\mathbf{s}}_i), \quad m_t-1, \cdots, 1, \quad (9b)$$

where we define the candidate subvector as $\check{\mathbf{s}}_i = [\check{s}_i, \cdots, \check{s}_{m_t}]$. Clearly, $J_i(\check{\mathbf{s}}_i)$ exhibits the following essential property

$$J(\check{\mathbf{s}}) = J_1(\check{\mathbf{s}}_1) > J_2(\check{\mathbf{s}}_2) > \cdots > J_{m_t}(\check{\mathbf{s}}_{m_t}) > 0 \quad (10)$$

for all possible realizations of $\hat{\mathbf{x}} \in \mathbb{C}^{m_t}$ and $\check{\mathbf{s}} \in \mathcal{M}^{m_t}$, where the space $\mathbb{C}^{m_t}$ contains all possible unconstrained MMSE estimates $\hat{\mathbf{x}}$ of the transmitted signal vector $\mathbf{s}$.

2.3 Bitwise System Model

It is evident that each phasor point c_m of an M-QAM constellation map may be represented as the inner product of a unique bit-based vector $\boldsymbol{d}_m = \{d_{ml} = -1, 1\}_{l=1,\cdots,b}$ and the corresponding *quantisation vector* $\boldsymbol{q}$. Specifically, we have

$$c_m = \boldsymbol{q}^T \boldsymbol{d}_m. \quad (11)$$

For instance, the quantization vectors corresponding to the modulation schemes of QPSK and 16-QAM are $\boldsymbol{q} = \frac{1}{\sqrt{2}}[1,j]$ and $\boldsymbol{q} = \frac{1}{\sqrt{10}}[1,j,2,2j]$, respectively.

Furthermore, we define a $(bm_t \times m_t)$-dimensional *quantization matrix* $\mathbf{Q} = \mathbf{I} \otimes \boldsymbol{q}$, where $\mathbf{I}$ is an $(m_t \times m_t)$-dimensional identity matrix and $\boldsymbol{q}$ is the aforementioned *quantization vector*, while $\otimes$ denotes the *matrix direct product* [6]. Consequently the M-QAM-modulated signal vector $\mathbf{s}$ may be represented as

$$\mathbf{s} = \mathbf{Q}\mathbf{t}, \qquad (12)$$

where $\mathbf{t} = [\boldsymbol{t}_1^T, \cdots, \boldsymbol{t}_{m_t}^T]^T$ is a $r = bm_t$-dimensional column supervector comprising the bit-based vectors $\boldsymbol{t}_i$ associated with each transmitted signal vector component s_i. Substituting Equation (12) into the system model of Equation (1) yields

$$\mathbf{y} = \mathbf{H}\mathbf{Q}\mathbf{t} + \mathbf{w}. \qquad (13)$$

Moreover, since $\mathbf{t}$ is a real-valued vector, we can elaborate a bit further and deduce a real-valued system model as follows

$$\tilde{\mathbf{y}} = \begin{bmatrix} \mathcal{R}\{\mathbf{y}\} \\ \mathcal{I}\{\mathbf{y}\} \end{bmatrix} = \begin{bmatrix} \mathcal{R}\{\mathbf{HQ}\} \\ \mathcal{I}\{\mathbf{HQ}\} \end{bmatrix} \mathbf{t} + \begin{bmatrix} \mathcal{R}\{\mathbf{w}\} \\ \mathcal{I}\{\mathbf{w}\} \end{bmatrix} = \tilde{\mathbf{H}}\mathbf{t} + \tilde{\mathbf{w}}, \qquad (14)$$

where $\tilde{\mathbf{H}}$ is a real-valued $(2n_r \times bm_t)$-dimensional bitwise channel matrix.

Substituting the system model of Equation (1) by the bitwise real-valued system model of Equation (14) and subsequently exploiting the properties of the corresponding objective function $J(\check{\mathbf{t}})$ of Equation (10) enables us to employ a highly efficient reduced-complexity search algorithm, which decreases the number of objective function evaluations of the minimization problem outlined in Equation (4) to a small fraction of the set $\mathcal{M}^{m_t}$. This reduced-complexity search algorithm is outlined in the next section.

2.4 Search Strategy

Firstly, we commence the recursive search process with the evaluation of the CSC function value $J_r(\check{t}_r)$ of Equation (9a). Secondly, at each recursive step i of the search algorithm proposed we stipulate two legitimate hypotheses -1 and 1 concerning the value of the transmitted bitwise symbol t_i and subsequently calculate the conditioned sub-cost function $J_i(\check{t}_i)$ of Equation (9b). Furthermore, for each tentatively assumed value of $\check{t}_i$ we execute a successive recursive search step $i-1$, which is conditioned on the hypotheses made in all preceding recursive steps $j = i, \cdots, r = bm_t$. Upon each arrival at the index $i = 1$ of the recursive process, a complete candidate vector $\check{\mathbf{t}}$ is hypothesized and the corresponding value of the cost function $J(\check{\mathbf{t}})$ formulated in Equation (7) is evaluated.

Observe that the recursive hierarchical search procedure described above may be employed to perform an exhaustive search through all possible values of the transmitted signal vector $\check{\mathbf{t}}$ and the resultant search process is guaranteed to arrive at the ML solution $\hat{\mathbf{t}}$, which minimizes the value of the cost function $J(\check{\mathbf{t}})$ of Equation (7). Fortunately however, as opposed to other ML search schemes, the search process described above can be readily optimized, resulting in a dramatic reduction of the associated computational complexity. Specifically, the potential optimization complexity gain accrues from the fact that most of the hierarchical search branches can be discarded at an early stage of the recursive search process. The corresponding optimization rules proposed may be outlined as follows.

Rule 1 We reorder the system model of Equation (1) as suggested in [7]. Specifically, we apply the *best-first* detection strategy outlined in [1], which implies that the transmitted signal vector components are detected in the decreasing order of the associated channel quality. More specifically, the columns of the bitwise channel matrix $\tilde{\mathbf{H}}$ are sorted in the increasing order of their norm. Namely, we have

$$\|(\tilde{\mathbf{H}})_1\|^2 \leq \|(\tilde{\mathbf{H}})_2\|^2 \leq \cdots \leq \|(\tilde{\mathbf{H}})_r\|^2, \tag{15}$$

where $(\tilde{\mathbf{H}})_i$ denotes the ith column of the bitwise channel matrix $\tilde{\mathbf{H}}$.

Rule 2 At each recursive detection step $i = r, \cdots, 1$, the potential candidate values $d_m = \{-1, 1\}$ of the transmitted bitwise signal component t_i are considered in the increasing order of the corresponding value of the sub-cost function $\phi_i(\check{\mathbf{t}}_i) = \phi_i(d_m, \check{\mathbf{t}}_{i+1})$ of Equation (8), where we have

$$\phi_i(d_1, \check{\mathbf{t}}_{i+1}) < \phi_i(d_2, \check{\mathbf{t}}_{i+1}) \tag{16}$$

Rule 3 We define a *cut-off* value of the cost function $J_{\min} = \min\{J(\check{\mathbf{t}})\}$ as the minimum value of the total cost function obtained up to the present point of the search process. Consequently, at each arrival at step $i = 1$ of the recursive search process, the *cut-off* value of the cost function is updated as follows

$$J_{\min} = \min\{J_{\min}, J(\check{\mathbf{t}})\}. \tag{17}$$

Rule 4 Finally, at each recursive detection step i, only the high probability search branches corresponding to the highly likely bitwise symbol candidates d_m resulting in particularly low values of the CSC function obeying $J_i(d_m) < J_{\min}$ are pursued.

3. Simulation results

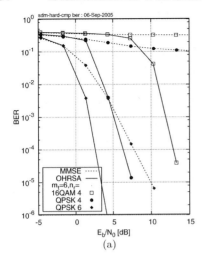 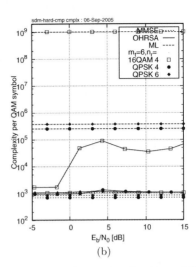

Figure 1. (a) **Bit Error Rate** and (b) the corresponding **complexity** exhibited by the rate-$\frac{1}{2}$ turbo-coded **SDM-OFDM** system employing the **OHRSA-ML** SDM detector. The abscissa represents the average E_b/N_0.

Our simulations were performed in the base-band frequency domain. The OFDM system considered utilises 128 QAM-modulated orthogonal subcarriers. For forward error correction (FEC) we use $\frac{1}{2}$-rate turbo coding [8] employing two constraint-length $K = 3$ Recursive Systematic Convolutional (RSC) component codes and the standard 124-bit WCDMA UMTS turbo code interleaver of [9]. The octally represented RCS generator polynomials of (7,5) were used. We assume a 9-tap CIR Rayleigh-fading multipath channel and stipulate the assumption of perfect channel knowledge, where the knowledge of the frequency-domain subcarrier-related coefficients $H[n,k]$ is deemed to be available in the receiver.

Figure 1(a) characterizes the achievable BER performance of the MIMO-OFDM system employing the OHRSA detector proposed, as well as that of the MMSE detector. We can see that as opposed to the MMSE detector, the proposed OHRSA detector performs equally well both in fully-loaded as well as in overloaded scenarios, where the number of transmit antennas exceeds that of the receive antennas and thus we have $m_t > n_r$. Furthermore, the OHRSA method exhibits an equally good performance, when employed in the overloaded 16-QAM-modulated MIMO-OFDM system.

On the other hand, Figure 1(b) illustrates the corresponding computational complexity quantified in terms of the total number of real addition and multiplication operations per detected QAM symbol.

Observe that the complexity exhibited by the OHRSA method in the QPSK scenario is only slightly higher than that exhibited by the MMSE detector. Moreover, the complexity exhibited by the OHRSA detector in the overloaded 16-QAM scenario is more than four orders of magnitude lower than that exhibited by the exhaustive ML detector, while their BER performance is expected to be similar. We note however that the BER performance of the ML detector is not shown since it would impose a more than 10^4-times higher complexity.

4. Conclusion

We proposed a novel OHRSA-ML space-time detector, which may be regarded as an advanced extension of the CSD method. We demonstrated that the MIMO-OFDM system employing the OHRSA-ML detector proposed is capable of achieving the near optimum ML performance in the overloaded scenario, where the number of transmit antennas exceeds that of the receive antennas.

References

[1] L. Hanzo, M. Münster, B.J. Choi, and T. Keller. *OFDM and MC-CDMA for Broadband Multi-user Communications, WLANs and Broadcasting*. John Wiley and IEEE Press, 2003. 992 pages.

[2] L. Hanzo, L-L. Yang, E-L. Kuan, and K. Yen. *Single- and Multi-Carrier DS-CDMA*. John Wiley and IEEE Press, 2003. 430 pages.

[3] A. Goldsmith, S. A. Jafar, N. Jindal, and S. Vishwanath. Capacity limits of MIMO channels. *IEEE Journal on Selected Areas in Communications*, 21(5):684–702, June 2003.

[4] D. Pham, K. R. Pattipati, P. K. Willet, and J. Luo. An improved complex sphere decoder for V-BLAST Systems. *IEEE Signal Processing Letters*, 11(9):748–751, September 2004.

[5] B. M. Hochwald and S. ten Brink. Achieving near-capacity on a multiple-antenna channel. *IEEE Transactions on Communications*, 51(3):389–399, March 2003.

[6] R. D. Schafer. *An Introduction to Nonassociative Algebras*. Dover, New York, 1996.

[7] M. K. Varanasi. Decision feedback multiuser detection: A systematic approach. *IEEE Transactions on Information Theory*, 45:219–240, January 1999.

[8] L. Hanzo, T. H. Liew, and B. L. Yeap. *Turbo Coding, Turbo Equalisation and Space-Time Coding*. John Wiley and IEEE Press, Chichester, UK; Piscataway, NJ, USA, 2002. 766 pages. (For detailed contents, please refer to http://www-mobile.ecs.soton.ac.uk/).

[9] H. Holma and A. Toskala, editors. *WCDMA for UMTS : Radio Access for Third Generation Mobile Communications*. John Wiley and Sons, Ltd., 2000.

ON CODING FOR OFDM AND THE BROADCAST CHANNEL

Martin Bossert

Dept. of Telecommunications and Applied Information Theory, TAIT
University of Ulm, Albert-Einstein-Allee 43, D-89081 Ulm, Germany
martin.bossert uni-ulm.de

Abstract   In this paper two fundamental aspects for wireless data transmission are addressed, namely the interrelation between multicarrier spread spectrum transmission and multilevel coding and the view of downlink transmission as *the broadcast channel* from information theory. In the first part it will be shown that multicarrier spread spectrum systems can be interpreted as a multilevel code with repetition codes in all levels and different mappings onto the modulation alphabet. This point of view might provide new insights into both, multicarrier spread spectrum systems as well as multilevel coding. In the second part we will describe the information theory point of view for the situation when data are transmitted from one sender to many receivers. This is called the broadcast channel. The opposite case when data are transmitted from many senders to one receiver is called the multiple access channel. It is known that the broadcast channel has a larger capacity than the multiple access channel which might be exploited in the downlink transmission. With the help of elementary examples these relations will be illustrated.

1.   Introduction

Orthogonal frequency division multiplexing (OFDM) is the transmission scheme for future high data rate wireless systems. OFDM with cyclic extension transforms any fading multipath channel into flat fading subcarriers. However, there is no built-in diversity using OFDM, neither in the time domain nor in the frequency domain. Multicarrier spread spectrum (MC-SS) offers frequency diversity using additional spreading in order to overcome severely faded subchannels. MC-SS systems based on OFDM have been first proposed in 1993 [1], [2]. In [2] the transmission of a single data symbol is organized using N replicas of the symbol whereas each copy of the data symbol is multiplied by a chip of a pseudo-random or some other code of length N. Already here specific MC-SS systems were described using replicas of the same symbol (repetition code). A similar description exists for direct sequence code division multiple access (DS-CDMA), which introduces additional time diversity. A DS-CDMA system can be interpreted as a binary modulator followed by an encoder of a repetition code and a complex valued scrambler [3]. Thus, it is possible, from a coding theory perspective, to view spreading as repetition code. We will analyze MC-SS systems from a coding perspective using multilevel codes (MLCs) with repetition codes in the different levels. This point of view might give insights into both MC-SS and MLC and hopefully helps to improve future systems.

For the notation and description we will follow [4], where also constellation coding is considered like *Linear Constellation Precoding* [5] and *Complex Field Coding* [6]. However, here we only will proof by construction, that it is possible to interpret a MC-SS system as a MLC with repetition codes in the levels. For general spreading matrices, the resulting MLC will use different signal constellations for the different subcarriers, resulting in so called polyalphabetic codes [7]. It can also be shown [4] that a simpler MLC encoder using just a single signal constellation might be used in case of Hadamard spreading matrices and modified Hadamard spreading matrices. The simplification is at the cost of some restrictions to the used signal constellations and cosets of the repetition codes. In Section 2 the considered MC-SS system is introduced and afterwards we recall the idea of multilevel coding. Then we show how general MC-SS systems can be analyzed using an MLC. For the discussion of the specialties using Hadamard type spreading matrices we refer to [4].

Concerning the second fundamental aspect we recall that already Shannon has defined the broadcast channel as a model in which one transmitter sends individual information to many receivers. Clearly each receiver has a different channel. In [8], [9] the broadcast channel was analysed and it was shown that the capacity region exceeds the one for the multiple access channel. Time division multiple access was assumed. An overview of the results of information theory until 1976 for different multiuser channels can be found in [10]; see also [11].

Figure 1 shows the broadcast case for several users. One and the same signal is sent to all users which contains individual information for all. Because the signal passes through different channels each user receives a differently disturbed signal. From this the user detects his individual information. Note, that from an information theoretic perspective the downlink of a wireless communication system is exactly the broadcast case. However, in existing systems this fact is not utilized and multiple access schemes are applied instead.

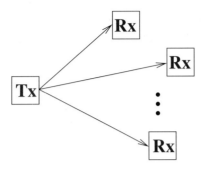

Figure 1. Broadcast transmission (downlink)

In order to illustrate the essential difference between broadcast and multiple access channels we look for example at UMTS (Universal Mobile Telecommunications System). This standard is based on code division multiple access (CDMA).The uplink, from the mobiles to the base station, is a multiple access channel in the sense of information theory because it consists of many transmitters and one receiver. The users transmit at the same time, in the same frequency band, and are separated by codes. In the downlink there is one transmitter and many receivers but also here the users are separated by codes. Thus, multiple access is intended for the downlink. However, the signal is constructed by adding the individual spread information signals together and transmit the sum signal. Clearly one single signal is transmitted to all users where the individual signals of all users are simultaneously included. This fact is the same as for the broadcast case. Indeed, a receiver has two possibilities. The first is to correlate the received (sum) signal with his individual spreading sequence. Since it is CDMA which is a multiple access scheme the individual information can be extracted. This is called single user detection. The second possibility is to perform joint detection which means that a receiver does not only detect his own data but all data contained in the sum signal. But this is as in the broadcast case. And it is commonly known that the performance is much better in case of joint detection than for single user detection [12].

We will recall the example by Cover [8] in order to illustrate the possible gain of broadcast over multiple access. In this simple case the region of possible data rates can be derived analytically. This principle will be used for an example where TDMA is compared to broadcast transmission using turbo codes. Both, theoretical bounds and simulation results show the effect that broadcast can achieve a higher data rate compared to TDMA. Again, despite this fact in all existing systems multiple access is used for downlink transmission.

2. Multicarrier Spread Spectrum and Multilevel Coding

In this part of the paper, we will show the interrelation between MC-SS systems and MLCs. We will first recall the system model for MC-SS and after that we will describe the principle of multilevel coding. Then we will show that MC-SS systems can be interpreted as MLCs.

2.1 Multicarrier Spread Spectrum

In Multicarrier spread spectrum (MC-SS) the spreading is performed in the frequency domain and usually orthogonal frequency division multiplexing (OFDM) is used. Figure 2, shows the block diagram of the MC-SS system. The vector

$$\underline{b} = \left(\underline{b}_1^T, \ldots, \underline{b}_N^T\right)^T$$

entering the mapper, consists of N binary vectors

$$\underline{b}_i^T = (b_{i,1}, \ldots, b_{i,m_i}), \quad \underline{b}_i \in \mathbb{F}_2^{m_i}, \quad b_{i,j} \in \mathbb{F}_2$$

with different lengths m_i, $i = 1 \ldots m$. Therefore, the length of vector $\underline{b}$ is

$$m = \sum_{i=1}^{N} m_i. \tag{1}$$

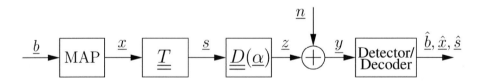

Figure 2. MC-SS transmission model

Each binary m_i-tuple is mapped to one point x_i from a complex valued signal constellation $\mathcal{X}_i$ with cardinality $|\mathcal{X}_i| = 2^{m_i} = M_i$. The mapping is therefore defined as:

$$\text{MAP} : \underline{b}_i \in \mathbb{F}_2^{m_i} \longrightarrow x_i \in \mathcal{X}_i. \tag{2}$$

Clearly, the mapper transforms the binary vector $\underline{b}$ of length m into the vector $\underline{x} = (x_1, \ldots, x_N)^T$ of length N with complex valued entries. Afterwards this vector $\underline{x}$ is linearly transformed by the $N \times N$ complex valued spreading matrix $\underline{\underline{T}}$ into the vector $\underline{s}$, $\underline{s} = \underline{\underline{T}} \cdot \underline{x}$. The vector $\underline{b}$ can be regarded as binary information word which is encoded to a codeword $\underline{s}$ of an N-dimensional complex space. Using OFDM transmission technique with N parallel subchannels, each component s_i of the codeword $\underline{s}$ is subsequently transmitted over a separate subchannel i. This OFDM transmission with N parallel subchannels is represented by the $N \times N$ diagonal channel matrix $\underline{\underline{D}}(\underline{\alpha})$ with channel coefficients α_i and the addition of a noise vector $\underline{n}$. All fading coefficients α_i in the vector $\underline{\alpha}$ are assumed to be complex valued and Gaussian distributed with variance σ_α^2 which means that the fading amplitudes $|\alpha_i|$ are Rayleigh distributed. Therefore, we have $\underline{z} = \underline{\underline{D}}(\underline{\alpha}) \cdot \underline{s}$. The noise vector $\underline{n}$ consists of N independent and Gaussian distributed complex valued random variables n_i with variance σ_n^2. So finally the received vector $\underline{y}$ is given by

$$\underline{y} = \underline{\underline{D}}(\underline{\alpha}) \cdot \underline{\underline{T}} \cdot \underline{x} + \underline{n}.$$

An example for a spreading transform $\underline{\underline{T}}$ is the well-known *Hadamard spreading matrix*, recursively defined by

$$\underline{\underline{T}}^{(1)} = \frac{1}{\sqrt{(2)}} \cdot \begin{pmatrix} 1 & 1 \\ 1 & -1 \end{pmatrix}, \underline{\underline{T}}^{(n)} = \underline{\underline{T}}^{(n-1)} \otimes \underline{\underline{T}}^{(1)}, \tag{3}$$

where $\otimes$ denotes the Kronecker product. With this spreading matrix, the energy of each symbol is equally spread over all subchannels. Furthermore, it is unitary $\underline{\underline{T}}^{-1} = \underline{\underline{T}}^H$, where $\underline{\underline{T}}^H$ denotes the Hermitian conjugate of $\underline{\underline{T}}$. In this case the Euclidean distances between each pair of vectors before and after spreading remain the same.

2.2 Multilevel Coding

The idea of multilevel coding [14] or generalized code concatenation [13] is that m outer codes $\mathcal{C}^{(i)}, i = 1, \ldots, m$ (here binary) protect the labeling of a partitioned signal constellation $\mathcal{B}$. The structure of such a multilevel encoder is depicted in Figure 3. The block **DEMUX** demultiplexes the binary vector $\underline{b} = (b_1 \ldots, b_m)^T$ into its components b_i which are assigned to separate levels of the MLC. For general multilevel coding schemes, the level codes $\mathcal{C}^{(i)}$ can have different complexity and rate. Here we restrict ourselves to a special case of multilevel coding, using only rate $1/N$ binary repetition codes in each level. Therefore, each bit b_i is repeated N times to a codeword $\underline{c_i} = (c_{i,1}, \ldots, c_{i,N})^T$ and the codewords $\underline{c_i}, i = 1, \ldots, m$ build a $m \times N$ codeword matrix $\underline{\underline{C}}$. Each

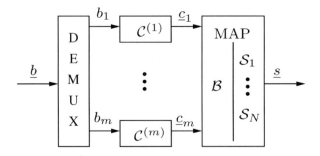

Figure 3. Structure of a multilevel encoder

column i of the codeword matrix $\underline{\underline{C}}$ is mapped to one point s_i from a complex valued signal constellation $\mathcal{S}_i$ with cardinality $|\mathcal{S}_i| = 2^m$. So each binary codeword matrix $\underline{\underline{C}}$ yields a codeword $\underline{s} = (s_1, \ldots, s_N)^T$ of length N in the N-dimensional complex space, $s_i \in \mathcal{S}_i$, $i = 1, \ldots, N$. This results in so-called polyalphabetic codes [7] and we get the following codeword matrix $\underline{\underline{C}}$:

$$\underline{\underline{C}} = \begin{pmatrix} \underline{c}_1^T \\ \vdots \\ \underline{c}_m^T \end{pmatrix} = \begin{pmatrix} c_{1,1} & \cdots & c_{1,N} \\ \vdots & \ddots & \vdots \\ c_{m,1} & \cdots & c_{m,N} \\ \underbrace{\phantom{c_{m,1}}}_{s_1 \in \mathcal{S}_1} & & \underbrace{\phantom{c_{m,N}}}_{s_N \in \mathcal{S}_N} \end{pmatrix}. \qquad (4)$$

Instead of varying the signal constellation for each column i, it is also possible to use a fixed constellation $\mathcal{B}$ for all columns of the matrix $\underline{\underline{C}}$.

2.3 Interrelation between MC-SS and MLC

Now we will show that a MC-SS system can be described by a MLC, first by a small example and then by construction which is a proof for this fact. In [4] further aspects are addressed, i. e., possible simplifications of the MLC for special signal constellations are considered as well as the use of rotated signal constellations. Furthermore in [4] it is shown that the spreading matrices are not restricted to Hadamard matrices but can be any nonsingular matrix. Here it is only important that the MC-SS system described in Section 2.1 can be interpreted as a MLC described in Section 2.2 using simple repetition codes with rate $R = 1/N$ as outer codes which protect the levels of the MLC. Let us start with the example.

Example. Assume the length $N = 2$ orthogonal Hadamard spreading matrix

$$\underline{\underline{T}} = \frac{1}{\sqrt{2}} \cdot \begin{pmatrix} 1 & 1 \\ 1 & -1 \end{pmatrix}$$

and the two signal constellations $\mathcal{X}_1 = \mathcal{X}_2 = \{\pm 1\}$, depicted in Figure 4. Each signal constellation represents one bit, thus $m_1 = m_2 = 1$

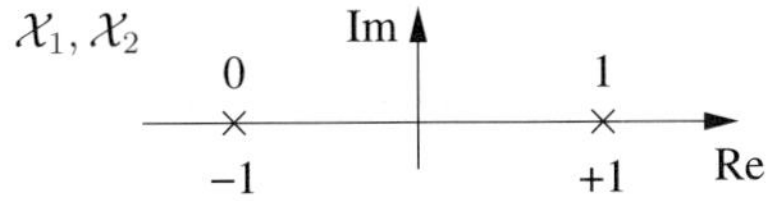

Figure 4. Signal constellations $\mathcal{X}_1$ and $\mathcal{X}_2$

and $M_1 = 2^{m_1} = M_2 = 2^{m_2} = 2$. The binary vector $\underline{b}$ has length $m = m_1 + m_2 = 2$ which results in $M = 2^m = 4$ binary information vectors $\underline{b}$, four different complex valued information vectors $\underline{x}$ and four different codewords $\underline{s}$, where the mapping $\underline{b} \to \underline{x} \to \underline{s}$ is as follows:

$$\begin{pmatrix} 0 \\ 0 \end{pmatrix} \to \begin{pmatrix} -1 \\ -1 \end{pmatrix} \to \frac{1}{\sqrt{2}} \begin{pmatrix} -2 \\ 0 \end{pmatrix} \quad \begin{pmatrix} 0 \\ 1 \end{pmatrix} \to \begin{pmatrix} -1 \\ +1 \end{pmatrix} \to \frac{1}{\sqrt{2}} \begin{pmatrix} 0 \\ -2 \end{pmatrix}$$
$$\begin{pmatrix} 1 \\ 0 \end{pmatrix} \to \begin{pmatrix} +1 \\ -1 \end{pmatrix} \to \frac{1}{\sqrt{2}} \begin{pmatrix} 0 \\ +2 \end{pmatrix} \quad \begin{pmatrix} 1 \\ 1 \end{pmatrix} \to \begin{pmatrix} +1 \\ +1 \end{pmatrix} \to \frac{1}{\sqrt{2}} \begin{pmatrix} +2 \\ 0 \end{pmatrix}. \quad (5)$$

Note, that the cardinality of $\{\underline{s}\}$ is 4 while the cardinality of the space $\frac{1}{\sqrt{2}} \begin{pmatrix} \{-2,0,2\} \\ \{-2,0,2\} \end{pmatrix}$ is 9. This can be interpreted as a code since a subset of a larger space is used. However, another selection of codewords would be possible.

To analyze this MC-SS system using a MLC like in Section 2.2 we first compute two signal constellations $\mathcal{S}_1$ and $\mathcal{S}_2$ which are depicted in Figure 5. The labeling is not the same for $\mathcal{S}_1$ and $\mathcal{S}_2$.

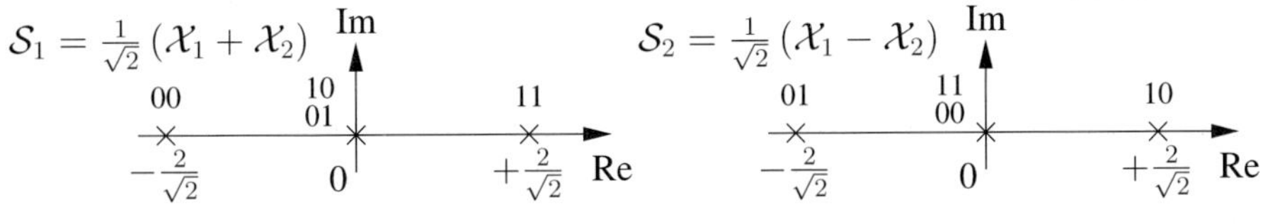

Figure 5. Signal constellations $\mathcal{S}_1$ and $\mathcal{S}_2$

Together with the binary code word matrix we get

$$\underline{\underline{C}} = \begin{pmatrix} b_1 & b_1 \\ \underbrace{b_2}_{s_1 \in \mathcal{S}_1} & \underbrace{b_2}_{s_2 \in \mathcal{S}_2} \end{pmatrix}$$

which is the same code word matrix as in (4). The MC-SS system can be obtained by the multilevel encoder depicted in Figure 6.

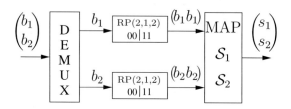

Figure 6. Equivalent multilevel encoder

For the case of a Hadamard spreading matrix and some simple restrictions on the constellations $\mathcal{X}_1$ and $\mathcal{X}_2$ and their labeling [4] we can simplify the corresponding multilevel encoder for the case of Figure 6. The simplification is that just one signal constellation can be used for all N symbols by using a coset of the second repetition code. The signal constellations depicted in Figure 5 differ in the labeling by just the single vector $\underline{a}_{S_2} = (0,1)$. Therefore, the labeling of constellation S_2 can be obtained by the modulo 2 addition of $\underline{a}_{S_2}$ to the labeling vector of the points from constellation S_1. Therefore we get the MLC shown in Figure 7.

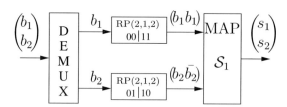

Figure 7. Simplified multilevel encoder

General Case. By comparing MC-SS systems from section 2.1 with the MLC from section 2.2, one can see that they are the same. Both encode a binary vector $\underline{b}$ of length m to a codeword $\underline{s}$ in the Euclidean space of length N. To show that a MLC can perform the same mapping from binary vectors $\underline{b}$ to complex valued codewords $\underline{s}$, we recall the functionality of the MAP block in Figure 2. Each binary m_i-tuple $\underline{b}_i$ is mapped to one point from the signal constellation $\mathcal{X}_i$, see equation (2). A mapping can only be organized if each point x_i has at least one binary m_i-tuple that labels it:

$$l(x_i) = \underline{b}_i. \tag{6}$$

Here the label of the signal x_i is denoted by $l(x_i)$, which is the binary m_i-tuple $\underline{b}_i$. As in Section 2.1 we introduce $\mathcal{X}$ as the set of all possible vectors $\underline{x} = (x_1, \ldots, x_N)$ with components x_i chosen from its own signal constellation $\mathcal{X}_i$ with cardinality $|\mathcal{X}_i| = 2^{m_i} = M_i$. The set $\mathcal{X}$ can be defined using the Cartesian product of the signal constellations $\mathcal{X}_i \subset \mathbb{C}$:

$$\mathcal{X} = \mathcal{X}_1 \times \ldots \times \mathcal{X}_N = \{\underline{x} = (x_1, \ldots, x_N) \,|\, x_i \in \mathcal{X}_i\}.$$

Given an $N \times N$ spreading matrix $\underline{\underline{T}}$ with row vectors $\underline{t}_i = (t_{i,1} \ldots t_{i,N})$ with entries $t_{ik} \in \mathbb{C}$, the 2^m codewords $\underline{s}$ can be computed by evaluating $\underline{s} = \underline{\underline{T}} \cdot \underline{x}$. The code symbols s_i from the codeword $\underline{s}$ are the scalar products of the row vectors $\underline{t}_i$ from $\underline{\underline{T}}$ and the vector $\underline{x}$ containing the information signals:

$$s_i = \underline{t}_i \cdot \underline{x}, \quad \underline{x} \in \mathcal{X}, s_i \in \mathbb{C}.$$

For each vector $\underline{x}$ we get one scalar s_i. The sets containing all the possible s_i are denoted by $\mathcal{S}_i$ with

$$\mathcal{S}_i = \{s_i | s_i = \underline{t}_i \cdot \underline{x}, \forall \underline{x} \in \mathcal{X}\}. \tag{7}$$

The cardinalities $|\mathcal{S}_i|$ of the sets $\mathcal{S}_i$ cannot exceed 2^m, $|\mathcal{S}_i| \leq 2^m$, $i = 1, \ldots, N$ because there are only 2^m different input vectors $\underline{x}$. However, it is possible that different vectors $\underline{x}$ result in the same symbol s_i. Hence, the cardinalities $|\mathcal{S}_i|$ could be strictly less than 2^m. The result of mapping and spreading can be viewed as a polyalphabetic code of length N, transmitting m bits per N channel uses.

According the labeling from (6), we label the resulting code symbols s_i by the juxtaposition of the corresponding labels of the components x_i:

$$l(s_i) = (l(x_1), l(x_2), \ldots, l(x_N)). \tag{8}$$

For the case that $|\mathcal{S}_i| < 2^m$ it is also possible that more than one bit tuple label the same symbol s_i. With the labeling in (8) we can organize the mapping $\underline{b} \longrightarrow \underline{s}$ via the binary codeword matrix $\underline{\underline{C}}$, $\underline{b} \longrightarrow \underline{\underline{C}} \longrightarrow \underline{s}$ with

$$\underline{\underline{C}} = \underbrace{(\underline{b}, \underline{b}, \cdots, \underline{b})}_{N \text{ copies}} = \begin{pmatrix} b_1 & \cdots & b_1 \\ \vdots & \vdots & \vdots \\ b_m & \cdots & b_m \end{pmatrix}_{s_1 \in \mathcal{S}_1 \quad s_N \in \mathcal{S}_N} = \begin{pmatrix} \underline{c}_1 \\ \vdots \\ \underline{c}_m \end{pmatrix}. \tag{9}$$

Repeating the binary information vector, $\underline{b}$, N times yields the binary codeword matrix $\underline{\underline{C}}$ and for each column i a signal s_i from signal constellation $\mathcal{S}_i$ is chosen. Comparing this with the multilevel coding scheme from Figure 3 and the corresponding mapping from the binary codeword matrix $\underline{\underline{C}}$ to the codeword $\underline{s}$ according to (4), we see that the above mentioned $\overline{\text{MC-SS}}$ can be encoded with a MLC having m identical binary *repetition codes (RPC)* of length N, dimension $k = 1$ and minimum distance $d = N$, $(RPC(N, 1, N))$ in each level, see Figure 8. The only

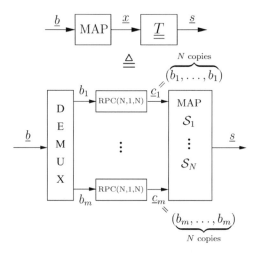

Figure 8. Interpreting MC-SS as MLC

specialty with the multilevel coding scheme is that N different signal constellations $\mathcal{S}_1$ up to $\mathcal{S}_N$ are used. These constellations are computed via (7) and labeled using (8).

In [4] it is shown that the different signal constellations can be represented by one constellation if they can be transformed by a single vector into each other. Furthermore, it is shown that the concept of diversity depends on the cardinalities of the signal constellations. All the theory for MLCs and generalized concatenation can be applied to improve the code properties on the one hand and decoding on the other. For example iterative multistage decoding could be used to decode MC-SS systems. However, there might be a complexity problem since the signal constellations are getting extremely large.

3. Transmission over the Broadcast Channel

In order to understand the approach of information theory to the broadcast channel we start with the simple example from Cover [8]. Figure 9 shows the broadcast channel for two users. They receive the same signal corrupted by two different channels. Note that individual information for each user is included in one signal X.

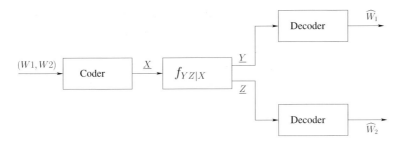

Figure 9. Broadcast transmission (downlink) for two users

Let us investigate the case of Figure 9. We assume that user 1 has an error free binary symmetric channel (BSC) with error probability $p_1 = 0$ and user 2 has a BSC with $p_2 = p$. This means that in case of exclusive use of the channel user 1 would have capacity $C_1 = 1$ whereas user 2 would have $C_2 = 1 - h(p)$. Here, $h(p)$ denotes the binary entropy function. For time division multiple access (TDMA) we split the transmission time into two parts, say λ is the part of user 1 and $1 - \lambda$ the one of user 2, where $0 \leq \lambda \leq 1$. We ignore guard times and assume that switching between users is perfect. Clearly, we can transmit at any rates $R_{1,T}$ and $R_{2,T}$ as long as the following conditions hold:

$$R_{1,T} \leq \lambda \cdot C_1 = \lambda \\ R_{2,T} \leq (1 - \lambda) \cdot C_2 = (1 - \lambda)(1 - h(p)). \tag{10}$$

Usually, the border line of the rate region is plotted over the rates $R_{1,T}$ and $R_{2,T}$ and the achievable rates are given by any point below and on the border line. The curve for our example is the dashed line depicted in Figure 11. Clearly, the value on the x-axis is 1 since for $\lambda = 1$ only data for user 1 are transmitted which can be done at capacity $C_1 = 1$. If only data for user 2 are transmitted the whole time, then we have $\lambda = 0$ and the according point on the y-axis is $1 - h(p)$ which corresponds to the capacity C_2 of user 2. By varying λ between 0 and 1 the points on the line are obtained in case of equality in (10).

For the broadcast case, we first consider the encoding of the information for user 2. However, we use a code which has a slightly higher error correcting capability than required for channel 2. This better code is then suitable for transmission over a cascade of two BSCs, one with error probability α and one with p, see Figure 10. Now the codewords of user 1 should be chosen in a way such that 1's occur with probability at most α. These codewords are added to the ones of user 2. This can be seen as an introduction of additional errors for user 2. Thus the effective channel of user 2 corresponds to the above described cascade of two BSCs. This effective channel has error probability $\varepsilon = \alpha \cdot (1-p) + (1-\alpha) \cdot p$.

At the receiver side, user 2 only has to decode the received word to get its information. Now let us look at user 1. We assume that the code of user 2 is also known to user 1. Therefore, user 1 can decode the received word according to the code of user 2. Since user 1 has an error free channel, subtracting the decoded word from the received one directly yields the codeword of user 1.

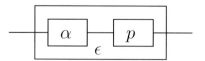

Figure 10. Cascade of two BSC

Therefore we can calculate the achievable rates for the broadcast case with two users

$$R_{1,B} \leq h(\alpha)$$
$$R_{2,B} \leq 1 - h(\varepsilon) = 1 - h\left(\alpha \cdot (1-p) + (1-\alpha) \cdot p\right) \quad (11)$$

This curve is plotted as bold line in Figure 11 and again each point can be realized below and on the curve. Thus, also points above the TDMA

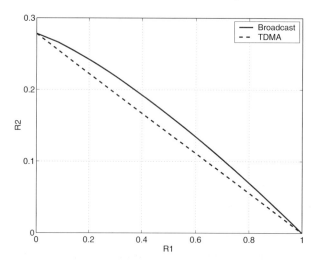

Figure 11. Rate regions for two BSCs

line can be reached which means that with broadcast higher rates than with TDMA can be achieved.

This example can be generalized to the case when not only one but both users have channels with errors. The principle idea is to use unequal error protection (UEP) codes (see e.g. [13]). To illustrate the UEP principle, in Figure 12 we see that the codewords of user 2 are the centers of the clouds and user 1 selects one satellite codeword inside a cloud.

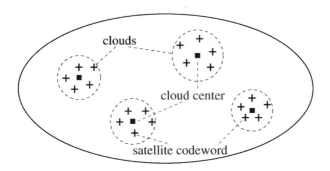

Figure 12. UEP codewords

In case we have additive white Gaussian noise (AWGN) channels instead of BSCs, the UEP can also be achieved by simple superposition of modulation alphabets with different energies. Thus, we obtain a *cloud modulation* as shown in Figure 13.

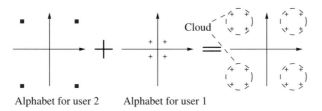

Figure 13. Cloud modulation

Clearly, this concept can be generalized to many users with different protection levels. However, here we only would like to give the principle ideas.

Example. As a more practical example we consider an AWGN channel. We assume that the signal to noise ratio (SNR) of user 1 is 22 dB and that of user 2 is 2 dB. We use cloud modulation with different alphabets as describe above. For user 1 we use 64QAM and for user 2 only QPSK whereas different energy corresponding to the channel SNR is utilized.

Additionally, the data of each user are encoded by a rate 1/3 turbo code with the component code's generator polynomials given by (31, 33). The encoding scheme is shown in Figure 14. For the TDMA scenario

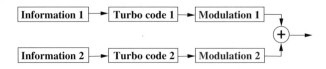

Figure 14. Encoding for broadcast transmission (Downlink)

we use 256 QAM for user one and QPSK for user 2. Again the data are encoded by the same turbo code. In order to obtain the different transmission rates the codes are punctured. In Figure 15 the simulation results and the theoretical limits are plotted.

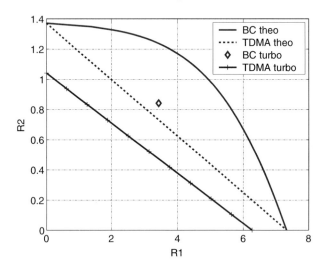

Figure 15. Comparison of TDMA and broadcast

The simulated broadcast case is a single point in the plot. Compared to the theoretical curve for TDMA, higher data rates can be achieved simultaneously for both users. By comparing the simulated broadcast rates with the simulated TDMA rates it can be observed that the rates could be more than doubled for user 1 and simultaneously almost doubled for user 2.

This example shows that with broadcast transmission an increase of achievable data rates can be obtained for some cases, especially, when the users have different channels which is not an unrealistic assumption. In existing wireless systems the feature of power control is implemented just because of this fact. But broadcast could also be used in this scenario. Unfortunately, there exist only few publications considering and comparing broadcast transmission with multiple access. In our group there exist two first results. In [15] the broadcast transmission was compared to FDMA and shown that the achievable data rates can be increased. Furthermore, in [16] it was shown analytically that for independent Rayleigh channels the delay limited capacity of broadcast transmission is finite while the one for single carrier FDMA is zero. Also in the case whith many subcarriers it could be shown that broadcast yields the bigger delay limited capacity compared to FDMA.

4. Conclusions

In this paper we introduced a special coding perspective of MC-SS systems. We gave a constructive proof, that each MC-SS system and similar schemes can be interpreted as a MLC with identical repetition codes in the levels. Hence, we revealed the interrelation between spreading transforms and repetition coding for the special case of MC-SS. This way of looking at MC-SS and MLC could result in insights for both, MC-SS as well as MLC. Particular parts from the existing theory for one system can be applied for the other one and vice versa. Of course it is up to the desired problem, whether the one or the other description method is more convenient.

In the second part we presented broadcast ideas and showed by means of simulation results that broadcast can be superior to multiple access schemes. It is still an open question if and how broadcast transmission can improve the data rates and the quality in future systems. A lot of work has to be done to transfer the theoretical advantage to practical methods and algorithms.

Acknowledgments

I would like to thank my PhD students for the help with this paper, namely Carolin Huppert, Dr.-Ing. Bernd Baumgartner, Axel Hof, and Aeman Mohammed.

Parts of this work were supported by Deutsche Forschungsgemeinschaft, Grants Bo867/8 and Bo867/13.

References

[1] K. Fazel and L. Papke. On the Performance of Convolutionally-Coded CDMA/OFDM for Mobile Communications Systems. In *Proc. IEEE PIMRC'93*, pages 468–472, Yokohama, Japan, September 1993.

[2] N. Yee, J. P. Linnartz, and G. Fettweis. Multi-carrier CDMA in Indoor Wireless Radio Networks. In *Proc. IEEE PIMRC'93*, pages 109–113, Yokohama, Japan, September 1993.

[3] T. Frey and M. Bossert. A First Approach to Concatenation of Coding and Spreading for CDMA Systems. In *Proc. Int. Symp. on Spreading Spectrum Techniques and Applications*, pages 667–671, Mainz, Germany, September 1996.

[4] B. Baumgartner, V. Sidorenko, and M. Bossert. Multicarrier Spread Spectrum: A Coding Perspective In *Proc. IEEE 8th International Symposium on Spread Spectrum Techniques and Applications*, pages 61-66, Sydney, Australia, September 2004.

[5] Z. Liu, Y. Xin, and G. B. Giannakis. Linear Constellation Precoding for OFDM with Maximum Multipath Diversity and Coding Gains. *IEEE Trans. on Comm.*, 51(3):416–427, March 2003.

[6] Z. Wang and G. B. Giannakis. Complex-Field Coding for OFDM Over Fading Wireless Channels. *IEEE Trans. on Inform. Theory*, 49(3):707–720, March 2003.

[7] M. Bossert and E. Gabidulin. Polyalphabetic Codes. Dept. TAIT, University of Ulm, Germany, preprint 2004.

[8] T. M. Cover. Broadcast channels. *IEEE Trans. Inf. Theory*, IT-18:2–14, January 1972.

[9] T. M. Cover. An achievable rate region for the broadcast channel. *IEEE Trans. Inf. Theory*, 21(4):399–404, July 1975.

[10] E. C. van der Meulen. A survey of multi-way channels in information theory: 1961–1976. *IEEE Trans. Inf. Theory*, IT-23(1):1–37, January 1977.

[11] T. M. Cover. Comments on broadcast channels. *IEEE Trans. Inf. Theory*, IT-44:2542–2530, October 1998.

[12] S. Verdu. *Multiuser Detection*. Cambridge University Press, 1988.

[13] M. Bossert. *Channel Coding*. John Wiley & Sons, New York, first ed., 1999, ISBN 0-471-98277-6.

[14] H. Imai and S. Hirakawa. A New Multilevel Coding Method Using Error-Correcting Codes. *IEEE Transactions on Information Theory*, IT-23(3):371–377, 1977.

[15] M. Bossert and A. S. Mohammed Downlink transmission as broadcast channel. *Proc. International Symposium on Information Theory and its Applications*, Parma, Italy, pp. 116–119, October 2004.

[16] C. Huppert and M. Bossert, Delay-Limited Capacity for Broadcast Channels, Proc. 11th European Wireless Conference, Nicosia, Cyprus, pp. 829–834, April 2005.

ADAPTIVITY IN MC-CDMA SYSTEMS

Ivan Cosovic
German Aerospace Center (DLR), Inst. of Communications and Navigation
Oberpfaffenhofen, 82234 Wessling, Germany
ivan.cosovic@dlr.de

Stefan Kaiser
DoCoMo Communications Laboratories Europe GmbH
Landsberger Strasse 312, 80687 Munich, Germany
kaiser@docomolab-euro.com

Abstract This paper investigates the performance of MC-CDMA with link adaptation in the uplink of a wireless communication system. Two types of link adaptation approaches can be distinguished: 1) Exploitation of channel state information (CSI) at the transmitter and 2) exploitation of CSI at both the transmitter and the receiver. Based on the theoretical lower bounds for the different approaches we design multi-user concepts and present their performances in fading channels with different system loads. Moreover, real channel estimation is considered in this investigations and the effects of different link adaptation techniques on the peak-to-average power ratio are shown.

1. Introduction

Multi-carrier communications realized by orthogonal frequency-division multiplexing (OFDM) have proven to be suitable for broadband wireless communications in several broadcasting and wireless LAN standards [1]. In addition, OFDM is currently under investigation for mobile radio systems beyond 3G (B3G). In order to further increase the capacity of wireless systems, the exploitation of link adaptation has become an interesting topic of research. Several publications cover link adaptation in orthogonal frequency-division multiple-access (OFDMA) systems, e.g., [2]. In OFDMA bit and power loading can be combined in order to increase the capacity of the system. Compared to OFDMA only few publications cover link adaptation for multi-carrier code-division multiple-access (MC-CDMA), e.g., [3]. With MC-CDMA each data symbol is spread over several subcarriers which can be faded independently. This requires more advanced link adaptation approaches compared to OFDMA.

In this paper we investigate the application of link adaptation in the uplink of an MC-CDMA system. Different approaches for link adaptation depending on the availability of channel state information (CSI) in the system are considered. When CSI is only available at the transmitter, the link adaptation is based on pre-equalization using power loading. If CSI is available at the transmitter and the receiver, the scheme is referred to as combined-equalization with link adaptation using power loading. These two link adaptation approaches differ significantly as the latter offers more degrees of freedom in the system design than the former. As reference, we include the performance of systems which have CSI available only at the receiver, i.e., post-equalization without link adaptation. The corresponding theoretical lower bounds are presented. Based on these, the achievable performance under real world conditions like channel fading, channel estimation inaccuracies, and different system loads is shown. Since the uplink is sensitive to high peak-to-average power ratios (PAPRs), the influence of the different combined-equalization techniques on the PAPR is compared.

The remainder of the paper is organized as follows. The basic principles of the investigated adaptivity methods for MC-CDMA are described in Section 2. In Section 3, we describe the considered uplink MC-CDMA signal model. The numerical simulations are provided in Section 4, and conclusions are drawn in Section 5.

2. Adaptivity in MC-CDMA

Transmission through the frequency-selective fading channel causes degradations of the transmission signal at the receiver. Considering MC-CDMA transmission, the signal-to-noise ratio (SNR) degrades and multiple-access interference (MAI) occurs. To eliminate or reduce the resulting performance degradations, equalization and detection of the received signal based on the CSI estimation can be performed at the receiver.

Alternatively, as illustrated in Fig. 1, CSI can be exploited prior to transmission, i.e., at the MC-CDMA transmitter, as inherently available in, for example, TDD and closed-loop systems. In these systems, the link adaptation can be performed in the form of pre-equalization such that SNR is maximized, MAI is eliminated, or a certain trade-off between these two is achieved. However, when SNR is maximized MAI is further enhanced and, vice-versa, when MAI is eliminated SNR is further degraded. Thus, it can be expected that, if a system is heavily loaded, a technique which compromises these two counteracting effects will not be capable to closely approach the corresponding single-user bounds.

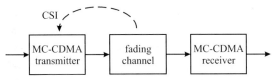

Figure 1. Block diagram of the MC-CDMA transmission system with CSI available only at the MC-CDMA transmitter.

As illustrated in Fig. 2, adaptivity can be used in conjunction with post-equalization at the receiver by exploiting CSI at both MC-CDMA transmitter and receiver. This approach is called combined-equalization and targets a common optimization of data transmission and reception. The main characteristics of combined-equalization are as follows:

- The classical matched filter (MF) bound is valid as the MC-CDMA performance bound for all concepts which are based on pre- and post-equalization applied separately. However, this does not hold for combined-equalization Combined-equalization is lower-bounded by the selection diversity (SD) combined-equalization bound that lies well-below the MF bound [4]. The reason for this is based on the fact that, observed from the receiver, the transmission signal is distorted by both link adaptation and fading channel. Hence, detection at the receiver is performed with respect to a *pre-equalized fading channel* which takes into account both link adaptation and fading channel influence and has different statistics from the fading channel observed alone.

- CSI at both transmitter and receiver ensures more degrees of freedom in the system design and helps to better resolve the problem of joint MAI avoidance/suppression and SNR maximization. In particular, pre-equalization coefficients are designed for efficient distribution of transmit power among the subcarriers with respect to the fading channel, whereas MAI cancellation is the main task of post-equalization. Ideally, the SD combined-equalization bound will be approached closely even if the MC-CDMA system is heavily-loaded.

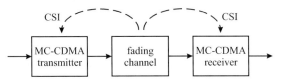

Figure 2. Block diagram of the MC-CDMA transmission system with CSI available at both MC-CDMA transmitter and receiver.

3. Uplink MC-CDMA Signal Model

We consider an uplink MC-CDMA system with a total number N_c of subcarriers. The system is divided into $M \cdot Q$ subsystems of L subcarriers.

This flexible scheme is better known as MC-CDMA with $M\&Q$-modification [1]. Note, M denotes the number of parallel data symbols transmitted per OFDM symbol for each user, while Q denotes the number of different, parallel user-groups per OFDM symbol. Without loss of generality we concentrate on a single subsystem of K users ($K \leq L$). The users are separated by spreading sequences of length L. Moreover, we assume that the corresponding L subcarriers are spread over the signal bandwidth in order to better exploit frequency diversity of the channel, i.e., they are frequency interleaved with the remaining $N_c - L$ subcarriers. We denote with l, $l = 1, 2, \ldots, L$, the subcarriers in the considered subsystem. The block diagram of the kth, $k = 1, 2, \ldots, K$, uplink MC-CDMA transmitter of the considered subsystem is shown in Fig. 3(a). The input bits are channel encoded and symbol mapped producing the complex-valued data symbols. Each data symbol $d^{(k)}$ is spread by a unit-energy spreading sequence $\mathbf{c}^{(k)} = (c_1^{(k)}, c_2^{(k)}, \ldots, c_L^{(k)})^{\mathrm{T}}$. The spreading process results in the sequence $\mathbf{s}^{(k)}$ given by

$$\mathbf{s}^{(k)} = \mathbf{c}^{(k)} d^{(k)} = (s_1^{(k)}, s_2^{(k)}, \ldots, s_L^{(k)})^{\mathrm{T}}. \tag{1}$$

Pre-equalization is applied resulting in a new sequence $\bar{\mathbf{s}}^{(k)}$,

$$\bar{\mathbf{s}}^{(k)} = \mathbf{G}_{\mathrm{pre}}^{(k)} \mathbf{s}^{(k)} = (\bar{s}_1^{(k)}, \bar{s}_2^{(k)}, \ldots, \bar{s}_L^{(k)})^{\mathrm{T}}, \tag{2}$$

where $\mathbf{G}_{\mathrm{pre}}^{(k)}$ is a diagonal $L \times L$ pre-equalization matrix with diagonal elements $G_{\mathrm{pre},l,l}^{(k)}$, $l = 1, 2, \ldots, L$. In this paper, pre-equalization is performed applying generalized pre-equalization (G-pre-eq) whose pre-equalization coefficients are given by

$$G_{\mathrm{pre},l,l}^{(k)} = \left|H_{l,l}^{(k)}\right|^p H_{l,l}^{(k)*} \sqrt{\frac{L}{\sum_{n=1}^{L} |H_{n,n}^{(k)}|^{2p+2}}}, \quad l = 1, 2, \ldots, L, \tag{3}$$

where $H_{l,l}^{(k)}$ represents the fading coefficient on the lth subcarrier for user k. G-pre-eq is a unified approach of pre-equalization and comprises several well-known pre-equalization techniques for different values of its design parameter p, e.g., pre-equalization maximum ratio transmission (pre-eq MRT) for $p = 0$, pre-equalization equal gain transmission (pre-eq EGT) for $p = -1$, and pre-equalization zero-forcing (pre-eq ZF) for $p = -2$. G-pre-eq is especially suitable for combined equalization since setting $p > -1$ and combining it with post-equalization enables better single-user bounds than the MF bound [4]. However, increasing p is expected to enhance MAI at the receiver as the selectivity of the pre-equalized fading channel is increased. Hence, as shown in Section 4,

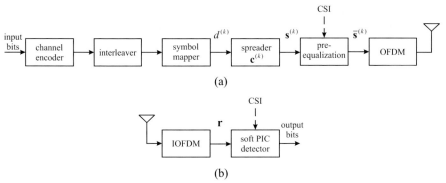

Figure 3. Block diagram of the uplink MC-CDMA transmission system (a) transmitter and (b) receiver.

the parameter p can be used to achieve a proper trade-off between SNR maximization and MAI elimination in a multi-user uplink MC-CDMA system.

Finally, the pre-equalized sequence $\bar{\mathbf{s}}^{(k)}$ is OFDM-modulated onto the corresponding L out of N_c subcarriers and transmitted. OFDM comprises inverse fast Fourier transform and addition of a guard interval by cyclic extension of the OFDM symbol.

The block diagram of an uplink MC-CDMA receiver of the considered subsystem is shown in Fig. 3(b). After the inverse OFDM (IOFDM) operation the received signal results in

$$\mathbf{r} = \sum_{k=1}^{K} \mathbf{H}^{(k)} \mathbf{G}_{\text{pre}}^{(k)} \mathbf{s}^{(k)} + \mathbf{n} = (r_1, r_2, \ldots, r_L)^{\mathrm{T}}, \qquad (4)$$

where $\mathbf{H}^{(k)}$ represents the $L \times L$ diagonal channel matrix for user k with the diagonal elements $H_{l,l}^{(k)}$, $l = 1, 2, \ldots, L$. The vector $\mathbf{n} = (n_1, n_2, \ldots, n_L)^{\mathrm{T}}$ represents the AWGN with the one-sided noise spectral density N_0 in each component. As shown in Fig. 3(b), if CSI is available at the receiver, in this paper, the received signal $\mathbf{r}$ is detected with a soft parallel interference detector (PIC) similar to the one described in [5]. The soft PIC is performed with respect to the pre-equalized fading channel which is defined as

$$\overline{\mathbf{H}}^{(k)} = \mathbf{H}^{(k)} \mathbf{G}_{\text{pre}}^{(k)}. \qquad (5)$$

Note, if CSI is available only at the transmitter, instead of soft PIC only symbol demapping and channel decoding are performed at the receiver.

4. Numerical Results

In this section, several numerical results are given that illustrate the performance of an adaptive uplink MC-CDMA system.

Table 1. MC-CDMA system parameters for simulations

carrier frequency	5.2 GHz
transmission bandwidth	50 MHz
number of subcarriers	$N_c = 512$
guard length	$1.8\,\mu$s
OFDM symbols per frame	64
OFDM frame duration	0.77 ms
signal constellation	QPSK
spreading sequences	Walsh-Hadamard; length $L = 8$
system load	full-load ($K = 8$) or half-load ($K = 4$)
channel coding	convolutional; rate $R = 1/2$; memory $m = 6$
channel model	ETSI BRAN Model E

Table 1 shows the main parameters of the uplink MC-CDMA simulation system. The underlying channel model is based on the Channel E defined in the ETSI BRAN HIPERLAN/2 standardization project [6]. Unless otherwise stated, we assume that CSI is perfectly known and that mobile speed equals $v = 0$ m/s. The SNR is given in E_t/N_0, i.e., the transmitted energy per bit E_t over the one-sided noise spectral density N_0.

First, we consider adaptive uplink MC-CDMA with combined equalization and determine the G-pre-eq parameter p such that the SNR required to achieve an average bit-error-rate (BER) of $P_b = 10^{-3}$ is minimized. Fig. 4 shows the required SNR as a function of the parameter p. Simulation results are given for fully- ($K = 8$) and half-loaded ($K = 4$) systems, whereas the parameter p of G-pre-eq is varied in order to find the optimum value which minimizes the required SNR. At the receiver soft PIC with 3 iterations is applied. It can be seen that a minimum SNR is achieved by setting $p = 1$ and $p \in [4, 9]$ in the case of full- and half-load, respectively. The reason for this difference is a considerably different amount of MAI in both cases.

In Fig. 5, the performance of adaptive uplink MC-CDMA is shown for the cases when CSI is available only at the transmitter and when CSI is available at both transmitter and receiver. In the case when CSI is available only at the transmitter, a pre-equalization technique which maximizes SINR is applied [3]. Adaptive MC-CDMA that exploits CSI at both transmitter and receiver employs G-pre-eq together with soft PIC (3 iterations). G-pre-eq with $p = 1$ and $p = 6$ is used in the case of full- and half-load, respectively. As references, the classical MF bound and the SD combined-equalization bound are given. In addition, performance results of uplink MC-CDMA with post-equalization applying soft PIC with 3 iterations (i.e., of uplink MC-CDMA without adaptivity) are shown. From Fig. 5 can be seen that adaptive uplink MC-CDMA

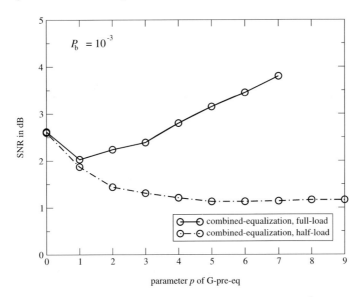

Figure 4. SNR required to achieve an average BER of $P_b = 10^{-3}$ versus the G-pre-eq parameter p for adaptive uplink MC-CDMA with combined-equalization.

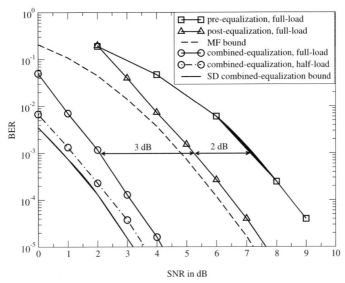

Figure 5. BER performance versus SNR for adaptive uplink MC-CDMA with pre-equalization and with combined-equalization.

that employs only pre-equalization performs worse than other considered techniques and is far from the MF bound. However, adaptive uplink MC-CDMA that employs combined-equalization outperforms others by several dB and approaches the SD combined-equalization bound closely.

The effects of imperfect CSI at the uplink transmitter on the system performance are considered in Fig. 6. We estimate CSI at the mobile

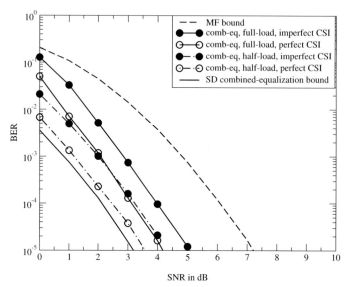

Figure 6. Effect of imperfect CSI at the mobile terminals on the BER performance of adaptive uplink MC-CDMA with combined-equalization.

terminal using a downlink channel estimation approach discussed in [7]. We exploit the estimates obtained from the pilot symbol inserted at the end of the downlink OFDM frame assuming that the power of this pilot symbol is boosted four times providing a more accurate CSI for the purposes of pre-equalization. We see that imperfect CSI induces an SNR loss of less than 1 dB irrespective of system load.

The impact of channel variations on the system performance is addressed in Fig. 7 where the SNR required to achieve a BER of $P_b = 10^{-3}$ is shown as a function of the mobile speed v. For pre- and combined-equalization the pre-equalization coefficients are computed at the beginning of each uplink frame and are kept fixed over the entire frame while the channel varies continuously according to the mobile speed v. As we can see, due to the soft PIC at the receiver, post- and combined-equalization are much more resistent to channel variations than pre-equalization. As expected the required SNR grows as v increases, but the performance loss is relatively low for the mobile speeds up to $v \leq 5$ m/s. Thus, the proposed scheme is well-suited for scenarios with low-mobility.

Finally, we address the effect of pre-equalization on the PAPR. Pre-equalization leads to power re-distribution at the transmitter. As shown in [4], the distribution of the transmission power strongly depends on the G-pre-eq parameter p. As p grows, more power is invested on subcarriers with high gains and less on others. Already for $p = 3$ more than 90% of the transmission power is concentrated on around 25% of available subcarriers. Considering the time domain representation most of the

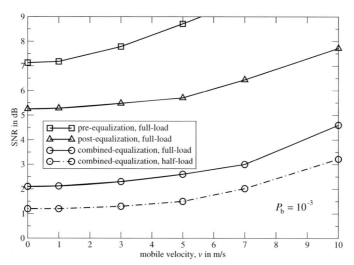

Figure 7. SNR required to achieve a BER of $P_b = 10^{-3}$ versus mobile speed v for uplink MC-CDMA with combined-equalization, post-equalization, or pre-equalization.

power is concentrated on only a part of available sinusoidal waveforms. Such a situation leads to a lower PAPR, since there are fewer possibilities for constructive addition of signals. This is illustrated in Fig. 8 where a cumulative distributive function (CDF) of the PAPR for uplink MC-CDMA applying different pre-equalization techniques is shown assuming, for simplicity reasons, that a system employs Q-modification. It can be seen that already for $p = 3$ the PAPR is significantly reduced in comparison to pre-eq EGT. However, as shown in Fig. 4 adjustment of this parameter towards lower PAPR can lead to a worse BER performance, which can be partly compensated by introducing additional complexity at the base station, e.g., by introducing successive interference cancellation instead of PIC. Thus, in realistic uplink MC-CDMA with combined equalization, a certain trade-off between PAPR reduction, receiver complexity, and achievable performance can be sought.

5. Conclusions

Different approaches to exploit link adaptation in MC-CDMA are investigated for the uplink. Under consideration are pre-equalization and combined-equalization. As reference, post-equalization is also included. It could be shown that gains of 3 dB in SNR could be achieved when applying link adaptation compared to MC-CDMA systems without link adaptation. The presented results include the theoretical lower bounds and the performance of full and half loaded systems. Channel fading as well as real channel estimation is taken into account. Since the focus is on the uplink, the influence of different link adaptation approaches on

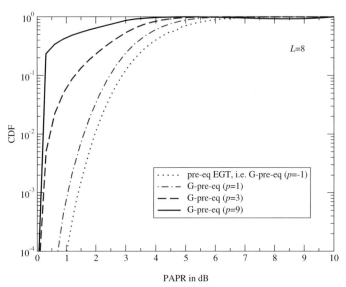

Figure 8. CDF for PAPR in dB for uplink MC-CDMA with pre-equalization for different G-pre-eq.

the PAPR is investigated and it could be shown that there is a trade-off between performance with link adaptation and reduction in PAPR.

References

[1] K. Fazel and S. Kaiser. *Multi-Carrier and Spread Spectrum Systems*. John Wiley & Sons, 2003.

[2] C. Y. Wong, R. S. Cheng, K. B. Letaief, and R. D. Murch. Multiuser OFDM with adaptive subcarrier, bit, and power allocation. In *IEEE J. Select. Areas Commun.*, pages 1747–1758, Oct. 1999.

[3] D. Mottier and D. Castelain. SINR-based channel pre-equalization for uplink multi-carrier CDMA systems. In *Proceedings IEEE International Symposium on Personal, Indoor and Mobile Radio Communications (PIMRC'02)*, pages 1488–1492, Sept. 2002.

[4] I. Cosovic and S. Kaiser. Exploitation of diversity in MC-CDMA systems. In *Proceedings IEE International Conference on 3G and Beyond (3G'05)*, Nov. 2005.

[5] S. Kaiser and J. Hagenauer. Multi-carrier CDMA with iterative decoding and soft-interference cancellation. In *Proceedings IEEE Global Telecommunications Conference (GLOBECOM'97)*, pages 6–10, Nov. 1997.

[6] J. Medbo. Channel models for HIPERLAN/2 in different indoor scenarios. In *ETSI EP BRAN 3ERI085B*, March 1998.

[7] L. Sanguinetti, M. Morelli, and U. Mengali. Channel estimation and tracking for MC-CDMA signals. In *European Transactions on Communications (ETT)*, pages 249–258, May/June 2004.

PARAMETER ESTIMATION FOR INTERFERENCE CANCELLATION IN POWER-CONTROLLED MC-CDMA DOWNLINK TRANSMISSIONS

Michele Morelli, Luca Sanguinetti and Umberto Mengali
University of Pisa
Department of Information Engineering
Pisa, Italy
{michele.morelli,luca.sanguinetti,umberto.mengali}@iet.unipi.it

Abstract We consider the downlink of a power-controlled MC-CDMA network in which the base station assigns the power to each user according to its location within the cell. In particular, more power is assigned to the users near the cell boundaries in order to improve the geographical fairness in data reception. This results in a near-far effect for the users located close to the base station, which enhances the multiple-access interference (MAI) and limits the performance of conventional single-user receivers. Interference cancellation (IC) schemes may be employed to mitigate the detrimental effects of MAI. However, they require knowledge of the channel responses as well as of the noise power and strengths of the interfering signals. In this paper, we address the problem of estimating all of the above parameters using a maximum likelihood (ML) approach and exploiting pilot blocks multiplexed with the transmitted data. The estimates are then used by a non-linear detector in which an interfering signal is cancelled provided that its estimated power exceeds a given threshold. Simulation results show that the proposed scheme outperforms other existing IC-based techniques.

1. Introduction

Multi-Carrier Code-Division Multiple-Access (MC-CDMA) is considered as a promising candidate for supporting high-speed transmissions in future cellular networks [1] due to its high spectral efficiency and robustness to frequency selective fading.

Recent publications show that MC-CDMA is particularly suitable for downlink transmissions, i.e., from the base station (BS) to the mobile terminals (MTs). In these applications orthogonal spreading codes are usually employed to provide protection against cochannel interference. In the presence of multipath propagation, however, the code orthogonality is lost and multiple-access interference (MAI) occurs.

A key issue in cellular communication networks is power control, which is used in downlink applications to improve the geographical fairness in data reception. Power control is commonly achieved by transmitting more power to the users located near the cell boundaries, which experience a much higher path loss compared with those close to the BS [2]-[3]. On the other hand, this makes the terminals near the BS more exposed to the near-far effect, which enhances the interference and limits the performance of conventional single-user receivers. Interference cancellation (IC) techniques [4] may be employed to improve the system performance. In order to work properly, however, IC-based schemes require knowledge of the number of active users and their corresponding channel responses and power levels.

In this work we address the problem of estimating all of the above parameters in the downlink of a power-controlled MC-CDMA network. The users' amplitudes as well as the channel responses and noise power level are estimated following a maximum likelihood (ML) approach and exploiting some training blocks inserted into the frame structure.

The estimates of the above parameters are then passed to an IC-based receiver, in which the interfering signals with amplitude above a specified threshold are cancelled out in parallel from the received waveform before detection of the desired user's data. Compared with the partial parallel interference cancellation (PPIC) receivers discussed in [5], our scheme is different in two aspects. First, an interfering signal is either totally cancelled or retained depending on its specific reliability measure (i.e., the estimated signal amplitude). Vice versa, in [5] all the interfering signals are partially subtracted from the received waveform, irrespective of their power levels. Second, our scheme has affordable complexity as it employs a single stage of cancellation whereas the detector in [5] is a multistage receiver (hence, it is not suited for downlink applications where system complexity and power consumption must be kept as low as possible).

The rest of the paper is organized as follows. The next section describes the signal model and introduces basic notation. In section 3 we discuss the proposed IC-based data detector. Channel and noise power estimation is addressed in section 4 while a scheme for estimating the amplitudes of the interfering signals is proposed in section 5. Simulation results are discussed in section 6 and, finally, some conclusions are drawn in section 7.

2. System model

We consider the downlink of an MC-CDMA network in which the total number of subcarriers, N, is divided into smaller groups of Q elements [1]. By exploiting the subcarriers of a given group, the BS simultaneously communicates with K active users ($K \leq Q$), which are separated by means of orthogonal Walsh-Hadamard (WH) codes of length Q. Without loss of generality, we concentrate on a single group and assume that the Q subcarriers are uniformly spread over the signal bandwidth so as to better exploit the channel frequency diversity. We denote $\{i_n\,;\ 1 \leq n \leq Q\}$ the subcarrier indexes in the group, with $i_n = 1 + (n-1)N/Q$ and call $a_k(m)$ the symbol transmitted to the kth user during the mth MC-CDMA block. The latter is spread over Q chips using the code sequence $\mathbf{c}_k = [c_k(1), c_k(2), \ldots, c_k(Q)]^T$, where $c_k(n) \in \{\pm 1/\sqrt{Q}\}$ and the superscript $(\cdot)^T$ means transpose operation. The contributions of the users are then summed chip-by-chip to form the multi-user signal

$$\mathbf{s}(m) = \sum_{k=1}^{K} \lambda_k a_k(m)\, \mathbf{c}_k \qquad (1)$$

where $\{\lambda_k; k = 1, 2, \ldots, K\}$ are non-negative real numbers that serve to perform power control. The elements of $\mathbf{s}(m)$ are finally mapped on Q subcarriers using an OFDM modulator, which comprises an inverse fast Fourier transform (IFFT) unit and the insertion of a cyclic prefix (CP).

At the receiver side the incoming waveform is passed to an OFDM demodulator, which eliminates the CP and performs a fast Fourier transform (FFT) operation. We concentrate on the mth received MC-CDMA block and denote $\mathbf{X}(m) = [X(m, i_1), X(m, i_2), \ldots, X(m, i_Q)]^T$ the demodulator output corresponding to the Q subcarriers in the considered group. Assuming ideal timing and frequency synchronization, we have

$$\mathbf{X}(m) = \sum_{k=1}^{K} \lambda_k a_k(m)\, \mathbf{u}_k(m) + \mathbf{w}(m) \qquad (2)$$

where $\mathbf{u}_k(m)$ is a Q-dimensional vector with entries

$$u_k(m, n) = H(m, i_n)\, c_k(n) \qquad 1 \leq n \leq Q \qquad (3)$$

and $H(m, i_n)$ is the channel frequency response over the i_nth subcarrier of the mth block. Also, $\mathbf{w}(m)$ is thermal noise, which is modelled as a white Gaussian process with zero mean and covariance matrix $\sigma^2 \mathbf{I}_Q$ ($\mathbf{I}_Q$ denotes the identity matrix of order Q).

3. Data detection

The need to keep the remote units as simple and power efficient as possible makes the PPIC receiver discussed in [5] quite unattractive for downlink applications. As an alternative, we propose a novel scheme employing a single stage of cancellation in which interfering signals are subtracted from the DFT output provided that they are sufficiently reliable. Without loss of generality, we concentrate on the jth MT. Then, the kth interfering signals is cancelled provided that

$$\frac{\lambda_k}{\lambda_j} \geq \gamma \qquad (4)$$

where the threshold γ is a design parameter. This leads to the following expurgated vector

$$\mathbf{Z}'_j(m) = \mathbf{X}(m) - \sum_{k \in I_c} \lambda_k \hat{a}_k(m) \mathbf{u}_k(m) \qquad (5)$$

where I_C is the set of indexes k (with $k \neq j$) satisfying (4) and $\hat{a}_k(m)$ is an estimate of $a_k(m)$. The entries of $\mathbf{Z}'_j(m)$ are then exploited to detect the useful symbol $a_j(m)$. To this purpose we observe that, even assuming ideal data decisions (i.e., $\hat{a}_k(m) = a_k(m)$), $\mathbf{Z}'_j(m)$ is still affected by residual MAI due to the (weak) users that are not cancelled. Therefore, instead of combining the elements of $\mathbf{Z}'_j(m)$ using the maximum-ratio-combining (MRC) strategy (which gives the best results in the MAI-free case), we adopt the minimum-mean-square-error (MMSE) approach. Hence, the decision statistic is given by

$$v_j(m) = \sum_{n=1}^{Q} \xi_j(m,n) c_j(n) [\mathbf{Z}'_j(m)]_n \qquad (6)$$

in which $[\mathbf{Z}'_j(m)]_n$ denotes the nth entry of $\mathbf{Z}'_j(m)$ and

$$\xi_j(m,n) = \frac{\lambda_j H^*(m, i_n)}{|\lambda_j H(m, i_n)|^2 + \sigma^2} \qquad n = 1, 2, \ldots, Q. \qquad (7)$$

Finally, passing $v_j(m)$ to a threshold device produces the estimate of $a_j(m)$. In the sequel, the proposed scheme is referred to as the reliability-based IC (RBIC) detector. To summarize, the RBIC operates through the following steps:

1) Select the interfering users that must be cancelled, i.e., those satisfying the constraint (4);

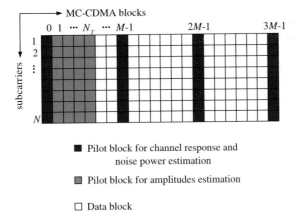

Figure 1. Downlink frame structure.

2) Compute the corresponding data decisions $\hat{a}_k(m)$ using the following single-user (SU)-MMSE decisions statistics

$$y_k(m) = \sum_{n=1}^{Q} \xi_k(m,n) c_k(n) X(m, i_n) \qquad k \in I_c. \qquad (8)$$

3) Compute the expurgated vector $\mathbf{Z}'_j(m)$ in (5);
4) Use $v_j(m)$ in (6) to estimate $a_j(m)$.

It is worth noting that RBIC has similarities with the detector proposed by Ochiai and Imai (OID) in [6]. The main difference is that in [6] the kth interfering signal is subtracted from the DFT output provided that $|y_k(m)| > |y_j(m)|$ whereas in our scheme we take the amplitude of the interfering signal as an indicator of reliability. As is shown later, this leads to a better error rate performance.

4. Estimation of the signal amplitudes

5. Estimation of the channel response and noise power

From (5)-(7) we see that RBIC requires knowledge of σ^2, $\{H(m, i_n)\}$ and the amplitudes $\{\lambda_k\}$ of the interfering signals. All of the above quantities must be estimated in some manner. In this section we extend the method discussed in [7] to jointly estimate the channel response and noise power. To this end, the transmission is organized in frames and pilot blocks are inserted into the frame structure as shown in Figure 1. We denote M the distance between two consecutive pilot blocks and assume that the channel variations are negligible over M blocks (slow fading). The amplitudes $\{\lambda_k\}$ of the users' signals are kept fixed over the frame duration.

For notational simplicity, we drop the block index m. Then, the DFT outputs corresponding to the pilot block takes the form [7]

$$\mathbf{X}^{(p)} = \mathbf{PH} + \mathbf{w}^{(p)} \qquad (9)$$

where $\mathbf{H} = [H(1), H(2), \ldots, H(N)]^T$ is the channel frequency response and $\mathbf{P} = \text{diag}\{p_1, p_2, \ldots, p_N\}$ with $\{p_n;\ n = 1, 2, \ldots, N\}$ pilot symbols taken from a PSK constellation. Also, vector $\mathbf{w}^{(p)}$ accounts for the noise contribution. Following [7], the joint ML estimates of $\mathbf{H}$ and σ^2 are

$$\hat{\mathbf{H}}_{ML} = \frac{1}{N}\mathbf{F}\mathbf{F}^H\mathbf{Y}, \qquad (10)$$

$$\hat{\sigma}^2_{ML} = \frac{1}{N-L}\mathbf{Y}^H\mathbf{F}^\perp\mathbf{Y} \qquad (11)$$

where $\mathbf{Y} = \mathbf{P}^H\mathbf{X}^{(p)}$, $\mathbf{F}$ is an $N \times L$ matrix with entries

$$[\mathbf{F}]_{n,\ell} = e^{-j2\pi(n-1)(\ell-1)/N} \qquad 1 \leq n \leq N,\ \ 1 \leq \ell \leq L \qquad (12)$$

and $\mathbf{F}^\perp = \mathbf{I}_N - \mathbf{F}(\mathbf{F}^H\mathbf{F})^{-1}\mathbf{F}^H$ is the orthogonal complement of $\mathbf{F}$. Also, L is the duration of the channel impulse response (CIR) in sampling periods.

The mean square estimation error (MSEE) of $\hat{H}_{ML}(n)$ is computed in [7] and reads

$$\mathrm{E}\left\{\left|\hat{H}_{ML}(n) - H(n)\right|^2\right\} = \frac{L}{N}\sigma^2. \qquad (13)$$

The variance of $\hat{\sigma}^2_{ML}$ can be found using standard calculations and it is given by

$$\text{var}\{\hat{\sigma}^2_{ML}\} = \frac{\sigma^4}{N-L}. \qquad (14)$$

Inspection of (5) and (7) reveals that RBIC requires knowledge of the amplitudes $\mathbf{\Lambda} = [\lambda_1, \lambda_2, \ldots, \lambda_Q]^T$ corresponding to the complete set $\{\mathbf{c}_1, \mathbf{c}_2, \ldots, \mathbf{c}_Q\}$ of potentially active codes ($\lambda_k = 0$ if the kth user is turned off). In the following we address the problem of estimating these parameters. For this purpose, training blocks are placed at the beginning of each frame (see Fig.1), just after the first pilot block (which is employed to estimate the channel response and noise power). We begin by rewriting (2) in the equivalent form

$$\mathbf{X}(m) = \mathbf{H}(m)\sum_{k=1}^{Q}\lambda_k a_k(m)\mathbf{c}_k + \mathbf{w}(m) \qquad m = 1, 2, \ldots, N_T \qquad (15)$$

where $\mathbf{H}(m) = \text{diag}\{H(m,i_1),\ H(m,i_2),\ldots,H(m,i_Q)\}$ while $\{a_k(m);\ m=1,\ 2,\ldots,N_T\}$ is the training sequence of the kth user. Note that the summation index in (2) counts the active users whereas in (15) it counts the potentially active codes and spans the interval $1 \le k \le Q$. The reason is that the generic MT has no a-priori knowledge of which codes are effectively employed in the current frame and, therefore, it assumes that all them are potentially active.

Equation (15) may be rewritten as

$$\mathbf{X}(m) = \mathbf{B}(m)\mathbf{\Lambda} + \mathbf{w}(m) \qquad m = 1,\ 2,\ldots,N_T \qquad (16)$$

where $\mathbf{B}(m)$ is the following $Q \times Q$ matrix

$$\mathbf{B}(m) = \mathbf{H}(m)[a_1(m)\,\mathbf{c}_1\ a_2(m)\,\mathbf{c}_2\ \cdots\ a_Q(m)\,\mathbf{c}_Q]. \qquad (17)$$

From (16), we see that the ML estimate of $\mathbf{\Lambda}$ is given by

$$\hat{\mathbf{\Lambda}}_{ML} = \mathbf{R}^{-1} \sum_{m=1}^{N_T} \Re\left\{\mathbf{B}^H(m)\mathbf{X}(m)\right\} \qquad (18)$$

with

$$\mathbf{R} = \sum_{m=1}^{N_T} \Re\left\{\mathbf{B}^H(m)\mathbf{B}(m)\right\}. \qquad (19)$$

In the sequel, equation (18) is referred to as the maximum likelihood estimator (MLE) of the signal amplitudes. It can be shown that $\hat{\mathbf{\Lambda}}_{ML}$ is unbiased and its covariance matrix is given by

$$\mathbf{C}_{\hat{\mathbf{\Lambda}}_{ML}} = \frac{\sigma^2}{2}\mathbf{R}^{-1}. \qquad (20)$$

The following remarks are of interest.

i) Inspection of (17) reveals that computing $\mathbf{B}(m)$ requires channel state information. In practice, an estimate of the channel response is obtained from the first pilot block (corresponding to $m=0$ in Figure 1) using the estimator discussed in the previous section. The channel estimate is exploited to compute an estimate of $\mathbf{B}(m)$, say $\hat{\mathbf{B}}(m)$, which is then used in (18)-(19) in place of the true $\mathbf{B}(m)$.

ii) From (18)-(19) we see that the crux in the calculations is the inversion of $\mathbf{R}$. The complexity of MLE can be greatly reduced by resorting to orthogonal training sequences. In these circumstances $\mathbf{R}$ reduces to $N_T E_H \cdot \mathbf{I}_Q$ with

$$E_H = \frac{1}{Q} \sum_{n=1}^{Q} |H(i_n)|^2 \qquad (21)$$

and (18) becomes

$$\hat{\mathbf{\Lambda}}_{ML} = \frac{1}{N_T E_H} \sum_{m=1}^{N_T} \Re\mathfrak{e}\left\{\mathbf{B}^H(m)\mathbf{X}(m)\right\}. \tag{22}$$

Also, substituting $\mathbf{R} = N_T E_H \cdot \mathbf{I}_Q$ into (20) and using the identity $E\{|\hat{\lambda}_k - \lambda_k|^2\} = [\mathbf{C}_{\hat{\mathbf{\Lambda}}_{ML}}]_{k,k}$ yields

$$\mathrm{E}\left\{\left|\hat{\lambda}_k - \lambda_k\right|^2\right\} = \frac{\sigma^2}{2N_T E_H}. \tag{23}$$

6. Performance evaluation

6.1 System Parameters

We consider a power-controlled cellular system in which the users are uniformly distributed within a cell of radius $R = 1$. We adopt the power allocation scheme discussed in [2], where a common power level is assigned to those users whose distance from the BS is less than a certain threshold value $d_{\min}$. This is tantamount to setting

$$\lambda_k = \begin{cases} (d_{\min}/R)^{n/2} & \text{if } d_k \leq d_{\min} \\ (d_k/R)^{n/2} & \text{if } d_k > d_{\min} \end{cases} \quad k = 1, 2, \ldots, K \tag{24}$$

where $d_{\min} = 0.55 \cdot R$ and $n = 5$. The information bits are mapped onto QPSK symbols using a Gray map. The total number of subcarriers is $N = 64$ and the spreading gain is $Q = 8$. The signal bandwidth is $B = 8$ MHz corresponding to a block duration $T = N/B = 8$ μs. A cyclic prefix of $T_G = 2$ μs is adopted to eliminate inter-block interference. Each frame has duration of 1.28 ms and comprises 128 blocks.

The channel impulse response has 16 paths ($L = 16$). The path delays are kept constant over a frame and are uniformly distributed within $[0, 2 \ \mu s]$. The path gains have powers

$$\sigma_\ell^2 = \sigma_h^2 \times \exp(-\ell) \quad \ell = 0, 1, \ldots, 15 \tag{25}$$

where σ_h^2 is chosen such that the average energy of the channel is normalized to unity. The gain of each path varies independently of the others within a frame and is generated by filtering a white Gaussian process in a third-order low-pass Butterworth filter. The 3-dB bandwidth of the filter is taken as a measure of the Doppler rate $f_D = f_0 v/c$, where $f_0 = 2$ GHz is the carrier frequency, v denotes the speed of the mobile terminal and $c = 3 \times 10^8$ m/s is the speed of light.

The distances d_k are kept fixed over a frame. The amplitudes λ_k are generated from (24) and are estimated using MLE in conjunction with

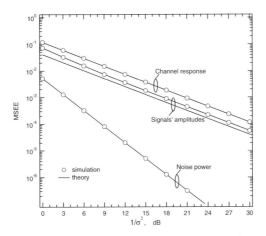

Figure 2. Accuracy in the estimation of the channel response, noise power and users' amplitudes.

WH training sequences of length $N_T = 8$. Pilot blocks are periodically inserted within the frame with period M. Estimates of the channel response and noise power are computed based on the observation of a pilot block and are kept fixed over the next $M-1$ data blocks. The system is fully-loaded ($K = 8$) and the mobile speed is 60 km/h (corresponding to $f_D = 110$ Hz).

The threshold γ in (4) must be carefully designed so as to achieve a reasonable trade-off between complexity and error-rate performance. Indeed, increasing γ reduces the complexity of RBIC since the number of interfering signals that are cancelled from the DFT decreases. At the same time, however, this leads to higher residual interference. In our simulations we set $\gamma = 1.2$.

6.2 Performance Assessment

We begin by assessing the accuracy achieved in the estimation of the channel response, noise power and signal amplitudes. Figure 2 illustrates the MSEE of $\hat{\mathbf{H}}_{ML}$, $\hat{\sigma}_{ML}^2$ and $\hat{\mathbf{\Lambda}}_{ML}$ vs. $1/\sigma^2$ with $N_T = 8$. Marks indicate simulations while solid lines represent analytical results as given by (13), (14) and (23). Perfect agreement between simulations and theory is observed in the estimation of the channel response and noise power while the MSEE of $\hat{\mathbf{\Lambda}}_{ML}$ is 1.5 dB worse than the predicted value. The reason is that the result (23) has been derived assuming perfect knowledge of $\mathbf{B}(m)$ while in practice $\hat{\mathbf{\Lambda}}_{ML}$ is computed from (22) after replacing $\mathbf{B}(m)$ with its estimate $\hat{\mathbf{B}}(m)$.

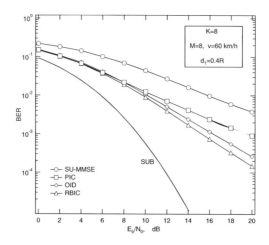

Figure 3. Performance of the various detection schemes with $K = 8$ and $d_1 = 0.4R$.

Figures 3-4 compare the BER performance of the SU-MMSE, brute-force PIC, OID and RBIC detectors vs. E_b/N_0 (E_b is the average received energy per bit and $N_0/2$ is the two-sided noise spectral density). For a fair comparison with OID and RBIC, the brute-force PIC employs a single stage of cancellation. The separation between pilot blocks is $M = 8$ while the distance d_1 is $0.4R$ in Figure 3 and $0.8R$ in Figure 4. The single user bound (SUB) is also shown as a benchmark. We see that RBIC gives the best results irrespective of d_1, even though the gain with respect to OID is limited to 1 dB at an error probability of 10^{-3}. Still, this gain comes for free since RBIC and OID have comparable complexity. It is worth noting that SU-MMSE performs poorly when $d_1 = 0.4R$. The reason is that the user of interest is relatively near the BS and strong interfering signals are likely. While they degrade the performance of SU-MMSE, they are reliably cancelled by the other detectors. Increasing d_1 reduces the near-far effect, thereby improving the error probability of SU-MMSE. When $d_1 = 0.8R$, SU-MMSE outperforms the PIC. This can be explained by observing that the user of interest is now located near the cell boundaries and $\lambda_k \ll \lambda_1$ with high probability. In these circumstances, the tentative decisions employed by PIC to remove the MAI are not reliable. If they are wrong, MAI is enhanced and PIC performs poorly.

Figure 5 illustrates the BER of RBIC vs. E_b/N_0 for $d_1 = 0.6R$ and some values of M. The SUB and the BER with perfect knowledge of the channel parameters (PKCP) (i.e. $\hat{\mathbf{H}} = \mathbf{H}$, $\hat{\sigma}^2 = \sigma^2$ and $\hat{\mathbf{\Lambda}} = \mathbf{\Lambda}$) are also indicated as benchmarks. At BER $= 10^{-3}$ we see that imperfect knowledge of the channel parameters entails a loss of approximately 1

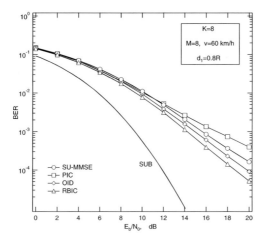

Figure 4. Performance of the various detection schemes with $K = 8$ and $d_1 = 0.8R$.

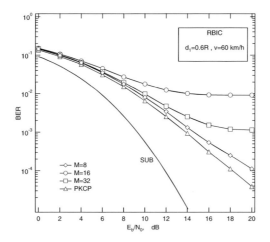

Figure 5. Performance of RBIC with $K = 8$, $d_1 = 0.6R$ and various M.

dB with $M = 8$ while an irreducible error floor is observed with either $M = 16$ or 32.

7. Conclusions

We have discussed channel parameter estimation for interference cancellation in power-controlled MC-CDMA downlink transmissions. Pilot blocks are periodically inserted in the data stream for joint ML estimation of the channel response and the noise power. A scheme based on ML reasoning has been proposed for estimating the users' amplitudes. It exploits orthogonal training sequences with length equal to the spreading gain. All of the above parameters are passed to a non-linear receiver (RBIC) which performs interference cancellation provided that the power of the interfering signal exceeds a given threshold. Computer simulations indicate that the proposed scheme is superior to other existing IC-based schemes. Imperfect knowledge of the channel parameters entails a loss of approximately 1 dB for a fully-loaded system and a mobile speed of 60 km/h.

Acknowledgement

This work was supported by the Italian Ministry of Education under the FIRB project PRIMO.

References

[1] K. Fazel and S. Kaiser. *Multi-Carrier and Spread Spectrum Systems*. John Wiley and Sons, 2003.

[2] W.C.Y. Lee. Overview of cellular CDMA. *IEEE Trans. on Vehicular Technology*, 40:291–302, May 1991.

[3] M. Zorzi. Simplified forward-link power control law in cellular CDMA. *IEEE Trans. on Vehicular Technology*, 43:1088–1093, Nov 1994.

[4] S. Moshavi. Multiuser detection for DS-CDMA communications. *IEEE Comm. Magazine*, pages 124–136, Oct. 1996.

[5] D. Divsalar, M.K. Simon, and D. Raphaeli. Improved parallel interference cancellation for CDMA. *IEEE Trans. on Communications*, 46(2):258–268, Feb. 1998.

[6] H. Ochiai and H. Imai. Performance of downlink MC-CDMA with simple interference cancellation. In *Proc. of the Int. Workshop on Multi-Carrier Spread-Spectrum (MC-SS'99)*, pages 211–218, Sept. 1999.

[7] L. Sanguinetti, M. Morelli, and U. Mengali. Channel estimation and tracking for MC-CDMA signals. *European Trans. on Telecommunications*, 15:249–258, March 2004.

PERFORMANCE EVALUATION AND PARAMETER OPTIMIZATION OF MC-CDMA

M. Guenach, H. Steendam
*Department of Telecommunications and Information Processing Ghent University,
St.-Pietersnieuwstraat 41, B-9000 GENT, Belgium
TEL: +32 9 264 8900; Fax: +32 9 264 4295*
{guenach,hs}@telin.UGent.be

Abstract   The performance of multicarrier systems depends on the propagation channel behavior. The latter is subject to time and/or frequency selectivity. The designer has to select properly the guard interval and the number of carriers for a given system bandwidth to combat the channel dispersiveness. In this paper we investigate the sensitivity of MultiCarrier Code Division Multiple Access (MC-CDMA) performance to these parameters in different environments. We derive closed form expressions of the useful and the different interference powers after Maximum Ratio Combining (MRC) and despreading. The optimum parameters correspond to the minimum of the Signal to Noise Ratio Degradation (SNRD). It turns out that the derivations in [1] restricted to the Orthogonal Frequency Division Multiplexing (OFDM) system, are a particular case of our results. Numerical evaluation of the analytical expressions reveals that the optimum parameters of the MC-CDMA and its corresponding OFDM system are similar and depend in the same way on the channel characteristics.

1. Introduction

Multicarrier systems have emerged as a powerful candidate to wireless communication systems. The OFDM technique was selected as a transmission standard technique in Digital Audio Broadcasting (DAB) from satellite or fixed terrestrial to mobile users [2] and in Digital Video Broadcasting in Europe [3]. Another important application of MC systems can be found in transmission of high data rate over twisted pair [4]. So far, the conventional CDMA technique that is the core wireless technology in the third generation is limited by channel dispersion, causing Inter Symbol Interference (ISI) which requires advanced detection algorithms to be removed. The combination of the OFDM and CDMA, i.e. MC-CDMA [5], has drawn a lot of attention due to its robustness to channel dispersion, hence ISI, and its ability to accommodate a higher number of users as compared to CDMA only. Basically we can distinguish three versions namely MC-CDMA, MC-DS-CDMA and MultiTone (MT)-CDMA [5].

The robustness of the MC systems to the channel frequency selectivity that induces Inter Carrier Interference (ICI) (which is caused by other carriers from the same MC symbol) and ISI (which is caused by other MC symbols) is obtained by transmitting the data over subchannels with a very low bandwidth compared to the total system bandwidth. If the subchannel bandwidth is smaller than the channel Coherence Bandwidth (CB), the interference can be removed completely after the FFT at the receiver. Otherwise, the interference can be reduced by adding a guard interval per MC symbol. The subchannel bandwidth is proportional to the inverse of the MC symbol period. Therefore increasing this parameter will reduce the interference level at the input of the decision device. However, there is another constraint that should be fulfilled in order to avoid interference caused by the time selectivity of the channel: the MC symbol duration has to be smaller than the channel Coherence Time (CT) duration related to the time variation of the channel. If this condition is not respected, ICI will be introduced.

Hence the designer has to select the optimum parameters according to a certain criterion. The few contributions found in the literature deal with the OFDM parameter optimization. In [6] and [3] the optimum number of OFDM subchannels is found in the absence of the ISI, i.e. the guard interval is sufficiently high to cope with the ISI. In [1] the authors extend the previous contributions to deal with the effect of the ISI in the OFDM system when the guard interval is smaller than the delay spread.

In this contribution we extend the work done in [1] to MC-CDMA systems equipped with the MRC equalizer and using Walsh Hadamard codes with a random overlay sequence as spreading codes. The results obtained in [1] for the OFDM system can be seen as particular case of our MC-CDMA system with a processing gain of one and will serve as a reference system. We derive analytical expressions for the useful and interference power and the SNRD after MRC equalization and despreading. The SNRD will serve as a cost function to be minimized in order to find the optimum number of carriers and guard interval. In the numerical evaluation, we consider an aeronautical channel with different Dopplers to come up with general optimization rules that are channel independent.

The outline of this paper is as follows: first the system model is described in section 2. In section 3, we derive analytical expressions for the useful power, the interference power and the SNRD. Numerical evaluations can be found in section 4 before we conclude in section 5.

2. System model

The composite signal to be transmitted is obtained as follows: after serial to parallel conversion of P symbols, the p-th data symbol $a_{i,p}$ of the i-th block is spread by the spreading sequence $c_{p,s}$ that repeats from frame to frame with $p = 0, ..., P - 1$ and $s = 0,..., N_s$ where N_s is the spreading factor. Each chip $a_{i,p}c_{p,s}$ is mapped to a carrier denoted $n_{p,s}$ where $0 \leq n_{p,s} \leq N - 1$. In total we have $N = PN_s$ carriers. In order to achieve frequency diversity, the assignment of the carriers to chips is made such that the frequency separation among carriers conveying the chips of the same data symbol is maximized. One possible mapping is as follows: the p-th data stream is transmitted on N_s carriers with frequencies $f_{p+sP} = f_c + (p + sP) / (NT)$ where $1/T$ is the system bandwidth and f_c is the RF carrier frequency of the MC system. After feeding the i-th block of spread data symbols to an Inverse Fast Fourier Transform (IFFT) of length N, the resulting block of N time domain samples is cyclically extended by a prefix of v samples. The m-th sample of the resulting MC block of $N + v$ samples is given by:

$$s_{i,m} = \sqrt{\frac{E_s}{N+v}} \sum_{p=0}^{P-1} a_{i,p} \sum_{s=0}^{N_s-1} c_{p,s} \exp\left[j2\pi m \frac{n_{p,s}}{N}\right], \quad m = -v, \ldots, N-1.$$

Considering a tapped delay line channel $h(l;k)$ with symbol spacing, the received samples $r(k)$ of the j-th frame (after removal of the cyclic prefix) are:

$$r(k) = \sum_{i=\infty}^{+\infty} \sum_{m=-v}^{N-1} s_{i,m} h(k - m - i(N+v); k) + n(k),$$

$$k = j(N+v), \ldots, j(N+v) + N - 1. \tag{2}$$

We concentrate on the MC symbol detection during the frame $j = 0$. The output $b_{0,p',s'}$ of the FFT corresponding to the chip $a_{0,p} c_{p',s'}$ can be written as:

$$b_{0,p',s'} = \sqrt{\frac{1}{N}} \sum_{k=0}^{N-1} r(k) \exp\left[-j2\pi k \frac{n_{p',s'}}{N}\right].$$

After the expansion of the received samples, it can be shown that

$$b_{0,p',s'} = \sqrt{\frac{N}{N+v}} E_s \sum_{i=-\infty}^{+\infty} \sum_{p=0}^{P-1} \sum_{s=0}^{N_s-1} a_{i,p} c_{p,s} \gamma(n_{p',s'}; n_{p,s}; i)$$

$$+ \frac{1}{\sqrt{N}} \sum_{k=0}^{N-1} n(k) \exp\left[-j2\pi k \frac{n_{p',s'}}{N}\right] \tag{3}$$

where

$$\gamma(n_{p',s'}; n_{p,s}; i) = \frac{1}{N} \sum_{m=-v}^{N-1} \sum_{k=0}^{N-1} \exp\left[-j2\pi \frac{n_{p',s'} k - n_{p,s} m}{N}\right]$$

$$h(k - m - i(N+v); k). \tag{4}$$

Note that the useful part in the FFT output $b_{0,p',s'}$ in (3) can be expressed as

$$\left(b_{0,p',s'}\right)_U = \sqrt{\frac{N}{N+v}} E_s a_{0,p'} c_{p',s'} \gamma\left(n_{p',s'}; n_{p',s'}; 0\right). \tag{5}$$

Therefore, the MRC equalizer multiplies the $n_{p',s}$-th FFT output with $d_{0,p',s'} = \gamma^*(n_{p',s};$ $n_{p',s}; 0)$. After MRC and despreading of the different chips related to the symbol index p', the decision variable $y_{0,p'}$ is as follows:

$$y_{0,p'} = \sum_{s'=0}^{N_s-1} c^*_{p',s'} d_{0,p',s'} b_{0,p',s'} \tag{6}$$

3. Performance analysis

Based on the decision variable in (6), we can identify on top of the useful component and the AWGN component two sources of interference: the Inter-Symbol-Interference (ISI) due to the frequency selectivity of the channel and the Inter-Carrier-Interference (ICI) due to both the time and frequency selectivity of the channel. The power at the output of the equalizer can be decomposed as $E_s \frac{N}{N+v}$ $(P_U + P_{ICI} + P_{ISI}) + P_{AWGN}$ where P_X, $X = U$, ICI, ISI, AWGN corresponds to the useful, ICI, ISI and AWGN powers, respectively:

$$P_U = E\left\{\left|\Gamma\left(n_{p'}; n_{p'}; 0\right)\right|^2\right\} \tag{7}$$

$$P_{ICI} = \sum_{\substack{p=0 \\ p \neq p'}}^{P-1} E\left\{\left|\Gamma\left(n_{p'}; n_p; 0\right)\right|^2\right\} \tag{8}$$

$$P_{ISI} = \sum_{\substack{i=-\infty \\ i \neq 0}}^{+\infty} \sum_{p=0}^{P-1} E\left\{\left|\Gamma\left(n_{p'}; n_p; i\right)\right|^2\right\} \tag{9}$$

$$P_{AWGN} = E\left\{\left|\sqrt{\frac{1}{N}} \sum_{s'=0}^{N_s-1} c^*_{p',s'} d_{0,p',s'} \sum_{k=0}^{N-1} n(k) \exp\left[-j2\pi n_{p',s} k/N\right]\right|^2\right\} \tag{10}$$

where $\Gamma(n_{p'};n_p;i) = \sum_{s',s=0}^{N_s-1} c_{p',s'}^* c_{p,s} d_{0,p',s'} \gamma(n_{p',s'};n_{p,s};i)$.

The averaging in P_X is with respect to the statistics of the data symbols, the spreading chips and the channel. In the following we derive closed analytical expressions for the useful and interference powers in terms of the Wide Sense Stationary Uncorrelated Scatterer (WSSUS) channel autocorrelation $R(l; k)$ function defined as E $\{h(l, k)h^*(l', k')\} = \delta(l-l') R(l; k-k')$. P_T is the total power defined as $P_T = P_U + P_{ICI} + P_{ISI}$, and can be expressed as

$$P_T = \sum_{i=-\infty}^{+\infty} \sum_{p=0}^{P-1} E\left\{\left|\Gamma(n_{p'};n_p;i)\right|^2\right\}. \qquad (11)$$

The derivations related to the total and useful powers are straightforward but quite lengthly and therefore are not included in the manuscript.

One important parameter in the system performance is the Signal to Noise Ratio (SNR) which is the ratio of the useful power and the other sources of interference power:

$$SNR = \frac{E_s \frac{N}{N+v} P_U}{E_s \frac{N}{N+v}(P_{ICI} + P_{ISI}) + P_{AWGN}} \qquad (12)$$

To tackle how the performance degrades in the fading channels as compared to the ideal case, i.e. the equivalent frequency-time flat channels, we define the SNR degradation as the SNR reduction compared to the ideal case:

$SNRD = 10\log\left(\frac{SNR_{AWGN}}{SNR}\right)$. It can be shown that

$$SNRD = -10\log\left(\frac{\frac{N}{N+v}P_U}{2\left[\frac{1}{N}\sum_{q=-\infty}^{+\infty}\sum_{r=-\infty}^{+\infty} w(q;r) R(q;r)\right] + \frac{E_s}{N_0}\frac{N}{N+v}(P_T - P_U)}\right) \qquad (13)$$

where the multivariate function $w(q; r)$ has the same definition as the function $w(q; r)$ in [1].

4. Numerical results

In this section we have evaluated the obtained analytical expressions to optimize the system parameters of the MC-CDMA system. The OFDM system will serve as a reference system and the optimum parameters of the MC-CDMA system will be compared to those of the OFDM system. We consider the aeronautical channel model of the parking scenario in [7] with two different mobile speeds $v = 20$ and $v = 100$ km/h respectively. The channel impulse response is a multipath Rayleigh fading channel with a maximum delay spread of $\tau_{max} = 7.10^{-6}$ s and a power delay profile that is exponentially decreasing with slope time $\tau_{slope} = 10^{-6}$ s. Computations are carried out for a MC-CDMA system using Walsh Hadamard spreading codes followed by an overlay random code of spreading factor $N_s = 8$, a system bandwidth $B = 2$ MHz, i.e. $T = 5\ 10^{-7}$ s, and a carrier frequency of $f_c = 1$ GHz. To compute channel Coherence Bandwidth (CB) and Coherence Time (CT) we use the following expressions [8] $B_c = \frac{1}{2\pi\tau_{max}}$ and $T_c = \frac{1}{16\pi f_{D_{max}}}$ respectively where $f_{D_{max}} = \frac{v}{c} f_c$ is the maximum Doppler spread. It turns out that the CB is $B_c = 22.74$ kHz and the maximum Doppler frequency is $f_{Dmax} = 92.6$ Hz (resp. 18.52 Hz) for a mobile speed of $v = 100$ km/h (resp. $v = 20$ km/h) which results in a CT of $T_c = 1.9$ ms (resp. $T_c = 9.7$ ms). We will refer to the scenario with $v = 100$ km/h (resp. $v = 20$ km/h) as scenario A (resp. scenario B).

The determinant factor in the system performance is the BER that should be minimized or equivalently we have to minimize the SNRD defined in (13). In figure 1 we plot the optimum number of carriers N (left) and guard interval v (right) for the MC-CDMA system with $N_s = 8$ and for the corresponding OFDM system with $N_s = 1$. It can be noticed that the minimization of the SNRD appears to result in different parameters N and v for the MC-CDMA system and its equivalent OFDM system per E_b/N_0. From the figure, it follows that for all E_s/N_0 operating points, the optimum carrier spacing $\left(\frac{1}{NT}\right)_{opt}$ ranges in the interval $\left[\frac{1}{600T}, \frac{1}{200T}\right]$ for mobile speed of $v = 100$ km/h and $\left[\frac{1}{200T}, \frac{1}{400T}\right]$ for a mobile speed of $v = 20$ km/h. It can easily be verified that these optimum carrier spacings are located within the interval determined by the maximum Doppler spread f_{Dmax} as a lower bound and the coherence bandwidth, which is inversely proportional to the maximum delay spread τ_{max} as an upper bound: $f_{D_{max}} \ll \left(\frac{1}{NT}\right)_{opt} \ll B_c$.

Performance Evaluation and Parameter Optimization of MC-CDMA

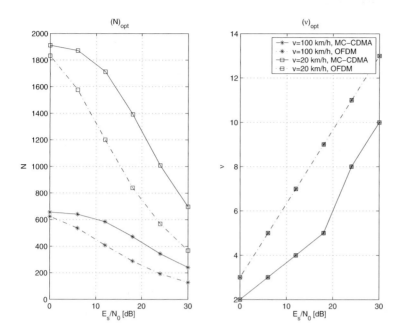

Figure 1. The optimum number of carriers (left) and guard of interval (right) as function of the E_b/N_0 in case of a mobile speed of 20 and 100 km/h.

This can be explained as follows: when the carrier spacing is larger than the Doppler spread, the MC symbol period NT will be smaller than the coherence time of the channel, such that the channel appears almost time flat to the MC system reducing the amount of time selectivity induced ICI. On the other hand if the subchannel bandwidth $\frac{1}{NT}$ is smaller than the coherence bandwidth of the channel, the channel is frequency flat per carrier which is a key factor in MC systems to reduce frequency selectivity induced ICI and ISI. It can be noticed that the scenario B can accommodate more carriers. This can be explained by the fact that the Doppler spread is smaller as compared to scenario A. On the hand, for all situations the optimum guard interval v_{opt} is not necessarily equal to the maximum delay spread. To eliminate the ISI completely the guard interval should be at least equal to the channel delay spread but this will be translated in a system throughput loss. Generally, the optimum v is a few samples below the maximum delay spread for both scenarios.

5. Conclusion

In this paper, we derived closed analytical expressions for the useful and interference power of MC-CDMA. Based on these derivations, we carried out computations in order to minimize the SNR degradation and to compute the optimum guard interval and number of carriers. The optimization results of the MC-CDMA were compared to the corresponding OFDM system. Numerical results in different environments confirm the following rule of thumb in selecting the optimum parameters: the carrier spacing should be higher (resp. smaller) as compared to the maximum Doppler (resp. channel coherence bandwidth) while the guard interval has to be few samples smaller than the channel delay spread. Comparing the results of the parameter optimization for OFDM and MC-CDMA, and evaluating the sensitivity of the performance on the choice of the system parameters, it turns out that the results obtained for both multicarrier systems are very similar.

References

[1] H. Steendam and M. Moeneclaey. "Analysis and optimization of the performance of the OFDM on frequency-selective time-selective fading channels". *IEEE Transactions on Communications*, 47:pp.1811–1819, December 1999.

[2] H. Sari, G. Karam and I. J. Claude. "Transmission techniques for digital terrestrial TV broadcasting". *IEEE Communications Magazine*, 36:pp.100–109, February 1995.

[3] M. Russel and G. L. Stüber. "Terrestrial digital video broadcasting for mobile reception using OFDM". *Wireless Personnal Communications*, 2:pp.45–66, 1995.

[4] J. S. Chow, J. C. Tu and J. M. Cioffi. "A discrete multitone tranceiver system for HDSL applications". *Journal on selected areas in communications*, 9:pp.895–908, August 1991.

[5] S. Hara and R. Prasad. "Overview of Multicarrier CDMA". *IEEE Communication Magazine*, 35:pp.126–133, December 1997.

[6] F. Tufvesson and T. Maseng. "Optimization of sub-channel bandwidth for mobile OFDM systems". In *IEEE ICC'94*, pages 103–114, 1998.

[7] E. Haas. "Aeronautical channel Modeling". *IEEE Transactions on Vehicular Technology*, 51:pp.254–264, March 2002.

[8] R. Steele. "*Mobile radio communications*". PENTECH PRESS, London, 1992.

A NOVEL ALGORITHM OF INTER-SUBCHANNEL INTERFERENCE CANCELLATION FOR OFDM SYSTEMS

Ming-Xian Chang
Department of Electrical Engineering and
Institute of Computer and Communication Engineering
National Cheng-Kung University, Tainan, TAIWAN
mxc@ee.ncku.edu.tw

Abstract Orthogonal frequency-division multiplexing (OFDM) transmission has inter-subchannel interference (ICI) in fast fading environment. Several algorithms have been proposed to reduce the ICI effects. Among these algorithms, the ICI self-cancellation schemes [3, 4] needs not to estimate the multipath responses, and its performance doesn't depend on the signal-to-noise ratio (SNR). In [4], the pre-processor in the transmitter extends the original symbol interval to obtain diversity that can be used by the post-processor of the receiver to make ICI self-cancelled. When the original symbol interval is doubled and the multipath channels vary linearly with time within the extended symbol interval, this algorithm can completely the ICI. However, the complexity of the post-processor in [4] is not low. It needs $N+1$ parallel fast-Fourier transforms (FFT). Moreover, there is no performance comparison between [3] and [4].

In this paper, we continue our work in [4] and propose an equivalent low-complexity post-processor that needs only one FFT. We also give analyze the variance of the residual ICI and derive an bit-error probability (BEP) upper bound. Furthermore, we compare the performance of our algorithm with Zhao's algorithm and give some related discussion.

1. Introduction

Orthogonal frequency-division multiplexing (OFDM) transmission is known to be effective in multipath fading channels. While the fading of each subchannel is non-selective, the insertion of a guard interval of cyclic prefix longer than the maximum channel delay helps remove the intersymbol interference (ISI). If the channel responses are invariant during one symbol interval, the received signal of the mth subchannel at one symbol interval can be written as

$$Y_m = H_m X_m + N_m, \qquad (1)$$

where H_m is the channel effect, X_m is the transmitted symbol, and N_m is the additive white Gaussian noise (AWGN). Eq. (1) indicates that the orthogonality between subchannels and the receiver needs only a one-tap equalizer to compensate the complex multiplicative channel effect H_m.

However, in a time-varying channel, the orthogonality between subchannels is distorted, and this incurs the effect of inter-subchannel interference (ICI). In this scenario, the signal model in (1) becomes

$$Y_m = H_m m X_m + \sum_{\substack{m'=1 \\ m' \neq m}}^{N} H_{mm'} X_{m'} + N_m, \qquad (2)$$

where $H_{mm'}$ denotes the channel effect that causes the interference from the m'th subchannel. Several ICI cancellation or reduction algorithms have been proposed. Some of them need to estimate the multipath channel impulse responses, and hence the complexity increases with the number of significant paths [1] [2]. On the other hand, a self-cancellation scheme was proposed by Zhao [3]. The self-cancellation scheme has the advantage of low complexity, and its performance doesn't depend on the signal-to-noise ratio (SNR). Zhao's algorithm modulates a data pair $(a, -a)$ onto two adjacent subchannels, and the received signals on each pair of adjacent subchannels are used to jointly make a symbol decision. Besides [3], in [4] an effective ICI self-cancellation algorithm is proposed. In transmitter a pre-processor is inserted after the inverse fast Fourier-transform (IFFT). The pre-processor extends the signal interval to adds diversity in it. At the receiver, after the guard interval is removed, a post-processor is designed to utilize the diversity in the signal such that the ICI can be self-cancelled. This algorithm shows that, if the extended interval is twice of the original and the variation of each multipath is linear within the extended interval, then the ICI can be completely removed.

However, the complexity of the post-processor in [4] is not low, due to several parallel FFTs. Moreover, there is no performance analysis and comparison with [3]. In this paper, we continue our work in [4] and propose an equivalent post-processor with much lower complexity than [4]. This equivalent post-processor needs only one FFT operation. We also analyze the variance of residual ICI, and the simulated results validate our analysis. Furthermore, we compare the performance of our algorithm with Zhao's algorithm and give some related discussion.

This paper is organized as follows. In Section 2, we introduce the algorithm in [4]. An equivalent post-processor with low complexity is

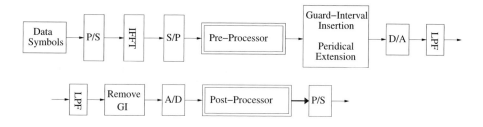

Figure 1. The system block diagram of the proposed algorithm.

proposed in Section 3. We analyze the variance of residual ICI in Section 3. Section 5 shows some the simulation results and the comparison with Zhao's algorithm. We conclude our paper in Section 6

2. The Algorithm in [4]

In this section, we introduce the algorithm in [4]. The system block diagram of this algorithm is shown in Fig. 1. Before inserting the guard interval, the pre-processor extends periodically the original signal

$$x'(t) = \sum_{m=0}^{N-1} X_m e^{i2\pi \frac{m}{T} t}, \quad 0 \leq t < T, \tag{3}$$

to a signal of $2T$ duration,

$$x_1(t) = \sum_{m=0}^{N-1} X_m e^{i2\pi \frac{m}{T} t}, \quad -\frac{T}{2} \leq t < \frac{3T}{2}, \tag{4}$$

where N is the number of subchannels and T is the original symbol interval. After the insertion of guard interval T_g by another periodic extension, we have the transmitted baseband signal

$$x(t) = \sum_{m=0}^{N-1} X_m e^{i2\pi \frac{m}{T} t}, \quad -\frac{T}{2} - T_g \leq t < \frac{3T}{2}. \tag{5}$$

Note that by sampling $x(t)$ with rate $\frac{N}{T}$ during $(-\frac{T}{2}, \frac{3T}{2})$, we obtain $2N$ samples $\{x_k\}_{k=-\frac{N}{2}}^{\frac{3N}{2}-1}$, which is just the periodic extension of the samples $\{x_k\}_{k=0}^{N-1}$. By performing FFT on any N successive samples, say $x_d, \ldots, x_{d+N-1}$, we will obtain $\{X_m e^{i2\pi \frac{md}{N}}\}_{m=0}^{N-1}$. Therefore, we see that the periodic extension in time provides diversity of the frequency domain symbols X_m's.

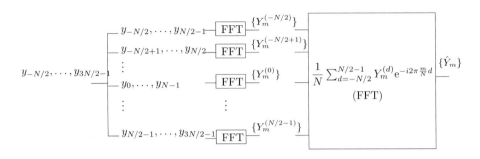

Figure 2. The post-processor in [4].

In the receiver, after removing the guard interval and sampling the received signal $y(t)$ with rate $\frac{N}{T}$ within one extended symbol interval, we obtain $2N$ samples, y_k's, $k = -\frac{N}{2}, \ldots, \frac{3N}{2} - 1$. The post-processor in Fig. 2 first partitions these $2N$ samples into N overlapped sequences, $\{y_k\}_{k=-\frac{N}{2}}^{\frac{N}{2}-1}$, $\{y_k\}_{k=-\frac{N}{2}+1}^{\frac{N}{2}}$, $\ldots$, $\{y_k\}_{k=0}^{N-1}$, $\ldots$, $\{y_k\}_{k=\frac{N}{2}-1}^{\frac{3N}{2}-2}$. Then the FFT is performed on each of these sequences, which resulted in

$$Y_m^{(d)} = \frac{1}{N} \sum_{k=0}^{N-1} y_{k+d} \, e^{-i2\pi \frac{mk}{N}}, \quad d = -\frac{N}{2}, \ldots, \frac{N}{2} - 1. \quad (6)$$

These $Y_m^{(d)}$'s are then combined into an $\hat{Y}_m$ by another FFT,

$$\hat{Y}_m = \frac{1}{N} \sum_{d=-N/2}^{N/2-1} Y_m^{(d)} e^{-j2\pi \frac{md}{N}}. \quad (7)$$

It is shown in [4] that if all multipath channels vary linearly with time withing one extended symbol interval $2T$, then the corresponding ICI in (7) is zero.

This algorithm extend the symbol interval from T to $2T$ to obtain the diversity for ICI self-cancellation in the receiver. However, this also reduces the system throughput. A general scheme in [4] is by extending the signal from T to $\frac{D}{N}T$, where D is a parameter in $(0, N)$. Shortening the extended interval loses the property of complete cancellation when channels vary linearly with time. However, this gives a trade-off between ICI reduction and throughput.

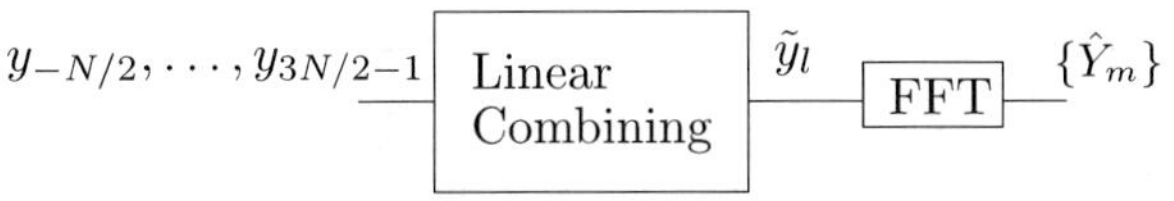

$y_{-N/2}, \ldots, y_{3N/2-1}$ → [Linear Combining] → $\tilde{y}_l$ → [FFT] → $\{\hat{Y}_m\}$

Figure 3. The proposed equivalent post-processor.

3. An Equivalent Post-Processor

The post-processor of the general scheme in [4] needs $D+1$ FFTs, as shown in Fig. 2 for $D = N$. We now propose an equivalent post-processor with much lower complexity. For the general scheme, (6) and (7) become

$$Y_m^{(d)} = \frac{1}{N} \sum_{k=0}^{N-1} y_{k+d}\, e^{-i2\pi \frac{mk}{N}}, \quad d = -\frac{D}{2}, \ldots, \frac{D}{2} - 1, \qquad (8)$$

$$\hat{Y}_m = \frac{1}{N} \sum_{d=-D/2}^{D/2-1} Y_m^{(d)} e^{-j2\pi \frac{md}{N}}. \qquad (9)$$

Substituting (8) into (9), we have

$$\hat{Y}_m = \frac{1}{ND} \sum_{d=-D/2}^{D/2-1} \sum_{k=0}^{N-1} y_{k+d}\, e^{-i2\pi \frac{m(k+d)}{N}}. \qquad (10)$$

After some manipulation, we can write $\hat{Y}_m$ as

$$\hat{Y}_m = \frac{1}{N} \sum_{q=0}^{N-1} \tilde{y}_q e^{-i2\pi \frac{mq}{N}}, \qquad (11)$$

with

$$\tilde{y}_q \triangleq \begin{cases} \frac{1}{D}\left[(\frac{D}{2}+q+1)y_q + (\frac{D}{2}-q-1)y_{q+N}\right], & \text{for } 0 \leq q < \frac{D}{2}, \\ y_q, & \text{for } \frac{D}{2} \leq q < N - \frac{D}{2}, \\ \frac{1}{D}\left[(N+\frac{D}{2}-q-1)y_q + (-N+\frac{D}{2}+q+1)y_{q-N}\right], & \\ & \text{for } N - \frac{D}{2} \leq q < N. \end{cases} \qquad (12)$$

Fig. 3 shows the block diagram of this equivalent post-processor for $D = N$, and the linear combination is according to (12). We see that the equivalent post-processor needs only one FFT and has much lower complexity. Moreover, when we substitute Fig. 3 into the post-processor of Fig. 1, we obtain a receiver that differs from a standard OFDM receiver only by an additional linear combining block. Therefore, the proposed algorithm can be readily combined with other algorithms of channel estimation, synchronization, coding, etc. for OFDM systems in fast fading.

4. Performance Analysis

In this section, we will derive the variance of residual ICI of our algorithm in the general scheme. The residual ICI of the mth subchannel can be written as

$$\text{ICI}_m = \frac{1}{ND} \sum_{l=1}^{L} \sum_{\substack{m'=0 \\ m' \neq m}}^{N-1} e^{-i2\pi \frac{m'}{T} \tau^{(l)}} X_{m'} \sum_{d=-D/2}^{D/2-1} \sum_{k=0}^{N-1} h_{k+d}^{(l)} e^{i2\pi \frac{(m'-m)(k+d)}{N}}, \quad (13)$$

where $h_k^{(l)}$ and $\tau^{(l)}$ are the kth sample and delay of the lth path, respectively. The variance of ICI_m is

$$E\left[|\text{ICI}_m|^2\right] = \frac{1}{N^2}|X|^2 \sum_{l=1}^{L} \sum_{m'=1}^{N-1} E\{|\tilde{H}_{m'}^{(l)}|^2\}, \quad (14)$$

where where $|X|^2$ denotes the average power of the symbol $X_{m'}$, and

$$\tilde{H}_m^{(l)} \triangleq \sum_{q=0}^{N-1} \hat{h}_q^{(l)} e^{-i2\pi mq/N}, \quad (15)$$

with

$$\hat{h}_q^{(l)} \triangleq \begin{cases} \frac{1}{D}\left[(\frac{D}{2}+q+1)h_q^{(l)} + (\frac{D}{2}-q-1)h_{q+N}^{(l)}\right], & \text{if } 0 \leq q < \frac{D}{2}, \\ h_q^{(l)}, & \text{if } \frac{D}{2} \leq q < N - \frac{D}{2}, \\ \frac{1}{D}\left[(N+\frac{D}{2}-q-1)h_q^{(l)} + (-N+\frac{D}{2}+q+1)h_{q-N}^{(l)}\right], \\ & \text{if } N - \frac{D}{2} \leq q \leq N-1. \end{cases} \quad (16)$$

We can write $E\{|\tilde{H}_{m'}^{(l)}|^2\}$ as

$$E\{|\tilde{H}_{m'}^{(l)}|^2\} = \sum_{q=0}^{N-1} \sum_{q'=0}^{N-1} F^{(l)}(q,q') e^{-i2\pi \frac{m'(q-q')}{N}}, \quad (17)$$

where $F^{(l)}(q,q')$ is the correlation function of $\hat{h}_q^{(l)}$. Therefore, start from the correlation function of $h_q^{(l)}$, say, $r_h^{(l)}(q)$, we can obtain $F^{(l)}(q,q')$ through (16), and then the variance of ICI by (17) and (14).

5. Simulation Results

In Fig. 4, we plot the analytical and simulated results of the variance of residual ICI of our algorithm for various $f_d T$'s and ER's, where ER is defined as $\frac{D}{N}$. Each path of channel has the Rayleigh correlation function.

A Novel Algorithm of Inter-Subchannel Interference Cancellation

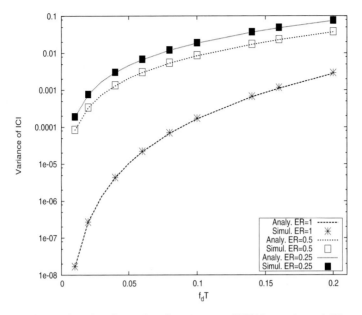

Figure 4. The analyzed and simulated variances of ICI for various f_dT's and ER's.

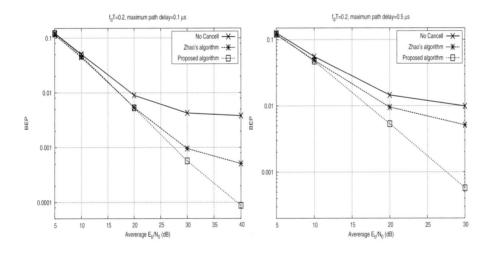

Figure 5. Performance comparison with Zhao's algorithm. The maximum path delays are 0.1 and 0.5 μs, respectively

The analytical results are obtained based on the approach in section 4. To obtain the simulated variance of ICI, we select one subchannel and transmit no symbol on it. The ICI of this subchannel is the received signal on it, assuming no other noise or interference. We see that the simulation results coincide with the analytical ones, and the variance of ICI increases with both $f_d T$ and $\frac{1}{ER}$, as expected.

We also compare our algorithm with Zhao's algorithm [3]. Following [3], we apply the frequency-domain differential coding to avoid channel estimation. Zhao's algorithm constrains the transmitted symbols such that $X_1 = -X_0, X_3 = -X_2, \ldots, X_{N-1} = -X_{N-2}$, and an ICI cancellation demodulation was proposed to make the ICI self-cancelled in the receiver. Zhao's algorithm can be considered as an approach that adds frequency-domain diversity in the transmitted signals. It's clear that the throughput of Zhao's algorithm is only half of the original system, and therefore it's fair to compare Zhao's algorithm with our proposed complete algorithm ($D=N$). In left part of Fig. 5, the maximum path delay is 0.1 μs, we see that our algorithm appears to have lower BEP than Zhao's when the SNR is larger than 30 dB, while both algorithms have much smaller BEPs at high SNR comparing with the system without ICI cancellation. In right part of Fig. 5, the maximum path delay increases to 0.5 μs. Comparing the BEPs at $\frac{E_b}{N_0} = 30$ dB, we see that Zhao's algorithm appears some degradation, due to the frequency-domain diversity is distorted by the reduced coherent bandwidth, which is the reciprocal of the delay spread. On the other hand, the proposed algorithm obtains almost the same BEP at SNR=30 dB.

6. Conclusion

In this paper, we continue our work of ICI self-cancellation in [4]. We propose an equivalent post-processor with much lower complexity and make the algorithm can be readily combined with other algorithms of coding, channel estimation, synchronization, etc. We also analyze the variance of the residual ICI of our algorithm and compare the BEP performance with Zhao's algorithm.

References

[1] W. G. Jeon, K. H. Chang, and Y. S. Cho, "An equalization technique for orthogonal frequency-division multiplexing systems in time-variant multipath channels," *IEEE Trans. Commun.*, vol. 47, pp 27-32, Apr. 1999.

[2] M. Münster and L. Hanzo, "Second-order channel parameter estimation assisted cancellation of channel variation-induced inter-subcarrier interference in OFDM systems," *EUROCON'2001, International Conference on Trends In Communications*, vol. 1 pp. 1-5, Jul. 2001.

[3] Y. Zhao and S.-G. Häggman, "Intercarrier Interference Self-Cancellation Scheme for OFDM Mobile Communication Systems," *IEEE Trans. Commun.*, vol. 48, pp 1185-1191, Jul. 2001.

[4] Ming-Xian Chang, "A Novel Algorithm of Inter-Subchannel Interference Cancellation in OFDM Systems,"in *Proc. IEEE 60th Vehicular Technology Conf.*, Los Angeles, pp. 460-464, Sep. 2004.

COMPARATIVE MULTICARRIER SYSTEM PERFORMANCE IN JAMMING AND FADING

David W. Matolak, Wenhui Xiong, Indranil Sen
School of Electrical Engineering & Computer Science Ohio University Athens, OH, USA 45701

Abstract: The performance of MC/MT-DS-SS and FH-OFDM in the presence of jamming and fading is investigated. Performance with both swept tone and partial band jamming is compared under conditions of identical data rate, transmit power, and occupied bandwidth.

Key words: MC-DS-SS, MT-DS-SS, FH-OFDM, Fading Channel, Jammer

1. INTRODUCTION

Continually increasing demands for mobile communications have caused methods related to orthogonal frequency division multiplexing (OFDM) [1] to see an enormous amount of study. Applying this modulation scheme to spread spectrum (SS) and code-division multiple access (CDMA), several multicarrier (MC) approaches have been proposed, including the direct sequence (DS) MC-SS techniques MC-DS-SS and MT-DS-SS [2]-[8]. Without spread spectrum, OFDM performance is very poor in the presence of interference [10]. In [11], the authors provided a performance analysis of the IEEE 802.11a system in the presence of a fading channel and pulsed-interference. Here, we employ a FH-OFDM system to incorporate spread spectrum into OFDM. Our motivations for analyzing the performance of MC systems in jamming are the possible future coexistence of wideband MC/MT-DS-SS systems and narrow band systems [9], and potential military anti-jam use of MC systems.

To date, the bit error ratio (BER) performance of MC-DS-SS systems in the presence of partial band interference has been extensively studied [2]-[8], but few references [7], [8], address MT-DS-SS performance. In [2] and [5], the authors investigated MC-DS-SS performance in the presence of a partial band jammer with rectangular power spectral density (PSD). In practice though, an unfiltered (or filtered) *M*-ary random modulated signal (MPSK, QAM) with PSD in the form of a $sinc^2(f)$ function may better represent an existing narrow band interference signal. Here, we investigate the effect of a partial band interferer with $sinc^2(f)$ PSD and variable center frequency, bandwidth, and duty cycle, on several MC/MT systems in a fading channel.

In Section 2 we describe the MC transmitter and receiver structures, and provide signal models used for the jammers and channels. Section 3 provides performance analysis in the presence of jamming. In Section 4, we provide a system performance comparison. Computer simulation results validate and complement the analysis. Finally, Section 5 summarizes our findings.

2. SYSTEM MODELS

2.1 Transmitter structure and channel model

The transmitter block diagram for MC/MT-DS-SS systems is similar to those in [2]-[4]. For MC-DS-SS, we replicate the same data bit across subcarriers, and for MT, we use a S:P conversion. Without loss of generality and for ease of analysis, BPSK is used on each subcarrier. Transmitted signal *s(t)* is

$$s(t) = \sqrt{\frac{2E_{b,x}}{T_{b,x}}} c(t) \sum_{i=0}^{M_x - 1} d_i(t) \cos[2\pi f_{x,c,i} t] \tag{1}$$

where *x* is either MC or MT. For MC, $d_i = d$; $E_{b,x}$ and $T_{b,x}$ are the bit energy and bit duration of each subcarrier; $d_i(t) = d_i(n) p_{Tb,x}(t - T_{b,x})$ is the transmitted data waveform with $d_i(n)$ the nth transmitted data bit of subcarrier *i* and $p_T(t)$ the rectangular pulse waveform; *c(t)* is the spreading waveform. Carrier frequencies for subcarrier *i* for MC, MT are $f_{c,i} = f_c + i/T_{c,MC}$ and $f_{c,i} = f_c + i/T_{b,MT}$.

The transmitter for OFDM [12] employs an inverse FFT followed by a complex modulator. The companion OFDM receiver essentially inverts these operations. The OFDM bandpass signal is given by $v(t) = \sum_{i=0}^{M_{OFDM}-1} \sqrt{2E_{b,OFDM}/T_{b,OFDM}} d_i(n) \cos(2\pi f_i t)$, where f_i is the ith subcarrier frequency, $f_i = f_c + i/T_{b,OFDM}$ and f_c is the center frequency, which is randomly hopped within the available bandwidth, *BW*. The number of possible hops is given by $N_{HOP} = BW/B_{OFDM}$, where B_{OFDM} is OFDM signal bandwidth. The OFDM signal center frequency hops randomly within the available bandwidth under the constraint that the difference between the center frequencies of two consecutive hops is equal to or greater than the channel coherence bandwidth B_c—this ensures independent fading on different hops. To fairly compare the systems in the presence of interference, the three systems utilize the same modulation scheme on each individual subcarrier (BPSK), and they all have the same bandwidth, data throughput, and transmission power. The numbers of subcarriers used for MC, MT and FH-OFDM are 3, 3, and 64, respectively. We assume B_c equals the chip rate of MC-DS-SS, thus the baseband impulse response of the channel for each subcarrier of MC and the entire OFDM spectrum is flat fading. The baseband channel impulse response for the ith subcarrier of MT-DS-SS is a tapped delay line model, given by $h_i(t) = \sum_{l=0}^{L-1} \alpha_{i,l} \exp(-j\varphi_{i,l}) \delta(t - \tau_{i,l})$, where $\alpha_{i,l}$, $\varphi_{i,l}$, $\tau_{i,l}$ are respectively the channel attenuation, phase shift and delay for the lth path of the channel between transmitter and receiver for subcarrier i.

2.2 Jammer model and receiver structure

The SPT is a sinusoidal tone with frequency periodically swept over a given bandwidth, and can be expressed as $J(t) = \sqrt{2J} \cos(2\pi f_{Jam} t)$, where J is the power of the jammer, and f_{Jam} is the tone frequency. The tone sweeps within a frequency range $[f_L, f_H]$ in one period, T_{swp}. For the linearly swept case, $f_{Jam} = f_L + \Delta f t$, with $\Delta f = (f_H - f_L)/T_{Jam}$ (Hz/sec) the sweeping rate.

The PBJ signal is $J(t) = \sqrt{2J/\rho} J(t) \cos(2\pi f_{Jam})$, with duty cycle ρ, and J and f_{Jam} defined as in the SPT case. The PBJ baseband bandwidth is $1/T_J$, and $J(t)$ is a random process with autocorrelation function $R_\tau(t) = 1 - |\tau|/T_J; |\tau| < T_J$.

The PBJ/SPT interferers and the communicator signal under investigation are put into independent fading channels with the same statistics. Figure 1 shows a frequency domain representation of two of the MC schemes in the presence of the jammers. We use diversity receivers for both MT-DS-SS and MC-DS-SS. For MT, a RAKE [2]-[4] is used in order to exploit the multipath diversity offered by the frequency selective channel, whereas for MC, an MRC exploits the system frequency diversity.

Each MC-DS-SS subcarrier incurs flat fading, so the received signal is

$$r(t) = \sqrt{2E_{b,MC}/T_{b,MC}} d(t) c(t) \sum_{i=0}^{M_{MC}-1} \alpha_i(t) \cos[2\pi f_{c,i} t - \varphi_i] + J_{REC}(t) + n(t), \quad (2)$$

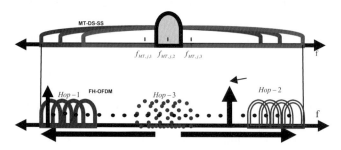

Figure 1. [MT-SS, and FH-OFDM spectra in presence of SPT and PBJ.]

where $\alpha_i(t)$ is the amplitude of the channel gain for subcarrier i, and φ_i is the phase, J_{REC} is the received jammer signal, and $n(t)$ is additive white Gaussian noise (AWGN) with two sided PSD $N_0/2$. We assume the receiver perfectly estimates the channel phase, and perfect synchronization and timing are maintained. When MRC is used to form the decision metric the correlator output for the jth subcarrier is $Z_j = D + J_j + N$, where D is the desired signal term $(\alpha_j)^2 \sqrt{E_{b,MC} T_{b,MC}/2} d_j(0)$, J_j is the jammer term, and N is the AWGN term, a zero mean Gaussian RV with variance $\alpha_j N_0 T_{b,MC}/4$.

The correlator output for OFDM is of the same form as for MC, but D is $\alpha_j^2 \sqrt{E_{b,OFDM} T_{b,OFDM}/2} d_j(0)$ and N has variance $N_0 T_{b,OFDM}/4$. Similar to MC, the received signal for MT can be written as

$$r(t) = \sum_{i=0}^{M_{MC}-1} \sum_{l=0}^{L-1} \alpha_{i,l} \sqrt{\frac{2E_{b,MT}}{T_{b,MT}}} d_i(t - \tau_{i,l}) c(t - \tau_{i,l}) \cos[2\pi f_{c,i}(t - \tau_{i,l}) + \varphi_{i,l}] + J_{REC}(t) + n(t) \quad (3)$$

The correlator output for the mth finger of the jth MT subcarrier is $D+J+N+I_1+I_2$, where I_1 is due to the multipath interference from the same subcarrier, and I_2 is contributed by the multipath from other subcarriers. The variances for I_1 and I_2 can be obtained as in [3], with results given in Table 1.

3. PERFORMANCE ANALYSIS

For brevity, we show performance analysis only for MC in SPT. The SPT sweeps the whole bandwidth within time period T_{swp}, which equals the input symbol duration. For the jth subcarrier, the jammer statistic is

$$J_j = \int_0^{T_{b,MC}} \beta \alpha_j \sqrt{2J} \cos(2\pi f_{Jam} t + \varphi_{Jam}) c(t) \cos[2\pi f_{c,j} t + \varphi_j] dt \quad (4)$$

where β is the amplitude of the jammer channel, which has the same distribution as α_j, and φ_{Jam} is the jammer channel phase. Simplifying,

$$J_j = \frac{\sqrt{2J}}{2}\beta\alpha_j \sum_{m=0}^{N_{MC}-1} c(m) A_m \qquad (5)$$

The A_m terms in (5) can be expressed via Fresnel integrals,

$$A_m = \int_{mT_{c,MC}}^{(m+1)T_{c,MC}} \cos[2\pi(B_j t + \Delta f t^2) + \theta] dt \qquad (6)$$

$$= \frac{1}{2\sqrt{\Delta f}}\left\{\cos\frac{\pi B_j^2 - 2\theta\Delta f}{2\Delta f}(C[g_1(j,m)]-C[g_0(j,m)]) + \sin\frac{\pi B_j^2 - 2\theta\Delta f}{2\Delta f}(S[g_1(j,m)]-S[g_0(j,m)])\right\}$$

where $g_k(j,m)=[2\Delta f T_{c,MC}(m+k)+B_j]/\sqrt{\Delta f}$, $k=0,1$; $B_j=f_L-f_{cj}$; $\theta = \varphi_{Jam} - \varphi_j$ is the random phase, uniform in $[0,2\pi)$; and $C(x)$, $S(x)$ are Fresnel integrals: $C(x) = \int_0^x \cos(\pi t^2/2) dt$, $S(x) = \int_0^x \sin(\pi t^2/2) dt$. J_j is zero mean due to the random code assumption, so after normalizing the channel power to 1, the variance of J_j is

$$\text{var}[J_j] = \frac{\alpha_j^2 E[\beta^2] J}{2}\sum_{m=0}^{N_{MC}-1} E(A_m)^2 = \frac{\alpha_j^2 J}{16\Delta f}\sum_{m=0}^{N_{MC}-1}\left\{\begin{array}{l}(C[g_1(j,m)]-C[g_0(j,m)])^2 + \\ (S[g_1(j,m)]-S[g_0(j,m)])^2\end{array}\right\}. \qquad (7)$$

The MT and OFDM systems are analyzed similarly. The main term of interest in BER performance is the jammer variance, listed in Table 1, along with resulting SNIR and BER expressions. For Table 1, the following definitions apply: $P_{MT}=T_{b,MT}/T_{swp}$; $N_{SWP}=T_{SWP}/T_{c,MT}$; $\Delta f_{j,Jam} = f_{c,j} - f_{Jam}$ is the frequency separation between subcarrier j and the PBJ center frequency; $\gamma_{\overline{Jam}}$ is the SNIR when the (PBJ) jammer is off; and $h_0(j,m)$ and $h_1(j,m)$ are

$$h_1(j,m) = [2\Delta f T_{c,MC}(pNswp + m + 1) + B_j]/\sqrt{\Delta f}$$
$$h_0(j,m) = [2\Delta f T_{c,MC}(pNswp + m) + B_j]/\sqrt{\Delta f} \qquad (8)$$

P_{hit} is the probability that the PBJ actually interferes with an OFDM symbol, given by $P_{hit}=1/N_{hop}$. To obtain the average BER, the BER of each subcarrier given in Table 1 should be averaged over all subcarriers.

Table 1. [Jammer statistics for different multicarrier schemes]

System	Variances	Instantaneous SNIR (γ) and Bit Error Rate for one subcarrier ($P_{b,j}$)
MC + SPT	$\dfrac{J\alpha_j^2}{16\Delta f} \sum_{m=0}^{N_{MC}-1} \{(C[g_1(j,m)] - C[g_0(j,m)])^2 + (S[g_1(j,m)] - S[g_0(j,m)])^2\}$	$\gamma_j = \left\{ \sum_{j=0}^{M_{MC}-1} \left[\dfrac{N_0}{E_{b,MC}} \alpha_j^2 + + \dfrac{4Var[J_j]}{E_{b,MC} T_{b,MC}} \right] \right\}^{-1}$ $P_b(\gamma) = \int\int\int_0^\infty Q(\sqrt{2\gamma_{MC}}) p_{\alpha_1}(\alpha_1) p_{\alpha_2}(\alpha_2) p_{\alpha_3}(\alpha_3) \, d\alpha_1 d\alpha_2 d\alpha_3$ $P_{b,j,PBJ} = \rho P_b(\gamma_{j,jam}) + (1-\rho) P(\gamma_{\overline{j,jam}})$
MC + PBJ	$\begin{cases} \dfrac{\alpha_j^2 J N_{MC} T_{c,MC}^2}{4\rho} \left[\dfrac{1 + (T_{c,MC}/T_j - 1)\cos(2\pi\Delta f_{j,jam} T_{c,MC}) + T_{c,MC}/T_j}{(2\pi\Delta f_{j,jam} T_{c,MC})^2} \right] & \Delta f_{j,jam} \neq 0 \\ \dfrac{\alpha_j^2 J N_{MC} T_{c,MC}^2}{4\rho} \left[1 - \dfrac{T_{c,MC}}{3T_j} \right] & \Delta f_{j,jam} = 0 \end{cases}$	
OFDM + SPT	$\dfrac{J}{16\Delta f} \sum_{p=0}^{P_{ODFM}-1} \left\{ \left[C\left(\dfrac{2\Delta f(p+1)T_{SWP} + B_j}{\sqrt{\Delta f}}\right) - C\left(\dfrac{2\Delta f p T_{SWP} + B_j}{\sqrt{\Delta f}}\right) \right]^2 + \left[S\left(\dfrac{2\Delta f(p+1)T_{SWP} + B_j}{\sqrt{\Delta f}}\right) - S\left(\dfrac{2\Delta f p T_{SWP} + B_j}{\sqrt{\Delta f}}\right) \right]^2 \right\}$	$\overline{\gamma}_{OFDM,j} = \left[\dfrac{N_0}{E_{b,OFDM}} + \dfrac{4Var[J_j]}{E_{b,OFDM} T_{b,OFDM}} \right]^{-1}$ $P_b(\overline{\gamma}) = \dfrac{1}{M} \sum_{j=0}^{M-1} \dfrac{1}{2} \left(1 - \sqrt{\dfrac{\overline{\gamma}_{OFDM,j}}{1 + \overline{\gamma}_{OFDM,j}}} \right)$ $P_{b,j,PBJ} = P_{hit} \rho P_b(\overline{\gamma}_{j,jam}) + (1 - P_{hit}\rho) P(\overline{\gamma}_{\overline{j,jam}})$
OFDM + PBJ	$\begin{cases} \dfrac{J}{2\rho} \dfrac{(T_{b,OFDM} + T_j) + (T_{b,OFDM} - T_j)\cos(2\pi\Delta f T_{b,OFDM})}{T_j (2\pi\Delta f)^2} & \Delta f_{j,jam} \neq 0 \\ \dfrac{J T_{b,OFDM}^2}{4\rho} \left[1 - \dfrac{T_{b,OFDM}}{3T_j} \right] & \Delta f_{j,jam} = 0 \end{cases}$	
MT + SPT	$I_1: \sum_{i=0,i\neq m}^{L-1} \dfrac{\alpha_{j,m}^2 \Omega_i E_{b,MT} T_{b,MT}}{6 N_{MT}}$ $I_2: \sum_{i=0,i\neq m}^{L-1} \sum_{j=0,i\neq j}^{M_{MT}-1} \dfrac{\Omega_i \alpha_{j,m}^2 E_{b,MT} T_{b,MT} N_{MT}}{4\pi^2 (i-j)^2} \{1 - \text{sinc}[2(i-j)/N_{MT}]\}$ $J_j: \dfrac{(\alpha_j)^2 J}{16\Delta f} \sum_{l=0}^{P_{MT}-1} \sum_{m=0}^{N_{MT}-1} \{(C[h_1(j,m)] - C[h_0(j,m)])^2 + (S[h_1(j,m)] - S[h_0(j,m)])^2\}$	$\gamma_j = \left\{ \sum_{l=0}^{L-1} \left[\dfrac{N_0}{E_{b,MC}} \alpha_{j,j}^2 + + \dfrac{4Var[J_{j,j} + I_1 + I_2]}{E_{b,MT} T_{b,MT}} \right] \right\}^{-1}$ $P_{b,j}(\gamma_j) = \int\int Q(\sqrt{2\gamma_{MTj}}) p_{\alpha_{j,1}}(\alpha_1) p_{\alpha_{j,2}} \, d\alpha_{j,1} d\alpha_{j,2}$ $P_{b,j,PBJ} = \rho P_{b,j}(\overline{\gamma}_{j,jam}) + (1-\rho) P_{b,j}(\overline{\gamma}_{\overline{j,jam}})$
MT + PBJ	$\begin{cases} \dfrac{J N_{MT} (\alpha_{j,j})^2}{\rho} \left[\dfrac{\pi \Delta f T_j + \pi \Delta f (T_{c,MT} - T_j)\cos 2\pi \Delta f T_{c,MT} + \pi \Delta f T_{c,MT} - \sin(2\pi \Delta f T_{c,MT})}{2 T_j (2\pi \Delta f)^3} \right] & \Delta f_{j,jam} \neq 0 \\ \dfrac{(\alpha_{j,j})^2 J N_{MT} T_{c,MT}^2}{4\rho} \left[1 - \dfrac{T_{c,MT}}{3T_j} \right] & \Delta f_{j,jam} = 0 \end{cases}$	

4. PERFORMANCE RESULTS

Using analytical results developed in section 3, we show performance plots in Figures 2-3. In Fig-2(a), we see good agreement between analytical and simulation results for MC and MT schemes for both jammer types. In Fig-2(b), we illustrate the effect of varying the PBJ duty cycle on the BER performance of the multicarrier schemes. It is evident that MC performs the best, followed by MT, and the worst performance is for the FH-OFDM scheme. This ranking of performance is visible for SPT jammer also.

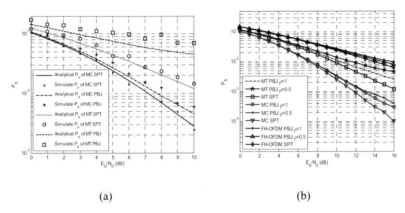

(a) (b)

Figure 2. (a) [BER vs. E_b/N_0, simulation vs. analytical, $N_{MC}=32$, $N_{MT}=64$, J/S=5dB, $B_J/BW=0.1$, $\rho=0.4$, T_{SP} is one bit duration (b) BER for MC schemes in SPT and PBJ (different ρ's), J/S=3 dB $B_J/BW=0.1$]

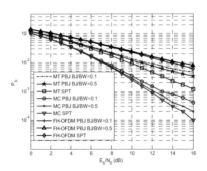

Figure 3. [BER vs. E_b/N_0, SPT and PBJ (different PBJ BW), J/S=3dB, $\rho=0.5$]

In Fig-3, we have plotted the BER performance of these MC schemes when we change the BW of the PBJ. Even in this case, the relative ranking remains: MC performs best, followed by MT, then by FH-OFDM.

5. CONCLUSIONS

In this work, we have analyzed the performance of three MC schemes in the presence of two types of jammer, a PBJ and a SPT, with different parameters. The performance of the MC schemes was compared under the condition that all systems have same transmit power and data rate. Our results show that MC-DS-SS with same data on each subcarrier outperforms the other two systems, and FH-OFDM has the worst performance.

6. REFERENCES

[1] IEEE 802.11 working group website, 7 October 2004.

[2] S. Kondo, L. B. Milstein, "Performance of Multicarrier DS CDMA Systems," *IEEE Trans. Comm.*, vol. 44, no. 2, pp. 238-246, February 1996.

[3] E. A. Sourour, et-al, "Performance of Orthogonal Multicarrier CDMA in a Multipath Fading Channel," *IEEE Trans. Comm.*, vol. 44, pp. 356-366, March 1996.

[4] L. Vandendorpe, "Multitone Spread Spectrum Multiple Access Communications System in a Multipath Rician Fading Channel," *IEEE Trans. Vehicular Tech.*, vol. 44, no. 2, pp. 327-337, May 1995.

[5] K. Cheun, et-al, "Antijamming Performance of a MC-DS-SS system," *IEEE Trans. Comm.*, vol. 47, no. 12, pp. 1781-1784, December 1999.

[6] S. Zhou, et-al, "Digital Multi-Carrier Spread Spectrum Versus Direct Sequence Spread Spectrum for Resistance to Jamming and Multipath," *IEEE Trans. Comm.*, vol. 50, no. 4, pp. 643-655, April 2002.

[7] D. W. Matolak, V. Deepak, F. A. Alder, "Performance of Multitone and Multicarrier Direct Sequence Spread Spectrum in the Presence of Narrowband Interference," *Proc. 12th Virginia Tech Symp. on Wireless Pers. Comm.*, Blacksburg, VA, pp. 135-144, June 5-7, 2002.

[8] D. W. Matolak, et-al, "Performance of Multitone and Multicarrier Direct Sequence Spread Spectrum in the Presence of Partial-Band Pulse Jamming/Interference," *Proc. IEEE VTC, Session 21*, Vancouver, BC, CA, Sept. 24-28, 2002.

[9] L. B. Milstein, et-al, "On the Feasibility of a CDMA Overlay for Personal Communications Networks," *JSAC*, vol. 10, pp. 655-668, May 1992.

[10] D. W. Matolak, et-al, "Potential Multicarrier and Spread Spectrum Systems for Future Aviation Data Links," *Proc. IEEE Aerospace Conf.*, Big Sky, MT, March 5-12, 2005.

[11] C. Kalogrias, C. Robertson, "Performance analysis of IEEE 802.11a WLAN standard optimum and sub-optimum receiver in frequency-selective fading Nakagami channels with AWGN and pulsed-noise interference," *Proc. MILCOM '04*, Monterey, CA Oct. 31-Nov. 3, 2004.

[12] R. Prasad, *OFDM for Wireless Communications Systems*, Artech House, Boston, MA, 2004.

CDMA IN THE CONTEXT OF OFDM AND SC/FDE – A COMPARATIVE STUDY

H. Witschnig[1], M. Huemer[2], R. Stuhlberger[3] and A. Springer[3]
1) Philips Austria, Mikronweg 1, 8101 Gratkorn, harald.witschnig@philips.com
2) Institute for Technical Electronics, University of Erlangen, Germany, huemer@lfte.de
3) Institute for Communications and Information Engineering, University of Linz, Austria, {a.springer, r.stuhlberger}@icie.jku.at

Abstract: Topic of this work is to characterize the frequency domain equalization concepts of OFDM (Orthogonal Frequency Division Multiplexing) and SC/FDE (Single Carrier Transmission with Frequency Domain Equalization) in the context of CDMA (Code Division Multiple Access) – as besides efficient equalization in the frequency domain, future communication systems will also be judged by their possibilities to be combined with multiple access techniques.
Due to specific characteristics of the investigated systems, also their behaviour in the context of CDMA is a different one and advantages and disadvantages of the underlying concepts will be pointed out. Aspects as preferred spreading codes, despreading in the frequency domain, possible performance in the context of multi path propagation, or multi user detection are investigated.

Keywords: Frequency Domain Equalization, CDMA

1. INTRODUCTION

Time dispersion caused by multi path propagation makes equalization necessary. One of the most powerful concepts are these which carry out the equalization in the frequency domain as there the effort grows significantly slower with increasing data rate as for the equalization in the time domain. Two well known concepts exist: OFDM (Orthogonal Frequency Division Multiplexing) and SC/FDE (Single Carrier System with Frequency Domain Equalization) [1]. Although SC/FDE is gaining more and more interest, it is still comparably rare investigated and therefore its principle is pointed out in brief.

Frequency domain equalization for single carrier systems is based on the equivalence between the convolution of two sequences in the time domain and the product of their Fourier transforms in the frequency domain. In [2] it has been proposed to perform a block wise transmission similar to OFDM and also to insert a Cyclic Prefix (CP) between successive blocks. Due to the cyclic extension of the transmitted blocks, the convolution of one cyclically extended transmitted block and the channel impulse response can be calculated by a circular convolution or equivalently by multiplications in the frequency domain, based on fast Fourier transformations. Figure 1 points out the underlying block structure in the time domain.

Besides efficient equalization algorithms it is the concept of CDMA (Code Division Multiple Access), which represents an almost unavoidable strategy for future communication systems [3][4][5].

From that point of view the question is obvious which transmission concepts (modulation schemes/physical layer concepts) are combined with CDMA in a more profitable way than others - concepts that combine the best performance with a low signal processing complexity. Therefore aspects like, integration of CDMA and optimal choice of spreading sequences, multiple users and multi-path propagation, multi user detection, asynchronous transmission or the near-far problem are investigated and are topic of research. Exemplary results and differences between OFDM and SC/FDE are pointed out in this paper.

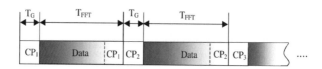

Figure 1. Transmitted data structure for SC/FDE.

2. INTEGRATION OF CDMA FOR SC/FDE AND OFDM

- Integration of CDMA for SC/FDE

Of primary interest in the context of SC/FDE is the despreading as it is to be combined with the concept of frequency domain equalization in the most advantageous way. Reception and despreading, which is formulated in the time domain for one symbol by

$$y(t) = \int_0^T r(t) \cdot c(t) dt, \tag{1}$$

can also be seen as the convolution of *r(t)* with *c(T-t)*,

$$y(t) = r(t) * c(T-t), \tag{2}$$

where *r(t)* describes a received value and *c(t)* describes the despreading code. Therefore despreading can also be formulated in the frequency domain by

$$Y(f) = R(f) \cdot C(f)^* e^{-j2\pi fT}. \tag{3}$$

This formulation points out obvious that despreading can be formulated as part of the optimal (frequency domain) reception for SC/FDE.

To keep the signal processing complexity as low as possible, the length of a spreaded symbol is defined by the FFT-size ($N=2^n$), which influences the choice of the spreading sequences. Although Walsh functions would fit for this scenario, they suffer from unfavorable auto-correlation properties (ACP), which lead to a severe influence on the performance by constructive or destructive interference - dependent on the actual radio channel. Compared to Walsh functions m- and Gold sequences fit significantly better for SC/FDE. These sequences have lengths of $J=2^n-1$ and must therefore be adapted for the SC/FDE system. The adaptation to a length of $J=2^n$ is carried out in that way that the sequences show an equal number of +1 and -1. Exemplary results are given by figure 2, demonstrating the strong dependence on the chosen spreading sequence – based on an arbitrarily chosen radio channel snapshot (the radio channel has been modeled as a tapped delay line, each tap with random uniformly distributed phase and Rayleigh distributed magnitude and with the power delay profile decaying exponentially). Note that PN32-52 stands for a PN Code of length 32 and the feedback is defined by $g_5=1$ and $g_2=1$.

Besides spreading it is the receive structure for multi path propagation – known as the Rake receiver which has to be adapted for SC/FDE. In [6] it has been demonstrated that an optimal Rake receiver is equivalent to a channel adapted matched filter, which is part of the receive structure for SC/FDE anyway. Nevertheless, a Rake receiver leads only to an optimal performance if the channel transfer function is known perfectly, the correlation properties of the spreading codes are ideal in terms of auto- and cross-correlation properties and if only one user is taken into account. As these requirements are not met in general, an additional equalizer is justified and necessary.

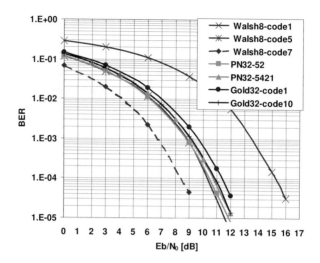

Figure 2. Performance due to different spreading codes.

For (frequency domain) equalization the following aspects have to be mentioned:

Due to channel adapted matched filtering and in particular due to the used equalizer the output noise sequence is colored, containing the characteristic of the actual radio channel. This leads to an additional, unwanted influence as a correlation with the despreading sequence results. This statement is underlined by figure 3, showing the overall transceiver structure, including the elements of CDMA as well as of SC/FDE.

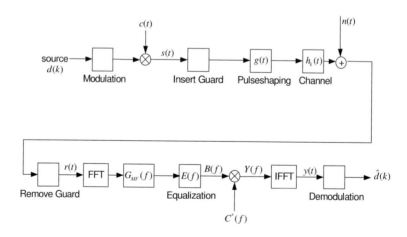

Figure 3. Block diagram of a CDMA-SC/FDE base band system.

- Integration for OFDM

Completely different appears the situation for multi carrier transmission. Our investigations/comparisons are based on the concept MC-CDMA (Multi-Carrier CDMA). It is well known that the spreading is carried out before the serial/parallel transformation [5] leading to the fact that every datasymbol is spreaded over all subcarriers. At the receiver, a parallel/serial transformation and a despreading in the frequency domain is carried out. This means that for MC-CDMA no adaptation of the length of the despreading sequence is necessary, compared to SC/FDE.

Due to the guard period the orthogonality of the subcarriers is ensured and equalization is reduced to a compensation of amplitude and phase of every subcarrier – therefore no colouring of the noise takes place and no unwanted correlation with the despreading sequence results. As a result of this, the performance rarely depends on the used sequence and the actual radio channel respectively. Based on that, Walsh Codes represent the optimal sequences for MC-CDMA.

Therefore it is concluded that one of the most significant difference between SC/FDE and OFDM in the context of CDMA is the choice of the used spreading sequences.

3. MULTIPLE USERS AND MULTIUSER DETECTION

For the multiple users case we assume a synchronous downlink. At the base station data of different users are spreaded with assigned codes and are sent at the same time in parallel, leading to an additive superimposement of several user signals.

All sent information reaches the mobile station over the same radio channel and has to be separated there by efficient methods. Exemplary this characteristics is given for two users in equation (4) where $r(t)$ describes the received sequence

$$r(t) = d_1(t)c_1(t) + d_2(t)c_2(t) + n(t) \qquad (4)$$

By carring out the despreading, based on sequence c_1, equation (5) results, characterizing the mutual influence of each user very well.

$$y(t) = \int_0^{T_D} d_1(t)c_1^2(t)dt + \int_0^{T_D} d_2(t)c_1(t)c_2(t)dt + \int_0^{T_D} n(t)c_1(t)dt \qquad (5)$$

It is obvious that the second term in equation (5) leads to an additional degradation if the used spreading sequences show not perfect cross correlation properties as it is the case for the used m- and Gold-sequences leading to a relative loss of SC/FDE-CDMA compared to MC-CDMA. Additionally the radio channel has to be taken into account, pointing out further differences between the investigated concepts. Therefore a two-path channel, whose echo is delayed by a value of τ, is taken into account, resulting in equation (6).

$$y(T_D) = \underbrace{\int_0^{T_D} d_1(t)dt}_{d_1(T_D)} + \underbrace{\int_0^{T_D} d_1(t-\tau)c_1(t-\tau)c_1(t)dt}_{I(T_D)} +$$
$$+ \underbrace{\int_0^{T_D} d_2(t)c_2(t)c_1(t)dt}_{S_1(T_D)} + \underbrace{\int_0^{T_D} d_2(t-\tau)c_2(t-\tau)c_1(t)dt}_{S_2(T_D)} + \qquad (6)$$
$$+ \int_0^{T_D} n(t)c_1(t)dt$$

Equation (6) indicates that besides the periodic auto-correlation function (PACF) and the periodic cross-correlation function (PCCF) at the time instant $t=T_D$ also their values at the time instants $t=T_D+\tau$ ($n\tau$ for an arbitrary channel respectively) are of significance.

For the SC/FDE system the terms $d_1(T_D)$ and $S_1(T_D)$ are of significance in particular, as $I(T_D)$ and $S_2(T_D)$ are compensated by ZF or MMSE equalization. As a consequence primarily the PCCF at time-instances $t=nT_D$ is responsible for a decreased performance.

In comparison to this, for MC-CDMA only $d_1(T_D)$ and $S_1(T_D)$ are of relevance as $I(T_D)$ and $S_2(T_D)$ are compensated completely. This means that only the PCCF values at the time instants $t=nT_D$ are of relevance.

Nevertheless, equation (5) and (6) point out the need of multi user detection. Different methods to compensate the degradation due to multiple users are known. The one used in this contribution is parallel interference cancellation (PIC) [5]. Here the degradation due to interference of every user is canceled in parallel by "subtraction" as shown in figure 4 for SC/FDE-CDMA (the principle for MC-CDMA is equivalent). The received, distorted sequence is being equalized in the frequency domain and despreaded by all possible user codes. After going back to the time domain the symbols are demodulated and decided. Again they are modulated, spreaded, summed, distorted by the estimated channel and finally subtracted from the originally received sequence. This signal, which is free of interferers is then processed as in the case of a single user. The efficiency of this method is demonstrated in figure 5: while the loss in terms of E_b/N_0 due to 6 users at a BER of 10^{-4} is more than 6 dB, this loss is reduced to less than 1 dB if PIC is used. Of course it is to mention that the processing complexity for this method is high, and grows linear with the number of interferers, but the concept of SC/FDE allows at least to reduce the implementation effort by carrying out the equalization and despreading in the frequency domain.

A comparable investigation based on simulations has been carried out for MC-CDMA, leading to comparable results as for SC/FDE CDMA, which may be summarized as follows:

a) For Walsh codes (and ideal assumptions of synchronity) no PIC is necessary as the degradation is prevented anyway.

b) For Gold codes the performance will decrease tremendously, leading even to a saturation effect.

c) PIC improves the performance enormous, leading to a BER comparable to the one of Walsh sequences. Based on these results it can be concluded that the behavior of SC/FDE CDMA and MC-CDMA in the context of PIC is comparable: A significant performance gain is possible at the cost of a high implementation effort.

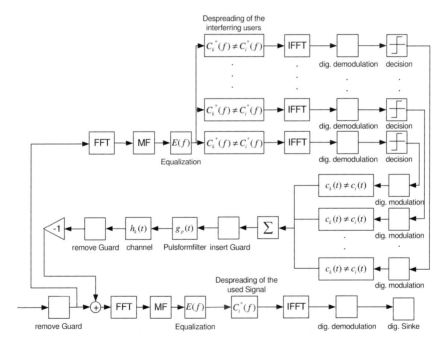

Figure 4. Block-Diagram of a PIC receiver.

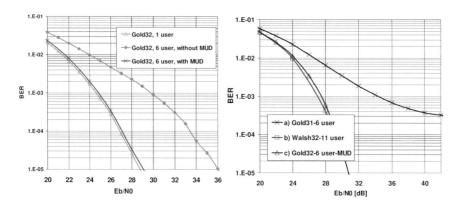

Figure 5. Performance Comparison due to PIC (SC/FDE left, MC-CDMA right).

4. COMPARISON OF SC/FDE-CDMA AND MC-CDMA - SUMMARY

SC/FDE and OFDM in combination with CDMA for high rate wireless systems have been compared in this work. The most significant differences are primarily based on the different spreading codes which have to be used (SC/FDE requires m- or Gold sequences instead of Walsh sequences which are optimal for OFDM). This leads to somewhat worse performance for SC/FDE, taking only one user into account. Nevertheless, it has been demonstrated that the necessary multi user detection leads finally to a comparable performance for both concepts. Therefore it is concluded that both concepts represent powerful strategies which can be combined with CDMA advantageously.

REFERENCES
[1] H. Witschnig, Frequency Domain Equalization for Broadband Wireless Communication, Dissertation, Linz, Austria, April 2004
[2] H. Sari, G. Karam, I. Jeanclaude, "An Analysis of Orthogonal Frequency-Division Multiplexing for Mobile Radio Applications", Proc. of the IEEE Vehicular Technology Conference (VTC '94), Stockholm, Sweden, pp. 1635-1639, June 1994
[3] A. Goiser, Handbuch der Spread-Spectrum Technik, Springer Verlag, Wien, 1998 (in German)
[4] S. Gliscic, B. Vucetic, Spread Spectrum CDMA Systems for wireless communications, Artec House Inc. 1997
[5] K. Fazel, S. Kaiser, Multi-Carrier and Spread Spectrum Systems, Wiley-Verlag, 2003
[6] H. Witschnig, R. Stuhlberger, A. Springer, "On the Use of CDMA for an SC/FDE System – an Overview", Proc. of the Int. Symposium on Signals, Systems and Electronics (ISSSE 04), Linz, Austria, August 2004

Section II

CELLULAR ASPECTS

IMPACT OF THE INTERCELL INTERFERENCE IN DL MC-CDMA SYSTEMS

Xenofon G. Doukopoulos and Rodolphe Legouable
France Telecom R&D RESA/BWA, 4 rue du Clos Courtel, 35512 Cesson-Sévigné, France.
Email: {xenofon.doukopoulos rodolphe.legouable}@francetelecom.com

Abstract: The impact of intercell interference for MC-CDMA systems is studied. We first model mathematically the pdf of the interfering signal, and its effects to the system's performance are verified via simulation results, where drastic performance degradation is observed especially at the cell borders. Then, we propose a simple intercell interference cancellation receiver in order to cope with this kind of interference. Simulations show the significant performance improvements that our receiver offers in all SIR values.

Key words: MC-CDMA, intercell interference, parallel interference cancellation.

1. INTRODUCTION

The multicarrier code division multiple access (MC-CDMA) scheme constitutes one of the most promising candidates for the downlink of B3G communication systems. It has been presented for the first time in[1,2] and combines the principles of both CDMA and orthogonal frequency division multiplexing (OFDM).

The intercellular interference constitutes one of the major handicaps for modern wireless communication systems. It has been verified in[3,4] that it affects the performance as a Gaussian noise of the same power. Moreover, it can be modeled as WGN for the case of unsynchronized base stations[5]. This actually means that the intercell interference degrades much more severely the performance compared to the multiple access interference (MAI). Thus, we focus on cancellation methods in order to ameliorate the performance.

The literature of Interference Cancellation (IC) techniques is very rich, but only for the single cell environment, where the only source of interference is the MAI. However, the same *does not* hold for the case of intercell interference, where the corresponding literature is very limited. Focusing to MAI suppression the successive interference cancellation (SIC) detector consists in estimating (and then subtracting) each interfering user in a serial way[6], while the parallel interference cancellation (PIC) detector estimates in a parallel manner the entire set of interfering users[7]. Moreover, hybrid techniques[8] have been proposed that combine both SIC and PIC schemes. Finally, for the rest of the article whenever the word "interference" is used, we refer to the "intercell interference", unless otherwise stated.

The rest of the paper is organized as follows. Section 2 introduces briefly the signal model of an MC-CDMA system. In Section 3, we analyze the nature of the intercell interference. In Section 4, we develop a parallel interference cancellation scheme specially tuned for the case of multiple cells. Simulation results are presented in Section 5, and finally Section 6 concludes our article.

2. SYSTEM MODEL

Consider an MC-CDMA transmitter with N_c subcarriers. Each user's bit vector is encoded by a UMTS-like turbocoder, being interleaved π, and then mapped according to the bit to symbol alphabet. Each symbol is spread with the corresponding Walsh-Hadamard signature of length N; the resulting signal is fed to a serial to parallel converter of size equal to the modulated carriers $N_{\text{mod}} \leq N_c$ to give $\mathbf{x}(n)$. If m denotes the number of CDMA symbols inside an OFDM symbol, i.e. $m = N_{\text{mod}}/N$, then $\mathbf{x}(n)$ has the form:

$$\mathbf{x}(n) = \begin{bmatrix} \mathbf{0} & \mathbf{Sb}_1(n) & \cdots & \mathbf{Sb}_m(n) & \mathbf{0} \end{bmatrix}^t,$$

where $\mathbf{S}$ is the Walsh-Hadamard matrix, $\mathbf{b}_i(n) = \begin{bmatrix} b_1(n) & \cdots & b_K(n) \end{bmatrix}^t$ the transmitted symbol vector for K users with $K \leq N$. Multiplication with the scrambling matrix $\mathbf{C}(n)$ follows before IDFT modulation of size N_c. Zero padding is added at both ends to match the filtering spectrum mask and after cyclic prefix insertion the signal is transmitted over a multipath channel.

At the receiver, after OFDM demodulation and descrambling, we have:

$$Y_j(n) = H_j(n)x_j(n) + w_j(n),$$

where $H_j(n)$ is the channel frequency response at the j-th subcarrier, $x_j(n) = s_1(j)b_1(n) + \cdots + s_K(j)b_K(n)$, and $w_j(n)$ AWGN.

If we consider an MC-CDMA model containing two synchronized cells, then the received signal will now have the form:

$$Y_j(n) = H_j(n)x_j(n) + c_j(n)c_j^I(n)H_j^I(n)x_j^I(n) + w_j(n),$$

where the I superscript refers to the signal from the interfering base station.

3. NATURE OF INTERCELL INTERFERENCE

Assuming that the interfering channel is known and since the scrambling coefficients $c_j(n), c'_j(n) = \pm 1$, then we may only focus on the discrete transmitted interfering signal $x'_j(n)$ to characterize the pdf of the interference. The latter is a random variable of the form $\omega = \alpha_1 + \cdots + \alpha_K$, where α_i are discrete uniform random variables depending on the modulation of the interfering base station. The pdf of the random variable ω is a staircase function that approaches the one of the normal distribution as K grows.

If we further consider BPSK modulation, then we can easily arrive to a closed form relation for this true pdf (coefficients of the Binomial Theorem). More specifically, $p(k)$ equals

$$p(k) = \binom{K}{(k+K)/2} a^K,$$

where $k = -K : 2 : K$ and $a = 1/2$. The latter can be approximated from the values of the Gaussian $\mathcal{N}(K/2, K/4)$ at points $0, 1, \ldots, K$ with one-to-one mapping, i.e.

$$p(k) \approx P(\omega = l) = \frac{1}{\sqrt{2\pi * (K/4)}} e^{-\frac{(l-K/2)^2}{2*(K/4)}}$$

Similarly for 16-QAM, we obtain the true pdf $p(k)$ of the random variable ω from the convolution of the pdf of a single symbol, i.e. $[1/4 \quad 1/4 \quad 1/4 \quad 1/4]$, with itself $K-1$ times. The discrete values that $p(k)$ takes are approximated from the values of the Gaussian $\mathcal{N}(3K/2, 5K/4)$ at points $0, 1, \ldots, 3K$ with one-to-one mapping, i.e.

$$p(k) \approx P(\omega = l) = \frac{1}{\sqrt{2\pi * (5K/4)}} e^{-\frac{(l-3K/2)^2}{2*(5K/4)}},$$

where $k = -3K : 2 : 3K$ and $l = \frac{k + 3K}{2}$.

Finally following the same logic for 64-QAM, we have the true (discrete) pdf $p(k)$ that can be approximated from the values of the Gaussian $\mathcal{N}(7K/2, 21K/4)$ at points $0, 1, \ldots, 7K$ with one-to-one mapping.

$$p(k) \approx P(\omega = l) = \frac{1}{\sqrt{2\pi * (21K/4)}} e^{-\frac{(l-7K/2)^2}{2*(21K/4)}},$$

where $k = -7K : 2 : 7K$ and $l = \dfrac{k + 7K}{2}$.

Let us verify the aforementioned facts with some graphs. In Figure 1, we present the discrete pdf of the transmitted signal as well as the corresponding Gaussian approximations for the BPSK, 16-QAM, and 64-QAM modulations. The processing gain equals the number of users and both take the value $N = K = 32$.

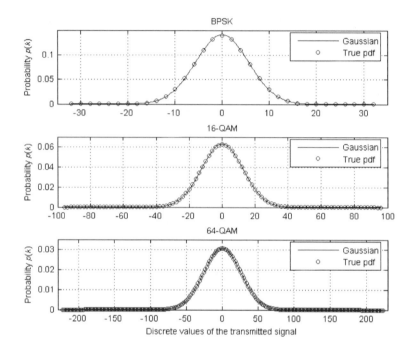

Figure 1. PDF of the transmitted symbols.

4. INTERFERENCE CANCELLATION

It has been verified in[4,5] that the intercell interference has a "Gaussian noise"-like effect in the performance of an MC-CDMA system. In simple words, the intercell interference behaves similarly to AWGN of the same power. Therefore, close to the cell borders, the SIR would be usually equal to 0dB, which degrades drastically the BER performance. It is thus of great interest to treat somehow this kind of interference, in order to achieve a unitary frequency reuse factor. The latter is very important, since it is widely known the scarcity of available frequency bands.

For OFDMA systems, it has been proposed in[9] to use LLR updates combined with some kind of time-frequency allocation patterns to avoid or reduce possible collisions[1] between users of different cells. However, as it has been shown in[10], in MC-CDMA we can not apply techniques that use LLR and MMSE equalization updates

$$G_j(n) = \frac{H_j^*(n)}{|H_j(n)|^2 + 1/SINR}.$$

The main reason is that in MC-CDMA systems just *a single* user covers *all* the available carriers, due to the use of spreading sequences, and thus it exists *no* interference-free region. Therefore, LLR updates *do not* produce a set of carriers with increased confidence, i.e. free of interference. Taking the aforementioned facts into consideration, we deduce that interference should be treated in a different way with respect to OFDMA systems.

4.1 Parallel Structures

Here, we mainly focus on parallel intercell interference cancellation (PIIC) schemes due to the fact that they present low latency times that respect real time data processing. The important characteristic of parallel architecture enables their practical implementations in contrast to serial schemes. Below we present the main blocks of a PIIC receiver, which provides an estimation of the signal emitted from base station-i. It also assumes that FFT and guard interval removal have been already performed.

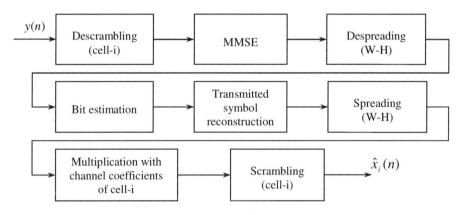

Figure 2. Structure of PIIC for cell-i.

---

[1] It should be noted here that if at least one cell works at full load, then collisions are unavoidable even for OFDMA systems.

It can be observed from Figure 2 that the PIIC needs to know the channel of the interfering cell. This step can be carried out thanks to known data signaling, i.e. during synchronization for example where each mobile terminal knows specific sequences transmitted by different cells and then performing some kind of adaptive algorithm. Finally, we should point out that the PIIC structure of Figure 2 *does not* use any coding scheme into the processing IC detection. The latter can sometimes improve the overall performance at the cost of an increased complexity.

5. SIMULATION RESULTS

In this section we would like to test the performance of a two cell MC-CDMA system. The parameters are N_c=1024, N_{mod}=704, cyclic prefix of size 64, Vehicular A channel, carrier frequency 2GHz, bandwidth 5MHz, QPSK R=2/3 modulation, and the noise is AWG.

5.1 No Interference Treatment

In Figure 3, BLER and BER curves are presented for the case where the intercell interference is not treated. The cell of interest operates always in full load, containing $N=K=32$ users, while a wide variety of 32, 16, 8, and 4 users is considered for the interfering cell. At each point the SNR is *equal* to

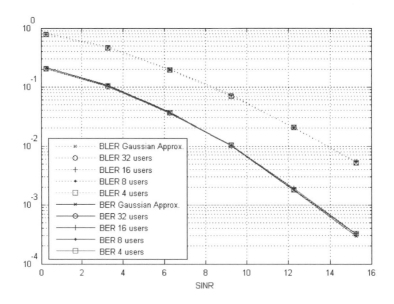

Figure 3. BLER and BER performance for different intercell loads.

the SIR, whatever the number of users in the interfering cell. Moreover, the curve of the Gaussian approximation case is also depicted. The performance, either in terms of BLER or BER, is actually identical for all curves. Thus, we see that the system's behavior does not actually depend on the number of the existing users in the interfering cell, as long as the SIR remains constant. Finally, it is true that the Gaussian noise constitutes an efficient approximation of the interference in MC-CDMA systems, whatever the interfering cell's load.

5.2 Interference Cancellation

In the present part we are interested to test the performance of the previously developed PIIC technique. We consider a two cell downlink MC-CDMA system. The processing gain of the system is equal to $N=16$ and the SIR levels vary from -10dB to 15dB, while the SNR is fixed to 15dB.

In Figure 4, we plot the BER and BLER performance for the case where both cells are working at half load, i.e. $K=7$ users. We can easily observe the superiority of the PIIC technique for all the tested SIR levels. For example, at SIR=0dB (which is a logical value for the cell borders) the PIIC is approximately 10dBs better than without IIC approach. We can note the fact that even for negative SIR values the PIIC scheme preserves a totally acceptable performance level (i.e. at 10^{-3} BER). In contrast by doing nothing the performance degrades by more than two orders of magnitude. Negative

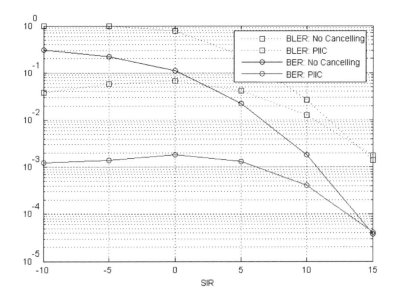

Figure 4. BLER and BER of intercell interference cancellation.

SIR values can appear in a real system, as for example when the user of interest has entered an interfering cell of full load. It should be also stressed here that the performance degrades as we move from SIR=-10dB till 0dB, since the quality of the estimated interfering signal deteriorates.

6. CONCLUSION

The present work focuses on the topic of intercell interference in the MC-CDMA context. It has been shown theoretically and by simulations results that this kind of interference behaves as Gaussian noise whatever the modulation. Moreover, we have proposed a simple parallel cancellation scheme that offers significant performance gains in all SIR levels, thus permitting a frequency reuse factor equal to unity. Future work will continue on the prominent area of intercell interference cancellation. Finally, we are very interested in studying the impact of the multicell interference in MIMO MC-CDMA systems.

7. REFERENCES

1. N. Yee, J.P. Linnartz, and G. Fettweis, "Multicarrier CDMA in Indoor Wireless Radio Networks," in Proc. of IEEE PIMRC Conference, Yokohama, Japan, Sep. 1993.
2. K. Fazel and L. Papke, "On the Performance of Convolutionally-Coded CDMA/OFDM for Mobile Communication System," in Proc. of IEEE PIMRC Conference, Yokohama, Japan, Sep. 1993.
3. G. Auer, S. Sand, A. Dammann, and S. Kaiser, "Analysis of the Cellular Interference for MC-CDMA and its Impact on Channel Estimation," *European Transactions on Telecommunications*, pp. 173-184, vol. 15, May-June 2004.
4. S. Plass, S. Sand, and G. Auer, "Modeling and Analysis of a Cellular MC-CDMA Downlink System," in Proc. of PIMRC'2004 Conference, Barcelona, Spain, Sep. 2004.
5. G. Auer, A. Dammann, S. Sand, and S. Kaiser, "On Modeling Cellular Interference for MultiCarrier Based Communication Systems Including a Synchronization Offset," in Proc. of WPMC'2003, Yokosuka, Japan, pp. 290-294, Oct. 2003.
6. J. G. Andrews and T.H.Y. Meng, "Performance of Multicarrier CDMA with Successive Interference Cancellation in a Multipath Fading Channel," *IEEE Trans. on Comm.*, vol. 52, pp. 811-822, May 2004.
7. R. Hofstad and M. J. Klok, "Performance of DS-CDMA Systems with Optimal Hard-Decision Parallel Interference Cancellation," *IEEE Trans. on Inf. Theory*, vol. 49, pp. 2918-2940, Nov. 2003.
8. D. Koulakiotis and A. H. Aghvami, "Evaluation of a DS/CDMA Multiuser Receiver Employing a Hybrid Form of Interference Cancellation in Rayleigh-Fading Channels," *IEEE Comm. Let.*, vol. 2, pp. 61-63, Mar. 1998.
9. J. Lainé, *Interference Estimation in a Multicellular OFDMA Environment*, M. Thesis, France Telecom R&D, Rennes, France, Jul. 2004.
10. France Telecom & Orange, "Inter-cell Interference Estimation for OFDMA and MC-CDMA on the DL," TSG RAN WG1 \#41 meeting, R1-050405, Athens, Greece, May 2005.

QUANTIFYING THE IMPACT OF THE MC-CDMA PHYSICAL LAYER ALGORITHMS ON THE DOWNLINK CAPACITY IN A MULTICELLULAR ENVIRONMENT

Abdel-Majid Mourad[1], Arnaud Guéguen[1], and Ramesh Pyndiah[2]
[1]*Mitsubishi Electric ITE-TCL, Telecommunications Research Laboratory, Rennes, France;*
[2]*Ecole Nationale Supérieure des Télécommunications de Bretagne, Brest, France*

Abstract: In this paper, we present an evaluation methodology for quantifying the impact of the MC-CDMA physical layer algorithms on the system capacity in the downlink of a multi-cellular environment. The methodology consists of a qualitative evaluation through a novel capacity indicator at the link level and a quantitative evaluation through a semi-analytical statistical approach at the system level. The qualitative and quantitative evaluations are complementary and lead to similar and consistent conclusions. They constitute an efficient tool that can be used for optimizing the MC-CDMA physical layer algorithms and identifying the most suitable configurations for a given environment.

Key words: MC-CDMA, Outage capacity, Multi-cellular environment, Downlink.

1. INTRODUCTION

A complete and realistic evaluation of the performance of a cellular system requires joint consideration of both microscopic and macroscopic aspects, from the microscopic level of binary information transmission to the macroscopic level of network control mechanisms. A typical cellular system usually includes several cells and a large number of mobile users, and therefore a combined approach where the microscopic and macroscopic aspects are modeled into one single simulator would lead to very complex simulations of large time consumption. Thus, for obvious complexity and feasibility reasons, the combined approach is usually discarded in practice and another splitting approach is used instead. The splitting approach carries out the performance evaluation in two simulation phases: at *link level* and at *system level*, with suitable interfacing between the two levels [1][2].

A *link level simulator* typically includes a transmitter, a multi-path fading channel, and a receiver. It operates with a microscopic time resolution equal to the sampling period so that the physical layer algorithms and multi-path channel can be modeled exhaustively. The main concern here is to investigate the impact of the physical layer algorithms on the quality of the binary information transmitted through the multi-path fading channel between the transmitter and receiver. This is with the aim of accounting for the effectiveness of the physical layer algorithms and their robustness to multi-path channel impairments. The transmission quality of the binary information is measured in terms of the average bit (BER) and frame (FER) error rates, which are usually determined from link level simulations spanning a large number of frames experiencing independent fading.

The *system level simulator* typically includes several base stations and all the mobiles that are connected to these base stations. The main interests here are to evaluate the cellular capacity and coverage of the system and to develop proper *Radio Resource Management* (RRM) algorithms for appropriate sharing of the system resources among the mobile users. The system level simulations are more or less complex depending on how accurate and how realistic the models represented are. Two approaches of system level simulations are usually used [1][2]: the *static* and *dynamic* approaches. The static approach requires low computational costs and does not have time dependency. It has been extensively used in the literature for preliminary system level studies as it can easily and efficiently provide meaningful statistical capacity and coverage estimates [2]. The dynamic approach is a much more sophisticated time-based approach, which is much more accurate and realistic than the static approach, but at the expense of much higher complexity [2]. In this paper, the static approach is adopted instead of the dynamic approach as our objective is to investigate only the capacity benefits of the MC-CDMA physical layer algorithms in the downlink of a cellular system.

Because of the separation between link and system level simulations, a suitable interface between the two levels needs therefore to be defined. The target of the interface is to enable the system level simulators to predict easily and accurately the actual transmission quality of the different links in the system. This is because the transmission quality of the different links cannot be measured online within the system level simulations. Usual procedures to interface link and system levels are to use a set of *Look-Up Tables* (LUTs) mapping the BER and FER transmission quality measures to an adequate *Signal to Interference plus Noise Ratio* (SINR)-based measure that can easily be calculated at the system level. Different LUTs generally need to be produced for different operating conditions, and different levels of accuracy can be targeted depending on the particular study carried out at the system level [3].

In this study, since we only focus on the performance evaluation of the physical layer algorithms through static system level simulations, we consider the conventional transmission quality measure of the physical layer, i.e. FER averaged over a large number of frames experiencing independent fading, in order to set the requirements for link satisfaction at the system level. The average FER measure is simply mapped to the *local mean SINR* measure through the so-called *multi-frame oriented link to system interface* [6]. The local mean SINR is the ratio between the average signal power and average interference plus noise power at the output of the detection module. The multi-frame oriented link to system interface considered here is simple and provides a degree of accuracy that is acceptable for the objective targeted in this study.

The rest of this paper is organized as follows. Section 2 presents briefly the system model and then Section 3 presents the evaluation methodology for quantifying the impact of the MC-CDMA physical layer algorithms on the downlink multi-cellular capacity. Numerical results are then presented in Section 4 and conclusions are finally drawn in Section 5.

2. SYSTEM MODEL

We consider the MC-CDMA physical layer in the downlink of an hexagonal regular macro-cellular system [6]. The system is made up of one central cell surrounded by N tiers of neighboring cells. The number of cells in the system is therefore equal to:

$$Q = 1 + 3N(N+1) \qquad (1)$$

Each cell has a centrally located BS fit with an omni-directional antenna. Each BS has at its disposal a maximum number of M spreading codes, and its total output power is limited by P_{max}. For the sake of simplicity and in order to avoid the border effects [4], the results are collected only from the central cell although the whole system is simulated, and the Q-1 neighboring BS are assumed to transmit at the same fixed power $P_o \le P_{max}$. The users are assumed uniformly distributed within the disk delimiting the hexagonal cells (see Fig. 1). The connectivity between users and BSs follows the minimum path loss criterion, i.e. a user is connected to the BS to which the path loss is minimum. A user is connected only to one BS, i.e. there is no handover, and all users are assumed to have the same physical layer configuration.

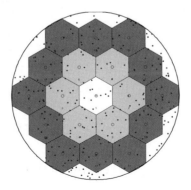

Figure 1. Cellular layout with $N = 2$ tiers and 250 users uniformly distributed within the disk delimiting the 19 hexagonal cells.

3. EVALUATION METHODOLOGY

Let us consider the problem of satisfying the target FER requirements of the users connected to the central BS (BS_1). Thanks to the direct mapping between the local mean SINR and average FER [6], the target FER can then be replaced by a target SINR. As discussed in [6], the target SINR value is specific to the given physical layer configuration and channel model. The problem of satisfying the SINR requirements under the constraint of limited BS power can then be formulated as:

$$SINR_k = \lambda_\phi, \quad \forall k = 1\ldots K \leq M, \quad p_k > 0, \quad P_c = \sum_{k=1}^{K} p_k \leq P_{max} \qquad (2)$$

where λ_ϕ denotes the target SINR that is specific to the physical layer configuration ϕ, K is the cell load of BS_1, P_c is BS_1 output power, and p_k is the power BS_1 should allocate to its k-th user in order to satisfy the SINR requirements. At last, $SINR_k$ denotes the local mean SINR at the output of the single user detection (SUD) module of the k-th user connected to BS_1 and it is given by [6]:

$$SINR_k = \frac{p_k g_{1k}}{g_{1k} \sum_{j=1,j\neq k}^{K} p_j \alpha_{kj} + \beta_k \left(\frac{1}{SF} \sum_{q=2}^{Q} P_o g_{qk} + P_n \right)} \qquad (3)$$

where g_{qk} stands for the path loss between BS_q and user k of BS_1. The quantities $\{\alpha_{kj}\}$ and β_k are respectively the mutual intra-cell interference and inter-cell interference plus noise factors at the output of the SUD module. These factors are derived analytically in [6] from the equalized channel coefficients correlation and family of spreading codes. At last, SF is the spreading factor and P_n is the thermal noise power.

Making use of (3), (2) can be rewritten as the following power allocation problem for BS_1:

$$\mathbf{p} = \lambda_\phi (\mathbf{Ap} + \mathbf{f}P_o + \mathbf{b}P_n), \quad \mathbf{p} > \mathbf{0}, \quad P_c = \mathbf{1}^T \mathbf{p} \leq P_{max} \qquad (4)$$

where $\mathbf{p}$ denotes the column vector of the K powers $\{p_k\}$, $\mathbf{A}$ is a $K \times K$ matrix representing the intra-cell interference, and $\mathbf{f}$ and $\mathbf{b}$ are two column vectors of length K representing respectively the inter-cell interference and thermal noise. The quantities $\mathbf{A}$, $\mathbf{f}$, and $\mathbf{b}$ are characterized by:

$$A[k,j] = \alpha_{kj}(1-\delta_{kj}), \quad f[k] = \frac{\beta_k}{SF} \sum_{q=2}^{Q} \frac{g_{qk}}{g_{1k}}, \quad b[k] = \frac{\beta_k}{g_{1k}} \qquad (5)$$

where δ_{kj} stands for the Kronecker symbol that is equal to 1 for $k = j$ and 0 otherwise. The matrix $\mathbf{A}$ is not strictly positive since its diagonal is null, but it is regular, i.e. its square is strictly positive, and so the Perron-Frobenius theory applies [5]. It is well known from Perron-Frobenius theory for non negative matrices that the form $\mathbf{p} = \mathbf{Ap} + \mathbf{b}$ has a positive solution

$\mathbf{p}^* = (\mathbf{I}-\mathbf{A})^{-1}\mathbf{b} > 0$ if and only if the maximum eigenvalue of $\mathbf{A}$ is strictly less than 1. Thus, the necessary and sufficient condition to obtain a positive and finite solution in (4) is that the maximum eigenvalue μ^* of $\mathbf{A}$ is less than $1/\lambda_\phi$. This condition is generally referred to as the *pole condition*, and the maximum cell load K satisfying the pole condition is referred to as the *pole capacity* K_{pole} [4]. The positive solution $\mathbf{p}^*$ in (4) can then be determined as:

$$\mathbf{p}^* = \lambda_\phi (\mathbf{x}P_o + \mathbf{y}P_n), \quad \mathbf{x} = (\mathbf{I}-\lambda_\phi \mathbf{A})^{-1}\mathbf{f}, \quad \mathbf{y} = (\mathbf{I}-\lambda_\phi \mathbf{A})^{-1}\mathbf{b} \qquad (6)$$

By taking into account the constraint of limited BS power, we define the *constrained capacity* as:

$$K_{const} = \arg\max_K \left\{ \Pr\left(\mu^* > \frac{1}{\lambda_\phi} \bigcup P_c > P_{max} \right) \leq \varepsilon \right\} \qquad (7)$$

where ε denotes the maximum tolerated outage threshold typically set to 5%. The cell throughput can then be derived from (7) by multiplying K_{const} by $(1-\text{FER})R_\phi$, where R_ϕ is the single user bit rate for the physical layer configuration ϕ.

3.1 Capacity Indicator at the Link Level

Let us consider the case where $\alpha_{kj} \approx \alpha$ and $\beta_k \approx \beta$. In this case, the power $P_c = \mathbf{1}^T \mathbf{p}^*$ that is necessary to satisfy the SINR requirements can simply be written as (cf. (6)):

$$P_c = \underbrace{\frac{\lambda_\phi \beta K}{1-\lambda_\phi(K-1)\alpha}}_{C_\phi} \underbrace{\left(\frac{1}{K}\sum_{k=1}^{K}\left(\frac{P_o}{SF}\sum_{q=2}^{Q}\frac{g_{qk}}{g_{1k}} + \frac{P_n}{g_{1k}} \right) \right)}_{T} \qquad (8)$$

Note that T in (8) is independent of the physical layer configuration ϕ. Furthermore, by applying the law of large numbers, T can then be assumed independent of the cell load K. Only the factor C_ϕ remains therefore specific to the physical layer configuration ϕ and cell load K. Note that the factor C_ϕ needs only to be evaluated at the link level since it is only function of λ_ϕ, α, and β, which are outputs of the link to system level interface.

We extend the expression of the link level capacity indicator C_ϕ to the case of multiple factors $\{\alpha_{kj}\}$ as follows:

$$C_\phi = \frac{\lambda_\phi \beta K}{1-\lambda_\phi \mu^*} \qquad (9)$$

Thus, from (9), we can determine the maximum cell load K_{max} at a given C_ϕ threshold. Note that the pole capacity can simply be determined from (9) as when C_ϕ tends to infinity.

The interest of this novel capacity indicator in (9) is that it allows to evaluate at the link level the impact of the physical layer algorithms on the system capacity, i.e. without performing system level simulations. This makes it an efficient and accurate tool at the link level for optimizing the physical layer algorithms and identifying the most appropriate physical layer configurations for a given environment.

3.2 Particular Case of MMSEC Equalization

In the particular case of MMSEC equalization [7], the equalization coefficients are functions of the useful and interference powers received by the k-th user. The n-th equalization coefficient for the n-th channel coefficient $h[n]$ is obtained as:

$$w_k[n] = \frac{\sqrt{P_k g_{1k}}\, \overline{h}[n]}{\dfrac{P_c g_{1k}}{SF}|h[n]|^2 + \left(\dfrac{1}{SF}\sum_{q=2}^{Q} P_o g_{qk} + P_n\right)} \tag{10}$$

Thus, in this case, the intra-cell interference factors $\{\alpha_{kj}\}$ and inter-cell interference plus noise factors $\{\beta_k\}$, which are derived from the correlations of the equalized coefficients $\{h[n]w_k[n]\}$, become functions of the power vector $\mathbf{p}$. Thus, the power allocation problem in (4) becomes nonlinear as:

$$\mathbf{p} = \lambda_\phi (\mathbf{A}(\mathbf{p})\mathbf{p} + \mathbf{f}(\mathbf{p})P_o + \mathbf{b}(\mathbf{p})P_n) \tag{11}$$

In order to solve (11), the following recursive algorithm is used:

$$\mathbf{p}^{(r+1)} = \lambda_\phi (\mathbf{x}(\mathbf{p}^{(r)})P_o + \mathbf{y}(\mathbf{p}^{(r)})P_n) \quad \text{for } r = 1\ldots N_r \tag{12}$$

where r denotes the r-th recursion. It is observed that when (11) has a solution, the recursive algorithm in (12) converges to this solution in few ($N_r \leq 5$) recursions for any positive and finite initial vector $\mathbf{p}^{(1)}$.

The factors $\{\alpha_{kj}\}$ and $\{\beta_k\}$ should therefore be evaluated online for each recursion for each snapshot within the system level simulator. This highly increases the computational costs since these factors require the computation of K equalized channel coefficients correlations [6]. In order to reduce the computational costs, we make the following approximation:

$$w_k[n] \approx \sqrt{\frac{p_k}{g_{1k}}} v[n], \quad v[n] = \frac{\overline{h}[n]}{\dfrac{P_c}{SF}|h[n]|^2 + s} \tag{13}$$

where s replaces the second term in the denominator in (10) by its average value taken over the K users:

$$s = \frac{1}{K}\sum_{k=1}^{K}\left(\frac{P_o}{SF}\sum_{q=2}^{Q}\frac{g_{qk}}{g_{1k}} + \frac{P_n}{g_{1k}}\right) \tag{14}$$

Thus, by using (13) instead of (10), only one correlation, i.e. that of the equalized coefficients $\{h[n]v[n]\}$, instead of K correlations is therefore needed in order to evaluate the factors $\{\alpha_{kj}\}$ and $\{\beta_k\}$. This approximation has been validated via simulations where it is observed that using (13) instead of (10) increases the power P_c only by less than 0.25 dBw. Thus, this approximation has negligible impact on the accuracy of the capacity estimates, however, it has the major advantage of significantly reducing the time consumption at the system level simulator.

At last, it is important to point out here that since the factors $\{\alpha_{kj}\}$ and $\{\beta_k\}$ are specific to each snapshot at the system level, the capacity indicator C_ϕ, evoked in sub-section 3.1 for simple equalization schemes, becomes then specific to each snapshot. The exact value of C_ϕ cannot therefore be evaluated at the link level. However, one can still evaluate another C_ϕ at the link level by making use of the simplified link level MMSEC equalization coefficient given by [7]:

$$w[n] = \frac{\overline{h}[n]}{\frac{K}{SF}|h[n]|^2 + \sigma^2_{AWGN}} \quad (15)$$

where σ^2_{AWGN} is the variance characterizing the inter-cell interference plus noise at the link level. Thus, from (15), we can evaluate C_ϕ at the link level for different values of σ^2_{AWGN}.

4. NUMERICAL RESULTS

This section presents an illustration of the evaluation methodology presented in the previous section. It quantifies the impact of chip mapping strategies and equalization techniques on the system capacity in the context of the urban ETSI BRAN E channel model [10]. We consider six physical layer configurations resulting from the combination of either adjacent (AFM) or interleaved (IFM) frequency domain chip mapping with either EGC, MRC, or MMSEC equalization. All six configurations use the same modulation and coding scheme, which consists of QPSK-Gray modulation and UMTS-like convolutional code of rate ½. The key other parameters of the MC-CDMA physical layer are summarized in Table 1 [10].

Table 1. Key parameters of the MC-CDMA physical layer.

Sampling frequency f_s	57.6 MHz
FFT size N_{fft}	1024 samples
Guard interval size N_g	216 samples
Number of data carriers N_c	736 carriers
Frame size N_f	32 OFDM symbols
Spreading factor SF	32 chips

Table 2 summarizes the target SINR values required to achieve 1% target FER for all the six physical layer configurations. These values are obtained from [6] for a cell load $K = 24$. As discussed in [6], the target SINR is invariant with respect to K in the IFM context, whereas in the AFM context, it is more or less invariant for K in the range between 16 and 32. Thus, in the sequel, we confine our analysis to K between 16 and 32.

Table 2. Target SINR (dB) for 1% target FER.

	IFM context	AFM context
MRC	4.25	6.2
EGC	4.5	5.85
MMSEC	4.85	5.3

Fig. 2 illustrates the novel link level capacity indicator C_ϕ as a function of the cell load K for all the six configurations. For MMSEC equalization, we consider two values of σ^2_{AWGN} in (15): −5 dB and −10 dB. From Fig. 2, we can observe that for K between 16 and 32, AFM-EGC outperforms IFM-EGC that in turn outperforms AFM-MRC and IFM-MRC. Moreover, IFM-MMSEC and AFM-MMSEC for both values of σ^2_{AWGN} have very close performance to AFM-EGC. Thus, from this link level study, we can conclude that AFM always outperforms IFM for any given equalization technique. Moreover, AFM-EGC, AFM-MMSEC, and IFM-MMSEC are similar and provide the highest system capacity.

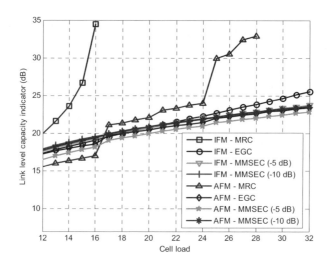

Figure 2. Link level capacity indicator (dB) versus the cell load.

Table 3 summarizes the most relevant system level parameters [10]. Note that we use the standard large-scale propagation model including path loss and log-normal shadowing [4].

Table 3. System level parameters.

Number of tiers	$N = 2$ ($Q = 19$ cells)
Cell radius	300 m
Thermal noise power density	−204 dBw/Hz
Propagation model	$L = -57.45$ dB, $\delta = 2.8$, $\sigma_s = 8$ dB, $\rho = 0.5$
Number of codes at BS	$M = 32$ codes
Outage threshold	$\varepsilon = 5\%$

Fig. 3 depicts the pole capacity and the constrained capacity for $P_{max} = 13$ dBw and $P_o = 3$ and 6 dBw. As we can see from Fig. 3, IFM-EGC, IFM-MMSEC, AFM-EGC, and AFM-MMSEC yield a full pole capacity of 32 codes. Moreover, for moderate inter-cell interference level (e.g. $P_o = 3$ dBw), IFM-MMSEC, AFM-EGC, and AFM-MMSEC yield almost the same and highest constrained capacity. However, for high and dominant inter-cell interference (e.g. $P_o = 3$ dBw), all configurations unless IFM-MRC are similar with a little advantage for AFM-EGC and AFM-MMSEC. These results match well those obtained from the previous analysis of the novel capacity indicator at the link level. Thus, for urban ETSI BRAN E channel model, AFM-EGC is the most suitable configuration since it provides the highest capacity and EGC is less complex than MMSEC equalization.

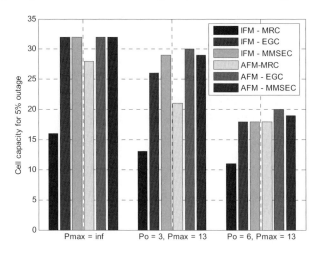

Figure 3. Pole and constrained capacity estimates.

5. CONCLUSIONS

This paper presented an evaluation methodology to quantify the impact of the MC-CDMA physical layer algorithms on the downlink capacity in a multi-cellular environment. The methodology consists of both qualitative and quantitative evaluations via link and system level analysis respectively. A very good match was shown between qualitative evaluation using a novel link level capacity indicator and quantitative evaluation using Monte Carlo statistical system level simulations. An illustration of this methodology showed that in particular adjacent frequency domain chip mapping always outperforms interleaved mapping for any given equalization technique in the context of the urban ETSI BRAN E channel model.

This methodology can further be applied to quantify the impact on the system capacity of different MC-CDMA physical layer configurations in different environments, which is crucial for system design.

6. ACKNOWLEDGMENT

The work presented in this paper was supported by the European IST project 4More (4G MC-CDMA multiple antenna system On chip for Radio Enhancements) [11].

7. REFERENCES

[1] J. Zander, S. L. Kim, "Radio Resource Management for Wireless Networks", *Artech House Publishers*, 2001.
[2] H. Holma, "A Study of UMTS Terrestrial Radio Access Performance", *Ph.D. dissertation, Helsinki University of Technology*, Oct. 2003.
[3] ETSI Technical Report 101 112, "Selection Procedures for the Choice of Radio Transmission Technologies of UMTS," *UMTS 30.03 version 3.2.0*, 1998.
[4] N. Enderlé, X. Lagrange, "Analyse de la capacité descendante d'un système WCDMA," *Actes du congrès DNAC*, Nov. 2001.
[5] E. Seneta, "Non-negative matrices and Markov chains," *Springer, 2nd edition*, 1981.
[6] A. Mourad et al., "Interface between Link and System Level Simulations for Downlink MC-CDMA Cellular Systems," *Proceedings of the 11th European Wireless Conference*, Nicosia, Apr. 2005.
[7] S. Hara, R. Prasad, "Design and Performance of Multi-carrier CDMA System in Frequency-Selective Rayleigh Fading Channels," *IEEE Transactions on Vehicular Technology*, vol. 48, no. 5, Sep. 1999.
[8] H. Atarashi et al., "Broadband Packet Wireless Access Based on VSF-OFCDM and MC/DS-CDMA," *IEEE PIMRC*, vol.3, Sep. 2002.
[9] K. Fazel, S. Kaiser, "Multi-Carrier and Spread Spectrum Systems," *John Wiley & Sons Ltd*, 2003.
[10] IST-MATRICE, *http://ist-matrice.org*.
[11] IST-4MORE, *http://ist-4more.org*.

RADIO RESOURCE MANAGEMENT FOR MC-CDMA OVER CORRELATED RAYLEIGH FADING CHANNELS

Simon Plass and Armin Dammann
German Aerospace Center (DLR)
Institute of Communications and Navigation
Oberpfaffenhofen, 82234 Wessling, Germany
simon.plass@dlr.de

Abstract This paper focuses on employing a radio resource management (RRM) over spreading codes for a coded downlink multi-carrier code division multiple access (MC-CDMA) transmission. We consider a cellular environment with correlated Rayleigh fading channels. For a degree of channel correlation, the investigations identify an improvement of performance offered by the RRM.

1. Introduction

We address a coded multi-carrier code division multiple access (MC-CDMA) downlink. Over the last decade, such downlink systems have been studied intensively for the single-cell scenario. However, it is necessary to extend the investigations to cellular structures. In [1], a radio resource management (RRM) for an orthogonal frequency division multiple access (OFDMA) scheme was investigated. By the allocation of different sub-carriers for neighboring interfering cells, the inter-cell interference could be reduced, and therefore, the performances enhanced. Since in an MC-CDMA system the data symbols of each user are spread over all sub-carriers, one idea of an RRM for MC-CDMA is using different spreading codes to avoid inter-cell interference in the managed cells.

Real channel scenarios show correlations in time and frequency direction. There are two extreme situations:

- The channel coefficients in an OFDM frame are uncorrelated in time and frequency:
  This provides maximal diversity but destroys the orthogonality between spreading codes of different cells [2]. Without any preequalization, that orthogonality cannot be recovered by equalization methods. Managing the resources between neighboring cells by using different spreading codes in different cells do not improve the performance for that reason.

- The channel coefficients in an OFDM frame are fully correlated (flat) in time and frequency:
  This provides minimal diversity, but maintains orthogonality, even between cells and an RRM over different spreading codes can be used.

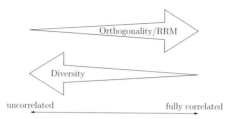

Figure 1. Contrariness of diversity and orthogonality regarding the correlation properties

Both orthogonality and diversity increase the system performance if they are increasing. However, with respect to multipath channel fading correlation orthogonality and diversity are counteracting (see the two extreme situations mentioned above and Figure 1). Therefore, it is interesting to investigate the system performance for channel scenarios, showing different correlation properties. For these investigations, a Rayleigh fading channel with adjustable correlation properties is used.

In this paper, we investigate the possibility of using an RRM for MC-CDMA if the channels have a degree of correlation. Therefore, we introduce the used MC-CDMA system and its cellular environment in Section 2. The following Section 3 describes the correlated Rayleigh fading channel and Section 4 handles the applied radio resource management over spreading codes in the cellular scenario. Finally, simulation results are given and discussed in Section 5.

2. System Model

In this section, we first give an outline of the used MC-CDMA down-link system. We then describe the cellular environment in which the MC-CDMA system in embedded.

2.1 MC-CDMA System

The block diagram of a transmitter using MC-CDMA is shown in Figure 2. The information bit streams of the N_u active users are convolutionally encoded and interleaved by the outer interleaver Π_{out}. With respect to the modulation alphabet, the bits are mapped to complex-valued data symbols. In the sub-carrier allocation block N_d symbols per user are arranged for each transmission scheme. In the case of MC-CDMA, the kth data symbol is multiplied by a user-specific Walsh-Hadamard spreading code which provides so-called chips. The spreading length L corresponds to the maximum number of active user $L = N_{u,\,max}$.

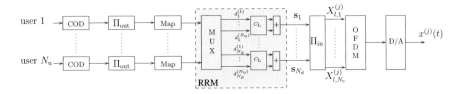

Figure 2 M C-CDMA receiver

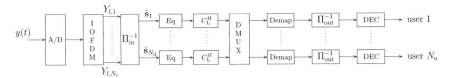

Figure 3 M C-CDMA receiver

An inner sub-carrier interleaver Π_{in} allows a better exploitation of diversity. The input block of the interleaver is denoted as one OFDM symbol and N_s OFDM symbols describe one OFDM frame. By taking into account a whole OFDM frame, a two-dimensional interleaving in frequency and time direction is possible.

Finally, an OFDM modulation is performed which includes an inverse fast Fourier transformation (FFT) and insertion of a guard interval to avoid inter-symbol and inter-carrier interference.

On the receiver side, the lth received OFDM symbol at sub-carrier i becomes

$$Y_{l,i} = X_{l,i}^{(j)} H_{l,i}^{(j)} + N_{l,i}, \qquad (1)$$

where $X_{l,i}^{(j)}$ denotes the value of the ith sub-carrier in the lth OFDM symbol at base station (BS) j, $H_{l,i}^{(j)}$ is the channel transfer function from BS j to the mobile terminal (MT), and $N_{l,i}$ is an additive white Gaussian noise (AWGN) process with zero mean. At the receiver, see Figure 3, the transmitter signal processing is inverted. Here, we describe the receiver for a single-cell scenario. In MC-CDMA the distortion due to the flat fading on each sub-channel is compensated by equalization. The received chips are equalized by using a linear minimum mean square error (MMSE) one-tap equalizer. The resulting MMSE equalizer coefficients are

$$G_{l,i} = \frac{H_{l,i}^{(j)*}}{|H_{l,i}^{(j)}|^2 + \frac{L}{N_u}\sigma^2}, \qquad i = 1,\ldots,N_c, \qquad (2)$$

where σ^2 is the actual variance of the AWGN process. Furthermore, N_c is the total number of sub-carriers. The operator $(\cdot)^*$ denotes the complex conjugate.

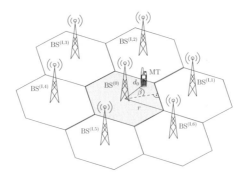

Figure 4. One-tier multi-cell environment

Then, the symbol demapper maps the data symbols to bits. In addition, it calculates the log-likelihood ratio for each bit based on the selected alphabet. The code bits are deinterleaved and finally decoded using soft-decision Viterbi decoding [3].

For multi-carrier schemes a resource load (RL) can be defined. The RL of the MC-CDMA system is defined by the ratio of the number of active users to the number of maximum users or spreading length, respectively:

$$\text{RL} = \frac{N_u}{L}. \qquad (3)$$

2.2 Cellular Environment

A typical hexagonal structure is assumed for the cellular network where all cell sizes are equal as depicted in Figure 4. A whole tier of interfering cells around the desired cell is assumed. The BS and the MT are perfectly synchronized in time and frequency. The distance between the desired BS and MT is denoted as d_0, and the cell radius r is normalized to 1. In this paper, the mobile is situated along a line from the desired BS to the intersection of the desired cell and two interfering cells, and therefore, $\beta = 30°$. A propagation model represents the locally averaged received energy from the jth BS at the MT. The slowly varying signal energy attenuation due to path loss is generally modeled as the product of the γth power of distance d_j and a log-normal component representing shadowing losses [4]. Therefore the resulting received signal energy is

$$E_j = E_{t,j} \cdot d_j^{-\gamma} \cdot 10^{\eta_j/10\text{dB}}, \qquad (4)$$

where $E_{t,j}$ is the transmitted signal energy from the jth BS. The path decay factor γ is assumed to be 4 and the standard deviation of the Gaussian-distributed shadowing factor η_j is set to 8 dB [5]. This scenario represents a power-controlled desired user at distance d_0 as well as power-controlled interfering cells, *i.e.*, all users are located at the same distance d_0 to their base station (BS), and therefore, they have the same E_b/N_0.

In the multi-cell scenario, the total received signal at the desired BS is

$$Y_{l,i} = X_{l,i}^{(0)} H_{l,i}^{(0)} + \sum_{j=1}^{m-1} \sqrt{E_j} X_{l,i}^{(j)} H_{l,i}^{(j)} + N_{l,i}, \qquad (5)$$

where m represents the total number of BSs. Then, the signals are passed to a MMSE equalizer after the deinterleaving process in the receiver. The coefficients in (2) have to be modified in such a way that the interfering signals are assumed to be an additional noise variance term in the denominator [6].

3. Correlated Rayleigh Fading Channel

We assume that the channel on each sub-carrier is a slowly-varying frequency-nonselective Rayleigh channel. Let α_l, $l = 1,\ldots, N_c$, denote the fade amplitude variable on the lth sub-carrier. The Rayleigh random variables α_l have zero mean and unit variance, $E\{|\alpha_l|^2\} = 1$, and uniform random variables over $(0, 2\pi)$. The fading correlation property between sub-carriers i and j can be described by the correlation coefficient

$$\rho_{i,j} = \frac{\text{Cov}\{\alpha_i, \alpha_j\}}{\sqrt{\text{Var}\{\alpha_i\}\text{Var}\{\alpha_j\}}}, \qquad 0 \leq |\rho_{i,j}| < 1, \qquad (6)$$

where the covariance is defined by [7]

$$\text{Cov}\{\alpha_i, \alpha_j\} = E\{(\alpha_i - E\{\alpha_i\})(\alpha_j - E\{\alpha_j\})^*\}. \qquad (7)$$

The sub-carriers are said to be uncorrelated if $\rho_{i,j} = 0$ and are said to be totally correlated if $|\rho_{i,j}| = 1$. For this paper, the correlation factor remains constant over all sub-carriers, and therefore, we set $\rho = \rho_{i,j}$. Thus, the channel and its correlation properties can be described by ρ.

4. Radio Resource Management over Spreading codes

For the applied RRM, perfect knowledge about the available and allocated spreading codes in the m_{RRM} managed cells is assumed. Each managed cell can assign up to $[L/m_{RRM}]$ spreading codes to N_u active users without allocating the same spreading code to different users. These users can transmit within the managed cells without disturbing each other if the orthogonality of the spreading codes at the receiver is preserved. By exceeding this number of spreading codes,

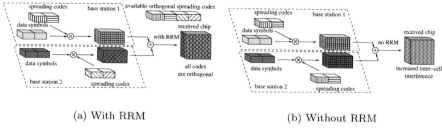

(a) With RRM (b) Without RRM

Figure 5. Radio resource management, over spreading codes with $m_{RRM} = 2$

the succeeding assignment of spreading codes is done randomly. Therefore, a total disturbance of one user is avoided. The procedure of using the available orthogonal spreading codes in different cells for the RRM is shown in Figure 5(a) for the case of two managed cells. In contrast, Figure 5(b) shows a scenario without RRM. Here, the codes of the superimposed received chips disturb each other, and therefore, the inter-cell interference is increased. In this paper, an RRM for $m_{RRM} = 3$ is assumed. Thus, no inter-cell interference can be guaranteed up to a resource load of 1/3.

5. Simulation Results

For the simulations, quadrature phase shift keying (QPSK) is used and perfect channel knowledge is assumed. Furthermore, a convolutional code with rate $R = 1/2$ and memory $M_{CC} = 6$ was selected as an outer channel code. The spreading length is set to $L = 16$ and the FFT size is equal to $N_c = 1024$.

Figure 6 shows results regarding the bit error rate (BER) performance versus E_b/N_0 with RL = 1/16. In the case of fully-correlated Rayleigh fading channels ($\rho = 1.0$), the RRM enables to enhance the performance by approximately 2 dB at the cell edge ($d_0 = 1.0$). Since the inter-cell interference in the inner area ($d_0 = 0.1$) of the desired cell is negligible [1], the performances with and without RRM are

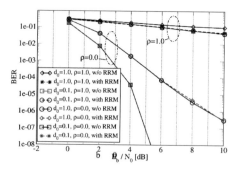

Figure 6. BER versus E_b/N_0 with RL = 1/16 for different MT locations and ρ

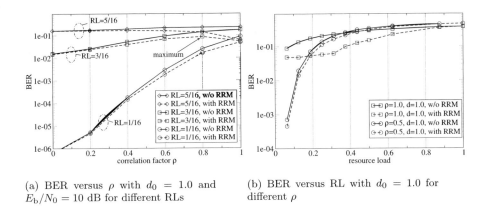

(a) BER versus ρ with $d_0 = 1.0$ and $E_b/N_0 = 10$ dB for different RLs

(b) BER versus RL with $d_0 = 1.0$ for different ρ

Figure 7. BER versus ρ and BER versus RL performances

equal and the performance at the cell edge with RRM matches the non-interfered case of the inner cell. For the uncorrelated channels, there is no difference between the managed system and the non-managed one due to the lost orthogonality. For all scenarios without RRM, the performances degrade for higher correlation factors because of less diversity. This can be also seen in Figure 7(a) where the figure shows the BER versus the correlation factor ρ. Furthermore, all three scenarios with different RLs (RL = [1/16, 3/16, 5/16]) show a gain by using the RRM for higher correlated Rayleigh fading channels. The performances can be enhanced for $\rho > 0.2$. Obviously, a higher RL degrades the performances because the desired spreading code is corrupted by the other active spreading codes including the codes from the interfering cells. In the case of RL = 5/16 inter-cell interference can be still avoided, and therefore, the RRM improves the performance up to the single-user scenario for fully-correlated Rayleigh fading channels. We would like to stress out that there is a maximum in the RRM performance for RL = 3/16 due to the contrariness of diversity and orthogonality.

The simulation results in Figure 7(b) show the BER versus the RL. Two different scenarios are investigated namely with $\rho = 0.5$ and $\rho = 1.0$. It is obvious that the performances with or without RRM are close-by for a less correlated Rayleigh fading channel ($\rho = 0.5$). In contrast, the RRM can ensure the maximum performance up to a 1/3-loaded system in each cell. Furthermore, the randomized assignment of spreading codes for higher RLs prevents a direct adaptation to the performance without RRM.

6. Summary and Conclusions

A coded downlink MC-CDMA system in a cellular environment is investigated regarding an introduced RRM of the used spreading codes within several neighboring cells. Simulations reveal the trade-off between diversity and orthogonality due to correlation characteristics of a channel using an RRM. Thus, we have the possibility of using an RRM for MC-CDMA over spreading codes if the fading channels coefficients have a degree of correlation, which is the case in multipath propagation environments.

References

[1] Simon Plass, Armin Dammann, and Stefan Kaiser. Error performance for MC-CDMA and OFDMA in a downlink multi-cell scenario. In *Proceedings IST Mobile & Wireless Communications Summit (IST Summit 2005), Dresden, Germany*, June 2005.
[2] John G. Proakis. *Digital Communications*. McGraw-Hill, 3rd edition, 1995.
[3] K. Fazel and S. Kaiser. *Multi-Carrier and Spread Spectrum Systems*. John Wiley and Sons, 2003.
[4] G. D. Ott and A. Plitkins. Urban path-loss characteristics at 820MHz. *IEEE Trans. Veh. Technol.*, VT-27:189–197, Nov. 1978.
[5] A. J. Viterbi, A. M. Viterbi, and E. Zehavi. Other-cell interference in cellular power-controlled CDMA. *IEEE Transactions on Communications*, 42(2/3/4):1501–1504, Feb./March/April 1994.
[6] Simon Plass, Stephan Sand, and Gunther Auer. Modeling and analysis of a cellular MC-CDMA downlink system. In *Proceedings IEEE International Symposium on Personal, Indoor and Mobile Radio Communications (PIMRC 2005), Barcelona, Spain*, volume 1, pages 160–164, Sep. 2004.
[7] Alberto Leon-Garcia. *Probability and Random Processes for Electrical Engineering*. Addison Wesley, 2nd edition, 1994.

THROUGHPUT OF HETEROGENEOUS MULTICELL MULTI-USER MIMO-OFDM SYSTEMS

Bartosz Mielczarek and Witold A. Krzymień
TRLabs/University of Alberta,
Edmonton, Alberta,
Canada T6G 2V4
{bmielcza, wak}@trlabs.ca

Abstract In this paper, we propose an architecture of a flexible multi-cell system that can support users with different numbers of receive antennas (heterogeneous system) on frequency-selective fading channels. The proposed solution uses orthogonal frequency division multiplexing (OFDM) to combat channel dispersion, and exploits multi-user diversity together with a form of multiple-input multiple-output (MIMO) approach to achieve capacity gains. We focus on the downlink transmission and assume that the feedback bandwidth from the mobile users is limited, and hence the base station has incomplete channel state information.

The proposed system divides the available transmission bandwidth into sub-bands and uses layered MIMO approach to transmit information to a selected group of users, which in turn employ their proprietary interference cancellation algorithms. The users are assigned to disjoint sub-bands according to their average supported rates, which facilitates the use of a form of proportionally fair scheduling.

We develop upper and lower bounds on the total achievable throughput using layered transmission, compare them with single-user scheduling capacity and discuss scheduling fairness of the system operating on an ITU-recommended vehicular channel.

1. Introduction

The development of the modern Internet-based data communication systems and ever increasing demand for bandwidth have spurred an unprecedented progress in fixed network capacity, enabling cabled users to access the data network at bit rates well in excess of 100 Mbits/sec. To catch up with that progress, current trends in wireless system design focus on the use of multiple antennas to provide capacity gains on fading channels [1], orthogonal frequency division multiplexing (OFDM) to facilitate the utilization of these capacity gains on frequency-selective channels [2], spread spectrum techniques to enhance robustness to multi-user interference in multi-cell systems [3] and strong channel codes (such as turbo codes [4]) to enable acceptable performance at high interference and noise levels.

In this paper, we propose a flexible multiple-input multiple-output (MIMO) OFDM system implementation that aims at providing very high spectral efficiency in a multi-user frequency selective mobile channel environment. Based on our assumptions, we calculate performance bounds for aggregate system throughput in a multi-cell scenario, taking into consideration the inter-cell interference. This paper is a direct extension of our previous work presented in [5, 6].

2. System Model

The system discussed in this paper consists of hexagonal cells with central base stations equipped with N antennas. All cells in the system divide their frequency channels into S sectors and within each sector there are K users with $n^{(k)}, k = 0, 1, ...K - 1$ antennas each. Users are assumed to be scattered uniformly over the cells at distances $r^{(k)}$ from the base station. The path losses $L^{(k)}$ for each user are specified as

$$L^{(k)}(r) = A\log_{10}(r^{(k)}) + B \quad (1)$$

where parameters A and B are given in [7]. In addition to path loss, signals for every user are subject to log-normal shadowing.

The signal from the base station is transmitted using OFDM format and any form of coded modulation. The number of sub-carriers N_{OFDM}, the guard interval and the OFDM symbol duration are adjusted to the known frequency selective channel model with the maximum Doppler shift so that inter-carrier interference (ICI) is eliminated.

For every user $k = 0, 1, ...K - 1$, we define mean received power (accounting for path loss and shadowing) as $P^{(k)}$. Moreover, for every sub-carrier $j = 0, 1, ..., N_{\text{OFDM}} - 1$, we define a matrix of complex small-scale channel gains $\mathbf{H}_j^{(k)}$ with dimensions $n^{(k)} \times N$ whose elements are independent circular complex Gaussian variables with unity variance. The total transmission power is equally divided between antennas so that mean power from each transmit antenna at the kth user's receiver is equal to $P^{(k)}/N$. Finally, the noise and out-of-cell interference are assumed to be Gaussian with variances equal to N_0 and I, respectively.

3. Sub-carrier Grouping

In principle, maximum frequency diversity is achieved by independent use of each of N_{OFDM} sub-carriers. However, large number of independent frequency transmission layers (sub-carriers) means that, for a fixed packet duration and total signal bandwidth, the codewords become shorter, which leads to higher error floors and higher susceptibility to interference. Moreover, the complexity of such a solution may be quite high.

Consequently, we propose grouping adjacent sub-carriers to form M sub-bands containing N_{OFDM}/M sub-carriers each. In this way, the length of the codeword in each sub-band can be increased. Moreover, we assume that fading within each sub-band is approximately flat and approximated by mean channel matrix $\overline{\mathbf{H}}_m^{(k)}$ as in [5].

4. Layered Transmission with Limited Feedback

In multiple antenna systems, optimal transmission (in sum-capacity sense) to many users at a time can be achieved by a special technique called *dirty paper coding* [8]. With such systems, depending on the channel state and the power region, the transmission should be simultaneously scheduled to multiple users, whose number may be sometimes even larger than the number of transmit antennas [9].

Unfortunately, dirty paper coding requires non-causal full channel state information at the transmitter side and in multi-user MIMO-OFDM systems, a very high feedback bit rate is needed to accommodate information about the full set of matrices $\mathbf{H}_j^{(k)}$ [10]. Such a high burden on the feedback link may not even be justified since, in reality, the throughput maximization is rarely the only objective from the system design point of view (user delays, fairness and hardware complexity being other crucial factors).

In this work, we assume that only receivers have perfect channel state information, including the noise and interference variance. To lower the feedback burden, layered transmission is used, where in each sub-band base station transmits N independent streams of data with maximum rate $R^{\max}$ per antenna. With such a setup, each receiver will only have to transmit information about its preferred base station antenna $l_m^{(k)} \in \{0, 1...N-1\}$ and the corresponding maximum supported link throughput $R_m^{(k)} \leq R^{\max}$ (compare to solution in [11]). Note that such a solution allows the base station to treat all receivers as 'black boxes' regardless of their number of receive antennas. In this way, the system can accept any receiver configuration and hence allow for true heterogeneity.

The base station will obtain receiver feedback regularly in order to track small-scale fading. In addition, we assume that the base station will have perfect knowledge of the mean received powers $P^{(k)}$ for each user, which should be relatively easy to obtain since the path loss and shadowing changes are much slower than small-scale fading and can span thousands of frames [7].

5. Sectoring

In order to limit the inter-cell interference, we assume that each cell is divided into S sectors with disjoint sets of frequency channels. Based on the above discussion, each sector will have at least $\lfloor M/S \rfloor$ sub-bands with resulting $N\lfloor M/S \rfloor$ channels available to the users. If the number of sectors is not a divisor of M, there will be additional sub-bands that can be assigned dynamically to the the cells experiencing increased traffic.

The mapping between sub-bands and sectors can be done in a variety of ways, with equal spacing being probably the best choice from the practical point of view. In addition to interference mitigation, mapping between sub-bands and sectors can be dynamically changed across the cellular system. This feature can be used to combat low mobility of the users which may cause some of them to experience bad quality of all available channels for an extended period of time. If the mapping of the sub-bands changes periodically, the effect is similar to opportunistic beamforming [12] and allows even the stationary users to improve their transmission quality.

6. Upper Bound on Capacity

In order to compute the upper bound on the system capacity, we assume that each base station antenna will transmit with rate equal to at most δ. To compute bounds, we will vary the parameter $\delta \in (0, R^{\max})$, although in a practical system, it would be fixed periodically by a base station, depending on the user distribution and signal powers. Moreover, we assume that each frame is long enough to employ capacity-achieving codes, so that the absolute minimum signal-to-noise and inter-cell interference threshold needed to perfectly decode the signal in a given layer is equal to Shannon limit $\theta(\delta) = 2^\delta - 1$. Finally, to make the presentation more clear, in the following sections we will ignore the sub-band index m.

We first define interference cancellation function $f^{(k)}(n)$ as

$$f^{(k)}(n) = \begin{cases} 0 & \frac{P^{(k)}}{N(N_0+I)} \| \left[\overline{\mathbf{H}}^{(k)} \right]_{\cdot n} \|^2 > \theta(\delta) \\ 1 & \text{otherwise} \end{cases} \quad (2)$$

which states that a signal from a layer n can be canceled if its signal-to-noise and inter-cell interference ratio, is larger than $\theta(\delta)$. Note that the condition in (2) is necessary but not sufficient for error-free decoding of layered signal n since it disregards interference from other layers' signal.

Using (2) to determine which layer signals can be canceled, the optimization algorithm at a receiver searches for antenna index $l^{(k)}$ that will maximize the upper limit to the instantaneous capacity of the user k receiving transmission from antenna l given as

$$C_{\text{up}}^{(k)}(l;\delta) = \min\left(\delta, \log_2\left[1 + \frac{\frac{P^{(k)}}{N}\|[\overline{\mathbf{H}}^{(k)}]_{\cdot l}\|^2}{N_0 + I + \frac{P^{(k)}}{N}\sum_{i\neq l} f^{(k)}(i)\|[\overline{\mathbf{H}}^{(k)}]_{\cdot i}\|^2}\right]\right), \quad (3)$$

which results in user's optimum set of parameters calculated as

$$C_{\text{up}}^{(k)}(\delta) = \max_l C_{\text{up}}^{(k)}(l;\delta), \quad l^{(k)} = \arg\max_l C_{\text{up}}^{(k)}(l;\delta). \quad (4)$$

Based on (4), the upper limit to the capacity of the simultaneous transmission of N layered signals is finally obtained as

$$C_{\text{up}} = \max_{\delta < R^{\max}}\left(\max_{\mathbf{k}\in\mathbb{K}^N}\sum_{n=0}^{N-1} C_{\text{up}}^{([\mathbf{k}]_n)}(\delta)\right) \quad (5)$$

where $\mathbb{K}^N$ is a vector containing a unique combination of N different values out of $\{0, 1, ...K-1\}$ with the constraint that no two users can be assigned to the same base station antenna $l^{(k)}$. A method of solving inner maximization in (5) is discussed in [5].

The value calculated in (4) is an upper bound on actual system capacity since it is based on weak condition in (2). Such a condition may lead to a situation when a user receives a signal with throughput equal to $NR^{\max}$ which is larger that the theoretical capacity of the link given by

$$C^{(k)} = \log_2\left[\det\left[\mathbf{I} + \frac{P^{(k)}}{N(N_0+I)}\overline{\mathbf{H}}^{(k)}\left(\overline{\mathbf{H}}^{(k)}\right)^\dagger\right]\right] \quad (6)$$

where † symbolizes conjugate transpose of the matrix [13]. Such a situation will occur when there are many users experiencing high quality channels at a given sub-band and the base station schedules the transmission with very high rates on all its antennas.

7. Lower Bound on Capacity

In this section, we develop a lower bound on the system capacity, taking into consideration the MIMO capacity limit in (6).

Similarly to the upper bound calculation, the receiver first searches for base station antenna index $l^{(k)}$ that will provide it with the best signal quality. In this case, however, the supported layer rate is reported to the base station in such a way, that the total rate of the signals received by any user may never exceed the limit in (6). We start by defining the *excess capacity* function $\Delta^{(k)}(l;\delta)$ as

$$\Delta^{(k)}(l;\delta) = \max\left(0, C^{(k)} - \delta\sum_{i\neq l}\left(1 - f^{(k)}(i)\right)\right), \qquad (7)$$

which represents the minimum excess capacity available to the given antenna, assuming that the signal from other antennas can be detected as specified by (2). If, for example, with a two-antenna base station, the total capacity of a user is $C^{(k)} = 1.5$ bps/Hz, the maximum rate per antenna is $\delta = 1$ bps/Hz, and the signal of one of the base station antennas can be canceled, the excess capacity for the other antenna is only 1.5-1=0.5 bps/Hz, since the user must assume that the signal on the first antenna will be equal to 1 bps/Hz (even though it can be lower). Based on (7) and (3), the lower limit to the instantaneous capacity of of the user k receiving transmission from antenna l is given as

$$C_{\text{lo}}^{(k)}(l;\delta) = \min\left(C_{\text{up}}^{(k)}(l;\delta), \Delta^{(k)}(l;\delta)\right), \qquad (8)$$

which gives

$$C_{\text{lo}}^{(k)}(\delta) = \max_{l} C_{\text{lo}}^{(k)}(l;\delta), \quad l^{(k)} = \arg\max_{l} C_{\text{lo}}^{(k)}(l;\delta) \qquad (9)$$

and finally

$$C_{\text{lo}} = \max_{\delta < R^{\max}}\left(\max_{\mathbf{k}\in\mathbb{K}^N}\sum_{n=0}^{N-1} C_{\text{lo}}^{([\mathbf{k}]_n)}(\delta)\right). \qquad (10)$$

Using the above calculations, no active user will receive the N layered signals from the base station which would together exceed the maximum theoretical capacity given by (6).

8. Fairness Control

Since users in any sector have differing long and short term channel statistics, aggregate throughput maximization will favor those users who are closer to the base station and/or experience improvement from the shadowing process. With such a performance criterion, users receiving weak signals (and hence having low average rates) may experience very long delays before assigned transmission slots.

To alleviate this problem, we propose that base station divides users in sub-bands, according to their mean supported rates and performs scheduling separately in each sub-band. In this way, only the users which have comparable rates are competing for resources in any subband and the problem with rate mismatch becomes much less severe. In this paper, we simply divide K users into G equal groups, which are then separately assigned $\lfloor M/S \rfloor /G$ sub-bands. In each sub-band, the base station can then adjust the transmitted power and/or the requested rate which, allows the operators to shape the traffic according to their customers needs. Note that such an approach can be interpreted as a form of Proportionally Fair (PF) scheduling [14] which equalizes average supported rates of the users grouped together. If their mean transmission rates are similar, the small scale variations of their instantaneous rates are then equivalent to the PF scheduling metrics.

9. Results

9.1 Simulation Parameters

We have simulated the system using the following set of OFDM parameters: carrier frequency $f_c = 2$ GHz; signal bandwidth $B = 5$ MHz; number of sub-carriers $N_{\text{OFDM}} = 256$, which are divided into $M = 32$ sub-bands of 8 sub-carriers each; OFDM symbol duration $T_{\text{OFDM}} = 53.8$ μs; OFDM frame length 41 symbols, which results in 2.2 ms frame duration. The channel model has been implemented using ITU-R M.1225 vehicular A channel model with the maximum velocity of the mobile receiver 120 km/h [7].

We have implemented a two-tier cellular system consisting of 19 hexagonal cells with a cell radius of D km. Each cell has been divided into $S = 6$ sectors, four of these sectors have 5 separate sub-bands and two of them have 6 sub-bands, resulting in the full use of the OFDM spectrum. The mapping of the sub-bands and sectors is periodically changed across the system.

System simulations are performed for the central cell of the system, using 100 random user drops between 0.03 km and D km distance from base station with each user location being traced for duration of 100 OFDM frames. The path loss and shadowing remain constant for each drop but small-scale fading changes independently from frame to frame. We vary the cell radius and normalize the transmitted power per sub-carrier so that mean $\text{SNR}(D = 0.5\text{km}) = 20$ dB.

9.2 Capacity Bounds and Outage Probability

Figures 1a) and 2a) present the capacity bounds calculated in (5) and (10) for $K = 20, 40$, $N = 4$, $G = 1$ and averaged over different channel realizations. All users are equipped with simple one-antenna

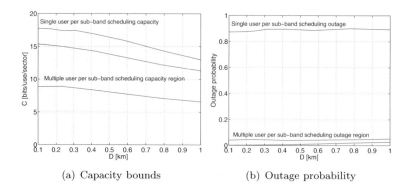

Figure 1. Ergodic capacity bounds and outage probability for $N = 4$, $K = 20$, $G = 1$ and one-antenna receivers.

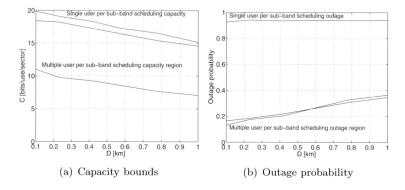

Figure 2. Ergodic capacity bounds and outage probability for $N = 4$, $K = 40$, $G = 1$ and one-antenna receivers.

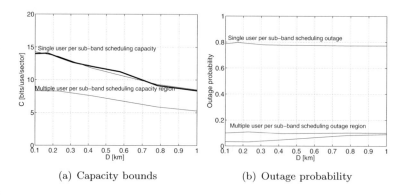

Figure 3. Ergodic capacity bounds and outage probability for $N = 4$, $K = 40$, $G = 2$ and one-antenna receivers.

receivers (the throughput gains would be higher for multiple-antenna receivers). For comparison, also ergodic capacity of the system where only one user at a time is scheduled in each sub-band is presented. One can see that the capacity provided by single-user scheduling is larger than the upper limit of capacity of the layered systems, although the difference between them decreases with increasing number of the users in the sector. However, as one can see in Figs. 1b) and 2b) presenting the averaged outage probability (defined as the fraction of users that the base station never transmits to during one epoch), the one-user scheduling effectively cuts off at least 90% of users. On the other hand, the layered multi-user scheduling, even though it does not explicitly take fairness issues into consideration, achieves much lower outage probabilities.

Although the outage probability of the proposed multi-layer system is much lower than one-user scheduling, it may not be enough in a realistic scenario. To decrease the outage probability, it may be necessary to trade-off some throughput for fairness as shown in Fig. 3. In this scenario, $K = 40$ users are divided into $G = 2$ groups: a 'weak' and a 'strong' one, and scheduling algorithm operates independently on both of them. Since only comparable users compete for resources, the outage probability is dramatically reduced, albeit at the cost of lowered throughput.

Due to the lack of space, we do not show the analysis of the influence of inter-cell interference on the system performance, however, it can be shown that with $S = 6$, the out-of-cell interference is negligible and no spreading techniques are necessary to combat it.

10. Conclusions

We have proposed a robust multi-cellular system able to support a variety of users characterized by different mobilities and hardware implementations. The proposed system is able to deal with frequency-selectivity and inter-cell interference without the use of spread-spectrum techniques. Moreover, it can be easily adjusted to balance throughput and fairness with changing number of users.

The future work will be concentrated on finding the tighter capacity bounds of the system, analytical functions for fairness and throughput trade-off and more advanced, adaptive scheduling algorithms for power and rate adjustment in sub-bands.

Acknowledgment

Funding for this work has been provided by TRLabs, the Alberta Ingenuity Fund, the Rohit Sharma Professorship Fund, the Natural Sciences and Engineering Research Council (NSERC) of Canada and the Italian Ministry of Education, Universities and Research (MIUR) through the FIRB-PRIMO project.

References

[1] A. Paulraj, D. Gore, and R. Nabar. *Introduction to Space-Time Wireless Communications*. Cambridge University Press, 2003.

[2] L. Hanzo, C. H. Wong, and M. S. Yee. *Adaptive Wireless Transceivers: Turbo-Coded, Turbo-Equalized and Space-Time Coded TDMA, CDMA, and OFDM Systems*. Canada: Wiley, 2002.

[3] A. J. Viterbi. *CDMA: Principles of Spread Spectrum Communication*. Prentice Hall, 1995.

[4] C. Berrou, A. Glavieux, and P. Thitimajshima. Near Shannon limit error control-correcting coding and decoding: Turbo-codes. In *Proc. ICC*, pages 1064–1070, Geneva, Switzerland, 1993.

[5] B. Mielczarek and W.A. Krzymień. Throughput of heterogeneous multi-user MIMO-OFDM systems. In *Proc. 16th International Conference on Wireless Communications*, Calgary, Canada, July 2004.

[6] B. Mielczarek and W.A. Krzymień. Throughput of realistic multi-user mimo-ofdm systems. In *Proc. IEEE Int. Symp. Spread Spectrum Techniques and Applications*, pages 434–438, Sydney, Australia, September 2004.

[7] Guidelines for evaluation of radio transmission technologies for IMT-2000. Recommendation ITU-R M.1225, 1997.

[8] M. Costa. Writing on dirty paper. *IEEE Trans. Inform. Theory*, 24:374–377, May 1978.

[9] D.J. Mazzarese and W.A. Krzymień. Throughput maximization and optimal number of active users on the two transmit antenna downlink of a cellular system. In *2003 IEEE Pacific Rim Conference on Communications, Computers and signal Processing, PACRIM.*, volume 1, pages 498–501, Victoria, Canada, August 2003.

[10] B. Mielczarek and W.A. Krzymień. Flexible channel feedback quantization in multiple antenna systems. In *Proc. IEEE Vehicular Technology Conference*, Stockholm, Sweden, May 2005.

[11] D. Aktas and H. El Gamal. Multiuser scheduling for MIMO wireless systems. In *Proc. IEEE Vehicular Technology Conference*, Orlando, FL, October 2003.

[12] N. Sharma and L.H. Ozarow. A study of opportunism for multiple-antenna systems. *IEEE Trans. Inform. Theory*, 51(5):1804–1814, May 2005.

[13] I. E. Telatar. Capacity of multi-antenna Gaussian channels. *Europ. Trans. Telecommun. (ETT)*, 10:585–595, November 1999.

[14] A. Jalali, R. Padovani, and R. Pankaj. Data throughput of CDMA-HDR a high efficiency-high data rate personal communication wireless system. In *Proc. IEEE Vehicular Technology Conference*, volume 7, pages 1854 – 1858, Tokyo, Japan, May 2000.

REAL-LINK PERFORMANCE OF A SS-MC-MA HIGH FREQUENCY RADIO MODEM

Héctor Santana-Sosa[1], Ivana Raos[2], Santiago Zazo-Bello[2], Iván A. Pérez-Álvarez[1] and Javier López-Pérez[1]

[1]*Dpto. Señales y Comunicaciones*
Universidad de Las Palmas de Gran Canaria
Las Palmas de Gran Canaria, Spain
{hector, ivan, javivi}@gic.dsc.ulpgc.es

[2]*Centro de Domótica Integral*
Universidad Politécnica de Madrid
Madrid, Spain
{ivana, santiago}@gaps.ssr.upm.es

Abstract   The HF band data communication link has been traditionally desired by many of the large range transmission systems although it is associated to unfavorable performances as low transmission rate, large delay and low confidence in terms of link establishment and maintenance. Although transmission rates may be high enough to transmit digital voice, delay, usually over several second, has been the main handicap to let the systems provide interactive digital voice links. Indeed, there is no unclassified equipment with this capability. The main achievement of this proposal is that we are able to guarantee digital voice transmission with low latency, around 135 msec (modem+codec), providing a full interactive digital voice link. Performances of two new 2460 bps. HF modems are presented versus the 39-tone 2400 bps MIL-STD-188-110A modem, working over an ITU-R moderate channel. Futhermore, these results are corroborated by real tests carried out in a 1800 Km. link.

1.   Introduction
1.1   The HF Channel

In the HF band, long distance communications are feasible thanks to the use of the ionosphere as passive reflector. However, the atmospheric nature of this reflector makes the systems to face a very hard commu-

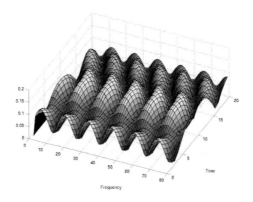

Figure 1. Time-Frequency representation of the HF channel

nication environment. Multipath effects are always present and have to be considered in depth, beside the very fast time-varying characteristics of the channel. The main parameters used to measure channel behavior are *frequency coherence* (Δf_c) and *time coherence* (Δt_c) [12]. The first one (Δf_c) gives information about how narrow must one modulated carrier be in order to consider the channel flat. Typical values in HF channel are close to 1 KHz or less. The second parameter (Δt_c) sets the time separation between two pulses with different attenuations. Its inverse is known as *doppler spread* (f_d) and in HF it is usually about 0.1 to 2 Hz. This multipath environment makes efficiencies over 0.5 bit/Hz very difficult to achieve. It is common to find channel characteristics as shown in Figure 1, where several deep nulls spread over the narrow transmission band and move arbitrarily over it. Usually, the bandwidth of HF transmissions are less than 3 KHz and the best performances in data communications are obtained usually by the combination of powerful codes and very long interleavers, which introduce an important delay in the communication and make interactive digital voice link impossible.

1.2 State of the Art

Single carrier schemes have been typically used for facing the hard conditions of the short-wave radio communications. Indeed, most of the systems designed for data transmission in the HF band use single carrier techniques beside powerful coding [6]. Optimum coding performance needs the use of long interleaving matrices in order to cope with burst errors introduced by the channel. Interactive digital voice communications are not feasible with this kind of systems due to this unacceptable delay. Most of present data modem applications are based on the standard MIL-STD-188-110A [9] as a military data transmission system, or in the standard STANAG 4285 [15]. The modem described by the standard MIL-STD-188-110A [9] specifies data rates range between 75 and 2400 bps. Transmission process uses an 8-states convolutional encoder with 8PSK modulation scheme with variable interleavers. This standard also describes in two appendixes, two multicarrier schemes: one of them is non-orthogonal 16 tones DPSK where channel estimation is not required, but also spectral efficiency is reduced. The other operating mode is an orthogonal (OFDM) 39-tones including a (14,10) Reed-Solomon encoder with a frequency diversity degree selected depending on the channel state. This modem has to be considered as one of the first HF-band military systems published in the open literature and thus, it is one reference for this kind of systems although it does not provide support for an interactive digital voice link. Indeed, nowadays, there are no unclassified systems to support this kind of links.

1.3 Multicarrier Modulations

One of the main problems to consider when a HF modem is being designed is related to the long impulse response length of the channel. The first strategy to deal with this problem is to reduce transmission rate reducing multipath distortion to a small symbol fraction. Required data transmission rate might be obtained by means of Multi-Carrier techniques, and in particular with *Orthogonal Frequency Division Multiplex* (OFDM), thus giving maximum spectral efficiency. The performances of OFDM schemes in HF channels have been deeply analyzed by C.Cook [4], also published by E.E. Johnson [6], considering a large set of parameters as interleaving length, robustness against impulsive noise, behavior in front of co-channel interference, relationship between average and peak power, equipment specifications, synchronization issues, and spatial diversity techniques applicability. In these studies, OFDM based modems have demonstrated to be more efficient than single carrier systems when long interleaving and powerful coding are avoided.

2. Data Modem Design

Two schequematic block diagrams of the developed modem are presented in Figure 2 and Figure 3. They represent the transmitter and the receiver respectively. The design of the new modem has been focused on three main topics: the use of continuos channel sounding distributed in the frequency-time grid, the use of MC-CDMA techniques to avoid deep-nulls effects, and completely avoid coding and interleaving to provide interactive digital voice links.

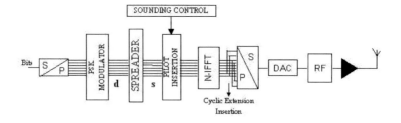

Figure 2. Block diagram of the transmiter

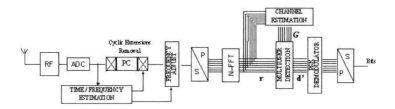

Figure 3. Block diagram of the receiver

2.1 SS-MC-MA Techniques

SS-MC-MA techniques provide a robust mechanism to avoid the effects of the deep-nulls introduced by the channel (Figure 1) without the need of powerful coding and long interleaving. Work showed in [14] and [11] firstly introduced these techniques in this kind of systems and demonstrated, in a simple simulated environment, that SS-MC-MA is a valid technique to provide interactive digital voice communications over ionospheric links. Using SS-MC-MA, the serial symbols are transmitted over all frequencies so the information affected by deep-nulls can be restored with the unaffected carriers using frequency diversity mechanism.

Taking [14] and [11] as a starting point, this paper presents a step forward in the field of SS-MC-MA techniques applied to short-wave radio modems. As it was seen in the block diagrams, coding has been completely removed from the previous versions as its benefits with no interleaving have been overcomed by MC-CDMA techniques. These techniques are common in researches related to new mobile communications systems (i.e. [7]) but have not been exploited in this kind of systems. The lack of coding and interleaving is not a problem as the new advanced reception patterns have increased performance thanks to the correct use of frequency diversity. Different detection strategies are used such as *Global Minimum Mean Square Error* (GMMSE) and *Interference Cancellation* techniques [2], all of them adapted in order to fit into new transmission schemes. Two new modems have been developed. One of them uses

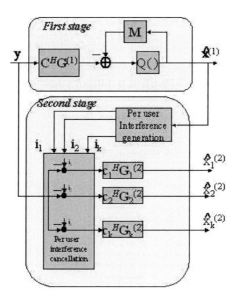

Figure 4. Interference Cancellation Process

GMMSE with a two-stages *Interference Cancellation* scheme. The second one performs a *Parallel Interference Cancellation* (PIC) following the hints given in [14] and [11] with *Equal Gain Combining* (EGC). The combination of GMMSE with standard PIC structures is straightforward as can be observed in the Figure 4; where $Q()$ is the standard symbol detector by minimum distance criteria and $\mathbf{M}$ represents the interference regeneration process:

$$\mathbf{M} = (\mathbf{C}^H \mathbf{GHC} - diag(\mathbf{C}^H \mathbf{GHC})) \quad (1)$$

where $diag()$ operator means a diagonal matrix whose inputs are the corresponding elements of the processed matrix, $\mathbf{G}$ is a matrix calculated following GMMSE [2] criteria, $\mathbf{C}$ represents the spreading matrix where $\mathbf{c}_k$ is its k-column and $\mathbf{i}_k$ is the interference of the whole users over the k-user.

3. Results Analysis

Simulated results are shown in the Figure 5, where MIL-STD 39 tones modem ([6], [4] and [9]) efficiency are presented beside the two new modems performances. These results were obtained with a Moderate Channel Watterson HF Channel (2 rays, time spread 1 msec, Doppler spread 0.5 Hz). The two new modems operate at 2460 bps and the 39-tone MIL-STD-188-110A operates at 2400 bps. It can be seen that

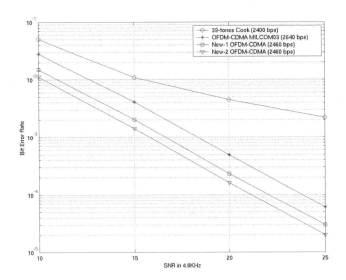

Figure 5. Performance of the two new modems vs. 39-tone MIL-STD-188-110A and [14]

the two new modems overcomes the MIL-STD performace. It is very important to recall that these results are obtained without any kind of interleaving matrix, so the data delay is kept extremely low. This delay does not goes over 135 msec Considering that the 2400 bps MELP vocoder [8] will operate satisfactorily with error probabilities below 10^{-2}, we can guarantee modem operability for SNR in 4.8 KHz above 10 dB.

It is also presented in the Figure 6 the performance of the two new modems in a five hours real link. This link was established between Canary Islands and Madrid in the 18 MHz band. It can be seen the parallelism between the simulated results and the real link performances. The main difference between the simulated and real link results is a SNR offset. This difference have its explanation in the fact that no real and *variable* channel will behave as a simulated CCIR channel. Deeper studies are considered in order to obtain actual channel conditions. Nevertheless, these results suppose a step forward in the way of demonstrating the system robustness.

4. Conclusions and Future Research Lines

Present work has shown the real feasibility of interactive digital voice transmission over the HF channel. This accomplishes a significant achievement in comparison with conventional analog transmission with a sig-

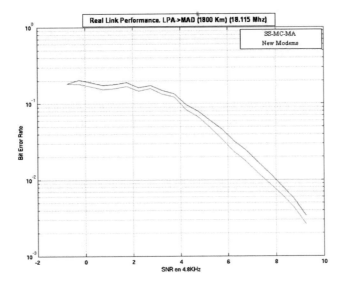

Figure 6. Performance of the two new modems in a 5 hours real link

nificant subjective improvement in terms of quality and intelligibility. The key issues are the following: the use of OFDM modulation and the use of MC-CDMA techniques inspired in mobile applications, increasing system robustness in front of deep spectral nulls and homogenizing performance over different subcarriers. Furthermore, once the transmission scheme has demonstrated to be efficient, complementary analysis, relaxing the delay constraint for data transmission [13], is a promising path to achieve new goals. This operation mode will increase data rate, by the increase of the constellation size and the use of combined interleaving and channel coding strategies. Updated information of this work can be found in the groups' ftp site.

Acknowledgments

The work presented in this contribution was supported by the Spanish National Project TEC2004-06915-C03-01/02 and AENA (Spanish Airports and Air Navigation) projects 00/49 and 00/50. It is also partially supported by the Spanish National Project TIC2003-09061-C03-01

References

[1] Khaled Fazel and Stefan Kaiser, *Multi-Carrier and Spread Spectrum Systems.* John Wiley and Sons, 2003.

Section III

SPREADING AND DETECTION

ANALYSIS OF ITERATIVE SUCCESSIVE INTERFERENCE CANCELLATION IN SC-CDMA SYSTEMS

Petra Weitkemper, Volker Kühn and Karl-Dirk Kammeyer
University of Bremen
Dept. of Communications Engineering
Otto-Hahn-Allee NW 1
28359 Bremen, Germany
{weitkemper, kuehn, kammeyer}@ant.uni-bremen.de

Abstract This paper analyzes convergence behavior and performance of iterative successive interference cancellation (SIC) for a CDMA system with random spreading using the so-called multi-user efficiency (MUE). The goal of such an analysis is the optimization of the detection scheme. Moreover, an optimized power allocation of the users at the transmitter is an important means for enhancing the convergence behavior of the detector and is based on the possibility of prediction. While this analysis has only been applied to parallel interference cancellation (PIC) we will apply it in this paper also to SIC. It will be shown that the achievable system load can be significantly increased.

1. Introduction

The turbo principle discovered in 1993 has been applied to nearly any concatenated system like channel estimation and equalization, coding and modulation. It was also applied to parallel and successive interference cancellation in a coded CDMA system. Both schemes exploit the soft information at the channel decoder output for improving interference cancellation in an iterative manner. The channel decoder and the multi-user detection can be regarded as serially concatenated systems. Analysis of the PIC scheme has been done by different approaches [1-3]. In this case the analysis reduces to a two-dimensional one as will be described in Section 3. For successive interference cancellation this simplification is not possible because the statistics of the users differ from each other. In this paper the analysis derived for PIC is generalized to SIC by taking the dependencies of the users into account. The goal is to predict the behavior not only of the whole system but of any particular user during iterations.

The paper is organized as follows: Section 2 introduces the system model of the considered CDMA system. In Section 3 the analysis based on multi-user efficiency (MUE) is described and applied to the parallel interference cancellation. The difference between SIC and PIC with respect to MUE is investigated and the MUE analysis of SIC is given in Section 4. Power optimization is done in Section 5 for both detection schemes. A conclusion is given in Section 6.

2. System Model

In order to simplify derivations and notation we assume a synchronous singel carrier- (SC-) CDMA system with a complex AWGN channel and pseudo-noise spreading sequences [4], but the analysis can be applied to MC-CDMA as well. The number of active users is denoted by U. The information bit vector of the u-th user is denoted by $\mathbf{d}_u$, which is encoded with a convolutional code of rate $R_c = 1/n$ identical for all users. The coded bit sequence is BPSK-modulated and interleaved by a user-specific interleaver Π_u of length L and then spread with random spreading codes $s_u(k) \in \{-1/\sqrt{N}, +1/\sqrt{N}\}$. k denotes the chip and l the symbol index. The length N of the sequences $s_u(k)$ is called spreading factor and the system load $\beta = U/N$ is an important parameter of the system. Assuming $\mathbf{b}(l)$ to be the vector comprising BPSK symbols of all users at time instance l and $\mathbf{C}(l)$ a $N \times U$ matrix containing the vectors of spreading sequences as columns each multiplied with an individual phase term of the channel, the received vector containing the superposition of the spread signals of all users and the noise can be described in vector-matrix notation

$$\mathbf{y}(k) = \mathbf{C}(l)\mathbf{b}(l) + \mathbf{n}(k) . \qquad (1)$$

The vector $\mathbf{n}(k)$ represents the complex additive white Gaussian noise with covariance matrix $\sigma_n^2 \mathbf{I}$. At the receiver a bank of matched filters (MF) is applied for despreading and the real-valued matched filter output can be written as

$$\mathbf{r} = \mathrm{Re}\left\{\mathbf{C}^H \mathbf{y}\right\} = \underbrace{\mathrm{Re}\left\{\mathbf{C}^H \mathbf{C}\right\}}_{\mathbf{R}} \mathbf{b} + \underbrace{\mathrm{Re}\left\{\mathbf{C}^H \mathbf{n}\right\}}_{\tilde{\mathbf{n}}} . \qquad (2)$$

For notational simplicity the time index k has been dropped. The off-diagonal elements of $\mathbf{R}$ contain the real part of the correlation coefficients $\varrho_{ij} = \mathrm{Re}\{\rho_{ij}\}$ between the i-th and the j-th user's signature sequence with $\mathrm{E}\{|\rho_{ij}|\} = 1/N$. Detection by individual decoding and hard decision will not be appropriate for systems with moderate to high loaded systems. The multi-user interference degrades the performance significantly. Considerable improvement can be obtained by the application of interference cancellation. The idea of interference cancellation schemes is to get some knowledge on the interference to substract it from the received signal before detection.

3. Multi-User Efficiency

The structure of the parallel interference cancelers depicted in Figure 1. The channel decoding is done by a Max-Log-MAP decoder

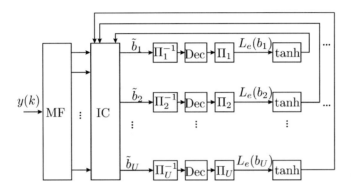

Figure 1. Receiver structure of PIC

deriving approximated extrinsic log-likelihood-ratios $L_e(b_u)$. The soft estimates of the coded symbols $\bar{b}$ are calculated as $\bar{b} = \tanh(L_e/2)$. The signal-to-interference-plus-noise-ratio (SINR) of each branch is relevant for the quality of the interference cancellation. It is defined as SINR $= 2\sigma_d^2/(\sigma_n^2 + \sigma_{mui}^2)$ and is equal to the SNR $= 2E_s/N_0$ in the case of perfect interference cancellation which is equivalent to the single-user bound (SUB). σ_d^2 is the variance of the desired signal and σ_{mui}^2 is the variance of the remaining multi-user interference after cancellation which can be calculated as $\sigma_{mui}^2 = \sigma_d^2 \cdot \mu(U-1)/N$. $\mu = E\{|\bar{b} - b|^2\}$ is the remaining mean squared error of the estimated symbols after decoding which is approximately the same for each user in the case of PIC. The ratio of SINR and SNR is called multi-user efficiency (MUE) and is denoted by η [2]. If η is equal to one there is no loss compared to the SUB and therefore we obtain perfect interference cancellation. For the large system limit $(N, U \to \infty)$ $(U-1)/N \approx U/N = \beta$, η can be written as

$$\eta = \frac{\text{SINR}}{\text{SNR}} = \frac{2\sigma_d^2/(\sigma_n^2 + \sigma_{mui}^2)}{2\sigma_d^2/\sigma_n^2} = \frac{1}{1 + \beta\mu E_s/N_0}. \qquad (3)$$

The parameter η can be used to visualize and predict the behavior of the

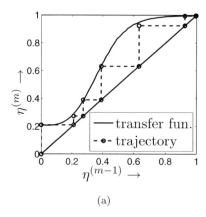 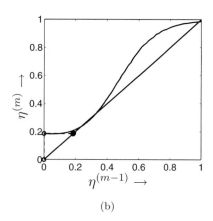

Figure 2. Predicted transfer function and simulated trajectory for PIC, $N = 8$, $U = 16$, $E_b/N_0 = 6$ dB (a) and $N = 8$, $U = 24$, $E_b/N_0 = 5$ dB (b)

iterative detection. In the first step, we have the matched filter outputs and no a-priori information from the channel decoder. The variance μ is therefore equal to 1 and the MUE becomes $\eta^{(1)} = 1/(1+\beta E_s/N_0)$. After simultaneously decoding all users, soft estimates $\bar{b}$ of the transmitted symbols are obtained which are used in the next iteration for interference cancellation. As channel decoding is generally a nonlinear process μ cannot be calculated analytically, but has to be predetermined. The output error $\mu^{(m)}$ of the decoder in the m-th iteration depends on the SINR at the input

$$\mu^{(m)} = g\,(\text{SINR}) = g\left(\eta^{(m-1)}\text{SNR}\right) \qquad (4)$$

and therefore on the MUE of the previous iteration $\eta^{(m-1)}$. Because $\eta^{(m)}$ depends itself on $\mu^{(m)}$ the behavior of the PIC at iteration m can be described by $\eta^{(m)} = f\left(\eta^{(m-1)}\right)$. This function is illustrated in a two-dimensional plot in Figure 2(a). The transfer function describes the theoretical behavior and the trajectory the measured values during simulation. The detection starts in the lower left corner and tends to the upper right corner. This point corresponding to $\eta = 1$ describes perfect interference cancellation. It can be seen that the behavior can be predicted very precisely. This plot corresponds to a system with a spreading factor $N = 8$, $U = 16$ users, an E_b/N_0 of 6 dB and a convolutional code with generator polynomials $[7\ 5]_8$ in octal representation. This system will converge to the SUB within 6 iterations. In Figure 2(b) a system with load of 3 and $E_b/N_0 = 5$ dB is depicted. There is an intersection between the transfer function and the bisecting line so the detection gets stuck at $\eta \approx 0.2$.

4. Analysis of Successive Interference Cancellation

The structure of successive interference cancellation is shown in Figure 3. The prediction in the same manner as for PIC is not possible. While

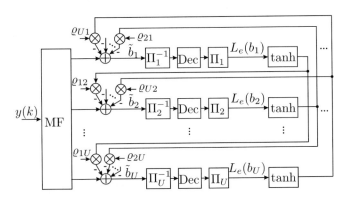

Figure 3. Receiver structure of SIC

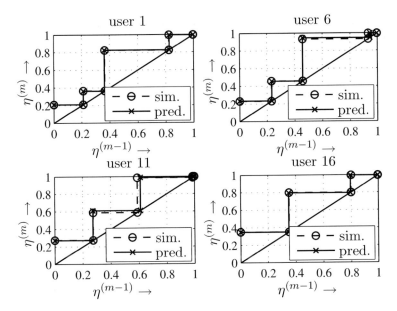

Figure 4. Predicted and simulated trajectories for SIC, $N = 8$, $U = 16$ and $E_b/N_0 = 6$ dB

for PIC the error variance μ is the same for all users in the large system limit, this is not the case for SIC. The U users have different variances $\mu_u^{(m)}$ at each iteration m. The remaining errors of the users are assumed to be independent. So a simple addition of their variances weighted with the corresponding correlation coefficient can be applied for calculating the resulting interference on the desired user signal. For that reason the MUE can still be calculated by

$$\eta_u^{(m)} = \frac{1}{1 + \frac{1}{N}\left(\sum_{i=1}^{u-1}\mu_i^{(m)} + \sum_{i=u+1}^{U}\mu_i^{(m-1)}\right)E_s/N_0} \quad (5)$$

To show how good the prediction works, Figure 4 depicts the predicted and the simulated trajectories for the same system parameters as in Fig. 2(a) in one diagram per user. It can be seen that the prediction works well also in the successive interference cancellation case. A kind of transfer function as shown in Fig. 2(a) cannot be drawn in these plots for clearness. The transfer function differs for each user and each iteration due to being conditioned on the current state of all the other users. In order to avoid a very complex diagram only the trajectories are depicted in Figure 4.

5. Power distribution optimization

Up to now analysis was based on uniformly distributed powers of the users. To describe an unequal power distribution with multi-user efficiency, a way to calculate a kind of average efficiency is necessary. So far the transfer characteristic was calculated by simply averaging over all users. For an average MUE in the case of different powers the individual μ's have to be obtained and combined. The error variance μ is defined as the residual symbol interference independent of the received power. But the effect of this error on other users' detection is indeed dependent on the receive power which is denoted as P_u. This fact is taken into account by weighting μ_u with this user's received power. Thus the resulting multi-user efficiency can be calculated more generally as

$$\eta_u = \frac{1}{1 + \frac{\bar{E}_s}{N_0}\frac{1}{N}\sum_{v \neq u}\mu_v \cdot P_v} \quad , \quad \sum_v P_v = U. \quad (6)$$

$\bar{E}_s/N_0$ is now defined as an average value over all users in order to get an appropriate criterion for fair comparison with the equal power case. μ_u depends on the SINR at the input of the decoder and is for that reason itself dependent on P_u

$$\mu_u = g(\text{SINR}) = g\left(\eta \frac{\bar{E}_s}{N_0} P_u\right) \quad . \tag{7}$$

The criterion for convergence is still reaching the point of $\eta = 1$ after an finite number of iterations. For the PIC this is fulfilled if $f(\eta) \geq \eta \; \forall \; \eta$. The number of iterations needed depends on the width of the tunnel. Whether the tunnel is open or not depends also on the power distribution. For PIC it turns out that equal power for all users is not the best choice, as presented e.g. in [1] and [5].

In this paper optimization is solved by means of *differential evolution* [6]. Especially in the case of SIC there exist many local optima in the cost function (multimodal function). For local optimization techniques a starting point is needed, which is already near to the global optimum. A better approach is a global optimization, especially algorithms motivated by evolution do not need any knowledge of the searching area.

The main difference between local search algorithms and evolutionary techniques is that instead of a single point a set of points called population is regarded at each time instance. Evolutionary algorithms combine elements of the current population and compare the generated children with the parents. If the child has a lower cost than the parents it will replace one parent. There exist many algorithms motivated by evolutionary processes, but differential evolution has the advantage of converging faster in the neighborhood of the optimal solution due to an adaptive step size. The optimization problem for PIC can be described by

$$\underbrace{\text{minimize}}_{P_1,\ldots,P_U} \sum_u P_u \quad \text{s.t.} \quad \begin{cases} f(\eta) \geq \eta \; (+\varepsilon) \; \forall \; \eta \in [0;1] \\ P_u > 0 \; \forall \; u \end{cases} \tag{8}$$

By ε the width of the tunnel can be increased in order to decrease the number of iterations at the cost of higher transmit power.

For the SIC this simple expression for the first condition is not possible. A more general condition for convergence used for SIC is to reach $\eta = 1$ after an finite number of iteration. The starting population is generated randomly and should be sufficiently large for a high diversity.

The performance improvement of power optimization can be seen in Figure 5(a) where the transfer functions for a system with load $\beta = 4$ at $E_b/N_0 = 6.5$ dB with equal and optimized power distribution is shown. For equal powers it will get stuck at $\eta \approx 0.1$ which is a loss of about 10 dB compared to the SUB. Optimized power levels enables the detection to converge to the SUB. With equal powers only a load of ≈ 3 is possible. In Figure 5(b) it is shown that the weakest user has only a loss of 2.5 dB

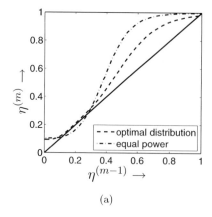
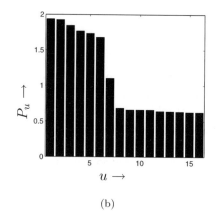

Figure 5. Transfer characteristic (a) and power profile (b) of optimized PIC at $E_b/N_0 = 6.5$ dB

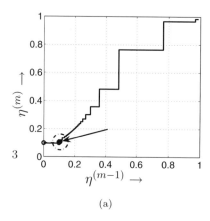
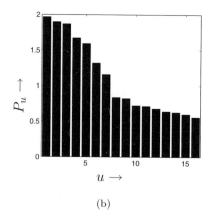

Figure 6. Trajectory (a) and power profile (b) of optimized SIC at $E_b/N_0 = 6.5$ dB

compared to the equal power case, so the gain is 7.5 dB.

In Fig. 6(a) a trajectory for optimized SIC is depicted and in Figure 6(b) the corresponding power profile. For the same system the results are nearly the same. If the tunnel of the transfer function is very small, the differences of the users are small and the SIC behaves more and more like the PIC. For that reason the optimized power levels are nearly the same, but the convergence speed is higher for SIC. To achieve an efficiency of $\eta = 0.98$ the PIC needs for this example 53 iterations, the SIC only 35.

6. Conclusion

In this paper, an analysis of iterative successive interference cancellation based on multi-user efficiency was presented. The quality of prediction is as good as for the PIC; only the graphical representation is not so simple and obvious. Unequal power distribution leads to significantly better performance. Optimization of the power distribution was applied by differential evolution for PIC and SIC based on the prediction of convergence. The convergence behavior could be improved significantly for high loads.

Acknowledgment

The authors would like to thank Dipl.-Ing. Karin Zielinski from the Institute for Electromagnetic Theory and Microelectronics at the University of Bremen for making her optimization tools available for this problem and for many fruitful discussions about evolutionary theory.

References

[1] C. Schlegel and Z. Shi. Optimal power allocation and code selection in iterative detection of random CDMA. In *Zurich Seminar on Digital Communications*, 2004.

[2] A. Nordio, M. Hernandez, and G. Caire. Design and performance of a low-complexity iterative multiuser joint decoder based on Viterbi decoding and parallel interference cancellation. In *ICC 2002 - IEEE International Conference on Communications*, volume 25, pages 298 – 302, April 2002.

[3] V. Kuehn. Analysis of iterative multi-user detection schemes with EXIT charts. In *IEEE International Symposium on Spread Spectrum Techniques and Applications (ISSSTA)*, Sydney, Australia, August 2004.

[4] S. Verdú. *Multiuser Detection*. Cambridge University Press, 1998.

[5] G. Caire, R. Mueller, and T. Tanaka. Iterative multiuser joint decoding: Optimal power allocation and low-complexity implementation. *IEEE Trans. Inform. Theory*, 50(9), Sep. 2004.

[6] K. Price. An introduction to differential evolution. In David Corne, Marco Dorigo, and Fred Glover, editors, *New Ideas in Optimization*, pages 79–108. McGraw-Hill, London, 1999.

BANDWIDTH AND POWER EFFICIENT DIGITAL TRANSMISSION USING SETS OF ORTHOGONAL SPREADING CODES

W.G. Teich, P. Kaim, and J. Lindner
Information Technology, University of Ulm, 89069 Ulm, Germany
werner.teich@uni-ulm.de

Abstract   A bandwidth and power efficient digital transmission is realized employing sets of orthogonal spreading codes. On each of the resulting subchannels antipodal transmission is used. Increasing the number of spreading codes will increase the data rate and thus the bandwidth efficiency of the transmission. The interference between the subchannels resulting from non-orthogonal codes is removed by an iterative vector detection algorithm (soft-decision iterative interference cancellation). Results for small and large spreading lengths are given. In the limit of an in nitely large spreading length, a variance transfer chart analysis of the iteration process is presented which utilises the speci c structure of the discrete-time channel matrix.

1. Introduction

Generally, in digital communications, a compromise between bandwidth efficiency and power efficiency of the underlying transmission scheme has to be obtained. In the third generation of mobile communication systems, e.g., the highest data rate modes are limited to small range and low mobility. Usually, bandwidth efficiency is increased by increasing the size of the complex transmit symbol alphabet. In the EDGE ("Enhanced Data Rates for GSM Evolution") system the symbol alphabet is extended in phase direction (8-PSK). In UMTS ("Universal Mobile Telecommunication System") and the family of WLAN ("Wireless Local Area Network") standards IEEE802.11x increased data rates are obtained by extending the alphabet in amplitude direction (16-QAM, 64-QAM). Both lead to a substantial decrease of the power efficiency.

Recently, Vanhaverbeke et al. [1] have reported an increased number of users in a code division multiple access (CDMA) system using multiple orthogonal spreading sequence sets combined with an iterative detection scheme at the receiver. They have shown, that this increase is substantially larger if sets of orthogonal spreading codes insted of pseudo noise (PN) codes are used. They call this method, in which each user is assigned a unique code out of m orthogonal sets, m orthogonal code division multiple access (m-OCDMA). They have also shown, that the capacity increase is the larger, the larger the spreading length N is. For the downlink and an additive white Gaussian noise (AWGN) channel they have reported that they could accomodate up to 50% more users in the system than orthogonal spreading codes are available without any substantial impairment of the bit error rate (BER) performance.

We propose to use m sets of orthogonal spreading codes in a single user communication scenario in order to generate K, not necessarily orthogonal, subchannels. On each subchannel antipodal transmission is used. Increasing the number of spreading codes will increase the bandwidth efficiency of the transmission. The interference between the subchannels resulting from non-orthogonal codes is removed by an iterative vector detection algorithm [2, 3] similar to the algorithm proposed by Vanhaverbeke et al. [1]. We call this an m-OCDM ("m orthogonal code division multiplexing") transmission. Note, that an m-OCDM transmission (one user is using K subchannels simultaneously) is very similar to an m-OCDMA transmission in the downlink (each user is assigned one of the K subchannels). The only difference being, that in m-OCDMA we are only interested in the information transmitted via the subchannel assigned to our user of interest, whereus in m-OCDM we have to detect all K subchannels.

2. m-OCDM: System Model

Basis for the generation of the m sets of orthogonal codes are Walsh-Hadamard (WH) codes with a spreading length of N. The orthogonal sets are then obtained by individual complex scrambling sequences for each set. Using a complex scrambling sequence each set contains at most $2N$ orthogonal spreading codes. The resulting signal space dimension is $2N$. The total number of spreading codes is given by $K = (m\ \ 1)N+M$ where $M\ \ N$ is the number of codes used in the last set. The load $L = \frac{K}{N}$ characterizes the bandwidth efficiency of the transmission. Using a sinc-function as chip waveform the resulting bandwidth efficiency is: $\eta = L = \frac{K}{N}$ [bit/s/Hz].

m-OCDM can be described in the general framework of a vector-valued transmission as introduced by Lindner, see, e.g. , [3, 4] and references therin. In this paper we restrict ourselves to an uncoded transmission over an AWGN channel with a two-sided noise power spectral density N_0. Assuming further a synchronous transmission, we obtain a very simple vector-valued discrete-time transmission model on symbol basis:

$$\underline{\tilde{x}} = R \ \underline{x} + \underline{\tilde{n}}_e \quad (1)$$

$\underline{x}$ is the K dimensional transmit vector and contains the transmit symbols of each code (subchannel). Since we use antipodal transmission, $x_i \epsilon \{-1, +1\}$. $\underline{\tilde{x}}$ is the K dimensional received vector obtained after the (channel) matched filter followed by a symbol spaced sampling. R is the KxK dimensional discrete-time channel matrix and in our case (AWGN channel) is simply the correlation matrix of the used codes, i.e. $R = U^H U$, where the K columns of matrix U are the various spreading codes. $\underline{\tilde{n}}_e$ is the k dimensional coloured noise vector with noise correlation matrix $\Phi_{n_e n_e} = N_0 \ R$.

Overloading of the system, i.e. using more codes than orthogonal spreading codes are available ($K > 2N$) inescapably leads to interference between the subchannels due to the non-orthogonality of the codes. This interference must be combated by suitable vector detection algorithm [2, 3].

2.1 Vector Detection

For small discrete-time channel matrices R, i.e. small K, a maximum likelihood (ML) vector detection (VD) can be applied. However, for large spreading length N, and therefore large matrices R, a ML-VD is far too complex. In this case VD with reasonable complexity and a performance as close as possible to the optimum must be applied. We use a VD which is based on soft-decision iterative interference cancellation [2, 5]. This detector is very similar to the detector proposed by Vanhaverbeke et al. [1]. The basic idea is to repeatedly use tentative decisions of the received symbol vector $\underline{\tilde{x}}$ in order to reconstruct the interference and subtract it subsequently. Important is, that the feedback is nonlinear, i.e. that a decision is taken, and that this is a soft decision. The interference cancellation process can then be either performed sucessive (sequential) or parallel or in a mixed form parallel/sucessive. In many applications, such as multiuser detection for CDMA systems [5, 6] or detection for multi-carrier CDMA systems [7] these iterative detectors have proven to be very powerful VD. The performance is often close to optimum while the complexity remains moderate. Note, that the structure of the soft-decision iterative interference cancellation VD is equivalent to the strucure of a recurrent neural network (RNN) [5]. The convergence behavior of soft-decision iterative interference cancellation can therefore be analysed using the theory of RNN [8]. Notably,

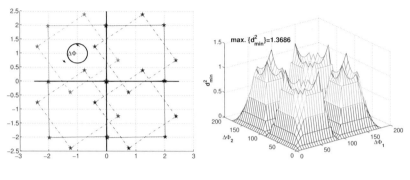

(a) Signal space constellation (b) Minimum distance in signal space

Figure 1. (a) Signal space constellation. Parameters: $N = 1$, $K = 4, \Delta\Phi = 0°$ resp. $35°$. (b) Minimum distance in signal space. Parameters: $N = 2, K = 6$.

if the iteration process converges to the global minimum of the energy function of the RNN, the VD has achieved the ML solution.

Finally, it should be remarked that the resulting discrete-time channel matrix R is not full-rank if the system is overloaded. This implies, that classical linear VD, which are based on the inversion of R, will not work in these cases or will lead to a large noise enhancement.

3. Results for Small Spreading Length

The smallest spreading length is $N = 1$. In this case we obtain a 2D signal space which can easily be depicted, cf. Fig. 1(a). For $K = 4$, the codes are given by:

$$u_1 = 1 \quad \text{and} \quad u_2 = j$$
$$u_3 = e^{j\Delta\Phi} \quad \text{and} \quad u_4 = je^{j\Delta\Phi}$$

Clearly, u_1 and u_2 are mutually orthogonal, as well as u_3 and u_4. In Fig. 1(a) for two cases the signal space constellation is shown. For a rotation $\Delta\Phi = 0°$ the two groups of orthogonal codes are identical. This leads to coinciding signal vectors in signal space, i.e. the minimum distance is zero, which results in a extremely bad BER performance. Increasing the rotation angle $\Delta\Phi$ leads to disjoint signal points. In order to optimize the BER performance, the rotation angle between the two sets must be chosen in such a way, that the minimum distance between any two signal points is maximum. This maximum is obtained for $\Delta\Phi = 30°$ or $\Delta\Phi = 60°$. The resulting minimum distance is $d_{min}^2 = 1.07$.

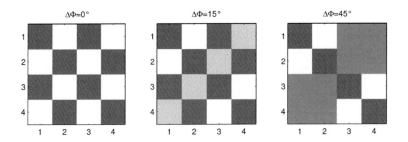

Figure 2. Discrete-time channel matrix R. Parameters: $N = 1, K = 4$.

Figure 1(b) shows the minimum distance for $N = 2$ and $K = 6$ as a function of two rotation angles which discriminate the two groups of codes. The maximum minimum distance has increased to a value of $d_{min}^2 = 1.37$. This is still much smaller than the maximum obtainable minimum distance of $d_{min}^2 = 4$, which is defined by the underlying antipodal transmission. Therefore a performance loss of up to 6dB compared to antipodal transmission can be observed even for the ML detector.

Figure 2 visualizes the discrete-time channel matrix R for $N = 1$ and $K = 4$ for different rotation angles $\Delta\Phi$. The diagonal terms of the matrix, R_{ii}, represent the total received energy for each symbols. For $\Delta\Phi = 0$, i.e. two identical groups of codes, there is only one interfering term for each transmit symbol. Interference power and signal power coincide in this case. Changing the rotation angle $\Delta\Phi$, or in general the complex scrambling sequence, will modify the distribution of the interference power. Note, however, that the total average interference power is not influenced by the complex scrambling code:

$$I = \sum_{i \neq j} |R_{ij}|^2 = \text{const.} \qquad (2)$$

The distribution of the interference power is important for the iterative VD algorithm. Generally, they perform better for a more evenly distributed interference power.

Allowing not only the phase, but also the amplitude of the scrambling codes to be varied, the maximum d_{min}^2 can be increased further. For specific values of phase and amplitude even the classical QAM constellations can be recovered. However, in these cases different symbols are transmitted with different power and therfore suffer from different error probabilities.

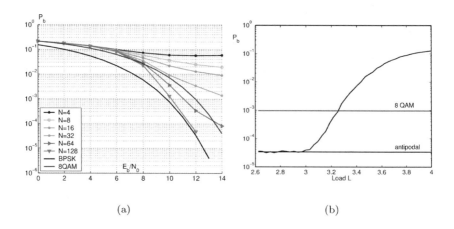

Figure 3. (a) Average BER as a function of E_b/N_0 for different spreading length N. Load $L = K/N = 3$ (b) Average BER as a function of the load $L = K/N$. Parameter: $N = 128, E_b/N_0 = 12$dB.

4. Results for Large Spreading Length

Figure 3(a) shows the run of the BER as a function of $\frac{E_b}{N_0}$ for different spreading length N. The load $L = K/N = 3$ is kept constant. For comparison the BER curves for antipodal and also for 8-QAM transmission are given as well. Soft-decision iterative interference cancellation as described in Section 2.1 has been used as VD algorithm. Fig. 3(a) shows that for $\frac{E_b}{N_0} > 6$dB the BER improves with increasing spreading length N. For $N = 128$ performance of antipodal transmission, which serves as a lower bound in the uncoded case, is practically reached for $\frac{E_b}{N_0} > 10$dB. This means, that we can increase bandwidth efficiency by a factor of three without sacrificing in power efficiency. Or, compared to 8-QAM transmission, which has also a bandwidth efficiency of 3bit/s/Hz, the power efficiency can be improved by more than 2dB for a BER of 10^{-4}. The only price to be paid is an increased complexity due to the VD.

Figure 3(b) shows the BER as a function of the load $L = K/N$ for a fixed spreading length $N = 128$ and a fixed $\frac{E_b}{N_0} = 12$dB. Above a load of $L = 3$ the BER performance of the system starts to degrade from the performance of antipodal transmission. Thus it is not possible to further increase the bandwidth efficiency substantially beyond 3bit/s/Hz without sacrificing in power efficiency. A similar behaviour has also been found for larger spreading length. It was not possible to increase the bandwidth efficiency beyond this point without sacrificing power

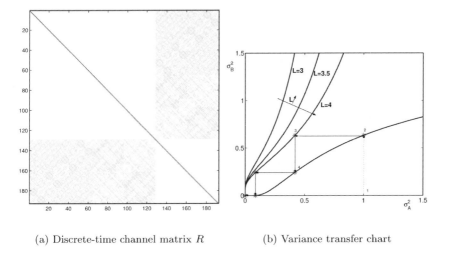

(a) Discrete-time channel matrix R (b) Variance transfer chart

Figure 4. (a) Discrete-time channel matrix R for $N = 64$ and $K = 192$. (b) Variance transfer chart for $N \to \infty$ and different loads $L = 3, 7/2$, and 4.

efficiency. This corresponds to the results of Vanhaverbeke et al. [1] for m-OCDMA in the downlink.

Figure 4(a) shows the discrete-time channel matrix R for $N = 64$ and $K = 192$. The scrambling codes have been obtained by randomly chosing phases $\{\frac{\pi}{4}, \frac{3\pi}{4}\}$ for each chip. For $2N < K \leq 4N$ the matrix R has a unique structure. We can distinguish two groups of codes, group "A", comprised of set 1 and set 2 (where the codes of set 2 are obtained by multiplying the codes of set 1 with "j") and group "B", comprised of set 3 and set 4. The codes of set 4 are obtained by multiplying the codes of set 3 with "j". Therefore codes within group "A" and codes within group "B" are mutually orthogonal. We have no intragroup interference. However, due to the overloading of the system, we have interference ("intergroup interference") between codes of group "A" and codes of group "B". In the iteration process of the VD, all symbols belonging to group "A" and all symbols belonging to group "B" can be updated in parallel. Only between the two groups a successive interference cancellation approach has to be taken.

4.1 Variance Transfer Chart Analysis

For large N, the specific structure of the discrete-time channel matrix for m-OCDM can be utilised to perform a convergence analysis of the iterative vector detection algorithm. This analysis is based on the varicance transfer charts as introduced, e.g., by Alexander et al. [9]. It is very similar to the extrinsic information transfer (exit) chart analysis of turbo decoding [10]. For simplicity we assume $N_0 = 0$ and we take the limit $N \to \infty$. In this limit the interference will follow a Gaussian distribution and the bit error probabilities $P_{b,A}$ or $P_{b,B}$ for each group can be approximated by the error function. The average interference power in group "A" can than be linked to the average interference power in group "B" in the previous iteration step and vice versa:

$$\sigma_A^2 = 4P_{b,B} \frac{K}{2N} \frac{2N}{2N} = \frac{K}{N} \frac{2N}{N} \mathrm{erfc}\left(\sqrt{\frac{1}{2\sigma_B^2}}\right) \quad (3)$$

$$\sigma_B^2 = 4P_{b,A} = 2\mathrm{erfc}\left(\sqrt{\frac{1}{2\sigma_A^2}}\right) \quad (4)$$

These relations are plotted in Fig. 4(b). This analysis explains the good convergence behavior of the soft-decision iterative interference cancellation for $L = 3$. However, Fig. 4(b) also indicates that the interference cancellation process should converge even for $L = 4$. A possible path is indicated in Fig. 4(b). However, this could not be varified by simulation. A more detailled analysis of the iteration process shows that after a few iterations the independence of the decision processes in group "A" and group "B" is lost. However, this is a critical assumption used to set up the relation between the interference powers (cf. Eq. 3 and 4).

5. Conclusions

We have proposed a new bandwidth and power efficient transmission scheme based on code division multiplexing. Using sets of orthogonal codes, separated by complex scrambling sequences, the bandwidth efficiency can be increased by up to a factor of three without loss in power efficiency. Resulting interference between different non-orthogonal codes can be combated by a VD algorithm based on soft-decision iterative interference cancellation. For large N, the structure of the discrete-time channel matrix for m-OCDM can be utilised to perform a convergence analysis of the iterative VD algorithm based on variance transfer charts. Results indicate that there is a bound in the increase of the bandwidth efficiency, even when the spreading length N is increased further.

Note, that the full potential of m-OCDM transmission can only be exploited for frequency selective channels and when channel coding is included. Using channel coding powerfull VD algorithms based on iterative detection and decoding can be employed [11]. Further analysis should also include the investigation of other sets of orthogonal spreading codes, such as Fourier spreading.

References

[1] F. Vanhaverbeke, M. Moeneclaey, and H. Sari, "Increasing CDMA Capacity Using Multiple Orthogonal Spreading Sequence Sets and Successive Interference Cancellation", *Proceedings ICC2002*, 28 April - 02 May 2002, New York, USA, pp. 1516-1520.

[2] A. Engelhart, W.G. Teich, J. Lindner, G. Jeney, S. Imre, and L. Pap, "A Survey of Multiuser/Multisubchannel Detection Schemes Based on Recurrent Neural Networks", Wireless Communications and Mobile Computing, Special Issue on Advances in 3G Wireless Networks, vol. 2, no. 3, May 2002, John Wiley & Sons, Ltd. Chichester, West Sussex, UK, pp. 269-284.

[3] J. Lindner, "Informationsübertragung - Grundlagen der Kommunikationstechnik", Springer-Verlag, Berlin, 2004.

[4] J. Lindner, "MC-CDMA in the Context of General Multiuser / Multisubchannel Transmission Models", European Transactions on Telecommunications (ETT), vol. 10, No. 4, July/August 1999, pp. 351-367.

[5] W.G. Teich and M. Seidl,"Code Division Multiple Access Communications: Multiuser Detection Based on a Recurrent Neural Network", *Proc. IEEE ISSSTA 1996*, Mainz, Germany, pp. 979-984.

[6] R.R. Müller and J.B. Huber,"Iterated Soft-Decision Interference Cancellation for CDMA", Broadband Wireless Communications, edited by Luise and Pupolin, Springer, London 1998, pp. 110-115.

[7] W.G. Teich, J. Egle, M. Reinhardt, and J. Lindner,"Detection Method for MC-CDMA Based on a Recurrent Neural Network Structure", *Proc. MC-SS 1997*, edited by K. Fazel and G.P. Fettweis, Kluwer Academic Publishers 1997, pp. 135-142.

[8] A. Engelhart, "Vector Detection Techniques with Moderate Complexity", Fortschr.-Ber. VDI Reihe 10, Nr. 724, VDI Verlag, Düsseldorf 2003.

[9] P.D. Alexander, A.J. Grant, and M.C. Reed,"Iterative Detection in Code-Division Multiple-Access with Error Control Coding", European Transactions on Telecommunications (ETT), vol. 9, no. 5, September/October 1998, pp. 419-425.

[10] S. ten Brink,"Convergence Behavior of iteratively Decoded Parallel Concatenated Codes ", IEEE Trans. Commun., vol. 49, no. 10, Oct. 2001, pp. 1727-1737.

[11] J. Lindner, M. Dangl, W.G. Teich, and D. Yacoub, "Comparison of Vector Detection Algorithms for MIMO-OFDM", Int. J. Electron. Commun. (AEÜ) vol. 59, 2005, pp. 137-146.

SIMPLIFIED REALIZATION OF PSEUDO-ORTHOGONAL CARRIER INTERFEROMETRY OFDM BY FFT ALGORITHM

K. Anwar, M. Saito, T. Hara, M. Okada and H. Yamamoto
Graduate School of Information Science, Nara Institute of Science and Technology (NAIST), 8916-5 Takayama, Ikoma, Nara, Japan 630-0192, Tel: +81-743-72-5348, Fax: +81-743-72-5349, E-mail: {anwar-k, saito, takao-ha, mokada, heiichi}@is.naist.jp

Abstract: This paper proposes new simplified design of Pseudo-orthogonal Carrier Interferometry Orthogonal Frequency Division Multiplexing (PO-CI/OFDM) system by utilizing IFFT/FFT algorithm as a substitute of PO-CI spreading and despreading. We also propose a separator design for guaranteeing that the resulted codes are still (pseudo) orthogonal to each other. This solution is capable of reducing the complexities of PO-CI/OFDM implementation significantly. Difficulties of generating PO-CI codes as complex spreading codes are successfully solved by employing two IFFT/FFTs with the proposed separator. In this paper, we show that PO-CI/OFDM using FFT algorithm (further called PO-CI-FFT/OFDM) is a very simple and low cost technique while providing good performance as the original PO-CI/OFDM. Additional multiplications and additions for PO-CI spreading/despreading and additional memory for storing the generated codes is not required when utilizing the proposed design. This solution is capable of making PO-CI/OFDM closer to the reality.

Key words: pseudo-orthogonal; carrier interferometry; OFDM; peak-to-average power ratio (PAPR); Fourier transform; complexity.

1. INTRODUCTION

Carrier Interferometry (CI) codes and its derivative, called Pseudo-Orthogonal Carrier Interferometry (PO-CI) codes, have been introduced for Orthogonal Frequency Division Multiplexing (called CI/OFDM and PO-

CI/OFDM) system to reduce the peak-to-average ratio (PAPR) level while providing full frequency diversity benefit [1]. CI/OFDM was tested in RF environment (a typical indoor office environment) for wireless LAN IEEE 802.11a [2]. The results showed that CI/OFDM can provides better BER performance of 5-7dB on BER level of 10^{-3} over current OFDM system.

When compared to CI codes, PO-CI codes have some additional advantages. First, PO-CI codes can increase the OFDM throughput double (2x) specifically for binary phase shift keying (BPSK) modulated data. It means that *2N* data stream are possible to be transmitted on *N* carriers. As presented in [1], PO-CI/OFDM system offers 8dB gain over traditional OFDM on Rayleigh fading channel at BER of 10^{-3}. Secondly, PO-CI codes provide lower OFDM PAPR as shown in [3]. With PO-CI, OFDM PAPR of BPSK can be reduced up to 8dB while with CI it is only 9dB from the original OFDM PAPR of 12dB at complementary cumulative distribution function (CCDF) of 10^{-4}. Third, a good interleaving architecture can be applied to obtain better BER performance on fading channel. The reason is that PO-CI codes have two code sets, so two kind of interleaving architectures exist and one of them was demonstrated has better performance by about 1dB over the first architecture [4]. These benefits lead us to realize PO-CI/OFDM with lower complexities while having good performance.

Recently, a simple method of how to generate CI codes and PO-CI codes does still not exist. Even though CI/OFDM has been tested in RF environment, the authors in [2] have used laptop computer to generate CI codes and to perform spreading process. In [5], we proposed the use of FFT algorithm for CI codes generation and at the same time doing a spreading process. Our results show that the proposed technique has some advantages such as low complexities and lower cost, because of the existing FFT processor can be used for performing both CI generation and CI spreading simultaneously.

In this paper, we extend our works in [5] to PO-CI codes for OFDM system. PO-CI codes are designed for doubling the OFDM throughput, so the length of codes is twice longer than the number of the codes. For covering this condition we propose two IFFTs at the transmitter and two FFTs at the receiver as substitute of PO-CI spreading and despreading, respectively. The most important part is the proposed separator for making sure that the signals resulted from IFFT1 and IFFT2 are not interference each other and possible to be detected.

Our results confirm that the proposed system is able provide the same performance as the original PO-CI/OFDM while providing lower complexities and high saving cost. Compared to partial transmit sequence (PTS) technique that has many IFFT/FFT processes for PAPR reduction [6], our technique resulted more lower complexities while able to increase the

OFDM throughput and offer lower PAPR level. If the improved FFT in [7] are realizable, our proposed system will be an excellent design with lower complexity and lower cost for PO-CI/OFDM realization.

2. PO-CI CODES AND IFFT/FFT MATRICES

PO-CI codes consist of two CI code sets, where the second set is the copy of the first code set. CI codes in set 1 are orthogonal to one another and CI code in set 2 are also orthogonal to one another. However, CI codes in set 1 are not orthogonal to CI code in set 2. This problem is then solved by seeking $\Delta\theta$ that is able to minimize the cross correlation value between these codes set. Paper [1] proposed $\Delta\theta = \pi N$ for obtaining minimum cross correlation, especially for high N. With this value, PO-CI codes for k-th data symbol on N subcarriers are then expressed as

$$C1(k) = \{e^{j(2\pi/N)k\cdot 0}, e^{j(2\pi/N)k\cdot 1}, \ldots, e^{j(2\pi/N)k\cdot(N-1)}\}$$
$$C2(k) = \{e^{j(2\pi/N)k\cdot 0+\Delta\theta\cdot 0}, e^{j(2\pi/N)k\cdot 1+\Delta\theta\cdot 1}, \ldots, e^{j(2\pi/N)k\cdot(N-1)+\Delta\theta\cdot(N-1)}\} \quad (1)$$

for C1, $k = 0, 1, 2, \ldots, N-1$ and for C2, $k = N, N+1, \ldots 2N-1$. We can represent PO-CI codes of (1) in a matrix as shown in (2).

$$C1 = \begin{pmatrix} 1 & 1 & \cdots & 1 & 1 \\ 1 & e^{j\cdot\frac{2\pi}{N}(1)\cdot 1} & \cdots & e^{j\cdot\frac{2\pi}{N}(1)(N-2)} & e^{j\cdot\frac{2\pi}{N}(1)(N-1)} \\ \vdots & \vdots & \ddots & \vdots & \vdots \\ 1 & e^{j\cdot\frac{2\pi}{N}(N-2)\cdot 1} & \cdots & e^{j\cdot\frac{2\pi}{N}(N-2)(N-2)} & e^{j\cdot\frac{2\pi}{N}(N-2)(N-1)} \\ 1 & e^{j\cdot\frac{2\pi}{N}(N-1)\cdot 1} & \cdots & e^{j\cdot\frac{2\pi}{N}(N-1)(N-2)} & e^{j\cdot\frac{2\pi}{N}(N-1)(N-1)} \end{pmatrix}$$

$$C2 = \begin{pmatrix} 1 & e^{j\left(\frac{2\pi}{N}(N+0)\cdot 1+\frac{\pi}{N}\cdot 1\right)} & \cdots & e^{j\left(\frac{2\pi}{N}(N+0)\cdot(N-1)+\frac{\pi}{N}(N-1)\right)} \\ 1 & e^{j\left(\frac{2\pi}{N}(N+1)\cdot 1+\frac{\pi}{N}\cdot 1\right)} & \cdots & e^{j\left(\frac{2\pi}{N}(N+1)\cdot(N-1)+\frac{\pi}{N}(N-1)\right)} \\ \vdots & \vdots & \ddots & \vdots \\ 1 & e^{j\left(\frac{2\pi}{N}(2N-2)\cdot 1+\frac{\pi}{N}\cdot 1\right)} & \cdots & e^{j\left(\frac{2\pi}{N}(2N-2)\cdot(N-1)+\frac{\pi}{N}(N-1)\right)} \\ 1 & e^{j\left(\frac{2\pi}{N}(2N-1)\cdot 1+\frac{\pi}{N}\cdot 1\right)} & \cdots & e^{j\left(\frac{2\pi}{N}(2N-1)\cdot(N-1)+\frac{\pi}{N}(N-1)\right)} \end{pmatrix}$$

(2)

The output signal for each subcarrier is then obtained by summing of data after multiplied with both code set 1 and 2. Number of row represents length of data while the number of column represents the number of subcarriers.

2.1 Existing PO-CI Codes Generation

The conventional method [1] of PO-CI codes generation is by saving a matrix of $2N \times N$ of equation (2) in a memory. Then, the input data stream are multiplied with these values and spread out over all subcarriers. Here, it is clear that $2N \times N$ number of multiplications; addition and memory size are required for performing a spreading process. This level of complexity is exponentially proportional to the number of subcarriers or $O(N^2)$. It shows that the conventional method is very complex and heavy.

2.2 Proposed PO-CI Codes Generation

Output signal for n-th subcarrier of (2) after spreading process can be expressed as

$$s(n) = \sum_{k1=0}^{N-1} S_{k1} \cdot e^{j\frac{2\pi}{N} \cdot k1 \cdot n} + \sum_{k2=N}^{2N-1} S_{k2} \cdot e^{j\left(\frac{2\pi}{N} \cdot k2 \cdot n + \frac{\pi}{N} \cdot n\right)} \tag{3}$$

These signals are periodic with period of N, so we can simplify the calculation of (3) as

$$s(n) = \sum_{k1=0}^{N-1} S_{k1} \cdot e^{j\frac{2\pi}{N} \cdot k1 \cdot n} + \left(e^{j\frac{\pi}{N} \cdot n}\right) \cdot \sum_{k1=0, k2=N}^{k1=N-1, k2=2N-1} S_{k2} \cdot e^{j\left(\frac{2\pi}{N} \cdot k1 \cdot n\right)} \tag{4}$$

The first part of (4) can be calculated by IFFT1 (called CI-IFFT1) and the second part can be calculated by IFFT2 (called CI-IFFT2) followed by one multiplication with a constant value per subcarrier which is called as *twiddle factor* [8] (further called as separator in this paper).

3. PROPOSED PO-CI-FFT/OFDM

Based on the (4), we propose a new structure of PO-CI-FFT/OFDM transmitter and receiver as described in Figs. 1 and 2. Data stream are modulated in BPSK and then converted from serial to parallel. Then the parallel data are spread out by PO-CI-IFFT (CI-IFFT2 and CI-IFFT2).

Spread signals are then converted to time domain by an IFFT with oversampling factor of L. After Guard Interval (GI) insertion, PO-CI-FFT/OFDM signals are transmitted. At the receiver GI is removed and FFT-ed to obtain frequency domain signals. On the multipath fading environment, a combiner is required.

Different with the proposal in [1] and [2], this design impacts to the simpler combiner implementation. The Multiplication part of minimum mean square error (MMSE) and threshold detection combining (TDC) [9] combiner's weight are located before the CI-FFT as shown in Fig. 2, while the conventional combiner design are located after the PO-CI despreading process. Then, the equalized signals are de-spread out using PO-CI-FFT (CI-FFT1 and CI-FFT2) and demodulated.

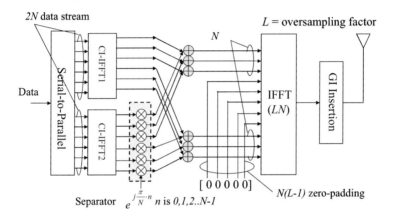

Figure 1. PO-CI-FFT/OFDM Transmitter

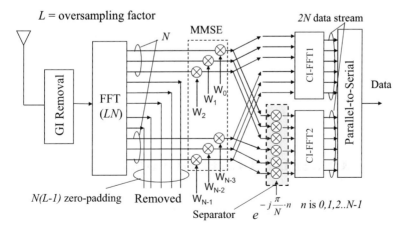

Figure 2. PO-CI-FFT/OFDM Receiver

4. COMPLEXITIES REDUCTION

Tables 1 and 2 shows the comparison of the number of arithmetic operation (multiplications and additions/subtraction) and required memory sizes. Approximately more than 90% complexities can be reduced by employing the proposed PO-CI-FFT (subcarriers > 64). The complexities values in these tables are calculated from existing IFFT/FFT algorithm [8].

Table 1. Computational Complexities of PO-CI Spreading Process

Subcarriers	PO-CI/OFDM		
	Multiplications	Additions	Memory Sizes
16	450	466	450
32	1922	1954	1922
64	7938	8002	7938
128	32258	32386	32258
256	130050	130306	130050
N	$2(N-1).(N-1)$	$2(N-1)(N-1)+N$	$2(N-1)(N-1)$

Table 2. Computational Complexities of Proposed PO-CI-FFT Spreading Process

Sub-carriers	Proposed PO-CI-FFT/OFDM					
	Mul.	Save (%)	Add.+Sub.	Save (%)	Mem.	Save (%)
16	80	82.2	144	69.1	46	89.8
32	192	90.0	352	82.1	94	95.1
64	448	94.4	832	89.6	190	97.6
128	1024	96.8	1920	94.1	382	98.8
256	2304	98.2	4352	96.7	766	99.4
N	$2(N/2)\log(N)+N$		$2N\log(N)+N$		$2(N-1)+N$	

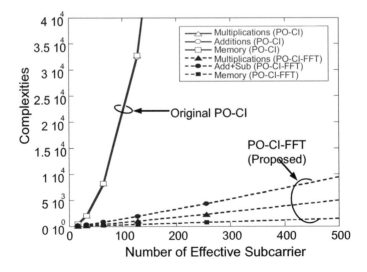

Figure 3. Complexity Reduction of PO-CI/OFDM using FFT as Spreading Codes

5. NUMERICAL RESULT

We show PAPR performance of the proposed system in Fig. 4 for BPSK modulation. For $L > 2$ the PAPR is by about 8dB or 4dB lower than the original 12dB of OFDM's PAPR level at CCDF of 10^{-4}. Fig. 5 shows the BER performance on BU COST-207 fading model [10] with MMSE and TDC combiner for equalizing and combining the received signals. The gain by about 12dB is obtained by employing PO-CI-FFT.

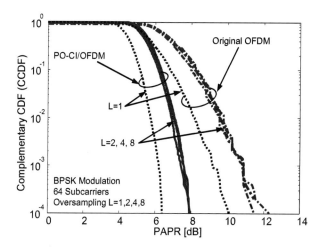

Figure 4. PAPR Performance of PO-CI-OFDM and PO-CI-FFT/OFDM

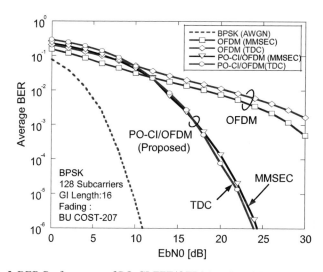

Figure 5. BER Performance of PO-CI-FFT/OFDM on BU-COST207 Fading Model

6. CONCLUSIONS

Simplified realization of PO-CI/OFDM using IFFT/FFT as a substitute of PO-CI spreading and despreading process is proposed. By utilizing PO-CI-IFFT/FFT, complexities of PO-CI/OFDM implementation are significantly reduced by about more than 90% for subcarriers > 64. PAPR performance and BER performance of the proposed system is completely similar to the original performance of PO-CI/OFDM system. Thus, PO-CI/OFDM will be easily realized by the proposed method.

ACKNOWLEDGEMENT

This research is supported by JSAT Corp., Tokyo, Japan.

REFERENCES

1. D. A. Wiegandt, Z. Wu and C. R. Nassar, "High-throughput, high-performance OFDM via pseudo-orthogonal carrier Interferometry spreading codes", IEEE Trans. On Comm., Vol. 51, No. 7, July 2003.
2. D. A. Wiegandt, Z. Wu, C. R. Nassar, "High-performance carrier interferometry OFDM WLANs: RF testing", IEEE ICC2003, vol. 26-1, pp. 203-207, May 2003.
3. Khoirul Anwar, "Peak-to-average power ratio reduction of OFDM signals using carrier interferometry codes and iterative processing", A Master Thesis of Nara Institute of Science and Technology (NAIST), March 2005.
4. D. A. Wiegandt and C. R. Nassar, "High-throughput, high-performance OFDM via pseudo-orthogonal carrier interferometry type 2", IEEE WPMC2002, Oct. 2002
5. K. Anwar, M. Saito, T. Hara, M. Okada and H. Yamamoto, "Simplified realization of carrier interferometry OFDM by FFT algorithm", IEEE VTS APWCS2005, Hokkaido, Japan, August 2005, pp. 199-203.
6. S. H. Muller and J. B. Huber, "OFDM with reduced peak-to-average power ratio by optimum combination of partial transmit sequences", Electronic Letters vol. 33, No. 5, February 1997.
7. S. Shatill and C. R. Nassar, "Improved Fourier transform for multi-carrier processing", Proceeding of SPIE vol. 4869, pp. 35-41, 2002
8. A. V. Oppenheim and R. W. Schafer, "*Discrete-time signal processing: 2nd Edition*", Prentice Hall, 1999.
9. T. Sao and F. Adachi, "Comparative study of various frequency equalization technique for downlink of a wireless OFDM-CDMA system", IEICE Trans. on Communications Vol. E86-B, No. 1 January 2003.
10. M. Patzold, "*Mobile Fading Channels*", John Wiley & Sons, 2002.

EXPLOITING ROTATED SPREADING FEATURES FOR CDM-OFDMA

Ronald Raulefs, Armin Dammann, Serkan Ayaz
German Aerospace Center (DLR)
Institute of Communications and Navigation
Mobile Radio Transmission Group
Oberpfaffenhofen, 82234 Wessling, Germany
{Firstname.Lastname}@dlr.de

Abstract   In this paper, we present a synchronous code division multiple scheme that uses orthogonal frequency division multiple access (CDM-OFDMA) to separate different users. The scheme is applicable for the up- or downlink. The CDM component generates a new alphabet that we compare with standard alphabet schemes. The potential of this alphabet is shown by EXIT-charts.

Keywords: CDMA, signal space diversity, rotated Walsh-Hadamard, EXIT charts

1. Introduction

Code division multiplexing (CDM) with orthogonal frequency division multiple access (OFDMA) is a flexible access scheme that can be applied in the down- as well as in the uplink. CDM-OFDMA is attractive as the complexity due to the multiplexing is only slightly higher compared to pure OFDMA in case a linear detector is applied [1]. The synchronous access is easily achieved as only a single user accesses the same time and frequency slot in the up- or downlink. The channel encoder is responsible for the main part of the overall complexity in a multicarrier system the therefore the flexibility for different code rates within a data frame is limited to adapt the needed performance to the channel capacity. This can be encountered in CDM-OFDMA by assigning in a flexible loading scheme the data symbols to a single spreading block. Therefore, the performance increases in case the load within a spreading block decreases as the multiple access interference (MAI) decreases as well. Alternatively, the receiver can apply iterative procedures to reduce the drawback of the MAI. Many solutions have been presented in the past decade for fully loaded systems to cope with the multiple access interference. One solution is to perform complex interference cancellation. Interference cancellation (IC) schemes enhance the system performance of multi-user systems [2], [3]. They apply in an iterative combination of channel coding and single user detectors, like the zero forcing or the minimum mean square error (MMSE), detector to keep the complexity relatively low.

In this paper, we focus on a CDM-OFDMA system, where we apply a different phase shift for each data symbol to be able to exploit signal space diversity [4]. We further extend the work presented in [5], [6], [7] by focusing on the performance of a single chip itself compared to the a M-ary PSK or QAM complex data symbol. Finally, we will show that performance improvements or complexity reduction is possible by applying soft interference cancellation.

2. Motivation and System Model

The system model of the investigated CDM-OFDMA scheme is depicted in Figure 1.

Walsh-Hadamard spreading codes are applied due to their simplicity to implement only a sign-function. The resulting spread data symbols are superposed and result as a chip sequence. The chip sequence is transmitted over a channel and the receiver uses the whole chip sequence to reverses the spreading. The drawback of the different chip sequences is the ambiguity of the possible chip values between different sequences. The Euclidean distance d between the possible chips is for some constellations 0. The resulting ambiguity diminishes the performance in a fading channel. Therefore, the goal is to find a set of chips $C = c_i, i = 0, ..., K-1$, where K defines the number of data symbols per spreading sequence, such that on the one hand the maximum pairwise error probability between different sequences is as small as possible, for the independent Rayleigh fading channel, on each complex chip:

$$d_{P,min}(C) = min_{c_i,c_j \in C, i \neq j} \prod_{n=0}^{N-1} |c_{i,n} - c_{j,n}|. \qquad (1)$$

On the other hand should $d_{P,min}(C) \neq 0$. In case $d_{P,min}(C) \equiv 0$ than the sequence can not exploit diversity. Boutros et al. [4] chose C such that $C = \Omega_i$, where Ω is a diagonal matrix with the elements $e^{j\omega_0}, e^{j\omega_1}, \ldots, e^{j\omega_{K-1}}$. Each data symbol that will be fed in the spreader is multiplied by another element of the elements of Ω. This guarantees that none of the data symbols has the same value and therefore

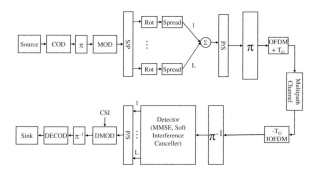

Figure 1. CDM-OFDMA system with rotated spreading (rot+spread).

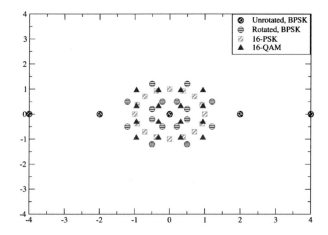

Figure 2. The different modulation schemes 16-PSK, 16-QAM and the (non-)rotated BPSK scheme for a spreading length of four.

minimizes

$$d_{P,min} = \left|\frac{min_{s_i,s_j \in C, i \neq j}|c_i - c_j|}{C}\right|. \quad (2)$$

In Figure 2 the 16-PSK, the 16-QAM constellation and the possible constellations of a single chip for the spreading output of four non-rotated and of four rotated BPSK data symbols that are spread with a Walsh-Hadamard spreading sequence of length four are drawn. The ambiguity for the non-rotated output of the spreading block is represented by the four different non-zero signal constellations and the zero itself. The rotated spreading output equals 16 different constellation points, where half of each different amplitudes circulate around the cen-

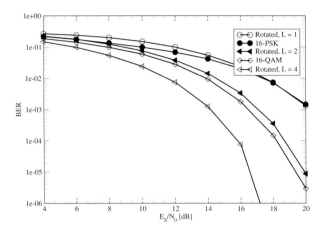

Figure 3. BER results for the AWGN channel and different modulation schemes.

center. The similarity between rotated chip alphabet and the PSK and the QAM alphabet are visible.

In Figure 3 we compared the different constellations using only a single carrier ($L = 1$, 16-PSK and 16-QAM). The used detector is a maximum likelihood soft-output demapper. The performance for the rotated constellation with spreading length four and BPSK as modulation alphabet is similar to the 16-PSK performance. To achieve the same performance as 16-QAM in an AWGN channel a repetition code is used. By this the effective data rate is divided in half for the rotated scheme. Four $L = 4$ the data rate is halved again, but equals now the normal spreading constellation. As in an AWGN channel no diversity gain is possible, the gains are based on the $3dB$ power amplification of using more subcarriers to detect the transmitted bits.

3. Iterative decoding

In the previous Section we compared the chip of a spread signal with well know 16-PSK and 16-QAM symbols for the AWGN channel. The output showed that it is not promising to use the chip as simple data symbol with a new modulation scheme. Now, we investigate the potential of the chip in an iterative decoding scheme, like it is presented in [8].

Figure 4 shows the iterative decoding and demapping scheme at the receiver. The received data is fed into the soft-output demapper, after deinterleaving the soft-out decoder generates the a priori log-likelihood

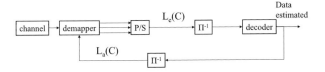

Figure 4. Iterative demodulation and decoding.

ratios (LLRs) $L_a(C_k(i)) = log\frac{(P(C_k(i)=0)}{(P(C_k(i)=1)})$ of the coded bits. The output of the extrinsic LLRs is

$$L_e(C_k(i)) = log\frac{\sum_{s_k \in \alpha_0^i} p(r_k|s_k) \prod_{j=1, j \neq i}^{m} e^{-L_a(C_k(j))c_k(j)}}{\sum_{s_k \in \alpha_1^i} p(r_k|s_k) \prod_{j=1, j \neq i}^{m} e^{-L_a(C_k(j))c_k(j)}} \qquad (3)$$

where α_b^i represents the subsets of chips whose original data symbol constellations have the values $b \in 0, 1$ at the $i'th$ position, where $i \in 0, \ldots, L$. The pdf of $p(r_k|s_k)$ is given by the channel model, an uncorrelated Rayleigh fading channel model, see e.g. [9]. After deinterleaving the extrinsic $L_e(C_k(i))$ is fed into the decoder. At the initial stage the a priori LLRs equal zero.

Figure 4 shows the EXIT charts for different mappings for different modulation alphabets with the cardinality of 16. The decoder curve is depicted for a convolutional channel code of $(5, 7)_8$ in the uncorrelated Rayleigh fading channel. The Gray mappings for 16-QAM and 16-PSK have a very promising start, but do not gain significantly from any a priori information. The MSP(modified set partitioning) mapping is optimized for iterative decoding schemes and outperforms all others. The chips which are created by the rotated spreader have a better start as the MSP mapping and have a similar steepness as the MSP mapping until the input information reaches 0.5. Then the MSP mapping improves significantly and outperforms the chips of the rotated spreading. The rotated spreading chips have a promising tunnel and will outperform the Gray mapping constellations already after three iterations. In comparison the MSP mapping has a very small tunnel to pass through and needs significantly more iterations to reach a similar performance as the outcome of rotated spreading.

4. Coded Results

As we have outlined in [7] the combination of rotated spreading and channel coding increases the overall Hamming distance between the different possible spreading vectors to the length of the spreading sequence.

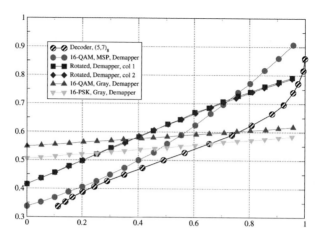

Figure 5. EXIT charts for 16-points modulation alphabets.

So far, we could show only improvements with a maximum likelihood detector. As the complexity of the ML detector limits the performance gains, especially in coded scenarios, the outlook was cheerless. In this section, we briefly motivate for further research and show results for an CDM-OFDMA downlink scheme [10] with a parallel interference canceller. We applied a convolutional channel code with memory six $(133, 171)_8$ and punctional scheme $(1110101010)_2$ to achieve a rate R of $\frac{5}{6}$ [11], [12] and used 4-QAM modulation. The channel is an uncorrelated Rayleigh fading channel, that offers on each subcarrier i.i.d. fading values to provide the maximum diversity to be exploited by the concatenated coded system. The soft bits of the decoder are used as a double value. At the first stage the MMSE is adopted to the number of data symbols here fully loaded, K equals four. At the further stages the MMSE is adopted to a single user. At the final stage, after the counted iterations, the maximum ratio combiner is applied.

The results in Figure 6 show that the performance for an MMSE detector performing only the first stage does not gain at all from rotated spreading. For further stages, here two and eight, the performance improves for a soft decided soft bit by approximately a dB.

5. Conclusions

In this paper we have shown that the scheme of rotated spreading performs similar to the well known modulation schemes, like PSK or QAM. The outcome shows that three different options at the receiver

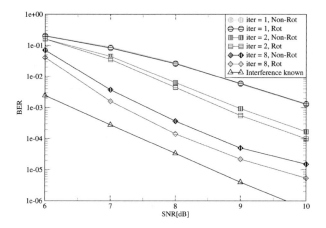

Figure 6. Different detectors exploiting rotated spreading.

remain possible to detect the transmitted bits without any prior knowledge about the possible receiver at the transmitter:

- standard single user detection, like the MMSE detector does not improve nor suffer due to rotated spreading,
- iterative decoding and demodulation does outperform the standard modulation alphabets,
- soft parallel interference cancellation allows to save iterations by using rotated spreading.

Further research about the possible adaptivity of the unique chips and its interconnections is needed. The repetition of different chips shows no differ in the BER performance. Nevertheless, the higher granularity could be a guide to a more flexible system, especially by taking into account automatic repeat requests schemes with variable load in the down- or uplink.

Acknowledgments

The work presented in this paper was partially supported by the European IST projects 4MORE (4G MC-CDMA multiple antenna system On chip for Radio Enhancements)[http://ist-4more.org/], WINNER (Wireless World Initiative New Radio) [https://ist-winner.org] and the German Aerospace Center, DLR.

References

[1] *IST Project Matrice*, Deliverable 5.2 2002-2005. http://www.ist-matrice.org.

[2] Volker Kuehn. Combined MMSE-PIC in coded OFDM-CDMA systems. In *Proceedings IEEE Global Telecommunications Conference (GLOBECOM 2001), San Antonio, TX, USA*, pages 231–235, November 2001.

[3] Xiaodong Wang and H. Vincent Poor. Iterative (turbo) soft interference cancellation and decoding for coded CDMA. *IEEE Transactions on Communications*, 47(7):1046–1061, July 1999.

[4] Joseph Boutros and Emanuele Viterbo. Signal space diversity: a power- and bandwidth-efficient diversity technique for the Rayleigh fading channel. *IEEE Transactions on Information Theory*, 44:1453–1467, July 1998.

[5] Dennis L. Goeckel and Ganesh Ananthaswamy. On the design of multidimensional signal sets for OFDM systems. *IEEE Transactions on Communications*, 50(3):442–452, March 2002.

[6] Andreas Bury, Jochem Egle, and Jürgen Lindner. Diversity comparison of spreading transforms for multicarrier spread sprctrum transmission. *IEEE Transactions on Communications*, 51(5):774–781, May 2003.

[7] Ronald Raulefs, Armin Dammann, Stephan Sand, Stefan Kaiser, and Gunther Auer. Rotated walsh-hadamard spreading with robust channel estimation for a coded MC-CDMA system. *EURASIP Journal on Wireless Communications and Networking — Special Issue on Innovative Signal Transmission and Detection Techniques for Next Generation Cellular CDMA Systems*, 2004(1):74–83, August 2004.

[8] Frank Schreckenbach, Norbert Goertz, Joachim Hagenauer, and Gerhard Bauch. Optimization of symbol mappings for bit-interleaved coded modulation with iterative decoding. *IEEE Communications Letters*, 7(12):593–595, December 2003.

[9] J. G. Proakis. *Digital Communications*. McGraw-Hill, 4th edition, 2001. ISBN 0-07-118183-0.

[10] K. Fazel and S. Kaiser. *Multi-Carrier and Spread-Spectrum Systems*. Wiley, 1st edition, 2003. ISBN 0470848995.

[11] L. H. C. Lee. New rate-compatible punctured convolutional codes for viterbi decoding. *IEEE Transactions on Communication*, 42(12):3073–3079, December 1994.

[12] M. Bossert. *Kanalcodierung*. B.G. Teubner, Stuttgart, 1998. ISBN 3-519-16143-5.

DETECTION OF PRE-CODED OFDM BY RECOVERING SYMBOLS ON DEGRADED CARRIERS

Sayaka Tamura, Masahiro Fujii, Makoto Itami and Kohji Itoh
Depatment of Applied Electronics, Tokyo University of Science
2641 Yamazaki, Noda, Chiba, 278-8510 Japan
saya@itlb.te.noda.sut.ac.jp

Abstract   In this paper, a detection scheme of precoded OFDM is proposed. The proposed scheme recovers the precoded data symbols correspond to the carriers whose reception powers are very degraded under frequency selective fading by using the data symbols after decision and the data symbols are detected using the recovered precoded data symbols. As the result of numerical analysis, it is confirmed that better performance under severe frequency selective fading channels is achieved.

1.   Introduction

OFDM (Orthogonal Frequency Division Multiplexing) is the modulation scheme that can utilize the frequency band very efficiently. Moreover, it is possible to eliminate influence of inter-symbol interference by adding a guard interval without much degrading transmission speed. Therefore, OFDM is applied to the many communication systems such as wireless LAN systems, digital terrestrial broadcasting systems, next generation mobile communication systems and so on.

One of the major factors of performance degradation in OFDM transmission is the influence of frequency selective fading channels. Under severe frequency selective fading channels, the reception power of a specific carrier often becomes very low and it becomes impossible to recover the symbol from the carrier under this situation. This affects the total performance of the system. In order to improve this problem, many schemes such as the use of error correcting codes, multiple antenna diversity reception and so on are proposed [1] [2] [3] [4] [5].

One of the effective schemes to improve this problem is the linear precoding scheme using an orthogonal transform [3]. This precoding

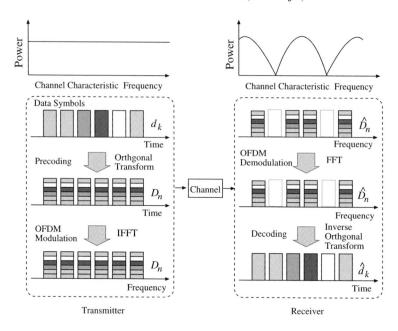

Figure 1. Precoded OFDM

scheme first transforms the block of data symbols using an orthogonal transform before OFDM modulation and each transformed symbol is transmitted on each carrier. By applying precoding, data symbols spread over the whole carriers and it becomes possible to recover data symbols even if several carriers are degraded by frequency selective fading. Different from the use error correcting codes, the precoding using orthogonal transform doesn't add redundant symbols and transmission speed is not reduced by precoding. However, performance of precoding scheme that uses an inverse orthogonal transform for detection is not always good under the case where influence of frequency selective fading is very severe.

In this paper, an iterative detection scheme that well improves the performance of the precoded OFDM scheme under severe frequency selective channels is proposed and its performance is examined by simulations.

2. System Model

In this paper, the linear precoded OFDM scheme shown in Fig. 1 is assumed and its detection scheme is discussed in the following section. In the precoded OFDM transmission, the block of data symbols, $d_k (k = 0, 1, 2, \cdots, N-1)$, are first precoded into the new block,

D_n ($n = 0, 1, 2, \cdots, N-1$), using a specific orthogonal transform. The original data symbol, d_k, is modulated by a specific modulation scheme such as QPSK, QAM, etc. The precoded symbol, D_n, is transmitted on each carrier of OFDM. In the receiver, the received precoded symbols are inversely transformed to recover original data symbols. Although some of the received precoded symbols are degraded by frequency selective fading, original data symbols can be recovered by an inverse orthogonal transform because the original data symbols are distributed over the whole carriers of OFDM. In this paper, it is assumed that the precoded data symbols are derived by the following formula.

$$D_n = \sum_{k=0}^{N-1} u_{n,k} d_k \tag{1}$$

where $u_{n,k}$ denotes a coefficient of the orthogonal transform. The precoded data symbol is used to generate OFDM signal, $s(t)$, as shown in the following formula.

$$s(t) = \operatorname{Re}\left[\sum_{n=0}^{N-1} D_n e^{j2\pi(f_c + nf_0)t}\right] \tag{2}$$

where N, f_0 and f_c denote number of sub-carriers, carrier interval frequency and lowest carrier frequency respectively. If it is assumed that $s(t)$ is transmitted under the channel whose transfer function is denoted by $H(f)$, X_n that corresponds to the received signal against each sub-carrier is derived after DFT processing in the receiver. The output of DFT, X_n, is shown by the following formula.

$$X_n = H_n D_n + Z_n \tag{3}$$

Where H_n denotes $H(f_c + nf_0)$ and Z_n denotes complex additive white Gaussian noise. X_n is the received symbol that is transmitted via n'th sub-carrier whose frequency is $(f_c + nf_0)$ and affected by channel transfer function and additive noise. The original data symbol, d_k, is detected from X_n according to the following procedure. In the first step, influence of channel transfer function is equalized. In this paper, zero forcing channel equalization shown in the following formula is assumed.

$$\hat{D}_n = D_n + Z_n/H_n \tag{4}$$

In this case, it assumed that H_n is known. In the second step, $\hat{d}_k$ is obtained by decoding $\hat{D}_n$ using an inverse orthogonal transform used in (1).

$$\hat{d}_k = \sum_{n=0}^{N-1} u_{n,k}^* \hat{D}_n = d_k + \sum_{n=0}^{N-1} u_{n,k}^* Z_n / H_n \qquad (5)$$

$\hat{d}_k$ is the detected data symbol and it is a combination of the desired symbol, d_k, and the additive noise term(the second term in (5)). Therefore signal to noise ratio of $\hat{d}_k$ is given by the following formula.

$$\gamma_k = \frac{E[|d_k|^2]}{E[|\sum_{n=0}^{N-1} \frac{u_{n,k}^* Z_n}{H_n}|^2]} \qquad (6)$$

$$= \frac{E[|d_k|^2]}{\sum_{n=0}^{N-1} \frac{|u_{n,k}|^2}{|H_n|^2} E[|Z_n|^2]} \qquad (7)$$

In (7), it is assumed that the average powers of d_k and Z_n are independent on their suffices and are denoted by E_s and N_0 respectively. And $E[Z_n Z_{n'}^*] = 0 (n \neq n')$ is also assumed. Moreover, if it is assumed that Walsh Hadamard transform is used for an orthogonal transform(i.e. $|u_{n,k}| = 1/\sqrt{N}$ for all n, k), (7) can be simplified as

$$\gamma_k = \frac{N}{\sum_{n=0}^{N-1} \frac{1}{|H_n|^2}} \cdot \frac{E_s}{N_0} \qquad (8)$$

As shown in (8), γ_k doesn't depend on index k. Therefore, all the symbols in the block have equivalent signal to noise ratio. Therefore, the influence of frequency selective fading is averaged and it is possible to improve bit error rate characteristics within the specific range of frequency selective fading as compared to the scheme without pre-coding. In the scheme without pre-coding, bit error rate of the data symbols on the carrier that is degraded by frequency selective fading can not improved while the pre-coding scheme can distribute the influence to the whole data symbols. However, the performance of pre-coding scheme degrades when the influence of frequency selective fading is very large.

3. Proposed Detection Scheme

In the proposed scheme, the received pre-coded data symbol, $\hat{D}_k$, whose reception power is very small is regenerated from the temporarily decoded data symbols and decoding is performed again using the regenerated pre-coded symbols. In this section, the concept of the proposed scheme is described.

Here it is assumed that $\tilde{d}_k(0)$ is derived by making a hard decision of $\hat{d}_k$ in (5). Therefore, $\tilde{d}_k(0)$ is a nearest signal point from $\hat{d}_k$ in the constellation surface. By re-transforming $\tilde{d}_k(0)$ using the equivalent transform used in (1), the following $\tilde{D}_k(0)$ is obtained.

$$\tilde{D}_n(0) = \sum_{i=0}^{N-1} u_{n,k} \tilde{d}_k(0) \tag{9}$$

If there are not many symbol errors in the block, it is possible to recover almost correct pre-coded symbol $\tilde{D}_n(0)$. For example, if only one symbol in the block is incorrectly made a decision, the influence of decision error against $\tilde{D}_n(0)$ is at most $1/N$ of the maximal distance between signal points. And this value is often smaller than the influence of additive noise to the received pre-coded symbol $\hat{D}_n$ whose received power is very small. Considering this characteristic, the new block of pre-coded symbols shown in the following formula is created.

$$\hat{D}_n(1) = \begin{cases} \tilde{D}_n(0) & \text{if} |H_n| < \text{Thr} \\ \hat{D}_n & \text{otherwise} \end{cases} \tag{10}$$

In (10), Thr is an appropriately determined threshold value. In this formula, the received pre-coded symbol whose corresponding channel transfer function H_n has a smaller magnitude than the predefined threshold value, Thr, is replaced by the recovered pre-coded symbol, $\tilde{D}_n(1)$. By applying an inverse orthogonal transform to the newly created block of symbols, $\hat{D}_n(1)$, the block of data symbols, $\hat{d}_k(1)$, is obtained.

$$\hat{d}_k(1) = \sum_{n=0}^{N-1} u_{n,k}^* \hat{D}_n(1) \tag{11}$$

As the result, the signal to noise ratio of $\hat{d}_k(1)$ can be improved as compared to that of $\hat{d}_k$ by applying an appropriate threshold value according to the channel characteristic.

4. Result of Simulation

In this section, the performance of the proposed scheme is shown as compare with the conventional detection scheme and the case without

Table 1. Parameters of OFDM

Modulation Scheme of d_k	QPSK
Number of Carriers(N)	1024
Carrier interval(f_0)	4 kHz
Minimum frequency(f_c)	100 MHz

pre-coding. The parameters of OFDM used here are shown in Table 1. In this case, a simple 2-path static multipath fading channel whose delay spread is 5µs is assumed. Moreover, the level of multipath is expressed by DUR(Desired to Undesired Ratio), which is the ratio of the power of the first path against the power of the delayed path. The smaller DUR indicates more severe multipath situation.

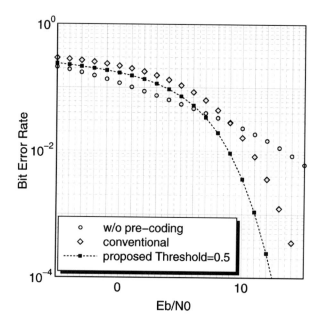

Figure 2. Bit Error Rate Characteristics against E_b/N_0 (DUR=2dB)

In Figure 2, the bit error rate characteristic of the proposed scheme against E_b/N_0 is shown using the threshold value of $Thr = 0.5$. In this case, received magnitude of the first path is assumed to 1.0 and the threshold value is determined relative to this value here. In the figure, 'w/o pre-coding' denotes the characteristic of the OFDM scheme without pre-coding, 'conventional' denotes the characteristic of the conventional pre-coded OFDM scheme and 'proposed' denotes the characteristic of the proposed detection scheme. In this case, bit error rate characteristic under the case where DUR=2dB is shown.

As shown in the figure, the proposed scheme can well improve the bit error rate characteristic by determining the threshold value, Thr appropriately. In Figure 3, the bit error rate characteristic against DUR is shown. The figure also shows that the proposed scheme can improve performance within the wide range of DUR.

In order to improve performance using the proposed scheme, it is necessary to determine the threshold value, Thr, appropriately and this depends on the actual channel characteristic. As the necessary condition, the value of Thr should be larger than that of minimal $|H_n|$. However, too large threshold value cannot improve performance. Therefore, a scheme of optimally determining the threshold value must be investigated in the further research.

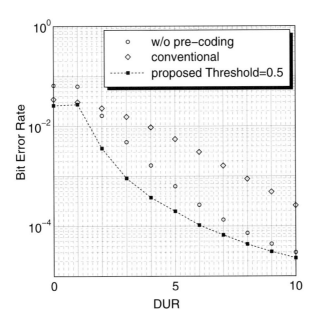

Figure 3. Bit Error Rate Characteristics against $DUR(E_b/N_0=10\text{dB})$

5. Conclusion

In this paper, a detection scheme of the pre-code OFDM is proposed and it is confirmed that better performance under severe multi-path channels can be achieved. And, it is necessary to determine the threshold value, Thr, appropriately for the performance. However, the value depends on E_b/N_0 and DUR and it is difficult to determine it optimally.

In the further researches, effect of channel estimation error and theoretical performance analysis are studied.

References

[1] B. Le Floch, M. Alard and C. Berrou, "Coded Othogonal Frequency Division Multiplex", Proceedings of the IEEE, 83, 6, pp. 982-996, 1995

[2] C. Tepedelenlioglu, "Maximum Multipath Diversity With Linear Equalization in Precoded OFDM Systems", IEEE Trans. Inform Theory, 50, 1, pp. 232-235, 2004

[3] M. Itami, T. Teramoto, K. Okada, H. Uesugi, S. Hatakeyama and K. Itoh, "Improving the Error Characteristics of OFDM by Distributing Data Symbols in Frequency Domain," ITE, 51, 9, pp. 1468-1475, 1997(Japanese).

[4] K. Kyomen, Y. Owaki, M. Itami and K. Itoh, "A Study on Data spread OFDM Scheme applying MAP Decoding", Proceedings of PIMRC 2003, 2, pp. 1085-1089, 2003

[5] S. Kanamori, M. Fujii, M. Itami and K. Itoh, "A Study on Sub-carrier Power and Phase Control of Data Spread OFDM with Transmission Diversity", Proceedings of the 7th International Symposium on Wireless Personal Multimedia Communications, 2, pp. 175-179, 2004

SEMI-BLIND MULTIUSER DETECTION IN ZERO-PADDED OFDM-CDMA SYSTEMS

Rafael Boloix-Tortosa, F. Javier Payan-Somet and Juan José Murillo-Fuentes
Dept. de Teoria de la Señal y Communicaciones
Universidad de Sevilla
Camino de los Descubrimientos sn, Sevilla 41092, Spain
{rboloix,jpayan,murillo}@us.es

Abstract   In this paper, we propose a new semi-blind multiuser detector for zero-padded OFDM-CDMA systems. We show that the downlink transmission in a wireless zero-padded OFDM-CDMA system can be modeled as a linear mixture of the transmitted data symbols. Therefore, blind source separation (BSS) algorithms based on the natural gradient are good candidates to blindly recover those symbols. Before applying BSS, we propose to reduce the effect of the additive noise performing a sub-space projection. On the other hand, we propose to take advantage of any available, may be rough, channel estimation to initialize the blind algorithm. The benefits of initializing the BSS approach are twofold: we avoid the source scaling and arrangement problem and we achieve a quite faster convergence.

1. Introduction

Multicarrier Code-Division Multiple Access (MC-CDMA) [1] refers to new multiple access schemes based on a combination of code division and orthogonal frequency division multiplexing (OFDM) techniques. MC-CDMA is, therefore, a multiuser and multicarrier communication system. In this paper we deal with the MC-CDMA scheme that first spreads the original data stream using a given spreading code, and then transmits the chips using an OFDM modulation. This scheme is usually referred to as OFDM-CDMA. In OFDM, blocks of symbols are transmitted in parallel over several narrowband subchannels at different orthogonal frequencies. If the channel is frequency-selective, the transmitted symbols will suffer both inter-block interference (IBI) and inter-channel interference (ICI). The most widely used version of OFDM is the one using a cyclic prefix (CP). In CP-OFDM, the CP acts as guard interval, so the IBI is avoided. Also, the ICI is removed if the transfer function of the channel can be estimated and does not cancel in any of the subcarriers. Recently, a zero padding (ZP) technique has been proposed as an alternative to the CP one [2],[3]. In ZP-OFDM, zero symbols are used as guard space and are appended before each transmitted block of symbols. The performance of a OFDM-CDMA system that uses this zero-padding alternative is analyzed in [4]. However, they proposed to use the overlap-and-add method to reduce the ZP-OFDM system to one equivalent to a CP-OFDM system [3]. Therefore, if a subcarrier is hit by a zero of the channel frequency response the chips sent through the corresponding subchannel would be lost. In this paper we will focus on the zero-padded OFDM-CDMA system that does not use the overlap-and-add method. We propose a new multiuser detector based on blind source separation (BSS). The application of BSS has been proved as a good alternative for blind receivers in MIMO OFDM [5] and MIMO MC-CDMA [6] systems. On the other hand, in [7] we showed that in ZP-OFDM BSS-based equalization is possible with a single antenna in both transmitter and receiver. However, as any blind solution, the algorithm proposed in that paper showed slow convergence and indeterminate scaling and arrangement of the recovered symbols. In [8] we tried to solve those problems proposing an initialization of the blind algorithm. An estimation of the channel impulse response is needed to obtain such an initialization. In this paper we will adapt the method in [8] to zero-padded OFDM-CDMA and show its good performance.

2. System Model

A single user zero-padded OFDM-CDMA system can be described by the generalized block-by-block discrete-time transceiver model shown in Figure 1. The data symbol sequence is divided into blocks of Q symbols, arranged in a vector $s(t)$. The user spreads the symbols in $s(t)$ using its spreading sequence. This spreading process is represented by $Cs(t)$, where $C = I_Q \otimes c$. Here, $\otimes$ denotes matrix Kronecker product, c is the spreading sequence of length P, with $N = PQ$, and I_Q is the $Q \times Q$ identity matrix. Chips are then multiplied by the IFFT matrix D_N^H, being D_N the $N \times N$ normalized FFT matrix with entries

$$D_N(k,n) = (N)^{-1/2} \exp(-j2\pi kn/N), \qquad (1)$$

for $0 \le k, n \le N-1$, and the operator $(\cdot)^H$ denotes conjugate trans-position. After the IFFT, L zero symbols are added at the end of the block as a guard space to avoid the inter-block interference. Finally, the block of $M = N + L$ transmitted symbols is

$$m(t) = TD_N^H Cs(t), \qquad (2)$$

where

$$T = \begin{bmatrix} I_N \\ 0_{(L \times N)} \end{bmatrix} \qquad (3)$$

and $0_{(L \times N)}$ is an all-zero $L \times N$ matrix. It is assumed that the length of the sampled version of overall impulse response of the continuous-time channel is equal or shorter than L, i.e., $h(k) = 0$, for $k < 0$ and $k > L$. The received block of M symbols with chip-rate sampling is given by

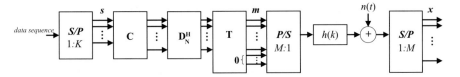

Figure 1. Discrete-time baseband equivalent model for the ZP-OFDM-CDMA system.

$$x(t) = HTD_N^H Cs(t) + n(t), \quad (4)$$

where $n(t)$ is the additive white Gaussian noise and H is the $M \times M$ convolution channel matrix, which is lower triangular Toeplitz with first column $[h(0) \ldots h(L) \; 0 \ldots 0]^T$.

If the system has K active users, multiuser interference will appear. When the base station transmits to a given user, i.e., in the downlink, the received block of symbols yields,

$$x(t) = \sum_{k=1}^{K} x_k(t) + n(t) = \sum_{k=1}^{K} HTD_N^H C_k s_k(t) + n(t), \quad (5)$$

where the Q_k-length vector $s_k(t)$ is the block of data symbols for the kth user and C_k its spreading process for the P_k-length spreading sequence c_k, with $N = Q_k P_k$.

In matrix notation (5) yields,

$$x(t) = HTD_N^H [C_1 \ldots C_K] S^T + n(t), \quad (6)$$

where $S^T = [s_1^T(t) \ldots s_K^T(t)]^T$. In the simplest case, only one data symbol is sent to each user per block, therefore $Q_k = Q = 1$ and $P_k = P = N$. In a most complex case, each user could received data at a different rate, using spreading sequences with different lengths.

3. Blind Source Separation

Blind Source Separation (BSS) involves the reconstruction of a set of n mutually independent unknown *sources*, $s(t)$, from the set of m observed *mixtures*, $x(t)$. Those mixtures are a linear instantaneous combination of the sources corrupted with additive white Gaussian noise $n(t)$:

$$x(t) = As(t) + n(t), \quad (7)$$

Blind stands for the lack of knowledge about the sources or the mixing matrix A. If $x(t)$ is a stationary ergodic random sequence, A is non-singular and there is no more than one Gaussian distributed source in the mixture, forcing the statistical independence of the outputs yields the sources as follows

$$y(t) = Bx(t) = BAs(t) + Bn(t) = Ps(t) + \tilde{n}(t), \tag{8}$$

where B is the separation matrix and P has one and only one nonzero element in each row and column. Notice that the original scaling and arrangement cannot be estimated from the mere independence assumption. Several algorithms have been proposed to solve the BSS problem based on statistical independence (see [9] and references therein). We will now introduce the BSS algorithms used in this paper.

3.1 The Natural Gradient

Adaptive or online BSS algorithms are based on optimizing an objective function using the relative [10] or natural gradient (NG) [11] as learning law. The steepest descent method updates matrix B according to the direction of the gradient $\nabla L(B)$ of a risk function $L(B)$. The natural gradient proposes to use $\tilde{\nabla} L(B) = \nabla L(B) B^T$ instead of the conventional gradient. Under the independence assumption, the maximum likelihood approach is widely used to derive the risk function [9] and obtain $\tilde{\nabla} L(B) = E[\varphi(y)y^T - I]$. Therefore, the normalized learning learning law using the NG yields:

$$B(i) = B(i-1) - \lambda \frac{E[\varphi(y)y^T - I]}{1 + \lambda | E[\varphi(y)^T y]|} B(i-1), \tag{9}$$

where $y = B(i-1)x$ and the entries of φ are the activation or score functions $\varphi_j(y_j) = -q_j'(y_j)/q_j(y_j)$, being $q_j(\cdot)$ the probability density function (pdf) of the j-th source. Since the sources are suppose to be unknown each author proposes his own activation function. The suitable selection of the activation functions $\varphi(\cdot)$ ensures the proper convergence of the method [11]. If the sources are complex, $(\cdot)^H$ replaces $(\cdot)^T$ in (9). The learning law in (9) is, at low noise, equivariant, i.e., it shows a good behaviour independent of the mixing matrix. Besides, under mild assumptions, it is endowed with the superefficiency property [12], i.e., we may achieve at time t, a covariance between two outputs of the order of $1/t^2$.

4. BSS applied to downlink ZP-OFDM-CDMA

The model of the downlink OFDM-CDMA system in (6) is a linear mixing process of the set of $n = \sum_{k=1}^{K} Q_k$ mutually independent unknown sources $[s_1(t) \ldots s_K(t)]^T$. The mixing matrix is given by

$$A = HTD_N^H [C_1 \ldots C_K].\tag{10}$$

The $M \times N$ matrix HT is Toeplitz and always guaranteed to be invertible [3], hence BSS can be applied to recover the sources. On the other hand, since the number of mixtures is always greater than the number of sources, we propose to perform a dimension reduction by whitening and projecting $x(t)$ onto the signal subspace. To do that we have to multiply $x(t)$ by the following matrix W

$$W = \Lambda^{-1/2} U^H,\tag{11}$$

where Λ is a diagonal matrix containing the n largest eigenvalues of $E[x(t)\, x(t)^H]$ and U is the matrix of the associated eigenvectors.

In most OFDM systems some knowledge of the channel is usually possible at the receiver, thanks to available pilot symbols or through signaling from the base station. That knowledge of the channel can be used to find an initialization of the separation matrix in (10). After that initialization, we will apply a NG-based BSS algorithm to blindly achieve a better solution of separation. Thanks to this initialization we can achieve a quite faster convergence. Also, that initial information helps to recover the sources scaling and arrangement. The whole process is as follows:

ALGORITHM 1 **NG-BSS based semi-blind multiuser detection.**
For each received frame:

1. *Channel estimation.* Using some fast pilot assisted procedure or any other method, obtain an estimate the channel.

2. *Mixing matrix estimation.* Calculate the estimated mixing matrix in (10) using the estimated channel.

3. *Whitening and projection.* Obtain $z = Wx$ using W in (11).

4. *BSS initialization.* Set the initial separation matrix as: $B_0 = [WA]^{-1}$.

5. *Natural Gradient.* Using the symbols z of the frame, and for It iterations, i.e., $i = 0$ to $It - 1$, update $B(i)$ using (9).

6. *Solution.* Compute the outputs $y(t) = Bz(t)$ using the last value of B.

5. Performance Evaluation

We consider here the performance of Algorithm 1 for a 4-QAM OFDM-CDMA system set over HIPERLAN/2 (HL2). HL2 is a multicarrier wireless system with $N = 64$ subcarriers. The length of the guard interval is set to $L = 16$, therefore the length of the block of zero-padded transmitted symbols is $M = N + L = 80$. Each frame in HL2 contains 500 blocks and the first 2 blocks are pilots known to the receiver that can be used for channel estimation. In the following, 400 frames are

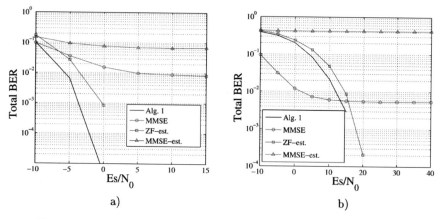

Figure 2. BER performance for: a) $K = 8$ users and $Q_k = 1$, b) $K = 16$ users and $Q_k = 1$.

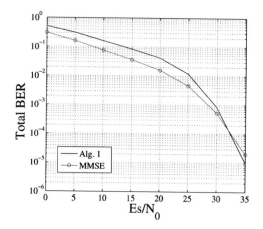

Figure 3. BER performance for $K = 8$ users and $Q_k = 8$.

sent through a typical HL2 channel model A. We suppose that the receiver knows the pilots and the spreading codes of all the active users. Then, the method proposed in [3] can be applied here to obtain the estimation of the channel in step 1 of Algorithm 1. The number of iterations in step 5 of the algorithm was set to $It = 100$. The adaptation coefficient was set to $\lambda = 0.05$ and $\varphi_j(y_j) = |y_j|^2 y_j$. In a first experiment orthoganal Walsh-Hadamard codes are used and each user receives a single symbol, i.e., $Q_k = 1$ and $P_k = 64$, $\forall k$. Figure 2 shows the average BER for all users with the proposed detector. SNR is the averaged chip energy over N_0. For the sake of comparison we include the performance of the MMSE detector [4] that knows the channel perfectly. We also include the MMSE and Zero-Forzing (ZF) detectors using the first estimation of the channel. In the second experiment $K = 8$ active users receive $Q_k = 8$ symbols, and the $P_k = 8$ length spreading codes are randomly generated. The results are shown in Figure 3, where the MMSE detector with perfect knowledge of the channel is included for comparison. It can be seen the good performance of the proposed Algorithm.

6. Conclusions

In this paper, we have faced the multiuser detection problem in zero-padded OFDM-CDMA as a new application of BSS techniques. We have first shown that the model for zero-padded OFDM-CDMA in a downlink transmission can be reduced to an instantaneous linear mixing matrix model and thus considered as a BSS problem. The application of a blind technique based on the natural gradient (NG) for solving the BSS problem allows to exploit its well-known benefits: stability, superefficiency, and equivariance. On the other hand, we have also proposed to apply a noise reduction method and find an initialization of the BSS algorithm based on the fact that in most OFDM-CDMA systems pilot symbols are available at the receiver. With the channel estimation we have obtained an initial value of a separation matrix and we have then used a NG-BSS algorithm to improve the solution. On the other hand, the pilots can be used to recover the order an scale of the original sources.

Acknowledgments

The authors would like to thank the Spanish government for funding TIC-2003-03781.

References

[1] Shinuke Hara and Ramjee Prasad. Overview of multicarrier cdma. *IEEE Communications Magazine*, pages 126–133, December 1997.

[2] A. Scaglione, G.B. Giannakis, and S. Barbarossa. Redundant filterbank precoders and equalizers -Part I: Unification and optimal designs -Part II: Blind channel estimation, synchronization and direct equalization. *IEEE Trans. on Signal Processing*, 47:1988–2022, 1999.

[3] Bertrand Muquet, Zhengdao Wang, Gerogios B. Giannakis, Marc de Courville, and Pierre Duhamel. Cyclic prefixing or zero padding for wireless multicarrier transmissions? *IEEE Trans. on Communications*, 50(12):2136–2148, December 2002.

[4] Xiaodong Ren, Shidong Zhou, and Zucheng Zhou. Performance of Zero-padded OFDM-CDMA signals. In *14th IEEE 2003 International Symposium on Personal, Indoor and Mobile Radio Communication Proceedings*, Beijing, China, 2003.

[5] Chiu Shun Wong, Drangan Obradovic, and Nilesh Madhu. Independent component analysis (ICA) for blind equalization of frequency selective channels. In *IEEE 13th Workshop on Neural Networks for Signal Processing, NNSP'03*, pages 419– 428, 17-19 Sept. 2003.

[6] Daniel I. Iglesia, Adriana Dapena, and Carlos J. Escudero. Bliind source separation detection in MIMO OFDM systems using blind source separation. In *15th International Symposium Personal, Indoor and Mobile Radio Communications. PIMRC 2004*, volume 4, pages 2925–2929, Sept. 2004.

[7] Rafael Boloix-Tortosa and Juan José Murillo-Fuentes. Blind source separation in the adaptive reduction of inter-channel interference for OFDM. *Lecture Notes on Computer Sciences 3195: Independent Component Analysis and Blind Signal Separation. ICA 2004.*, pages 1142–1149, Granada, Sept. 2004.

[8] Rafael Boloix-Tortosa, F. Javier Payan-Somet, and Juan José Murillo-Fuentes. Semi-blind equalization of one-antenna OFDM systems based on higher-order statistics. In *10th International OFDM-Workshop*, pages 70–74, Hamburg, Germany, August-September 2005.

[9] J. F. Cardoso. Blind signal separation: Statistical principles. *Proc. of the IEEE*, 86(10):2009–2025, Oct 1998.

[10] S. I. Amari. Natural gradient works efficiently in learning. *Neural Computation*, 10(2):251–276, Jan 1998.

[11] J. F. Cardoso and B. H. Laheld. Equivariant adaptive source separation. *IEEE Trans. on Signal Processing*, 44(12):3017–3030, Dec 1996.

[12] S. Amari. Superefficiency in blind source separation. *IEEE Trans. on Signal Processing*, 47(4):936–944, Apr 1999.

A STUDY ON ADAPTIVE SUCCESSIVE DETECTION USING M ALGORITHM BASED ON ML CRITERION FOR DOWN-LINK MC-CDMA SYSTEMS

Yoshihito Morishige, Masahiro Fujii, Makoto Itami and Kohji Itoh
Department of Applied Electronics, Tokyo University of Science, 2641 Yamazaki, Noda, Chiba 278-8510, Japan
{morishig,fujii,itami,itoh}@ itlb.te.noda.sut.ac.jp

Abstract In this paper, we propose a new multiuser detection scheme using Maximum Likelihood (ML) criterion and the M algorithm for Multi Carrier (MC)-Code Division Multiple Access (CDMA) systems in the down-link channel. We propose a tree search algorithm based on the linear filters using M algorithm in searching surviving paths which extend to next stage in order to reduce the computational complexity of ML detection. Although the complexity of ML detector as a function of the number of users is exponentially increasing, the proposed scheme requires linear complexity. We demonstrate that the proposed scheme achieves equivalent Bit Error Rate (BER) performance with lower complexity in comparison with ML detector by computer simulations.

1. Introduction

In order to achieve high speed and high capacity multiple access digital data transmission in the future mobile communication systems, MC-CDMA has drawn a lot of attention, and has been widely investigated [1] [2]. In sharp contrast to Orthogonal Frequency Division Multiplexing (OFDM), MC-CDMA makes use of frequency diversity effect by combining the received signals transmitted by spreading the data symbol over frequency carriers according to the signature sequences [3]. However, it suffers from the inherent influence of the Multiple Access Interference (MAI) caused by code division multiplexing of the data sequence in order to achieve higher spectrum efficiency. It is well known that ML detection is the optimum multiuser detection [4]. While ML detector minimizes the BER, it requires a complexity that grows exponentially with the increasing number of simultaneous users. Accordingly, several studies have been made on sub-optimum receivers for CDMA system which exhibit fair trade-off between the BER performances and the computational complexity. There are schemes using M algorithm for CDMA system [5]. In this paper, we propose a sub-optimum multiuser detection using M algorithm based on ML criterion for down-link MC-CDMA system. We propose a tree search algorithm based on the linear filters such as Maximum Ratio Combining (MRC), Zero Forcing (ZF) and Minimum Mean Square

Error (MMSE) using M algorithm in searching surviving paths which extend to next stage in order to reduce the computational complexity of ML detection. We show that these linear receivers help to perform ML detections in In- and Quadrature-phases separately. In the proposed algorithm, we generate a tree structure, whose depth is equal to the number of simultaneous users, according to the CDMA data sequence whose components are sorted in ascending order of the Log Likelihood Ratio (LLR)'s on the assumption that MAI is Gaussian random variable. And we adopt the best M likely paths in the tree structure as survivors. We demonstrate that the proposed scheme achieves equivalent Bit Error Rate (BER) performance with lower complexity in comparison with ML detector by computer simulations.

2. System Model

Figure 1 is the assumed transceiver of the MC-CDMA system. In a down-link MC-CDMA for K simultaneous users, the data symbols are transmitted by OFDM modulation with N carriers after spreading and code division multiplexing by $K \times K$ Walsh Hadamard matrix S and $N/K \times K$ block interleaving at the transmitter. $K \times 1$ received signal vector $\underline{y}$ after OFDM demodulation and the block deinterleaving at the receiver is represented by

$$\underline{y} = \boldsymbol{HS}\underline{d} + \underline{n} \qquad (1)$$

where $\boldsymbol{H} = \text{diag}[h_0, \ldots, h_{K-1}]$ is the $K \times K$ diagonal channel frequency response matrix, $\underline{d} = [d_0, \ldots, d_{K-1}]^T$ is the transmitted data symbol vector and $\underline{n} = [n_0, \ldots, n_{K-1}]^T$ is the additive white Gaussian noise vector with covariance matrix $\sigma_n^2 \boldsymbol{I}$, respectively.

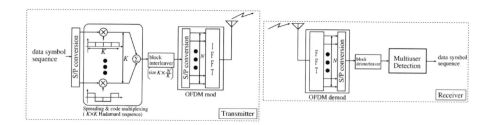

Figure 1. Transceiver for MC-CDMA

3. ML detector

Let W^H denote a $K \times K$ weight matrix of linear filter, then the output signal vector of the linear filter is given by

$$\underline{r} = W^H \underline{y} = W^H HS\underline{d} + W^H \underline{n} = G\underline{d} + \underline{\eta} \tag{2}$$

where $G = W^H HS$, $\underline{\eta} = W^H \underline{n}$, and covariance matrix R_η of $\underline{\eta}$ is written by $R_\eta = \sigma_n^2 W^H W$.

The ML detector is known to be optimum in the sense of minimizing the probability of the decision error and determines the symbol vector $\underline{d}$ which minimizes the Euclidean distance $\|\underline{y} - HS\underline{d}\|^2$. We can replace the distance with

$$\Lambda(\underline{d}) = \min_{\underline{d}} (\underline{r} - G\underline{d})^H R_\eta^{-1} (\underline{r} - G\underline{d}) \tag{3}$$

using the output of the linear filter $\underline{r}$. This is equivalent to maximizing the likelihood function. While ML detector minimizes the BER, it requires complexity that grows exponentially with the increasing number of simultaneous users K. In case of Q^2-ary modulation for the data symbol d_k, ML detector requires distance calculations for $(Q^2)^K$ possible vectors of $\underline{d}$.

We describe an implementation of ML detection by separating IQ components. If G is a real matrix and R_η^{-1} is a Hermitian matrix, $\Lambda(\underline{d}) = \Lambda(\underline{d}_I) + \Lambda(\underline{d}_Q)$ holds where

$$\Lambda(\underline{d}_i) = (\underline{r}_i - G\underline{d}_i)^T R_\eta^{-1} (\underline{r}_i - G\underline{d}_i), (i = I, Q) \tag{4}$$

where $\underline{d} = \underline{d}_I + j\underline{d}_Q$ and $\underline{r} = \underline{r}_I + j\underline{r}_Q$. The conventional linear filters such as MRC, ZF, MMSE (in case fully loaded) can satisfy the above matrix conditions. Therefore, we can separately implement ML detection on each phases. The ML detector requires distance calculations for $2 \cdot Q^K$ possible vectors.

4. Proposed scheme

In order to reduce the computational complexity of ML detector, we propose a tree search method using M algorithm. Figure 2 shows an example of the tree structure in binary data symbol, i.e. QPSK case.

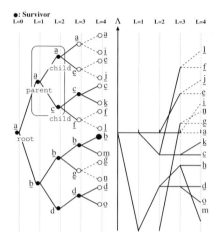

Figure 2. searching space for I or Q channel in case QPSK modulation, K = 4, M = 4

Each label depicts a possible data symbol vector. The label of the root is generated by hard decision vector which is provided by the output of the linear filter employed. A parent has 2-children. One of them has the same label as the parent and the other is different from the parent in one data symbol element of the data symbol vector. The proposed algorithm progresses starting from the hard decision vector based on the linear filter outputs making a tree by expanding, at each assumed decision stage, each of the decision vectors on the survivor paths into 2-children decision vectors, the one with one of the decisions inverted and the other retaining the decision, and the best M decision vectors on the extended paths of the same depth, regarding the likelihood or distance function, are kept as survivors. We assume that the reliability of hard decision on the linear filter output is lower as smaller the value of the log-likelihood-ratio be. Therefore, we can expect, at stages, improvement of the distance function takes place by inversion of the data symbol hypotheses more likely as smaller the LLR. We terminate the tree search algorithm at a predetermined number L of stage, so that the number of distance calculation is reduced in comparison with ML algorithm. We describe in the following the proposed algorithm in detail. Let M denote the maximum permissible number of the survivor paths. And let M' denote the number of the surviving paths.

- **Initialization**

  - Obtain the output of the combining filter $\underline{r} = W\underline{y}$ (W is the liner filter, e. g. MRC, ZF, MMSE)
  - Obtain the hard decision set $\hat{\underline{d}}_0 = \text{sgn}[\underline{r}] = [\hat{d}_{0,0}, \ldots, \hat{d}_{0,K-1}]^T$.
  - Sort $\{\hat{\underline{d}}_0\}$ in the ascending order of LLR, which is placed at the root of the assumed decision tree.
  - Let $M' = 1$

- $l = 0, \ldots, L-1$ (the number of stages $L(\leq K)$)
  - $m = 0, \ldots, M'-1$
    * Create a child decision vector from $\underline{\hat{d}}_m$ by replacing $\hat{d}_{m,l}$ by $\tilde{d}_{m,l} \neq \hat{d}_{m,l}$ as $\underline{\tilde{d}}_m^{(l)} = [\hat{d}_{m,0}, \ldots, \tilde{d}_{m,l}, \ldots, \hat{d}_{m,K-1}]^T$, including it into the candidate vectors for the following selection as well as the other child decision vector with the same label $\underline{\hat{d}}_m^{(l)}$ as the parent.
  - From the set of decision vectors $\{\underline{\hat{d}}_m^{(l)}, \underline{\tilde{d}}_m^{(l)}; 0 \leq m \leq M'-1\}$, select, in the ascending order of distance function $\Lambda(\underline{d})$, up to M vectors, and let them designated by $\{\underline{\hat{d}}_{m'}^{(l+1)}, 0 \leq m' \leq \min(M-1, 2(M'-1))\}$, letting M' be replaced by $2M'$ if $2M' < M$ or M if $2M' \geq M$

- Determine the symbol vector $\underline{\hat{d}}_m$ which minimizes the distance function $\Lambda(\underline{d})$ at the final stage.

Figure 2 shows an example of search tree structure of the proposed sub-optimum detection and transition of the value of distance function at every stage in case of QPSK modulation (i. e. $Q = 2$), $K = 4$, $L = 4$ and $M = 4$. In respect to the portion surrounded by solid line in Figure 2, let $\underline{a}$ be parent, its children are $\underline{a}$ and $\underline{c}$. And if $\underline{a} = [+1, -1, -1, +1]^T$, $\underline{c} = [+1, +1, -1, +1]^T$ whose second element is inverted $\underline{a}$ and the other elements are the same as $\underline{a}$. If we let all the paths survive (i. e. $M = 2^K$), the algorithm is equivalent to ML algorithm. The proposed algorithm requires the number of distance calculations for $2 \cdot L \cdot (Q-1) \cdot M$ possible vectors of $\underline{d}$.

4.1 Log Likelihood Ratio for ordering of testing of the data symbol

In this section, we describe a log likelihood ratio (LLR) criterion for ordering of testing of the data symbol in the proposed scheme with the tree structure. The linear filter output r_i for the ith user is written by

$$r_i = g_{ii} d_i + \underbrace{\sum_{l \neq i}^{K-1} g_{il} d_l}_{\xi_i} + \eta_i \qquad (5)$$

where $g_{ii} d_i$ is the desired signal term, ξ_i is the MAI term and η_i is the noise term which is a zero mean Gaussian random variable with variance σ_η^2, g_{il} is the (i,l) element of $\mathbf{G}$. Applying the central limit theorem, we assume in approximation that ξ_i is a zero mean

Gaussian random variable with variance σ_η^2. Note that σ_ξ^2 and σ_η^2 are, respectively, independent of the user index i. Since ξ_i and η_i are statistically independent, the conditional pdf of d_i is expressed as

$$p(r_i \mid d_i) = \frac{1}{\pi\left(\sigma_\xi^2 + \sigma_\eta^2\right)} \exp\left(-\frac{|r_i - g_{ii} d_i|^2}{\sigma_\xi^2 + \sigma_\eta^2}\right) \qquad (6)$$

and the hard decision $\hat{d}_i$ based on ML criterion is given by $\hat{d}_i = \arg\min_{d_i} \Omega(d_i)$ where

$$\Omega(d_i) = |r_i - g_{ii} d_i|^2 \qquad (7)$$

Since g_{ii} is also independent of the user index i, provided we employ MRC, ZF and MMSE in the linear filter, we replace g_{ii} by g. Let $\tilde{d}_i$ denote the possible hypothesis symbol which causes the second minimum $\Omega(d_i)$. Then we define LLR as

$$LLR_i = \log \frac{p(r_i \mid d_i = \hat{d}_i)}{p(r_i \mid d_i = \tilde{d}_i)} = \Omega(\tilde{d}_i) - \Omega(\hat{d}_i) \qquad (8)$$

LLR can be interpreted as a reliability of the hard decision $\hat{d}_i$ and is directly proportional to the reliability. Next, we show a merit of introducing LLR for ordering of testing of the data symbol. Figure 3 depicts an example of conditional pdf in case of BPSK. Let us assume that r_i and r_l are, respectively, given as the outputs of the linear filter for the ith and lth users. As Figure 3 indicates, it is clear that r_i and r_l result in same hard decision $\hat{d}_i = \hat{d}_l = +1$. Even if $\Omega(\hat{d}_i) = \Omega(\hat{d}_l)$ in Figure 3, the reliability of the hard decision $\hat{d}_i$ is greater than that of $\hat{d}_l$ because $LLR_i > LLR_l$. The proposed algorithm reconstructs the hard decisions in the order of reliability (i.e. LLR) from smallest to greatest.

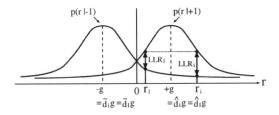

Figure 3. pdf of two signals (BPSK)

5. Numerical Results

In this section, we investigate, by computer simulations, BER performances and computational complexity of the proposed detection scheme. We employ $K \times K$ Walsh Hadamard sequence matrix as the spreading code matrix S (fully loaded) and assume $K = 8$. The data symbol $\{d_k\}$ is modulated by QPSK. In analyzing frequency diversity effect for MC modulation, we use signal bandwidth to coherence bandwidth ratio (*SCR*) defined in [6] as the parameter representing the effective number of uncorrelated scatters. As *SCR* gets larger, the channel fades more selectively in the frequency domain and the correlation decays faster as a function of the frequency separation. We assume Rayleigh fading channel environment with $SCR = 8.0$ and we can use ideal estimate of channel state information at the receiver.

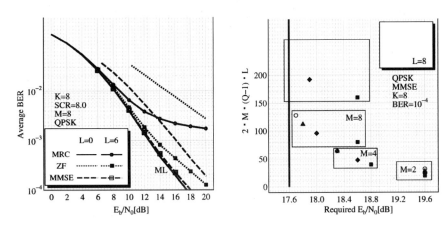

Figure 4. BER performances (L=6, M=8, QPSK) *Figure 5.* Computational Complexity (MMSE, QPSK)

Firstly, we evaluate BER performances and complexity of the proposed detection scheme. In Figure 4, we depict BER performances versus E_b/N_0 in case the number of stages L is equal to 6 and the number of survivor paths M is equal to 8 in comparison with those of MRC, ZF, MMSE (indicated as the case $L = 0$) and ML receivers. A significant improvement of the BER performance is provided by the proposed algorithm (i.e. $L = 6$) in comparison with the case in which only the initializing linear filters are used (i.e. $L = 0$). It is noteworthy that the proposed algorithm using MMSE receiver can achieve the BER performance of ML detector in spite of a reduction of the computational complexity from $(Q^2)^K = 65536$ to $2 \cdot L \cdot (Q-1) \cdot M = 96$. Next, we compare the proposed algorithm using MMSE receiver with ML detector from the viewpoint of the receiver computational complexity. In Figure 5, we show maximum number of calculation of distance functions as a function of the required E_b/N_0 for BER performance 10^{-4}. The proposed algorithm reduces the number of calculations considerably compared with ML detector in case $L = 6$, $M = 8$. The proposed algorithm results in a significant reduction of the computational complexity with a small amount of increase in the required E_b/N_0.

6. Conclusion

In this paper, in order to reduce the complexity of ML receiver with a slight degradation of BER performance, we investigated the multiuser detection algorithm in the down-link MC-CDMA system. We proposed the tree search algorithm based on the linear filters using M algorithm in searching surviving paths. The proposed algorithm using MMSE receiver achieved BER performance very close to ML detection in spite of much smaller number of calculations of distance function. Although the complexity of ML detector as a function of the number of users is exponentially increasing, the proposed scheme requires linear complexity. We demonstrated that the proposed scheme achieves equivalent Bit Error Rate (BER) performance with lower complexity in comparison with ML detector by computer simulations. We are studying the proposed algorithm for higher order modulations. And we will compare the proposed detection scheme with QRD-M algorithm [7] which is based on QR decomposition combined with M algorithm.

References

[1] S. Hara, "Overview of Multicarrier CDMA", IEEE Commun. Magazine, pp. 126–133, Dec. 1997.
[2] K. Fazel, S. Kaiser, "Multi-Carrier and Spread Spectrum Systems", John Wiley and Sons Ltd, 2003.
[3] S. Kaiser, "OFDM Code-Division Multiplexing in Fading Channels", IEEE Trans. on Commun., vol. 50, no. 8, pp. 1266–1273, Aug. 2002.
[4] Sergio Verdu, "Multiuser detection," Cambridge University Press, 2001.
[5] Lei Wei, "Near Optimum Tree-Search Detection Schemes for Bit-Synchronous Multiuser CDMA Systems over Gaussian and Two-Path Rayleigh-Fading Channels", IEEE Trans.on Commun., vol. 45, no. 6, pp. 691–700, June 1997.
[6] M. Fujii, M. Itami, K. Itoh, "Performance evaluation of diversity gain and coding gain in coded orthogonal multi-carrier modulation systems", Euro. Trans. Telecomms, vol. 15, no. 3, pp. 201–206, May-June 2004.
[7] Kyeong Jin Kim, Ronald A. Iltis, "Joint Detection and Channel Estimation Algorithm for QS-CDMA Signals Over Time-Varying Channels", IEEE Trans. on Commun, vol. 50, no. 5, pp. 845–855, May 2002.

Section IV

CHANNEL ESTIMATION

CHANNEL ESTIMATION IN THE PRESENCE OF TIMING OFFSETS FOR MC-CDMA UPLINK TRANSMISSIONS WITH COMBINED EQUALIZATION

Tiziano Mazzoni, Luca Sanguinetti and Michele Morelli
University of Pisa
Department of Information Engineering
Pisa, Italy
{tiziano.mazzoni,luca.sanguinetti,michele.morelli}@iet.unipi.it

Abstract   In this paper we consider the uplink of a multi-carrier code-division multiple-access (MC-CDMA) system employing combined equalization and we address the problem of channel estimation in the presence of residual timing offsets. In particular, we propose a method based on the maximum-likelihood (ML) principle that jointly estimates the pre-equalized channel response and timing offset of each user. The resulting scheme operates in the frequency-domain and exploits some training blocks (carrying known symbols) placed at the beginning of the uplink time-slot. Numerical results are given to highlight the effectiveness of the proposed method and to make comparisons with other existing solutions.

1. Introduction

Multi-carrier code-division multiple-access (MC-CDMA) has been proposed as a viable multiplexing technique for future wireless multimedia transmissions [1]. The main impairment of MC-CDMA is represented by the multiple-access interference (MAI), which occurs in the presence of multipath propagation. In conventional MC-CDMA systems, MAI mitigation is accomplished at the receiver using well-known single-user or multi-user detection schemes [2]. Alternatively, pre-filtering techniques based on the zero-forcing (ZF) or minimum-mean-square-error (MMSE) criterion can be employed at the transmit side [3], [4]. The idea behind pre-filtering is to vary the complex gain assigned to each subcarrier so that interference is reduced and the signal at the receiver appears undistorted.

A promising solution has recently been proposed in [5] in the form of combined equalization. This technique applies pre-filtering at the transmitter in conjunction with post-equalization at the receiver and results in enhanced performance with respect to conventional systems where pre-filtering or post-equalization are used solely. In order to work properly, however, combined schemes require explicit knowledge of the *pre-equalized* channel response of each active user.

In this work we consider the uplink of an MC-CDMA system employing combined equalization and we address the problem of channel estimation in the presence of residual timing offsets. The proposed scheme is based on the maximum-likelihood (ML) principle and requires knowledge of the transmitted symbols. For this purpose, some training blocks (carrying known symbols) are placed at the beginning of each uplink time-slot.

The rest of the paper is organized as follows. The next section describes the signal model and introduces basic notation. In section 3 we discuss the proposed channel estimation scheme. Simulation results are discussed in section 4 and some conclusions are drawn in section 5.

2. System Model

2.1 Transmitter structure

We consider the uplink of an MC-CDMA network in which the total number of subcarriers, N, is divided into I smaller groups of Q elements ($I = N/Q$). Several users within a group are simultaneously active and are separated by orthogonal Walsh-Hadamard (WH) spreading codes. Without loss of generality, we concentrate on the ith group and assume that the Q subcarriers are uniformly distributed over the signal bandwidth. The subcarriers indexes in the considered group are denoted $\{i_q; 1 \leq q \leq Q\}$ with $i_q = i + (q-1)I$, while $a_k^{(i)}(m)$ is the symbol transmitted by the kth user during the mth MC-CDMA block.

Each $a_k^{(i)}(m)$ is spread over Q chips using the spreading sequence $\mathbf{c}_k = [c_k(1), c_k(2), \ldots, c_k(Q)]^T$, where $c_k(q) \in \{\pm 1/\sqrt{Q}\}$ and the superscript $(\cdot)^T$ denotes the transpose operation. The resulting sequence $\mathbf{s}_k^{(i)}(m) = a_k^{(i)}(m)\mathbf{c}_k$ is then fed to the pre-filtering unit to produce the Q-dimensional vector

$$\mathbf{v}_k^{(i)}(m) = \mathbf{G}_k^{(i)}(m)\mathbf{s}_k^{(i)}(m) \qquad (1)$$

where $\mathbf{G}_k^{(i)}(m)=\mathrm{diag}\{G_k^{(i)}(m,q)\,;\,q=1,2,\ldots,Q\}$ is a diagonal matrix containing the pre-filtering coefficients. After frequency interleaving, the entries of $\mathbf{v}_k^{(i)}(m)$ are finally mapped on Q subcarriers using an OFDM modulator.

The idea behind conventional pre-filtering schemes is to design $\mathbf{G}_k^{(i)}(m)$ so as to mitigate MAI and channel distortions at the receiver end. An alternative solution has recently been proposed in [5] in the form of combined equalization. This technique applies pre-filtering at the transmitter together with post-equalization at the receiver. In particular, pre-filtering is employed to efficiently distribute the transmit power among the available subcarriers rather than as a method to cancel the MAI. This produces the following generalized pre-filtering (GPF) coefficients [5]

$$G_k^{(i)}(m,q) = \frac{|H_k(m,i_q)|^p H_k^*(m,i_q)}{\sqrt{\frac{1}{Q}\sum_{\ell=1}^{Q}|H_k(m,i_\ell)|^{2p+2}}} \qquad 1\le q \le Q \qquad (2)$$

in which $H_k(m,\ell)$ is the channel frequency response of the kth user over the ℓth subcarrier of the mth MC-CDMA block, while p is a design parameter that must be suitably chosen to optimize the system performance [5].

From (2), it is seen that GPF requires channel knowledge at the transmit side. In time-division-duplex (TDD) networks, this can be easily achieved by exploiting the channel reciprocity between alternative downlink and uplink transmissions. If channel variations are sufficiently slow (as occurs in indoor applications), the channel estimate derived at a given mobile terminal (MT) during a downlink time-slot can be reused for pre-filtering in the subsequent uplink time-slot. For simplicity, in the sequel we assume that the channel response of each user is practically constant over a time-slot (i.e., we let $H_k(m,\ell) = H_k(\ell)$) and ideal channel state information (CSI) is available at each MT.

2.2 Receiver structure

At the base station (BS), the incoming waveforms are implicitly recombined by the single receive antenna and passed to an OFDM demodulator. We call $\mathbf{X}^{(i)}(m) = [X(m,i_1), X(m,i_2),\ldots,X(m,i_Q)]^T$ the DFT output corresponding to the considered group of subcarriers during the mth MC-CDMA block and assume that each user has achieved timing synchronization at the beginning of the downlink slot by exploiting a broadcast synch channel as explained in [6]. In this way, each signal arriving at the BS is affected by a timing offset due to the two-way propagation delay and timing adjustment is necessary at each MT to avoid inter-block interference. Although this operation can be performed as indicated in [7], residual timing errors between the uplink signals and the BS time reference are expected to occur due to estimation errors. In the following we assume that the residual timing errors are smaller than the difference between the cyclic prefix (CP) duration and the length of the channel impulse responses. In these circumstances, $\mathbf{X}^{(i)}(m)$ takes the form

$$\mathbf{X}^{(i)}(m) = \sum_{k=1}^{K} a_k^{(i)}(m) \mathbf{C}_k \mathbf{H}_k^{(GPF)} + \mathbf{w}^{(i)}(m) \qquad (3)$$

where $\mathbf{C}_k = \mathrm{diag}\{c_k(1), c_k(2), \ldots, c_k(Q)\}$, $\mathbf{w}^{(i)}(m)$ is a Gaussian vector with zero-mean and covariance matrix $\sigma^2 \mathbf{I}_Q$ ($\mathbf{I}_Q$ denotes the identity matrix of order Q) and $\mathbf{H}_k^{(GPF)} = [H_k^{(GPF)}(i_1), H_k^{(GPF)}(i_2), \ldots, H_k^{(GPF)}(i_Q)]^T$ is a Q-dimensional vector with entries

$$H_k^{(GPF)}(i_q) = e^{-j(2\pi \tau_k i_q / NT_s - \theta_k)} A_k(i_q) \qquad 1 \leq q \leq Q. \qquad (4)$$

In the above equation, T_s is the sampling interval of the OFDM demodulator, τ_k denotes the residual timing offset of the kth uplink signal and $\theta_k = 2\pi f_0 \tau_k$, f_0 being the carrier frequency. Also, $A_k(i_q) = G_k^{(i)}(q) H_k(i_q)$ represents the *pre-equalized* channel frequency response of the kth MT over the i_qth subcarrier. As is seen from (2), these quantities are real-valued and non negative. Clearly, in conventional MC-CDMA systems with no signal pre-filtering $A_k(i_q)$ is replaced by the complex-valued quantity $H_k(i_q)$.

Vector $\mathbf{X}^{(i)}(m)$ is exploited to detect the data symbols of all users within the ith group of the mth MC-CDMA block. For this purpose, we employ the partial parallel interference cancelation (PPIC) receiver proposed in [8]. As occurs with all interference-cancelation (IC) schemes, its performance depends heavily on the quality of the channel estimates. In the following we discuss a method for jointly estimating the channel frequency response and the residual timing offset of each user at the BS. For notational simplicity we neglect the superscript $(\cdot)^{(i)}$ indicating the considered subcarrier group.

3. Joint Estimation of The Channel Responses and Timing Offsets

3.1 Problem formulation

We begin by decomposing the timing error τ_k into an integer part δ_k plus a fractional part ε_k, i.e.,

$$\tau_k = (\delta_k + \varepsilon_k) Q T_s \qquad (5)$$

where $\varepsilon_k = \mathrm{saw}\{\tau_k/(QT_s)\}$ is a sawtooth non-linearity that reduces $\tau_k/(QT_s)$ to the interval $(-1/2, 1/2]$ and $\delta_k = \tau_k/(QT_s) - \varepsilon_k$. Substituting (5) into (4) and recalling that $i_q = i + (q-1)I$ yields

$$H_k^{(GPF)}(i_q) = e^{-j[2\pi(q-1)\varepsilon_k - \varphi_k]} A_k(i_q) \qquad (6)$$

where $\varphi_k = \theta_k - 2\pi\tau_k i/NT_s$ and we have taken into account that δ_k is integer-valued. Letting $\mathbf{\Gamma}(\varepsilon_k) = \mathrm{diag}\{e^{-j2\pi(q-1)\varepsilon_k}\,;\; q=1,2,\ldots,Q\}$ and $\mathbf{A}_k = [A_k(i_1), A_k(i_2), \ldots, A_k(i_Q)]^T$, we may rewrite (6) in matrix form as

$$\mathbf{H}_k^{(GPF)} = e^{j\varphi_k} \mathbf{\Gamma}(\varepsilon_k) \mathbf{A}_k. \qquad (7)$$

Our goal is the estimation of the channel vectors $\{\mathbf{H}_k^{(GPF)};\ 1 \le k \le K\}$. For this purpose, we assume that N_T training blocks are placed at the beginning of each uplink time-slot. As in [9], we employ WH training sequences of length $N_T = Q$. In this way the users' signals can be easily separated at the BS by computing the quantities

$$\mathbf{Z}_k = \mathbf{C}_k \sum_{m=1}^{N_T} a_k(m) \mathbf{X}(m) \qquad k = 1, 2, \ldots, K. \qquad (8)$$

Substituting (3) into (8) and bearing in mind (7) yields

$$\mathbf{Z}_k = e^{j\varphi_k} \mathbf{\Gamma}(\varepsilon_k) \mathbf{A}_k + \mathbf{n}_k \qquad (9)$$

where $\mathbf{n}_k$ is a Gaussian vector with zero-mean and covariance matrix $\sigma^2 \mathbf{I}_Q$. Vectors $\{\mathbf{Z}_k;\ k=1,2,\ldots,K\}$ are next exploited to obtain the ML estimates of ε_k, φ_k and $\mathbf{A}_k$ (say $\hat{\varepsilon}_{k,ML}$, $\hat{\varphi}_{k,ML}$ and $\hat{\mathbf{A}}_{k,ML}$), which are finally used to compute the corresponding estimate of channel responses $\mathbf{H}_k^{(GPF)}$ in the form

$$\hat{\mathbf{H}}_{k,ML}^{(GPF)} = e^{j\hat{\varphi}_{k,ML}} \mathbf{\Gamma}(\hat{\varepsilon}_{k,ML}) \hat{\mathbf{A}}_{k,ML}. \qquad (10)$$

Note that vectors $\{\mathbf{X}(m);\ m=1,2,\ldots,N_T\}$ and $\{\mathbf{Z}_k;\ k=1,2,\ldots,K\}$ are related through a reversible operation. This means that passing from $\{\mathbf{X}(m)\}$ to $\{\mathbf{Z}_k\}$ does not produce any loss of information and, accordingly, the ML estimates of ε_k, φ_k and $\mathbf{A}_k$ based on $\mathbf{Z}_k$ are the same as those obtained from the observations $\{\mathbf{X}(m)\}$. Also, from (7) and (9) we see that $\mathbf{Z}_k$ represents an unbiased estimator of $\mathbf{H}_k^{(GPF)}$ that is equivalent to the unstructured channel estimator (UCE) proposed in [9].

3.2 Derivation of the channel estimation scheme

Without loss of generality, we concentrate on the kth user. From (9) we see that, for a given ε_k, φ_k and $\mathbf{A}_k$, vector $\mathbf{Z}_k$ is Gaussian distributed with mean $e^{j\varphi_k}\mathbf{\Gamma}(\varepsilon_k)\mathbf{A}_k$ and covariance matrix $\sigma^2 \mathbf{I}_Q$. Thus, the joint ML estimates of ε_k, φ_k and $\mathbf{A}_k$ based on the observations $\mathbf{Z}_k$ are found looking for the maximum of the following function

$$\Lambda(\tilde{\varepsilon}_k, \tilde{\varphi}_k, \tilde{\mathbf{A}}_k) = 2\tilde{\mathbf{A}}_k^T \Re e\{e^{-j\tilde{\varphi}_k}\mathbf{\Gamma}^H(\tilde{\varepsilon}_k)\mathbf{Z}_k\} - \tilde{\mathbf{A}}_k^T \tilde{\mathbf{A}}_k \qquad (11)$$

with respect to the trial values $\tilde{\varepsilon}_k$, $\tilde{\varphi}_k$ and $\tilde{\mathbf{A}}_k$. Note that in writing (11) we have taken into account that $\mathbf{A}_k$ is real-valued. Keeping $\tilde{\varepsilon}_k$ and $\tilde{\varphi}_k$ fixed and maximizing $\Lambda(\tilde{\varepsilon}_k, \tilde{\varphi}_k, \tilde{\mathbf{A}}_k)$ with respect to $\tilde{\mathbf{A}}_k$ produces

$$\hat{A}_{k,ML}(i_q) = \Re e\left\{Z_k(q)\,e^{j[2\pi(q-1)\tilde{\varepsilon}_k - \tilde{\varphi}_k]}\right\} \qquad (12)$$

where $Z_k(n)$ denotes the nth entry of $\mathbf{Z}_k$. Substituting (12) into (11) and bearing in mind that $[\Re e\{x\}]^2 = \tfrac{1}{2}\Re e\{x^2\} + \tfrac{1}{2}|x|^2$ yields

$$\Lambda_k(\tilde{\varepsilon}_k, \tilde{\varphi}_k) = \frac{1}{2}\sum_{q=1}^{Q} \Re e\left\{Z_k^2(q) e^{j[4\pi(q-1)\tilde{\varepsilon}_k - 2\tilde{\varphi}_k]}\right\} + \frac{1}{2}\sum_{q=1}^{Q} |Z_k(q)|^2 \qquad (13)$$

or, equivalently,

$$\Lambda_k(\tilde{\varepsilon}_k, \tilde{\varphi}_k) = \frac{1}{2}\left|\sum_{q=1}^{Q} Z_k^2(q) e^{j4\pi(q-1)\tilde{\varepsilon}_k}\right| \times \cos[\psi(\tilde{\varepsilon}_k) - 2\tilde{\varphi}_k] + \frac{1}{2}\sum_{q=1}^{Q} |Z_k(q)|^2 \qquad (14)$$

where

$$\psi(\tilde{\varepsilon}_k) = \arg\left\{\sum_{q=1}^{Q} Z_k^2(q) e^{j4\pi(q-1)\tilde{\varepsilon}_k}\right\}. \qquad (15)$$

From (14), we see that the ML estimates of ε_k and φ_k are given by

$$\hat{\varepsilon}_{k,ML} = \arg\max_{\tilde{\varepsilon}_k}\left\{|\lambda(\tilde{\varepsilon}_k)|\right\}, \qquad (16)$$

$$\hat{\varphi}_{k,ML} = \frac{1}{2}\arg\{\lambda(\hat{\varepsilon}_{k,ML})\} \qquad (17)$$

where $\lambda(\tilde{\varepsilon}_k)$ is defined as

$$\lambda(\tilde{\varepsilon}_k) = \sum_{q=1}^{Q} Z_k^2(q) e^{j4\pi(q-1)\tilde{\varepsilon}_k}. \qquad (18)$$

Finally, substituting (16) and (17) into (12), we obtain

$$\hat{A}_{k,ML}(i_q) = \Re\left\{Z_k(q)\, e^{j[2\pi(q-1)\hat{\varepsilon}_{k,ML} - \hat{\varphi}_{k,ML}]}\right\} \qquad 1 \leq q \leq Q. \qquad (19)$$

At this stage we are left with the problem of finding $\hat{\mathbf{H}}_{k,ML}^{(GPF)}$. To this end, we use the ML estimates of ε_k, φ_k and $A_k(i_q)$ in (16), (17) and (19) to obtain

$$\hat{H}_{k,ML}^{(GPF)}(i_q) = \hat{A}_{k,ML}(i_q)\, e^{-j[2\pi(q-1)\hat{\varepsilon}_{k,ML} - \hat{\varphi}_{k,ML}]}. \qquad (20)$$

In the sequel $\hat{H}_{k,ML}^{(GPF)}(i_q)$ is called the maximum likelihood estimator (MLE).

4. Simulation results

4.1 System Parameters

We consider an uncoded QPSK transmission with Gray mapping. The total number of subcarriers is $N = 64$ and WH codes of length $Q = 8$ are used for spreading purposes. The signal bandwidth is $B = 20$ MHz, so that the useful part of each MC-CDMA block has duration $T = N/B = 3.2$ μs while the sampling period is $T_s = T/N = 50$ ns. Each frame has 64 blocks and is preceded by $N_T = 8$ training blocks carrying WH sequences of length N_T. The channel of each user has 12 paths with exponential power delay profile. It is kept fixed over the frame (static channel) but varies from frame to frame.

4.2 Performance Assessment

We begin by designing the GPF parameter p in (2). As a design criterion we choose the minimization of the average bit-error-rate (BER) of the active users. In doing so, we resort to computer simulations since the optimal p is hard to get theoretically. Figure 1 shows the BER as a function of p for $E_T/N_0 = 6$ dB (E_T is the transmitted energy per information bit while $N_0/2$ is the two-sided noise power spectral density). The number K of active user is either 4 or 8 and perfect channel knowledge (PCK) is assumed at the receive end. As expected, the optimal p depends on K but we see that $p = 0$ represents a good choice in both cases. This value is adopted in all subsequent simulations.

Figure 2 illustrates the channel mean square estimation error (MSEE) vs. $1/\sigma^2$ as obtained with MLE. The curve labelled ITI (Ideal Timing Information) corresponds to perfect knowledge of the timing offsets and serves as a benchmark. Results obtained with UCE [9] are also shown for comparison. We see that MLE gives the best results. In particular, a

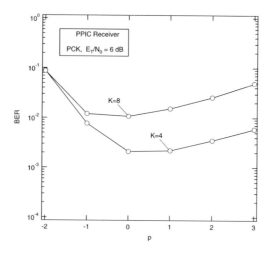

Figure 1. BER performance vs. p for $E_T/N_0 = 6$ dB and $K = 4$ or 8.

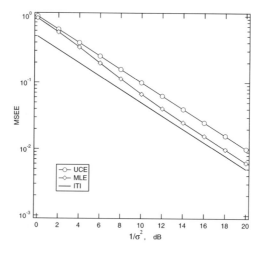

Figure 2. Accuracy of MLE vs. $1/\sigma^2$.

gain of approximately 2 dB is achieved with respect to UCE for $1/\sigma^2 > 8$ dB.

Figure 3 illustrates the BER vs. E_T/N_0 of a PPIC receiver endowed with the proposed channel estimation scheme in the presence of 4 active users. The curve labeled UCE has been obtained using the channel estimator discussed in [9] with no signal pre-filtering (only post-equalization is present), whereas GPF+UCE and GPF+MLE correspond to a system that employs the GPF coefficients in (2) in conjunction with UCE and MLE, respectively. The curve labeled PCK corresponds to a re-

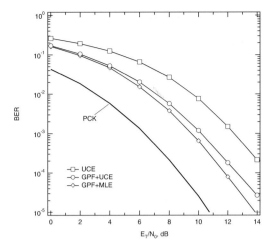

Figure 3. BER performance over a static channel with $K=4$.

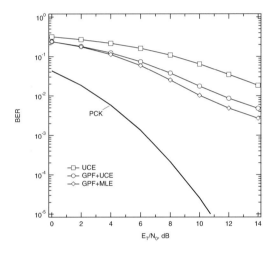

Figure 4. BER performance over a static channel with $K=8$.

ceiver with perfect knowledge of the pre-equalized channel responses and residual timing offsets and serves as a benchmark. We see that GPF significantly improves the system performance and achieves a gain of nearly 3 dB compared to a system without any signal pre-filtering. The best results are obtained with GPF+MLE. In this case, the gain with respect to GPF+UCE is approximately 1 dB at an error rate of 10^{-3}.

The results of Figure 4 have been obtained in the same operating conditions of Figure 3, except that eight users are now simultaneously active. As expected, the system performance deteriorates as the number of active users increases but GPF+MLE is still superior to other solutions.

5. Conclusions

We have discussed the problem of channel estimation in the presence of residual timing offsets for TDD MC-CDMA uplink transmissions with combined equalization. Training blocks placed at the beginning of each uplink frame are exploited for joint ML estimation of the pre-equalized channel response and timing offset of each user.

The performance of a PPIC receiver endowed with the proposed channel estimation scheme has been assessed by simulation and compared to other existing alternatives.

Acknowledgement

This work was supported by the Italian Ministry of Education under the FIRB project PRIMO.

References

[1] K. Fazel and S. Kaiser. *Multi-Carrier and Spread Spectrum Systems.* John Wiley and Sons, 2003.

[2] S. Verdú. *Multiuser Detection.* Cambridge University Press, 1998.

[3] D. Mottier and D. Castelain. SINR-based channel pre-equalization for uplink multi-carrier CDMA systems. In *Proc. of IEEE Int. Symposium on Personal, Indoor and Mobile Radio Communications (PIMRC'02)*, pages 1488–1492, Sept. 2002.

[4] I. Cosovic, M. Schnell and A. Springer. On the performance of different channel pre-compensation techniques for uplink time division duplex MC-CDMA. In *Proc. of IEEE Vehicular Technology Conference (VTC'03, Fall)*, volume 2, pages 857–861, Oct. 2003.

[5] Ivan Cosovic, Michael Schnell, and Andreas Springer. Combined-equalization for uplink MC-CDMA in rayleigh fading channels. *IEEE Trans. on Communications.* accepted for publication.

[6] S. Kaiser and W.A. Krzymien. An asynchronous spread spectrum multi-carrier multiple access system. In *Proc. IEEE Global Telecommun. Conf. (GLOBECOM'99)*, pages 314–319, Rio de Janeiro, Brasil, Dec. 1999.

[7] M. Morelli. Timing and frequency synchronization for the uplink of an OFDMA system. *IEEE Trans. on Communications*, 52:296–306, Feb. 2004.

[8] D. Divsalar, M.K. Simon, and D. Raphaeli. Improved parallel interference cancellation for CDMA. *IEEE Trans. on Communications*, 46(2):258–268, Feb. 1998.

[9] L. Sanguinetti, M. Morelli, and U. Mengali. Channel estimation and tracking for MC-CDMA signals. *European Trans. on Telecommunications*, 15:249–258, March 2004.

ITERATIVE CHANNEL ESTIMATION FOR MIMO MC-CDMA

Stephan Sand, Ronald Raulefs, and Armin Dammann
German Aerospace Center (DLR)
Institute of Communications and Navigation
Oberpfaffenhofen, 82234 Wessling, Germany
stephan.sand@dlr.de

Abstract   This paper investigates a downlink multiple-input multiple-output (MIMO) multi-carrier code division multiple access (MC-CDMA) system with iterative channel estimation (ICE) in the receiver. In MC-CDMA, each user's data symbols are spread with a Walsh-Hadamard code. Due to the superposition of different users' spread data signals, zero-valued subcarriers can occur, which cannot be used to compute least-squares (LS) estimates in ICE. For multiple transmit antennas, the data signals coincide non-orthogonal at each receive antenna. Hence, we propose two MIMO channel estimation methods to overcome these problems and avoid noise enhancement in ICE.

1.   Introduction

Orthogonal frequency division multiplexing (OFDM) [1] is a suitable technique for broadband transmission over multipath fading radio channels achieving high data rates. In addition to the coherent OFDM modulation [2], spreading in frequency or time direction is introduced for multi-carrier code division multiple access (MC-CDMA). MC-CDMA is a promising candidate in the downlink of future mobile communication systems and has been implemented in an experimental system by NTT DoCoMo [3]. High data rate MC-CDMA systems can additionally employ multiple-input and multiple-output (MIMO) techniques, e.g., the Alamouti space-time block code (STBC) [4].

Coherent OFDM systems require channel state information (CSI) at the receiver. Thus, pilot symbols are often periodically inserted into the transmitted signal to support channel estimation (CE). CE is performed by interpolating the time-frequency pilot grid and exploiting the correlations of the received OFDM signal [2], which are introduced by the time-

and frequency-selective radio channel. Exploiting these correlations, the pilot aided channel estimation (PACE) is performed by cascading two one-dimensional (1-D) finite-impulse response filters whose coefficients are based on the minimum mean square error (MMSE) criterion [2].

To further improve PACE, [5], [6], and [7] use previously decided data symbols as reference in iterative channel estimation (ICE). In [7], the authors propose an ICE algorithm for OFDM that feeds back information from the output of the channel decoder to the estimation stage to reduce decision feed-back errors. Since the CE gets additional information from the estimated data symbols, ICE achieves a further reduction of the bit error rate (BER).

In this paper, we extend the idea of ICE to a MIMO MC-CDMA system with Walsh-Hadamard spreading codes. In [8], we proposed the modified LS (MLS) and MMSE CE methods to solve the problem of zero-valued subcarriers in ICE. When using multiple transmit antennas, the transmitted data signals coincide non-orthogonal at each receive antenna. Thus, we extend the MLS and MMSE CE methods to cope with the non-orthogonal superposition of data symbols in ICE. We further examine the robustness of an MC-CDMA receiver with ICE if the channel statistics are not perfectly known.

2. System Model

Figure 1 represents the block diagram of the downlink MIMO MC-CDMA system with ICE. At the transmitter, a binary signal of a single user out of K active users is encoded by a channel coder and interleaved by a random code-bit interleaver. The bits b_k (Figure 1(b)) are modulated and serial-to-parallel converted to M data symbols per user in an OFDM symbol. After rotating [9] and spreading with a Walsh-Hadamard sequence of length L ($L \geq K$), the spread signals are combined, serial-to-parallel converted, and interleaved by a frequency interleaver to form the data symbol frame $S_{n,l}$ ($\{n,l\} \in \mathcal{D}$). Here, n denotes the subcarrier index, l the OFDM symbol number, and $\mathcal{D}$ the set of data symbol positions in a frame. Next, $S_{n,l}$ (Figure 1(a)) is space-time coded and each $S_{n,l}^m$ ($\{n,l\} \in \mathcal{D}^m$, $m = 1, \ldots, N_{\mathrm{TX}}$) is multiplexed together with pilot symbols $S_{n',l'}^m$, $\{n',l'\} \in \mathcal{P}^m$, $m = 1, \ldots, N_{\mathrm{TX}}$. $\mathcal{P}^m$ denotes the set of pilot symbol positions in an OFDM frame at transmit antenna m. The pilot symbols are inserted on a rectangular grid with N_L subcarriers and N_K symbols appart in frequency and time direction. For each transmit antenna, they are placed on disjoint positions ($\mathcal{P}^{m_1} \cap \mathcal{P}^{m_2} = \emptyset \ \forall m_1 \neq m_2$). The set $\bar{\mathcal{D}}$ denotes the union of all sets $\mathcal{D}^m$ ($m = 1, \ldots, N_{\mathrm{TX}}$). Here, we assume $\mathcal{D}^1 = \ldots = \mathcal{D}^{N_{\mathrm{TX}}} = \bar{\mathcal{D}}$ and

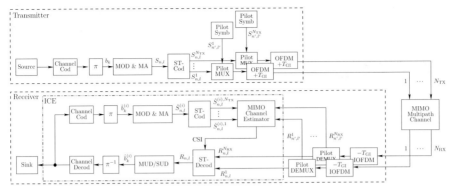

(a) MIMO MC-CDMA: transmitter and receiver with ICE

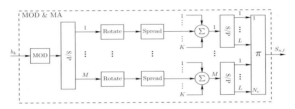

(b) Modulation and multiple access: MOD & MA

Figure 1. Downlink MIMO MC-CDMA system with ICE

$\mathcal{P}^m \cap \bar{\mathcal{D}} = \emptyset \; \forall m$, i.e., the pilot and data sets are disjoint. The resulting N_{TX} OFDM frames with N_{C} subcarriers and N_{S} OFDM symbols are OFDM modulated and cyclically extended by the guard interval (GI) before they are transmitted over a time-variant MIMO multipath channel, which adds white Gaussian noise.

The received symbols are shortened by the GI and OFDM demodulated for each of the N_{RX} receive antennas. Then, the received pilot symbols $R_{n',l'}^{m,p}$ ($\{n', l'\} \in \mathcal{P}^m$, $m = 1, \ldots, N_{\text{TX}}$, $p = 1, \ldots, N_{\text{RX}}$) are separated from the received data symbols $R_{n,l}^p$ ($\{n, l\} \in \bar{\mathcal{D}}$), and fed into the CE. In the initial ICE iteration ($i = 0$), the CE only uses pilot symbols to estimate the CSI.

After demultiplexing and space-time decoding the received data symbols $R_{n,l}^p$ ($\{n, l\} \in \bar{\mathcal{D}}$), either one multi-user detector (MUD) or K single-user detector (SUD) blocks return soft-coded bits $\hat{b}_k^{(i)}$ of all users [8]. Subsequently, the bits are used to reconstruct the transmit signal $\check{S}_{n,l}^{(i),m}$, $m = 1, \ldots, N_{\text{TX}}$, in ICE (Figure 1(a)).

In the ith iteration of ICE ($i > 0$), the CE exploits both the received pilot symbols $R_{n',l'}^{m,p}$ ($\{n', l'\} \in \mathcal{P}^m$) and the reconstructed transmit signal $\check{S}_{n,l}^{(i),m}$ ($\{n, l\} \in \bar{\mathcal{D}}$, $m = 1, \ldots, N_{\text{TX}}$) to improve the accuracy of the

CSI estimates. The newly obtained CSI estimates are fed back to the MUD/SUD block to improve the estimates of the transmitted bits.

3. Pilot Aided and Iterative Channel Estimation

This section investigates how ICE can further improve PACE in the downlink if decided data symbols are used as additional pilot symbols.

3.1 Pilot Aided Channel Estimation (PACE)

Since the pilot symbols for different transmit antennas are transmitted on disjoint positions and we assume independent and identically distributed (i.i.d.) subchannels, the MIMO channel estimation reduces to estimating $N_{\text{TX}} \cdot N_{\text{RX}}$ SISO channels.

In the initial iteration of ICE, only received pilot symbols are used to obtain the CSI in two steps [2]:

1. The initial estimate $\check{H}^{m,p}_{n',l'}$ of the channel transfer function between transmit antenna m and receive antenna p at pilot symbol positions is obtained by dividing the received pilot symbol $R^{m,p}_{n',l'}$ by the originally transmitted pilot symbol $S^{m}_{n',l'}$, i.e.,

$$\check{H}^{m,p}_{n',l'} = \frac{R^{m,p}_{n',l'}}{S^{m}_{n',l'}} = H^{m,p}_{n',l'} + \frac{Z^{p}_{n',l'}}{S^{m}_{n',l'}}, \quad \forall \{n',l'\} \in \mathcal{P}^m, \qquad (1)$$

   where $Z^{p}_{n',l'}$ denotes the additive white Gaussian noise component at the pth receive antenna.

2. The final estimates of the channel transfer function are obtained from the initial estimates $\check{H}^{m,p}_{n',l'}$ by two-dimensional (2-D) filtering:

$$\hat{H}^{m,p}_{n,l} = \sum_{\{n',l'\} \in \mathcal{T}^{m}_{n,l}} \omega^{m}_{n',l'} \check{H}^{m,p}_{n',l'}, \quad \mathcal{T}^{m}_{n,l} \subseteq \mathcal{P}^m, \; \forall \{n,l\} \in \bar{\mathcal{D}}, \qquad (2)$$

   where $\omega^{m}_{n',l'}$ is the 2-D filter impulse response. The subset $\mathcal{T}^{m}_{n,l} \subseteq \mathcal{P}^m$ is the set of initial estimates $\check{H}^{m,p}_{n',l'}$ used to obtain $\hat{H}^{m,p}_{n,l}$.

The optimum solution of (2) in the MSE sense is the 2-D Wiener filter [2]. Assuming the delay and Doppler power spectral densities (PSDs) to be statistically independent, the 2-D filter can be replaced by two cascaded 1-D filters, one for filtering in frequency and one for filtering in time direction. Since in practice the channel statistics are not perfectly known at the receiver, the CE filters are designed robust using a uniform delay PSD ranging from 0 to τ_{filter} and a uniform Doppler PSD ranging from $-f_{\text{D}_{\text{filter}}}$ to $f_{\text{D}_{\text{filter}}}$. Note, the maximum delay τ_{max} and Doppler frequency $f_{D_{\max}}$ of the channel can be different from the maximum delay τ_{filter} and Doppler frequency $f_{D_{\text{filter}}}$ of the CE filters.

3.2 Least-Squares Estimation in Iterative Channel Estimation

In the following the LS estimate in (1) is investigated in detail for ICE. [9] demonstrates that due to the superposition of Walsh-Hadamard spread data signals zero-valued subcarriers can occur in the transmit signal. For instance, a zero-valued subcarrier occurs with 27% probability for a binary phase shift keying (PSK) symbol alphabet and a Walsh-Hadamard spreading code of length 8 [8]. Consequently, the LS estimates in (1) can only be computed for some subcarriers and a method must be found to avoid a division by zero in (1) and noise enhancement.

Therefore, we propose the following MIMO MLS CE method: If the magnitude of the reconstructed subcarrier is zero, equal to, or below the threshold $\rho_{\text{th}} \geq 0 \in \mathbb{R}$, the initial estimate $\check{H}_{n,l}^{(i),m,p}$ is set to zero, i.e.,

$$\check{H}_{n,l}^{(i),m,p} = \begin{cases} \frac{R_{n,l}^{(i),m,p}}{\check{S}_{n,l}^{(i),m}} & \{n,l\} \in \bar{\mathcal{D}} \text{ and } |\check{S}_{n,l}^{(i),m}| > \rho_{\text{th}}, \\ 0 & \{n,l\} \in \bar{\mathcal{D}} \text{ and } |\check{S}_{n,l}^{(i),m}| \leq \rho_{\text{th}}, \end{cases} \quad (3)$$

where $R_{n,l}^{(i),m,p}$ is the received signal only comprised of the mth transmit signal $S_{n,l}^m$. In Section 3.3, we explain how to obtain $R_{n,l}^{(i),m,p}$. The threshold ρ_{th} should be chosen so that noise enhancement caused by subcarriers with small amplitudes is avoided and only a few possible constellation points of a subcarrier are below the threshold [8].

Alternative, we propose the MMSE MIMO CE method:

$$\check{H}_{n,l}^{(i),m,p} = \frac{\left(\check{S}_{n,l}^{(i),m}\right)^* R_{n,l}^{m,p}}{\left|\check{S}_{n,l}^{(i),m}\right|^2 + \frac{1}{\gamma_c}} \quad \{n,l\} \in \bar{\mathcal{D}}, \quad (4)$$

where γ_c is the actual SNR per subcarrier. In contrast to (3), it is not necessary to choose a threshold, which requires a trade-off between additional subcarriers used by ICE and noise enhancement due to subcarriers with small amplitudes. However, the MMSE estimate makes it necessary to compute a different set of Wiener filter coefficients taking into account the initial estimates from (4). To avoid a complexity increase [8], we use the same robust Wiener filter coefficients as for the MLS method. With (3) and (4), the initial estimates will not cause noise enhancement and degrade the channel estimates in any iteration of ICE.

3.3 Extension of Iterative Channel Estimation to MIMO MC-CDMA system

Since the pilot symbols for different transmit antennas are transmitted on disjoint positions and assuming i.i.d. subchannels, PACE for an MIMO MC-CDMA system is a straight forward extension of PACE for a SISO MC-CDMA system. In ICE, however, the data symbols from different transmit antennas are not placed on disjoint positions or generally orthogonal, which results in the following superposition

$$R_{n,l}^p = \sum_{r=1}^{N_{\text{TX}}} H_{n,l}^{r,p} S_{n,l}^r, \quad \{n,l\} \in \bar{\mathcal{D}}. \tag{5}$$

Hence, (3) cannot be directly applied when using the reconstructed transmit signal $\check{S}_{n,l}^{(i),m}$.

To obtain an interference reduced received signal $R_{n,l}^{(i),m,p}$ for (3), we subtract the estimated transmit signals except for the mth antenna, i.e.,

$$R_{n,l}^{(i),m,p} = R_{n,l}^p - \sum_{\substack{r=1 \\ r \neq m}}^{N_{\text{TX}}} \hat{H}_{n,l}^{(i-1),r,p} \check{S}_{n,l}^{(i),r}, \quad \{n,l\} \in \bar{\mathcal{D}}. \tag{6}$$

Thus, we initially estimate the CSI between transmit antenna m and receive antenna p by canceling the current estimates of the received signals from the other transmit antennas and applying then (3) or (4).

3.4 Iterative Channel Estimation (ICE)

After reconstructing the transmitted signal from the estimated information bits, the estimated data symbols and the transmitted pilot symbols form the set $\mathcal{P}_{\text{ICE}}^m = \mathcal{P}^m \cup \bar{\mathcal{D}}$ of reference symbols known at the receiver. The set $\mathcal{P}_{\text{ICE}}^m$ defines the complete frame of pilot and data symbols from transmit antenna m.

For ICE with one or more iterations ($i > 0$), the following steps have to be executed in each iteration:

1. Reconstruct the transmit signal $\check{S}_{n,l}^{(i),m}$ $\forall \{n,l\} \in \bar{\mathcal{D}}$ and $m = 1, \ldots, N_{\text{TX}}$ from the estimated information bits.

2. Compute the interference reduced received signal $R_{n,l}^{(i),m,p}$ for transmit antenna m and receive antenna p in (6).

3. Calculate the initial estimates $\check{H}_{n,l}^{(i),m,p}$ $\forall \{n,l\} \in \bar{\mathcal{D}}$ according to (3) or (4).

4. Obtain the final estimate of the channel transfer function between transmit antenna m and receive antenna p through filtering the initial estimates over the set $\mathcal{P}_{\text{ICE}}^m$ of all reference symbols, i.e.,

$$\hat{H}_{n,l}^{(i),m,p} = \sum_{\{n',l'\}\in\mathcal{T}_{n,l}^m} \omega_{n',l'}^{(i),m} \check{H}_{n',l'}^{(i),m,p}, \quad \mathcal{T}_{n,l}^m \subseteq \mathcal{P}_{\text{ICE}}^m, \quad \forall \{n,l\} \in \bar{\mathcal{D}}, \quad (7)$$

where $\omega_{n',l'}^{(i),m}$ is the Wiener filter coefficient obtained for the ith iteration.

5. Use the newly estimated CSI $\hat{H}_{n,l}^{(i),m,p}$ from (7) in the subsequent space-time decoder, MUD/SUD block, and decoder to obtain a new estimate of the transmitted information bits.

4. Simulation Results

This section presents simulation results for a downlink MIMO MC-CDMA system applying Walsh-Hadamard spreading and the Alamouti STBC [4] with perfect CE, PACE, and ICE. At a carrier frequency of $f_c = 5$ GHz, the MC-CDMA systems transmits 64 OFDM symbols per frame divided into 768 subcarriers over a bandwidth of 101.25 MHz resulting in a subcarrier spacing of 131.836 kHz. The length of the fast Fourier transforms in the OFDM modulation and demodulation is 1024. Thus, the sampling duration T_{spl} is 7.4 ns. The guard interval T_{GI} is set to $226\,T_{\text{spl}}$. The system uses a $R = 1/2$ convolutional code with generator polynomial $(133,171)_8$, QPSK symbols with Gray mapping, and Walsh-Hadamard spreading codes of length $L = 8$. Except for the single-user bound (SUB), $K = 8$ users transmit in parallel $M = 96$ data symbols per OFDM symbol that are randomly frequency interleaved. Pilot symbols are inserted on a rectangular grid with a frequency distance $N_{\text{L}} = 3$ or 4 and time distance $N_{\text{K}} = 9$ or 12.

The channel is modeled as a tapped delay line with exponential decaying PSD. The non-zero tap spacing $\Delta\tau$ of the PSD is $16\,T_{\text{spl}}$ and the maximum delay $\tau_{\max}$ is $176\,T_{\text{spl}}$. The channel employs $N_{\text{p}} = 12$ taps with a power decrement $\Delta P = 1$ dB. The Doppler frequencies are distributed according to a Jakes' spectrum and the maximum Doppler frequency $f_{D_{\max}}$ is 1500 Hz ($\approx 0.01\,\Delta f$), which is related to a mobile terminal speed of about 300 km/h at $f_c = 5$ GHz.

For robust PACE and ICE, two 1-D Wiener filters are cascaded with 15 filter coefficients in frequency and 4 in time direction to compute the CSI estimates in (2) and (7). The delay and Doppler PSDs are uniformly distributed with a maximum delay $\tau_{\text{filter}} = T_{\text{GI}}$ and Doppler frequency $f_{D_{\text{filter}}} = 1500$ Hz. In case of perfect CE, no pilot symbols are transmitted. Thus, the E_b/N_0 loss for transmitting pilot symbols for

PACE or ICE is 0.1 dB or 0.18 dB in a SISO system with $N_\mathrm{L} = 4$ and $N_\mathrm{K} = 12$ or $N_\mathrm{L} = 3$ and $N_\mathrm{K} = 9$. When employing Alamouti's STBC, the E_b/N_0 losses of the two pilot spacings increase slightly to 0.19 dB and 0.11 dB. In ICE and the parallel interference canceler (PIC) MUD, one iteration is computed. Simulation results indicate that more iterations only yield little improvements and hence, are omitted for clarity.

In Figure 2(a), we examine the influence of ρ_th in ICE with MMSE SUD for rotated and non-rotated Walsh-Hadamard spreading [9]. The pilot spacing and the robust PACE are optimized for this channel model to achieve the best BER perfromance. Hence, the results for the proposed ICE are very promising since it still gains up to 1 dB for MMSE SUD compared to PACE at a BER of 10^{-4} and reduces the loss to perfect CE to 1 dB. Note, the curves for $\rho_\mathrm{th} = 0.35$ for rotated and $\rho_\mathrm{th} = 0.4$ for non-rotated spreading with MLS CE as well as for MMSE-CE in ICE are omitted for clarity as they coincide with the curve for robust ICE with MLS CE and $\rho_\mathrm{th} = 0$. In the case of non-rotated spreading, we set ρ_th to 0, 0.4, 0.5 or 0.8 ignoring 8% in the first two cases and 32% or 50% in the latter two. For $\rho_\mathrm{th} < 0.5$, only zero-valued subcarriers are ignored in ICE. Thus, the curves for $\rho_\mathrm{th} = 0.4$ and $\rho_\mathrm{th} = 0$ coincide and yield the best BER performances as ICE exploits all information from data chips that cause no noise enhancement. In the case of rotated spreading (curves with solid markers), no zero-valued subcarriers occur in the constellation points. However, noise enhancement of small amplitude values decreases the performance. Increasing ρ_th from 0 over 0.2 to 0.35 significantly reduces noise enhancement in the CE. On the other hand, the MLS-CE ignores 0%, 4%, or 11% of the possible constellation points. Here, ICE with MLS CE and $\rho_\mathrm{th} = 0.35$ reaches the lowest BER and thus, the best trade-off between noise enhancement and exploiting all the information from data chips that cause no noise enhancement. Consequently, the optimal choice of the threshold depends on the minimum amplitude value that can occur in the subcarrier constellation.

In Figure 2(b), the Alamouti scheme is applied in time direction over two consecutive OFDM symbols. The coherence time of the channel is about 120 μs. So, the channel is approximately constant over two consecutive OFDM symbols, which is a necessary prerequisite for the Alamouti scheme. Additionally, the receiver employs two antennas, whose receive signals are maximum ratio combined. The Alamouti scheme and two receive antennas improve the performance of the SISO MC-CDMA system by 4.3 dB and 6.4 dB for SUB and MMSE SUD with perfect CE at a BER of 10^{-4}. These gains are mainly due to the increased receive power and the higher diversity. Additionally, the maximum ratio combining of the STBC and two receive antennas reduces the multiple access interfer-

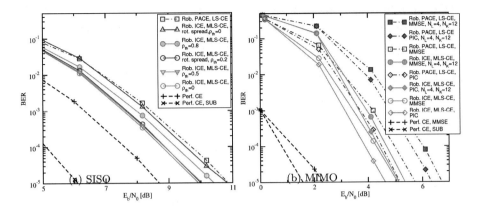

Figure 2. Comparison of perfect CE, robust PACE, and robust ICE for (a) MMSE SUD and fully-loaded SISO MC-CDMA system varying the threshold $\rho_{\rm th}$ or (b) MMSE SUD, PIC MUD, and fully-loaded MIMO MC-CDMA system employing Alamouti STBC and two receive antennas and varying the pilot spacings $N_{\rm L}$, $N_{\rm K}$

ence (MAI) in MMSE SUD. In the MIMO system all pilot distances for different transmit antennas are the same as those in the SISO system. So, in total, the MIMO system employs twice as much pilots as the SISO system. To keep the total transmitted pilot symbol energy constant, the pilot symbol energy for each transmit antenna is scaled by $1/N_{\rm TX}$ allowing a fair comparison between the SISO and MIMO channel estimation. This explains the reduced performance gains for MMSE SUD and robust PACE or ICE, which are only 4.8 dB or 4.5 dB at a BER of 10^{-4}. The differences between PACE and ICE for MMSE SUD reduces to 0.6 dB in the MIMO system as the interference cancellation in (6) is not perfect. When $N_{\rm L} = 4$ and $N_{\rm K} = 12$, the gain for ICE versus PACE is increased to 1.4 dB because ICE additionally uses $\check{S}_{n,l}^{(1),m}$ as reference symbols. Hence, ICE is less susceptible to interpolation errors because of larger pilot spacings. The BER curves of the PIC MUD demonstrate similar performances and gain 0.6 dB or 1.2 dB with ICE versus PACE and $N_{\rm L} = 3$ and $N_{\rm K} = 9$ or $N_{\rm L} = 4$ and $N_{\rm K} = 12$. In the MIMO system, the PIC MUD only improves by 0.4 dB and 0.3 dB compared to the MMSE SUD as MAI is already reduced by the MIMO algorithms.

5. Summary and Conclusion

In this paper, we present a downlink MIMO MC-CDMA system with Walsh-Hadamard spreading codes applying ICE at the receiver. Since the superposition of Walsh-Hadamard spread data signals can result in zero-valued subcarriers, LS channel estimates within ICE cannot be computed. Further, data signals from different transmit antennas coincide non-orthogonal at the receiver. Thus, we propose the MLS and MMSE MIMO channel estimation methods to resolve these problems and avoid noise enhancement. Simulation results indicate that a SISO MC-CDMA receiver with robust ICE and the MLS or MMSE method can outperform robust PACE by 1 dB for scenarios that were optimized for robust PACE. Applying Alamouti's scheme and two receive antennas, robust ICE and the MLS method gain up to 1.4 dB SNR versus robust PACE if the total pilot energy is constant in SISO and MIMO systems.

To conclude, the MLS and MMSE channel estimation can be applied with ICE to any other MC-CDMA, multi-carrier, CDMA, or spread spectrum system with general spreading codes or symbol alphabet whose superposition or constellation yields values below a certain threshold.

Acknowledgment

The here presented work was supported by the European IST projects 4MORE and WINNER [http://ist-4more.org/ and http://ist-winner.org].

References

[1] J. Bingham. Multicarrier modulation for data transmission: An idea whose time has come. *IEEE Commun. Mag.*, May 1991, pp. 5–14.

[2] K. Fazel and S. Kaiser. *Multi-Carrier and Spread Spectrum Systems*. John Wiley and Sons, 2003.

[3] Y. Kishiyama, et al. Experiments on throughput performance above 100-Mbps in forward link for VSF-OFDCDM broadband packet wireless access. In *Proc. of IEEE VTC'F03*. Orlando, USA, Oct. 2003, pp. 1863–1868.

[4] S. M. Alamouti. A simple transmit diversity technique for wireless communications. *IEEE J. Select. Areas Commun.*, Oct. 1998, pp. 1451–1458.

[5] V. Mignone and A. Morello. CD3-OFDM: A novel demodulation scheme for fixed and mobile receivers. *IEEE Trans. Commun.*, Sept. 1996, pp. 1144–1151.

[6] D. Kalofonos, et al. Performance of adaptive MC-CDMA detectors in rapidly fading rayleigh channels. *IEEE Trans. Wireless Commun.*, Mar. 2003, pp. 229–239.

[7] F. Sanzi, et al. A comparative study of iterative channel estimators for mobile OFDM systems. *IEEE Trans. Wireless Commun.*, Sept. 2003, pp. 849–859.

[8] S. Sand, et al. Iterative channel estimation for MC-CDMA. In *Proc. of IEEE VTC'S05*. Stockholm, Sweden, May 2005.

[9] A. Bury, et al. Diversity comparison of spreading transforms for multicarrier spread spectrum transmission. *IEEE Trans. Commun.*, May 2003, pp. 774–781.

PERFORMANCE INVESTIGATION OF IMPROVED CHANNEL ESTIMATION EXPLOITING LONG TERM CHANNEL PROPERTIES

Tobias Weber, Ioannis Maniatis and Michael Meurer
University of Kaiserslautern
Research Group for RF Communication
P.O. Box 3049, 67653 Kaiserslautern, Germany
{tweber,maniatis,meurer}@uni-kl.de

Wolfgang Zirwas
Siemens AG
Information & Communication Mobile Networks
Sankt-Martinstrasse 76, 81541 Munich, Germany

Abstract   MIMO transmission based on spatial multiplexing has been shown to offer great performance improvements and, therefore, will be included in future mobile radio systems. Unfortunately, the number of channel coefficients grows quadratically with the number of antennas on the transmitter and receiver sides whereas the channel capacity in the best case grows only linearly with the number of antennas [1]. Consequently, the performance improvement achievable by MIMO transmission will in practice be limited by the requirement to estimate the channel coefficients, which requires significant overhead for training signal transmission especially in high mobility scenarios. The present paper introduces a new subspace based channel estimation technique which can partially overcome this problem by exploiting long term channel properties to improve the snapshot based channel estimates.

1.   Introduction

State of the art channel estimation is based on the transmission of a priori known training signals, sometimes also referred to as pilots. In this case the received signal resulting from the training signal transmission is a known linear function of the unknown channel coefficients. These channel coefficients may be either thought of as samples of the channel impulse responses in the time domain, samples of the channel transfer functions in the frequency domain, or, more generally, as coefficients describing the channel properties with respect to a chosen set of basis functions. There is nothing special about the MIMO case except that in this case the channel coefficients stem from different physical channels.

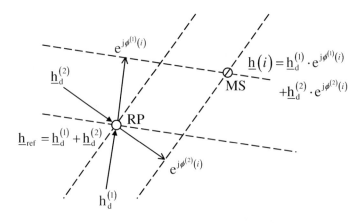

Figure 1. Directional channels

In the following the received vector $\underline{e}(i)$ describes the received signal resulting from the i-th training signal transmission with respect to a set of basis functions, e.g., with respect to time shifted sinc-functions resulting in a vector of samples in the time domain. Using the system matrix $\underline{G}$ and the Gaussian noise vector $\underline{n}(i)$, the received vector $\underline{e}(i)$ can be written as a linear function of the vector $\underline{h}(i)$ of length L of channel coefficients valid at time instant i as follows:

$$\underline{e}(i) = \underline{G} \cdot \underline{h}(i) + \underline{n}(i). \qquad (1)$$

For the sake of simplicity, white Gaussian noise is assumed in the following, i.e., the covariance matrix reads

$$\underline{R}_n = E\left\{\underline{n}(i)\underline{n}(i)^H\right\} = \sigma^2 \underline{I}. \qquad (2)$$

We will focus our investigations on the estimation of the channel coefficients of one of the SISO subchannels of the MIMO channel. The results can be easily extended to MIMO channel estimation, which is mathematically the same as SISO channel estimation.

2. Channel subspaces

Typically, the wavefronts impinge at the moving mobile station (MS) from rather few directions of arrival. Fig. 1 shows the most simple case of two directions of arrival. The two radio channels corresponding to the two directions of arrival can be described by directional channel impulse responses $\underline{h}_d^{(d)}$, $d = 1, 2$. At the reference point (RP) the channel impulse response is just the sum

$$\underline{\mathbf{h}}_{\text{ref}} = \underline{\mathbf{h}}_d^{(1)} + \underline{\mathbf{h}}_d^{(2)} \quad (3)$$

of the two directional channel impulse responses. At a position not too far away from the reference point, i.e., at a position where the wavefronts are still the same as at the reference point, the channel impulse response is a superposition

$$\underline{\mathbf{h}}(i) = \underline{\mathbf{h}}_d^{(1)} \exp\left(j\phi^{(1)}(i)\right) + \underline{\mathbf{h}}_d^{(2)} \exp\left(j\phi^{(2)}(i)\right) \quad (4)$$

of the two directional channel impulse responses. For the more general case of D directions of arrival the superposition of (4) can be equivalently written as

$$\underline{\mathbf{h}}(i) = \underbrace{\left(\underline{\mathbf{h}}_d^{(1)} \ldots \underline{\mathbf{h}}_d^{(D)}\right)}_{\underline{\mathbf{H}}_d} \cdot \underbrace{\begin{pmatrix} \exp\left(j\phi^{(1)}(i)\right) \\ \vdots \\ \exp\left(j\phi^{(D)}(i)\right) \end{pmatrix}}_{\underline{\mathbf{h}}_d(i)}, \quad (5)$$

i.e., the channel vector $\underline{\mathbf{h}}(i)$ lies in a rather low dimensional subspace spanned by the columns of the $L \times D$ matrix $\underline{\mathbf{H}}_d(i)$. It is important to notice that the matrix $\underline{\mathbf{H}}_d$ made up of the directional channel impulse responses $\underline{\mathbf{h}}^{(d)}_d$, $d = 1 \ldots D$, and representing the subspace does not change if the mobile station only moves in a small area where the wavefronts do not change, whereas the channel impulse response $\underline{\mathbf{h}}(i)$ and the subspace based channel vector $\underline{\mathbf{h}}_d(i)$ change quickly due to fast fading. Substituting (5) into (1) one obtains the subspace based linear system model

$$\underline{\mathbf{e}}(i) = \underline{\mathbf{G}} \cdot \underline{\mathbf{H}}_d(i) \underline{\mathbf{h}}_d(i) + \underline{\mathbf{n}}(i). \quad (6)$$

3. Concept of subspace based channel estimation

The basic idea of subspace based channel estimation is that in a certain snapshot i only the subspace based channel vector $\underline{\mathbf{h}}_d(i)$ needs to be estimated, which typically results in a significant reduction of the number of unknowns to be estimated as compared to the conventional snapshot based channel estimator which directly estimates the channel vector $\underline{\mathbf{h}}(i)$.

Fig. 2 shows the resulting block diagram of a subspace based channel estimator. In a first step a low quality initial snapshot based channel estimate

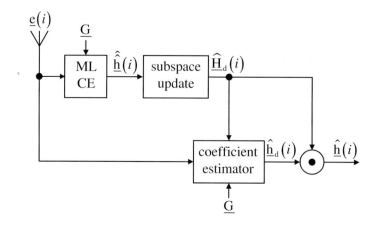

Figure 2. Block diagram of subspace based channel estimator

$$\hat{\underline{h}}(i) = \left(\underline{G}^H \underline{G}\right)^{-1} \underline{G}^H \underline{e}(i) \quad (7)$$

is obtained from the received signal $\underline{e}(i)$ at time instant i by conventional maximum likelihood channel estimation. By observing these initial channel estimates $\hat{\underline{h}}(i)$ over a longer time one can estimate the subspace represented by the matrix $\underline{H}_d(i)$, more precisely the subspace estimate $\hat{\underline{H}}_d(i)$ is updated at each time instant i using the past I initial channel estimates $\hat{\underline{h}}(j)$, $j = i - I + 1 \ldots i$. In the following I is also called the observation duration. This allows a tracking of slowly time variant subspaces. Due to the possibility to observe the channel for a longer time for subspace estimation the quality of the initial channel estimates $\hat{\underline{h}}(i)$ is not that critical. Using the subspace estimate $\hat{\underline{H}}_d(i)$ one can estimate the subspace based channel vector $\hat{\underline{h}}_d(i)$ at each time instant i. Following the maximum likelihood rationale [2],

$$\hat{\underline{h}}_d(i) = \left(\hat{\underline{H}}_d^H(i) \underline{G}^H \underline{G} \hat{\underline{H}}_d(i)\right)^{-1} \cdot \hat{\underline{H}}_d^H(i) \underline{G}^H \underline{e}(i) \quad (8)$$

holds. Finally, the estimate

$$\hat{\underline{h}}(i) = \hat{\underline{H}}_d(i) \cdot \hat{\underline{h}}_d(i) \quad (9)$$

of the channel vector $\underline{h}(i)$ at time instant i is obtained.

4. Subspace estimation

For the covariance matrix of the channel vector $\underline{h}(i)$

$$\begin{aligned}\underline{R}_h(i) &= E\left\{\underline{h}(i)\underline{h}(i)^H\right\} \\ &= \underline{H}_d(i) E\left\{\underline{h}_d(i)\underline{h}_d(i)^H\right\} \underline{H}_d(i)^H \end{aligned} \quad (10)$$

holds, i.e., the subspace spanned by the columns of the matrix $\underline{\mathbf{H}}_d(i)$ is equal to the subspace spanned by the covariance matrix $\underline{\mathbf{R}}_h(i)$. As we are only interested in any set of basis vectors of the subspace it is sufficient to find one orthonormal basis of the subspace spanned by the covariance matrix $\underline{\mathbf{R}}_h(i)$. In practical applications the covariance matrix $\underline{\mathbf{R}}_h(i)$ is not known but has to be estimated based on the initial channel estimates $\hat{\underline{\mathbf{h}}}(i)$ as follows:

$$\begin{aligned}
\underline{\mathbf{R}}_h(i) &\approx \mathrm{E}\left\{\hat{\underline{\mathbf{h}}}(i)\hat{\underline{\mathbf{h}}}(i)^H\right\} \\
&= \mathrm{E}\left\{\underline{\mathbf{h}}(i)\underline{\mathbf{h}}(i)^H\right\} + \frac{1}{\sigma^2}\left(\underline{\mathbf{G}}^H\underline{\mathbf{G}}\right)^{-1} \\
&\approx \frac{1}{I}\sum_{j=i-I+1}^{i}\hat{\underline{\mathbf{h}}}(j)\hat{\underline{\mathbf{h}}}(j)^H.
\end{aligned} \quad (11)$$

One can clearly see that the noise $\underline{\mathbf{n}}(i)$ is not really disturbing the subspace estimation as its influence can be eliminated, provided that one can average over a sufficiently large number I of initial channel estimates $\hat{\underline{\mathbf{h}}}(i)$.

Due to the time variance of the subspace and the disturbances in the initial channel estimates $\hat{\underline{\mathbf{h}}}(i)$ resulting from the noise $\underline{\mathbf{n}}(i)$ the estimated covariance matrix

$$\hat{\underline{\mathbf{R}}}_h(i) = \frac{1}{I}\sum_{j=i-I+1}^{i}\hat{\underline{\mathbf{h}}}(i)\hat{\underline{\mathbf{h}}}(j)^H \quad (12)$$

always has rank min (I, L) in real world scenarios. Concerning subspace estimation it is crucial

- to choose the observation duration I properly to allow sufficient averaging of the noise influence but also to avoid averaging over a significantly changing subspace, and

- to consider only the significant part of the subspace.

The second point needs some further clarification. To simplify the following basic considerations it shall be assumed that the influence of noise $\underline{\mathbf{n}}(i)$ can be neglected, e.g., due to perfect averaging and subtracting its contribution $\frac{1}{\sigma^2}(\underline{\mathbf{G}}^H\underline{\mathbf{G}})^{-1}$ when estimating the covariance matrix $\underline{\mathbf{R}}_h(i)$, see (11). In this case

$$\hat{\underline{R}}_h(i) = \frac{1}{I} \sum_{j=i-I+1}^{i} \underline{h}(j)\underline{h}(j)^H \quad (13)$$

holds. A suitable method for determining a orthonormal matrix $\hat{\underline{H}}_d(i)$ consists in eigenvalue decomposition of the hermitian covariance matrix

$$\hat{\underline{R}}_h(i) = \hat{\underline{U}}(i) \cdot \begin{pmatrix} \hat{\lambda}^{(1)}(i) & & & 0 \\ & \ddots & & \\ & & \hat{\lambda}^{(\min(I,L))}(i) & \\ 0 & & & 0 \end{pmatrix} \cdot \hat{\underline{U}}(i)^H . \quad (14)$$

The columns of $\hat{\underline{U}}(i)$ shall be ordered in such a way that

$$\hat{\lambda}^{(l)} \geq \hat{\lambda}^{(l+1)}, l = 1 \ldots \min(I,L) - 1, \quad (15)$$

holds for the eigenvalues, i.e., that the eigenvalues are ordered in a decreasing way. A first solution would be to choose the columns of the matrix $\hat{\underline{H}}_d(i)$ equal to the first min (I, L) columns of $\hat{\underline{U}}(i)$ corresponding to nonzero eigenvalues $\hat{\lambda}^{(l)}$. This will result in unbiased subspace based channel estimates $\hat{\underline{h}}(i)$. However, the performance improvement due to noise suppression by subspace based channel estimation will be typically rather small due to the large min (I, L). Considering only the first

$$S < \min(I, L) \quad (16)$$

columns and choosing the columns of the matrix $\hat{\underline{H}}_d(i)$ equal to the first S columns of $\hat{\underline{U}}(i)$ may improve the potential of noise suppression significantly. In this case, in general biased subspace based channel estimates $\hat{\underline{h}}(i)$ will be obtained. The average energy of the systematic channel estimation error due to the bias in the noise free case reads

$$E\left\{\left\|\hat{\underline{h}}(i) - \underline{h}(i)\right\|^2\right\} = \sum_{l=S+1}^{\min(I,L)} \hat{\lambda}^{(l)}, \quad (17)$$

i.e., the error will be small if the non considered eigenvalues $\hat{\lambda}^{(l)}$, $l = S + 1 \ldots \min(I, L)$, are small. One has to find a compromise among noise suppression and bias.

It should be kept in mind that the noise in the case of perfect averaging only leads to an additive term $\sigma^2 \mathbf{I}$ in the estimated covariance matrix $\hat{\mathbf{R}}_h(i)$ in the case of white noise (2) and optimum training signals with

$$\underline{\mathbf{G}}^H \underline{\mathbf{G}} = \mathbf{I}. \tag{18}$$

This additive term needs not to be subtracted as it does not influence the eigenvectors and shifts the eigenvalues only by an additive constant σ^2. In the end one should only consider the eigenvectors corresponding to eigenvalues which are larger than a certain threshold, which will be larger than σ^2.

5. Performance investigations

In the following a UMTS like scenario characterized by

- chip duration 244ns,
- channel impulse response length $L = 57$,
- D directions of arrival per transmitter,
- constant envelope low SNR-degradation training signals of length 456 obtained by the method described in [3], and
- random directional channel impulse responses $\underline{\mathbf{h}}_d^{(d)}$, $d = 1 \ldots D$, with power delay spectrum according to the COST 207 typical urban model [4]

is investigated.

Fig. 3 depicts the average mean square error $\mathrm{E}\left\{\left\|\underline{\hat{\mathbf{h}}}(i) - \underline{\mathbf{h}}(i)\right\|^2\right\}$ of the estimated channel impulse response as a function of the number D of directions of arrival for a noise variance

$$\sigma^2 = 1 \tag{19}$$

and different numbers S of considered largest eigenvalues when determining $\hat{\mathbf{H}}_d(i)$. Perfect averaging and time invariant subspace are assumed. One can clearly see the great performance gains achieved by the novel subspace based channel estimation especially for small numbers D of directions of arrival. In scenarios with large numbers D of directions of arrival considering not the full subspace, i.e., by choosing

$$S = \frac{D}{2} \tag{20}$$

improves the performance as the performance gain due to the better noise suppression in the small subspace is larger than the performance loss due to the bias of the channel estimates.

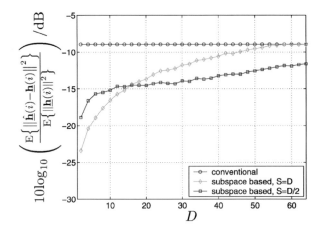

Figure 3. Performance comparison of subspace based channel estimation and conventional channel estimation, $\mathrm{E}\{\|\underline{\mathbf{h}}(i)\|^2\}=1, \sigma^2=1,$ perfect averaging

6. Conclusions

The paper presents the principles of a novel subspace based channel estimation technique which can improve the performance of channel estimation significantly as compared to conventional maximum likelihood channel estimation. The performance gains stem from the exploitation of long term channel properties in form of the subspace in which the channel vector lies. In contrast to minimum mean square error estimation the novel subspace based channel estimation technique can deliver unbiased estimates.

References

[1] Michael Meurer and Tobias Weber. Imperfect channel knowledge: An insurmountable barrier in Rx oriented multi-user MIMO transmission? In *Proc. 5th ITG Conference on Source and Channel Coding (SCC'04)*, pages 371–379, Erlangen, January 2004.

[2] Anthony D. Whalen. *Detection of Signals in Noise*. Academic Press, New York, 1971.

[3] Bernd Steiner and Paul Walter Baier. Low cost channel estimation in the uplink receiver of CDMA mobile radio systems. *Frequenz*, 47(11–12):292–298, 1993.

[4] COST 207: Digital land mobile radio communications. Final report, Office for Official Publications of the European Communities, Luxemburg, 1989.

PILOT SYMBOL-AIDED CHANNEL ESTIMATION FOR MC-CDMA SYSTEMS

Yusung Lee, Yeon-soo Kang, and Hyuncheol Park
Information and Communications University (ICU)
119 Munjiro, Yuseong-gu, Daejeon, 305-732, Korea
{diotima, yskang, hpark}@icu.ac.kr

Minseok Noh
Mobile Communication Technology Research Lab, LG Electronics Inc.
533 Hogye-1dong, Dongan-gu, Anyang-shi, Kyongki-do, 431-749, Korea

Abstract   In this paper, we investigate and propose channel estimation techniques for multi-carrier code division multiple access (MC-CDMA) system based on pilot arrangement. A low complexity linear minimum mean square error (LMMSE) channel estimation for pilot frequency is proposed. By partitioning channel autocorrelation matrix in LMMSE estimator into small sub-matrices whose size is determined by channel coherence bandwidth, the complexity is significantly reduced in spite of slight performance degradation. We also propose an improved discrete Fourier transform (DFT)-based interpolation for data subcarriers. It can achieve good estimation accuracy by ignoring non-significant channel taps which contain more noise than channel power, in which significant taps are detected by using threshold related to noise power. The proposed algorithms are verified by computer simulation.

1. Introduction

Multi-carrier code division multiple access (MC-CDMA) is a multiplexing technique that combines orthogonal frequency division multiplexing (OFDM) with CDMA [1]. Due to its advantages such as high spectral efficiency and robustness to frequency selectivity, MC-CDMA has been considered as one of the most promising candidates for mobile radio communications. The channel estimation technique is required before the demodulation of MC-CDMA signals for reliable data detection. For wideband mobile communication systems, the radio channel is usually frequency selective and time-variant. Channel estimation should be able to track these channel characteristics rapidly. Pilot symbol-aided channel estimation technique could be a good solution [2].

Pilot symbol-aided channel estimation can be achieved by two stages in general. A receiver first estimates the channel response at pilot locations and then, by using this information, obtains channel responses at data locations through interpolation. For the channel estimation at pilot locations, the least square (LS) or the linear minimum mean square error (LMMSE) estimate can be used [3]. The estimates at data locations can be performed by linear, polynomial, cubic spline or discrete Fourier transform (DFT)-based interpolation (or time domain interpolation) schemes with respect to complexity and performance [2] [4].

In this paper, we propose channel estimation techniques for MC-CDMA system based on pilot arrangement. To reduce the complexity of LMMSE estimator, we propose low complexity LMMSE channel estimates. By partitioning channel autocorrelation matrix in LMMSE estimator into small sub-matrices, the complexity is significantly reduced. We also propose an improved DFT-based interpolation.

This paper is organized as follows. The system model and channel model are described in Section 2. In Section 3, we introduce the proposed low complexity LMMSE channel estimators for channel estimation at pilot locations and modified DFT-based channel interpolation. In Section 4, we show our simulation results. Finally, we conclude the paper in Section 5.

2. System Model

We consider a downlink MC-CDMA system in which the total number of subcarriers is split into the spreading code length, SF [5]. At the transmitter, each user transmits simultaneously M data symbols per OFDM symbol and each individual data symbol is spread by frequency spreading code, $\mathbf{c}^{(k)}$, for which Walsh-Hadamard code is used. For obtaining channel response, we assume that N_p pilots are inserted in the OFDM block at known locations.

At the receiver, the incoming signal is first sampled with period T_s and the cyclic prefix is removed. The DFT output of m-th block, $\mathbf{Y}_m = [Y_m(1), Y_m(2), \cdots, Y_m(SF)]^T$, can be represented as:

$$\mathbf{Y}_m = \mathbf{X}_m \mathbf{H}_m + \mathbf{n}_m, \tag{1}$$

where $\mathbf{n}_m$ is a white Gaussian noise with zero mean and variance σ_n^2 and $\mathbf{H}_m$ is the channel frequency response over each subcarrier of the m-th block in which channel response, $H_m(k)$, over the k-th subcarrier of the

m-th block:

$$H_m(k) = \sum_{l=1}^{L_p} h_m(l) e^{-j2\pi(l-1)i_k/N}, \qquad (2)$$

where L_p represents the maximum channel length sampled by T_s, $h_m(l)$ is the complex fading envelope of the l-th path, and i_k, $\{i_k : 1 \leq k \leq SF\}$, is the subchannel index in which k-th spread data symbol is transmitted. Finally $\mathbf{X}_m$ is a diagonal matrix with the multiplexed data symbols:

$$\mathbf{X}_m = diag\{\mathbf{C}\mathbf{d}_m\} \qquad (3)$$

where $\mathbf{C} = [\mathbf{c}^{(1)}, \mathbf{c}^{(2)}, \cdots, \mathbf{c}^{(K)},]$ is $(SF \times K)$ matrix that consists of spreading code vector of each user, $\mathbf{c}^{(k)}$, and $\mathbf{d}_m$ is a K-dimensional vector consisting of the data symbol of K users.

3. Pilot Symbol-aided Channel Estimation

3.1 Channel Estimation at Pilot Locations

Denoting by $\mathbf{Y}_n^{(p)} = [Y_n^{(p)}(i_1), Y_n^{(p)}(i_2), \cdots, Y_n^{(p)}(i_{N_p})]^T$, the N_p-dimensional vector containing the DFT output at the pilot locations $\{i_n : 1 \leq n \leq N_p\}$ in OFDM symbol interval n can be represented as:

$$\mathbf{Y}_n^{(p)} = \mathbf{X}_n^{(p)} \mathbf{H}_n^{(p)} + \mathbf{n}_n^{(p)}, \qquad (4)$$

where $\mathbf{X}^{(p)}$ is a diagonal matrix with the pilot symbol, $\mathbf{H}^{(p)}$ is the channel frequency response at pilot location, and $\mathbf{n}^{(p)}$ is a white Gaussian noise.

For notation simplicity, we drop the time index n and the pilot index p. The LMMSE estimate can be expressed as [3] [6]:

$$\mathbf{H}_{LMMSE} = \mathbf{R_{HH}} \left(\mathbf{R_{HH}} + \frac{\beta}{SNR} \mathbf{I}_{N_p} \right)^{-1} \mathbf{H}_{LS} \qquad (5)$$

where $\mathbf{R_{HH}} = E[\mathbf{H}\mathbf{H}^H]$ is the channel autocorrelation matrix in which $(\cdot)^H$ denotes conjugate transpose, $\mathbf{I}_{N_p}$ denotes the identity matrix of order N_p, $\mathbf{H}_{LS} = \mathbf{X}^{-1}\mathbf{Y}$ is the LS estimates of $\mathbf{H}$, and $\beta = \frac{E[|X(k)^2|]}{E[1/X(k)^2]}$ is a constant depending on the type of modulation.

The average MSE of LMMSE channel estimate is:

$$MSE = \frac{1}{N_p} tr \left[\mathbf{R_{HH}} - \left\{ \mathbf{R_{HH}} \left(\mathbf{R_{HH}} + \frac{\beta}{SNR} \mathbf{I}_{N_p} \right)^{-1} \mathbf{R_{HH}} \right\} \right] \qquad (6)$$

where $tr(\cdot)$ is a trace operator.

From (5) we can see that LMMSE estimate requires knowledge of the channel correlation and the operating SNR. As the operating SNR varies,

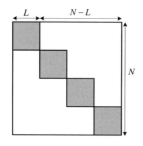 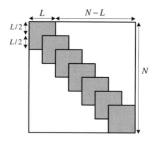

(a) $\mathbf{R_{HH}}$ by nonoverlap technique (b) $\mathbf{R_{HH}}$ by overlap technique

Figure 1. The decomposition of channel autocorrelation matrix $\mathbf{R_{HH}}$

the inverse matrix should be changed for reliable estimation. Therefore, the LMMSE channel estimation needs the matrix inversion and complex multiplication, which is the main drawback of LMMSE estimate.

The elements of channel correlation matrix within the coherence bandwidth that defines a range of high correlation among adjacent frequencies of the channel frequency response [8], have lager value than others. Therefore, the components outside the coherence bandwidth could be ignored. Consequently, channel auto-correlation matrix can be partitioned into sub-matrices of the size of coherence bandwidth, as shown in Figure 1-(a). By applying this concept, we can reduce the complexity of conventional LMMSE estimate.

We subdivide the channel vector using the coherence bandwidth:

$$\mathbf{H} = \begin{bmatrix} \mathbf{H}_1^T, \mathbf{H}_2^T, \cdots, \mathbf{H}_W^T \end{bmatrix}^T \quad (7)$$

where $W = \frac{N_p}{L}$, in which L is the coherence bandwidth normalized by pilot spacing D_f in frequency domain, and $\mathbf{H}_w = [H_{L(w-1)+1}, H_{L(w-1)+2}, \cdots, H_{L \cdot w}]^T$ is the w-th subvector of $\mathbf{H}$, for $w = 1, 2, \cdots, W$.

Using (5), the w-th subvector of size $L \times 1$, $\mathbf{H}_{LMMSE,w}$, is derived as:

$$\mathbf{H}_{LMMSE,\ w} = \mathbf{R}_{\mathbf{H}_w \mathbf{H}_w} \left(\mathbf{R}_{\mathbf{H}_w \mathbf{H}_w} + \frac{\beta}{SNR} \mathbf{I}_L \right)^{-1} \mathbf{H}_{LS,w}, \quad (8)$$

where $\mathbf{R}_{\mathbf{H}_w \mathbf{H}_w}$ is a auto-correlation matrix with size of $L \times L$, and $\mathbf{H}_{LS}$ is the w-th LS estimate.

Thus, the LMMSE estimate using the nonoverlap technique is:

$$\mathbf{H}_{nonoverlap} = \begin{bmatrix} \mathbf{H}_{LMMSE,1}^T, \mathbf{H}_{LMMSE,2}^T, \cdots, \mathbf{H}_{LMMSE,W}^T \end{bmatrix}^T. \quad (9)$$

The error covariance matrix of sub-block, $\mathbf{R}_{e_w e_w}$ is:

$$\mathbf{R}_{e_w e_w} = \mathbf{R}_{\mathbf{H}_w \mathbf{H}_w} - \mathbf{R}_{\mathbf{H}_w \mathbf{H}_w} \left(\mathbf{R}_{\mathbf{H}_w \mathbf{H}_w} + \frac{\beta}{SNR} \mathbf{I}_L \right)^{-1} \mathbf{R}_{\mathbf{H}_w \mathbf{H}_w}^H. \quad (10)$$

Finally, the averaged MSE of nonoverlap case is:

$$MSE_{nonoverlap} = \frac{1}{N_p} \sum_{w=1}^{W} Tr\left(\mathbf{R}_{\mathbf{e}_w \mathbf{e}_w}\right). \tag{11}$$

Note that the nonoverlap channel estimator is equivalent to the concept of partitioning into subsystem proposed in [3]. However, the MSE of estimator employing sub-matrices is high at the edge subcarriers of each subblock. So, we introduce overlap technique to compensate for this problem as shown in Figure 1-(b).

In the overlap technique, we divide the autocorrelation matrix into $\tilde{W}$ submatrices, where $W' = \frac{2N}{L} - 1 = 2W - 1$. The LMMSE estimate using the overlap technique is:

$$\mathbf{H}_{overlap} = \left[\tilde{\mathbf{H}}_{LMMSE,1}^T, \tilde{\mathbf{H}}_{LMMSE,2}^T, \cdots, \tilde{\mathbf{H}}_{LMMSE,W'}^T\right]^T, \tag{12}$$

where $\tilde{\mathbf{H}}_{LMMSE,1}$ and $\tilde{\mathbf{H}}_{LMMSE,W'}$ are sub-vector corresponding to the first $\frac{3L}{4}$ and the last $\frac{3L}{4}$ elements of $\mathbf{H}_{LMMSE,1}$ and $\mathbf{H}_{LMMSE,W'}$, respectively, and for $w = 2, 3, .., W'-1$, $\tilde{\mathbf{H}}_{LMMSE,w}$ is sub-vector of length $\frac{L}{2}$ corresponding to the center $\frac{L}{2}$ elements of $\mathbf{H}_{LMMSE,w}$. Finally, the MSE of overlap case is:

$$MSE_{overlap} = \frac{1}{N_p} \sum_{w=1}^{W} \Gamma(w), \text{ and} \tag{13}$$

$$\Gamma(w) = \begin{cases} \sum_{l=1}^{\frac{3}{4}L} \mathbf{R}_{\mathbf{e}_w \mathbf{e}_w}(l,l), & w = 1 \\ \sum_{l=\frac{1}{4}L+1}^{L} \mathbf{R}_{\mathbf{e}_w \mathbf{e}_w}(l,l), & w = W' \\ \sum_{l=\frac{1}{4}L+1}^{\frac{3}{4}L} \mathbf{R}_{\mathbf{e}_w \mathbf{e}_w}(l,l), & w = 2, 3, \cdots, W'-1. \end{cases}$$

The total complexity for proposed estimators as compared with conventional LMMSE is shown in Table 1.

Table 1. Complexity comparison

Operation	LMMSE	Nonoverlap	Overlap
Multiplication	$N_p^3 + 2N_p^2$	$L^3 + 2WL^2$	$L^3 + 2(2W-1)L^2$
Addition	$2N_p(N_p - 1)$	$2WL(L-1)$	$2(2W-1)L(L-1)$

3.2 Channel Estimation at Data Subcarriers

After the channel responses at pilot locations are obtained, the channel responses at data subcarriers can be estimated by interpolation. The DFT-based interpolation is a high resolution interpolation based on zero-padding and DFT/IDFT. After obtaining the channel information in pilot locations, it is transformed into time domain by IDFT and then interpolated by expanding the N_p points into N points by inserting $N-N_p$ zeros [2].

The performance of conventional DFT-based interpolation depends on the estimation accuracy of the number of channel taps and the number of pilots [6] [7]. Estimation accuracy degrades as the number of channel taps increases and the number of pilots decreases. Therefore, DFT-based interpolation can be improved by estimating the maximum channel taps accurately, or by ignoring the non-significant channel taps. By adapting significant channel tap detection and noise power estimation process, we can improve the DFT-based interpolation.

First, we decide the significant channel taps by following decision rule:

$$\hat{h}_{Proposed}(k) = \begin{cases} \hat{h}(k), & \text{if } |\hat{h}(k)|^2 \geq \lambda \\ 0, & \text{otherwise} \end{cases} \quad (14)$$

where $\hat{h}$ is the time-domain representation of estimated channel at pilot location and λ is the threshold value related to noise variance. The threshold can be obtained by comparing the individual subcarrier MSE before and after zero substitution of a specific channel tap. If k-th channel tap is ignored, the MSE is increased as much as the power of k-th channel tap, $|h(k)|^2$, but on the other hand, it is decreased by $\sigma_{\tilde{n}}^2 \cdot E\left\{|\frac{1}{X(k)}|^2\right\}$ in which $\sigma_{\tilde{n}}^2$ is the variance of time-domain noise. Finally, the consequent threshold can be easily derived as follows:

$$\lambda = 2\sigma_{\tilde{n}}^2 E\left\{|1/X(k)|^2\right\}. \quad (15)$$

4. Simulation Results

We consider downlink MC-CDMA system in the Rayleigh fading channel. The system parameter is given in Table 2.

Figure 2 shows the BER performance of various channel estimation methods in pilot location. For this simulation, we use cubic-spline interpolation in time and frequency grids and the channel autocorrelation matrix is partitioned into submatrix of size 16×16. For BER of 10^{-4}, the overlap LMMSE estimation is 3.4 dB better in E_b/N_0 than LS estimator and 0.5 dB worse than full LMMSE. Although the low complex LMMSE estimator has slight performance degradation as compared with

Table 2. System parameter

Parameters	Values
Bandwidth @ carrier frequency	20MHz @ 2.4GHz
FFT/IFFT size, N	1024
OFDM symbol duration	$51.2 + 6.4 \ \mu s$
Data modulation	QPSK
Spreading code	Walsh-Hadamard sequence (SF = 64)
Channel model	Rayleigh fading (Jakes' spectrum)
Normalized Doppler frequency	$f_d T_s = 0.0154$ (velocity = 120Km/h)
Multipath	Exponential decay channel ($\tau_{max} \Delta f = 0.0293$)
Pilot arrangement	Rectangular pilot ($D_f = 8, \ D_t = 8$)

LMMSE estimator, its complexity is significantly reduced. In Figure 3, we compare the BER performance of the modified DFT-based interpolation with others. In this simulation, we use LS method for pilot location. When BER is 10^{-4}, the modified DFT interpolation has 2 dB and 3.7 dB better performance than conventional one and 2-D cubic-spline interpolation, respectively. Also note that there is no error floor in the proposed scheme.

5. Conclusions

In this paper, we proposed channel estimation technique for MC-CDMA systems. For the channel estimation at pilot locations, we proposed low complexity LMMSE channel estimates, which use a small submatrix by non-overlap or overlap techniques applying the concept of coherent bandwidth to reduce the complexity of the full LMMSE estimate. The proposed LMMSE estimates have slight performance degradation as

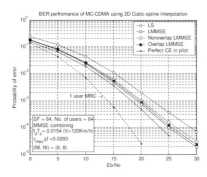

Figure 2. BER performance of channel estimates at pilot location

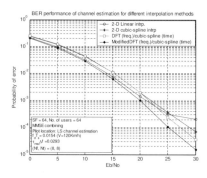

Figure 3. BER performance of different interpolations

compared with the full LMMSE estimate, however, the complexity of the full LMMSE is significantly reduced.

The proposed DFT-based channel interpolation can achieve good estimation accuracy by ignoring non-significant channel taps which contain more noise than channel power. The effective channel taps are detected by using threshold calculated from noise power level.

References

[1] S. Hara and R. Prasad, "Overview of multicarrier CDMA," *IEEE Commun. Mag.*, vol. 12, pp. 126-133, 1997.

[2] S. Coleri, M. Ergen, A. Puri, and A. Bahai, "A study of channel estimation in OFDM systems," in *Proc. IEEE 56th Vehicular Technology Conf.*, vol. 2, pp. 894-898, Sept. 2002.

[3] O. Edfors, M. Sandell, J.J van de Beek, S.K. Wilson, and P.O. Borjesson, "OFDM channel estimation by singular value decomposition," *IEEE Trans. Commun.*, vol. 46, pp. 931-939, July 1998.

[4] J. Choi and Y. Lee, "Design of the optimum pilot pattern for channel estimation in OFDM systems," in *Proc. IEEE Global Telecommun. Conf. (GLOBECOM'04)*. vol. 6, pp. 3661-3665, Nov./Dec. 2004.

[5] S. Kaiser, *Multicarrier CDMA mobile radio system-analysis and optimization of detection, decoding, and channel estimation*, Ph.D. thesis, VDI-Verlag, 1998.

[6] M. Morelli and U. Mengali, "A comparision of pilot-aided channel estimation methods for OFDM systems," *IEEE Trans. Signal Processing*, vol. 49, pp. 3065-3073, Dec. 2001.

[7] R. Negi and J. Cioffi, "Pilot tone selection for channel estimation in a mobile OFDM system," *IEEE Trans. Consum. Electron.*, vol. 44, pp. 1122-1128, Aug. 1998.

[8] T.S. Rappaport, *Wireless Communication: Principles and Practice*, Prentice Hall, 1996.

[9] M. Noh, Y. Lee, and H. Park, "A low complexity LMMSE channel estimation for OFDM," submitted to *IEEE Proc. Commun.*, Jan. 2005.

SUPERIMPOSED PILOT-BASED CHANNEL ESTIMATION FOR MIMO OFDM CODE DIVISION MULTIPLEXING UPLINK SYSTEMS

L. Cariou and J-F. Hélard
Electronics and Telecommunications Institute of Rennes (IETR)
INSA, 20, avenue des buttes de Coesmes, 35043 Rennes, France
laurent.cariou@ens.insa-rennes.fr, jean-francois.helard@ens.insa-rennes.fr

Abstract In this paper, we focus on the design of channel estimation techniques well suited to both STBC FH adjacent SS-MC-MA and STBC interleaved SS-MC-MA schemes. We present here the interest of the embedded pilot channel estimation technique in the SISO case and its adaptation to MIMO systems. A comparison with the recently proposed spread pilot channel estimation technique is performed. A loss of only 2dB compared to perfect channel estimation is obtained for both schemes.

1. Introduction

Nowadays, some of the most promising technologies for the air interface of the fourth generation (4G) terrestrial systems, concerned by flexibility and very high spectrum efficiency, are multi-carrier spread-spectrum techniques (MC-SS). At the European level, they have been studied during three years within the IST FP5 project MATRICE. The European IST FP6 project 4MORE [1] aims at enhancing MATRICE by taking advantages of the potentialities offered by multiple input multiple output (MIMO) techniques, and advancing one step towards an optimized implementation.

In downlink, multi-carrier code division multiple access (MC-CDMA) combines the merits of orthogonal frequency division multiplex (OFDM) with those of CDMA [2] by spreading users' signals in the frequency domain, offering a strong robustness to multipath propagation and inter-cell interference. However, in uplink, to reduce channel estimation complexity, spread spectrum multi-carrier multiple access (SS-MC-MA) has been proposed in [3]. SS-MC-MA assigns each user exclusively its own subset of subcarriers according to an additive FDMA scheme. Thanks to that, the BS only has to estimate, for each subcarrier, one channel compared to N_u for uplink MC-CDMA. Moreover, self interference (SI) can easily be cancelled by single user detection (SD).

The reduction, for each user, of the number of subcarriers with SS-MC-MA, compared to MC-CDMA, can lead to a reduction in the exploitation of the frequency diversity. In order to compensate for this, one can introduce a frequency multiplexing on the spread signals, *i.e.* distribute the chips of a spread symbol on the whole bandwidth in order to maximize the frequency separation between subcarriers inside a subset. As an alternative, we recently proposed in [4] to use subsets of adjacent subcarriers and to allocate them to the different users by applying a frequency hopping pattern (FH). In that case, it has been demonstrated that each user benefits as well from the frequency diversity linked to the total bandwidth, while offering a strong robustness to user's oscillators frequency shifts [5] and intercell interferences. Combined with Alamouti space-time block code (STBC) [6], the efficiency of the novel scheme STBC FH adjacent SS-MC-MA as a promising system for the uplink of the future wideband wireless networks has been successfully demonstrated in the case of perfect channel estimation [4].

In this paper, we focus on the design of channel estimation techniques well suited to both STBC FH adjacent SS-MC-MA and STBC interleaved SS-MC-MA schemes presented above. We present here the interest of the embedded pilot channel estimation technique in the SISO case and compare it to the recently proposed spread pilot channel estimation technique [7]. These techniques are then optimized in order to estimate a MIMO channel.

The article is organized as follows. In Section 2, we present the STBC SS-MC-MA scheme in a general way. Section 3 describes the proposed channel estimation technique based on embedded pilots in a SISO case and its optimization in the MIMO case. Section 4 presents the performance of STBC FH adjacent SS-MC-MA with embedded and spread pilot estimation over a 3GPP-like MIMO channel model, before drawing conclusions in Section 5.

2. STBC SS-MC-MA scheme description

Figure 1 shows a simplified STBC SS-MC-MA system for user j based on Alamouti's STBC with $N_t = 2$ transmit antennas and $N_r = 2$ receive antennas. User j simultaneously transmits N_L symbols $x_{j,l}^0$ and $x_{j,l}^1$ respectively from antenna 1 and 2 at time t, and the symbols $-x_{j,l}^{1*}$ and $x_{j,l}^{0*}$ at time $t+T_s$ where $l = 1, \ldots, N_L$ and T_s is the length of the SS-

Superimposed Pilot-based Channel Estimation for MIMO OFDM-CDM 249

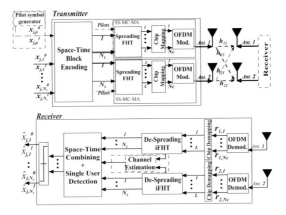

Figure 1. SS-MC-MA transmitter and receiver for user j with two transmit and two receive antennas.

MC-MA symbol. The number of subcarriers N_c is fixed to $N_u.L$ with the number of users N_u and the spreading length L. The *chip mapping* component determines the subset of subcarriers on which the spread symbols' chips from user j are distributed. Frequency multiplexing can be performed to achieve interleaved SS-MC-MA scheme. Subset of adjacent subcarriers, combined with a frequency hopping process can also be selected to achieve FH adjacent SS-MC-MA. In this last case, the Frequency hopping pattern respects a simple law which allows each user to experiment equiprobably each frequency subset, while avoiding collisions between users: $s_{i,j} = (s_{i-1,j} + inc) \mod N_s$ where $s_{i,j}$ is the subset index for OFDM symbol i and user j, N_s is the number of subsets per OFDM symbol and inc the subset index increment selected by a BER optimization to optimally take advantage of the frequency diversity. The optional two-dimensional spreading, more detailed in section 3, can also be activated.

At the reception, in the base station, the OFDM demodulation is carried out by a simple and unique fast fourier transform (FFT) applied to the sum of the N_u different users' signals. After equalization, for each receive antenna r, the two successive received signals are combined. The resulting signals from the N_r receive antennas are then added to detect symbols $x_{j,l}^0$ and $x_{j,l}^1$. After de-spreading and threshold comparison, the detected data symbols $\hat{x}_{j,l}^0$ and $\hat{x}_{j,l}^1$ for user j are:

$$[\hat{x}_{j,l}^0 \ \hat{x}_{j,l}^0]^T = (\mathbf{I}_2 \otimes \mathbf{c}_l^T)\mathcal{Y} = (\mathbf{I}_2 \otimes \mathbf{c}_l^T) \sum_{r=1}^{N_r} \mathcal{G}_r \mathcal{R}_r \quad \text{with} \quad \mathcal{G}_r = \begin{bmatrix} G_{1r} & G_{2r}^* \\ G_{2r} & -G_{1r}^* \end{bmatrix}$$

(1)

where $\mathbf{I}_2$ is the 2×2 identity matrix, $\otimes$ is the Kronecker product, $\mathcal{Y} = [y_1^0 \ldots y_k^0 \ldots y_L^0 \, y_1^1 \ldots y_k^1 \ldots y_L^1]^T$ is the vector of the received signals equalized and combined from the N_r antennas, G_{tr} is a diagonal matrix containing the equalization coefficients for the channel between the transmit antenna t and the receive antenna r. $\mathbf{c}_l = [c_{l,1} \ldots c_{l,k} \ldots c_{l,L}]^T$ is the Walsh-Hadamard orthogonal spreading code. Time invariance during two SS-MC-MA symbols is assumed to permit the recombination of symbols when STBC is used. In order to detect the $N_L \times 2$ transmitted symbols $x_{j,l}^0$ and $x_{j,l}^1$, per pair of SS-MC-MA symbols, for the desired user j, zero forcing (ZF) or minimum mean square error (MMSE) SD schemes are applied to the received signals in conjunction with STBC decoding.

3. Proposed channel estimation techniques

Both interleaved SS-MC-MA and FH adjacent SS-MC-MA schemes present very good results in a perfect channel estimation context, as they take advantage of the frequency diversity linked to the whole bandwidth with channel coding. Concerning channel estimation however, the two solutions are hardly compatible with frequency and time interpolation algorithm respectively. Classical time-multiplexed pilot based channel estimation is then not well suited.

Embedded pilot channel estimation Taking into consideration the sprectral and power efficiencies constraints, we took interest in the semi-blind pilot embedding channel estimation technique recently proposed for OFDM [8]. As it consists of transmitting low level overlay pilot-sequences concurrently with the data, there is no bandwidth consumption. Thanks to its very high flexibility, this solution can be used for both schemes. However, channel estimation errors due to cross-interferences between pilots and data symbols results in an error floor that reduces the performances in term of bit error rate (BER). Complex iterative solutions then have to be implemented to cancel the interferers. In order to simply reduce those interferences, we propose to select the pilot sequence orthogonal with the data sequence by allocating one of the WH orthogonal code $\mathbf{c}_p = [c_{p,1} \ldots c_{p,L}]^T$ to the pilots on every subset of subcarriers. This way, the interferences of the pilot sequence on the data sequence are strongly reduced. A small loss in terms of spectrum efficiency has to be paid. The power loss can be expressed by $10 log(\frac{N_L}{N_L+\eta})$, assuming the power ratio η between pilot P_p and data P_d.

Spread pilot channel estimation

Following the same idea but applying equalization after the despreading process, spread pilot channel estimation was recently proposed in [7] for STBC FH adjacent SS-MC-MA. In that case, only one equalization coefficient is used for each subcarriers' subset, while in the case of the embedded pilot channel estimation, one coefficient is used for each subcarrier. The spread pilot solution then requires non-selectivity on the subset of subcarriers on which the pilot is spread. Assuming the knowledge of the second-order statistics of the channel, chip mapping techniques such as two-dimensional spreading can be applied to SS-MC-MA in the same way it has already been applied to MC-CDMA to obtain Orthogonal Frequency and Code Division Multiplexing (OFCDM) [9]. This technique, can be simply expressed as a way of mapping the chips of the spread signal on adjacent sub-carriers (frequency dimension SF_{freq}) and on consecutive OFDM symbols (temporal dimension SF_{time} with $L = SF_{freq} \times SF_{time}$) in order for the channel to be as flat as possible on the whole spread symbol.

The spread pilot solution only works with FH adjacent SS-MC-MA and exhibits very good results [7], as it leads to a performance degradation of only 2 dB compared to perfect channel estimation in the MIMO case with 1 and 3 bits/s/Hz spectrum efficiencies systems. However, the performances decrease when the variance of the channel on the subset increases. We try to prove here that, at a price of a little increase of complexity, the embedded pilot channel estimation will be more flexible and will be more robust to the channel selectivity. As the SISO and MIMO spread pilot channel estimation scheme principles are presented in details in [7], this paper only describes the embedded pilot principles.

3.1 Embedded pilot channel estimation principles in the SISO case

Using QPSK for pilot and data symbols, x_p and x_l are selected from a set of fixed amplitude alphabet. x_p is deterministic and we assume x_l to be a zero mean random variable. Let the variance of x_p and x_l be the same, i.e. $\sigma_{x_p}^2 = \sigma_{x_l}^2$. η can be expressed as the amount of power allocated to the pilots when the data power is normalized to 1. At the reception point after OFDM demodulation and 2D-demapping, the received signal on subcarrier k can be expressed by:

$$r_k = \frac{\sum_{l=0}^{N_L-1} x_l.(h_k\, c_{l,k}) + \sqrt{\eta}.x_p.(h_k\, c_{p,k})}{\sqrt{N_L - 1 + \eta}} + n_k \qquad (2)$$

Knowing the value of the transmitted symbol $x_p \times c_{p,k}$, one can easily find the least square (LS) normalized channel coefficient on each sub-carrier:

$$\hat{h}_{kLS} = \frac{r_k \cdot \sqrt{N_L - 1 + \eta}}{\sqrt{\eta}.(x_p\, c_{p,k})} = h_k + h_k \cdot \underbrace{\sum_{l=0}^{N_L-1} \frac{x_l\, c_{l,k}}{\sqrt{\eta}.x_p\, c_{p,k}}}_{V_k} + \underbrace{\frac{n_k \cdot \sqrt{N_L - 1 + \eta}}{\sqrt{\eta}.x_p\, c_{p,k}}}_{U_k} \quad (3)$$

$\hat{h}_{kLS}$ appears in the form of a sum a desired signal h_k corrupted by a multiplicative noise $h_k.V_k$ and AWGN U_k as follows $\hat{h}_{kLS} = h_k + h_k.V_k + U_k$. Knowing that we use Walsh-Hadamard sequences, $c_{p,k} = \pm\frac{1}{\sqrt{L}}$, the variances $\sigma_{V_k}^2 = \frac{(N_L-1)\sigma_{x_l}^2}{\eta \sigma_{x_p}^2}$ and $\sigma_{U_k}^2 = \frac{N_L-1+\eta}{L\eta} \frac{\sigma_{n_k}^2}{\sigma_{x_p}^2}$ of the multiplicative noise V_k and the AWGN U_k respectively can be easily calculated. The optimum tap weights obtained in [8] can be applied on SF_{time} time pilots and SF_{freq} frequency pilots. The MMSE estimate is then

$$\hat{h}_{kMMSE} = w_k^T.\hat{h}_{kLS} \quad (4)$$

with $w_k = [R_{HH} + (\sigma_V^2 + \sigma_U^2)I]^{-1} r_{HH_k}$ and R_{HH} the channel autocorrelation.

3.2 Realistic application in a MIMO case

One great interest of the spread and embedded pilot estimation schemes is that it can easily be applied in a MIMO case. When combining Alamouti code in time to FH SS-MC-MA as in section 2, one can easily achieve the estimation of the MIMO channel by applying as well Alamouti code to the spread pilots.

At the receiver, on each receive antenna r, the embedded channel estimation block applies the Alamouti code detection and generates the least square (LS) estimate of the coefficients $\hat{h}_{k,tr,LS}$ from transmit antenna t as follows

$$\hat{h}_{k,1r,LS} = \frac{r_{k,t}\, x_{p_1}^* - r_{k,t+T}\, x_{p_2}}{(|x_{p_1}|^2 + |x_{p_2}|^2)} \quad (5)$$

$$\hat{h}_{k,2r,LS} = \frac{r_{k,t+T}\, x_{p_1} - r_{k,t}\, x_{p_1}^*}{(|x_{p_1}|^2 + |x_{p_2}|^2)} \quad (6)$$

$\hat{h}_{k,tr,LS}$ appears in the form of the following sum $\hat{h}_{k,tr,LS} = h_{k,tr} + h_{k,tr}.V_{k,tr} + U_{k,tr}$. The MMSE estimates can then be found as in the SISO case.

4. Simulation Results

4.1 System parameters

The system parameters are chosen according to the time and frequency coherence of the channel in order to match the requirements of a 4G mobile cellular system. The carrier and sampling frequencies are set to 5 GHz and 57.6 MHz respectively. The FFT size equals 1024 and the number of modulated subcarriers reaches 736. The guard interval duration is equal to 3.75 μs and the total symbol duration 21.52 μs. Channel coding is composed of a rate 1/3 UMTS turbo-coding followed by puncturing pattern defined to achieve a global coding rate R of 1/2. The time interleaving depth has been adjusted to the frame duration made of 32 OFDM symbols. With a QPSK modulation, this leads to an asymptotical spectrum efficiency of 1 bit/s/Hz. We use a link level MIMO channel model which has been specifically developed within the European IST MATRICE project and based on the 3GPP/3GPP2 proposal for a wideband MIMO channel exploiting multipath angular characteristics. The model parameters, adapted to the 5 GHz carrier frequency, uses the BRAN E channel average power delay profile (APDP), which refers to a typical outdoor urban multi-path propagation. The measured coherence bandwidth mono-sided at 3 dB is roughly 1.5 MHz. Spatial correlation is inferior to 0.1 for an antenna spacing of 10 wavelengths (λ) at the BS and close to 0.3 for 1 λ at the mobile terminal (MT). Finally, when the MT velocity is 72 km/h, the time correlation remains close to the frame duration.

4.2 Numerical results

In order to compare the two channel estimation schemes, we only present results with adjacent SS-MC-MA. Figure 2 presents the bit error rate (BER) performance of turbo-coded adjacent SS-MC-MA with MMSE SD. Embedded pilot channel estimation is performed. In spite of a small spectrum efficiency loss compared to pseudo-noise (PN) overlay pilot solution from [8], a very significant gain can be observed when one of the WH orthogonal codes is allocated to the pilots on every subset of subcarriers. The orthogonality reduces the interferences of the pilots on the data and suppresses the resultant error floor.

Figure 3 presents, for E_b/N_0=12dB, the BER performance of turbo-coded adjacent SS-MC-MA with MMSE SD for different 2D-spreading configurations over a realistic BRAN E SISO channel. The spreading length $L = SF_{freq} \times SF_{time} = 32$.

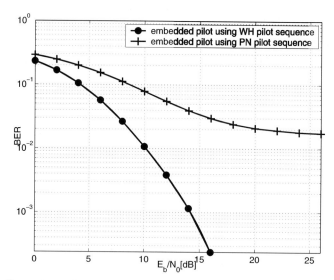

Figure 2. Comparison of embedded pilot channel estimation using WH or pseudo-noise (PN) pilot superimposed sequences; full load turbo-coded SS-MC-MA systems; realistic SISO model; MMSE single user detection, QPSK, channel coding rate R=1/2, 2D-spreading with $SF_{time} = 8$, MT velocity of 72 km/h.

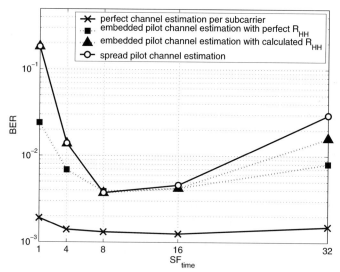

Figure 3. Comparison of different channel estimation techniques for different 2D-spreading configurations; realistic SISO model; full load turbo-coded SS-MC-MA; MMSE, QPSK, R=1/2, MT velocity of 72 km/h, $E_b/N_0 = 12dB$.

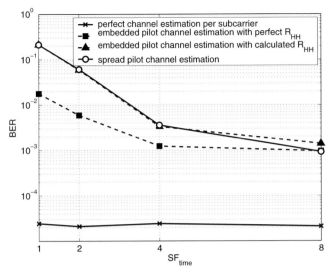

Figure 4. Comparison of different channel estimation techniques for different 2D-spreading configurations; realistic MIMO model; full load turbo-coded STBC SS-MC-MA; 2 transmit and 2 receive antennas, MMSE, QPSK, R=1/2, MT velocity of 72 km/h, $E_b/N_0 = 2dB$.

Embedded pilot and spread pilot channel estimations are compared, taking into account the pilot insertion power losses. In both cases, the optimum value $\eta_{opt} = 7$ of the power ratio in terms of BER has been determined thanks to Monte-Carlo simulations. The perfect channel estimation per subcarrier performance are also presented in reference. The behavior of the different curves confirms the assumptions concerning the tight link between the channel variance and the SI.

For the best 2D-spreading configurations, *i.e.* when the channel variance on the subset of subcarriers on which the spread data are distributed is low, spread and embedded pilot channel estimations have similar results. When this channel variance increases, for 2D-spreading configurations with SF_{time} equal to 1, 4 and 32, the embedded pilot outperforms the spread pilot technique when the correlation matrix R_{HH} is perfectly known. It is then clearly more robust against the selectivity of the channel. In a realistic system however, this correlation matrix R_{HH} has to be measured or calculated. The best results have been found for R_{HH} calculated with an exponentially decaying power delay profile (PDP) which approaches the BRAN E PDP, knowing the delay spread τ, the root mean square delay spread τ_{RMS} and the maximum doppler frequency $f_{D,max}$. The gain on the spread pilot channel estimation, using this calculated R_{HH}, is no longer significative. Only a slight gain is observed with SF_{time} equal to 32, enlightening the fact that the correlation matrix calculation is better along the time axis ($SF_{time} = 32$) than along the frequency one ($SF_{time} = 1$).

This behavior is emphasized by figure 4 which presents, for E_b/N_0=2dB, the BER performance of turbo-coded STBC adjacent SS-MC-MA with two transmit and two receive antennas, with MMSE SD for different 2D-spreading configurations over a realistic BRAN E MIMO channel.

5. Conclusion

Two different uplink channel estimation techniques based on embedded and spread pilot respectively are presented and compared. Leading to a loss of only 2dB compared to perfect channel estimation with Alamouti STBC adjacent SS-MC-MA systems, the efficiency of the two schemes as promising channel estimations for the uplink of the future wideband wireless networks is successfully demonstrated in SISO and MIMO cases. On one side, the spread pilot channel estimation presents, with a reduced complexity, similar optimum performance than the embedded solution. On the other side, the embedded pilot channel estimation appears to be more robust against the selectivity of the channel.

References

[1] IST 4MORE project. http://www.ist-4more.org.

[2] S. Hara and R. Prasad. Overview of multicarrier CDMA. *IEEE communication magazine*, 35(12):126–133, Dec. 1997.

[3] S. Kaiser and W.A. Krzymien. Performance effects of the uplink asynchronism in a spread spectrum multi-carrier multiple access system. *Europeen Transactions on Telecomunications*, 10(4), July/August 1999.

[4] L. Cariou and J.F. Helard. MIMO frequency hopping OFDM-CDMA: a novel uplink system for B3G cellular networks. *Proc. of IEEE International Conference on Networking (ICN'05)*, Saint-Denis, France, April 2005. Lecture Notes in Computer Science, volume 3421, pp. 8–17, April 2005.

[5] A. Arkhipov, M. Schnell, "The influence of user frequency offset on the uplink performance of SS-MC-MA", *Proc. of European Conference on Wireless Technology (ECWT'03)*, Munich, Germany, pp. 451–454, Oct. 2003.

[6] S.M. Alamouti. A simple transmit diversity technique for wireless communication. *IEEE Journal on Selected Areas in Communications*, 16:1451–1458, Oct. 1998.

[7] L. Cariou and J.F. Helard. A simple and efficient channel estimation for MIMO OFDM code division multiplexing uplink systems. *Proceedings of SPAWC 2005*, June 2005.

[8] C. K. Ho, B. Farhang-Boroujeny, and F. Chin. Added pilot semi-blind channel estimation scheme for OFDM in fading channels. *GLOBECOM 2001 - IEEE Global Telecommunications Conference*, (1):3075–3079, Nov. 2001.

[9] H. Atarashi, N. Maeda, S. Abeta, and M. Sawahashi. Broadband wireless access based on VDF-OFCDM and MC/DS-CDMA. *Proc. of IEEE PIMRC 2002*, 3:992–997, Sept. 2002.

PILOT TIME-FREQUENCY LOCATION ADJUSTMENT IN OFDM SYSTEMS BASED ON THE CHANNEL VARIABILITY PARAMETERS

Faouzi Bader
Centre Tecnologic de Telecomunicacions de Catalunya (CTTC)
Av. del Canal Olimpíc, 08860 Castelldefels, Barcelona, Spain.
faouzi.bader@cttc.es

Raúl Gonzalez
Universitat Politècnica de Catalunya (UPC)
ETSETB, c/Jordi Girona, 2-4, 08034 Barcelona, Spain.
rgon7822@alu-etsetb.upc.es

Abstract This work explores the possibility to adapt the pilot location in an OFDM frame, by choosing the more appropriate pilot pattern, the rectangular or the hexagonal one. Based on the prediction of some channel parameters and for a given pilot density, the transmitter seeks the most appropriate time-frequency pilot spacing which minimize the MSE of the channel estimated by the use of a general interpolator.

1. Introduction

When facing the wireless fading channel in the multiple carrier systems based on the Orthogonal Frequency Division Multiplexing (OFDM) scheme two types of modulation are usually used, differential or coherent. It is well known that in differential communication there is no need for channel estimation, since the information is encoded in the difference between two consecutive symbols. Therefore, in the coherent case the use of pilots symbols to aid the channel estimation (PSACE) is very suitable for tracking and acquiring the channels state information (CSI) changes during all the communication process. Results obtained in [1], [2] have demonstrated the direct influence of the pilot insertion within the frames of the OFDM on the system performance of channel estimation. The hexagonal pilot pattern has appeared as the most efficient strategy for the pilot distribution [1], [3], [5]. Several evaluations over different scenarios have demonstrated its good performances compared with other pilot patterns [5]. Usually, a simple interpolator such as the linear, or the spline interpolation is often used in the practice, but when these interpolators are used, the performance of the channel estimation is often affected by the pilot pattern form. Almost all of the previous research results on the design of pilot patterns for channel estimation in OFDM systems have been obtained based on the computer simulation results as in [2], [3], and [4]. It is more than desirable to design the PACE pattern in an analytical form without a need of exhaustive computer simulation. Based on the two-dimensional sampling and the time-frequency pilot location in the OFDM frames, a basic algorithm which consist in optimizing the time-frequency spaces of the pilots within the different OFDM frames has been developed in accordance with the maximum frequency Doppler f_{Dmax}, and the delay profiles of the channel τ_{max}. The remainder of the paper is outlined as follows; Section 2, describes the OFDM system over the wireless channel. In Section 3, the optimum time-frequency location is derived. The use of the proposed pilot distribution and its performances are verified in Section 4. Finally, conclusions are given in Section 5.

2. System Description

The system considered is based on an OFDM scheme, where different data information are transmitted over a set of N_c sub-carriers in the frequency domain, and modulated using the inverse fast Fourier transform (IFFT). A detailed description of the OFDM system is presented in [1], and the transmission of the signal over a wireless channel is,

$$h(t,\tau) = \sum_{p=0}^{P} h(t)\delta(\tau - \tau_p) \qquad (1)$$

where P is the total number of multi-paths, $\delta(.)$ is the Kronecker delta function, τ_p and $h(t)$ denote the delay and the complex -valued of the channel impulse response respectively. A cyclic prefix (CP) is inserted to prevent the orthogonality loss effects between the sub-carriers. At the receiver side, the CP is removed before applying the FFT process [1]. A perfect synchronization is assumed at the receiver, and the information related with the l-th time OFDM symbol at the k-th sub-carrier is given by,

$$R_{k,l} = H_{k,l}S_{k,l} + N_{k,l} \qquad (2)$$

where $H_{k,l}$ is the frequency response of the channel at the $k-th$ sub-carrier ($k = 1, \ldots N_c$ and $l = 1, \ldots N_s$), $N_{k,l}$ means the noise (additive white Gaussian noise) term with a variance σ^2. Usually, the pilots are introduced within the OFDM frame during each predefined N_t time slots, and N_f sub-carriers in the frequency domain. The structures of the pilots patterns are based on the theorem of the two-dimensional sampling described in [6], where any pilot pattern can be represented using two vectors, for instance; **u** and **g** (*in the Cartesian coordinate*). These two vectors have each one two components that compose the pilot spacing in the time direction (*sub-index 1*), and in frequency direction (*sub-index 2*). These two vectors of size (2×1) have each one the following structure,

$$\mathbf{u} = [u_1, u_2]^T, \mathbf{g} = [g_1, g_2]^T \qquad (3)$$

the vectors **u** and **g** can be combined and represented by a matrix **V** of size (2×2) as,

$$\mathbf{V} = [\mathbf{u} : \mathbf{g}] \qquad (4)$$

the matrix **V** represent the two dimensional distribution of the pilots within an OFDM frame. The determinant of **V** will give us the pilot density dp as,

$$dp = |det(\mathbf{V})|^{-1} = |u_1 g_2 - u_2 g_1| \qquad (5)$$

Although, u_1, u_2, g_1, and g_2 can be any value. For particular geometries the sampling matrix **V** could have the following values,

$$\mathbf{V} = V_{f_s} = \begin{bmatrix} N_t & 0 \\ 0 & 1 \end{bmatrix}, \mathbf{V} = V_h = \begin{bmatrix} N_t & -N_t \\ N_f & N_f \end{bmatrix} \qquad (6)$$

V_{f_s} and V_h denote the block type and the hexagonal pattern respectively. Forward it is assumed that the value of $u_2 = 0$ without the loss of generality, since any parallelogram can be made to $u_2 = 0$ by rotating the pattern and the pilot density in (5) can be rewritten as,

$$dp = |det(\mathbf{V})|^{-1} = |u_1 g_2| \qquad (7)$$

it can be emphasized from (5) and (7), that the pilot density dp is inversely proportional to the pilot spacing. The frequency response of the channel corresponding to the pilot location within the OFDM frame which is first estimated at the receiver is,

$$\hat{H}_{\acute{k},\acute{l}} = \frac{R_{\acute{k},\acute{l}}}{S_{\acute{k},\acute{l}}} = H_{\acute{k},\acute{l}} + \frac{N_{\acute{k},\acute{l}}}{S_{\acute{k},\acute{l}}}, \forall \{\acute{k}, \acute{l}\} \in \Omega \qquad (8)$$

the received pilot symbol is divided by the originally transmitted pilot symbol known at the base station, note that this operation is only valid at the pilot positions, $\acute{k}$ and $\acute{l}$. The final estimates of the complete channel transfer function belonging to the desired OFDM frame is obtained from the initial estimates $\hat{H}_{\acute{k},\acute{l}} \forall \{\acute{k},\acute{l}\} \in \Omega$, and by a two-dimensional (2D) interpolation process. Using the 2D Wiener filter, the total frequency response of the channel is given by,

$$\hat{H}_{k,l} = \sum_{\{\acute{k},\acute{l}\} \forall \Omega} w_{(k,l,\acute{k},\acute{l})} \hat{H}_{\acute{k},\acute{l}}, \ k=1,\ldots,N_c, l=1,\ldots,N_s \quad (9)$$

the value $\sum_{\{\acute{k},\acute{l}\}} w_{k,l,\acute{k},\acute{l}}$ denotes the coefficients of the Wiener filter for the desired estimate channel, and is closely linked with each pilot position $\{\acute{k},\acute{l}\}$, $\hat{H}_{k,l}$ is the estimated channel at any $\{k,l\}$ position within the OFDM-frame. Note that, the set of the pilot positions in the OFDM frame is Ω. The filter coefficients are obtained by applying the orthogonal principle in the linear mean square estimation (MSE). The two dimensional Wiener filter coefficients (*if the correlation matrix exist*) are calculated and given by,

$$\mathbf{w}_{k,l,\acute{k},\acute{l}} = \Theta_{k,l}^T \Phi_{\acute{k},\acute{l}}^{-1} \quad (10)$$

Where $\Phi_{\acute{k},\acute{l}}$ is the autocorrelation matrix of size $(N_{tap} \times N_{tap})$ which depends only on the distance between the different pilots positions. The vector $\Theta_{k,l}$ of size $(N_{tap} \times 1)$ contains the values of the cross-correlation function between the different pilots at the positions $(\acute{k},\acute{l}) \forall \{\acute{k},\acute{l}\} \in \Omega$, and the actual channel at the position (k,l). The vector $\mathbf{w}_{k,l,\acute{k},\acute{l}}$ of length N_{tap} contains the Wiener filter coefficients required for the channel estimation.

3. Optimum Pilot Location

Assuming an ideal interpolator, therefore (8) can be rewritten as,

$$H_{k,l} = \sum_{\{\acute{k},\acute{l}\} \forall \Omega} w_{P(k,l,\acute{k},\acute{l})} \hat{H}_{\acute{k},\acute{l}}, \ k=\{1,\ldots,N_c\}, l=\{1,\ldots,N_s\} \quad (11)$$

where in this case $w_{P(k,l,\acute{k},\acute{l})}$ denotes the 2D filter coefficient used when the channel is perfectly known (*i.e. when there is not errors in the channel estimation*). An ideal interpolator could be as that used in [7] and is,

$$w_{P(k,l,\acute{k},\acute{l})} = \frac{\sin\left(\frac{\pi \acute{k}}{u_1}\right) \sin\left(\frac{\pi \acute{l}}{g_2}\right)}{\left(\frac{\pi \acute{k}}{u_1}\right)\left(\frac{\pi \acute{l}}{g_2}\right)} \quad (12)$$

On the basis of (9) and (11), the error between the actual estimated channel and the ideal one is calculated using the MSE which is,

$$\begin{aligned}J &= E|\hat{H}_{k,l} - H_{k,l}|^2 \\ &= \frac{1}{(2\pi)^2} \int_{-\pi}^{\pi} \int_{-\pi}^{\pi} DFT(\Theta_{k,l})|DFT(w_{(k,l,\acute{k},\acute{l})} - w_{P(k,l,\acute{k},\acute{l})})|^2 \, df_1 \, df_2 \\ &+ \frac{\sigma_0^2 \, dp}{(2\pi)^2} \int_{-\pi}^{\pi} \int_{-\pi}^{\pi} |DFT(w_{(k,l,\acute{k},\acute{l})})|^2 \, df_1 \, df_2 = J_1 + J_2 \end{aligned} \qquad (13)$$

where f_1 and f_2 are the new variables in the frequency domain such that $\frac{\pi}{u_1} \geq |f_1|$ and $\frac{\pi}{g_2} \geq |f_2|$ [6]. The term J_1 depends on the pilot pattern. Therefore, the term J_2 is due to the interference caused essentially by the own 2D interpolator. We assume that J_2 is an irreducible interference term. However, we focus our analysis only upon J_1 with the purpose to reduce it effect. We Apply upon the interpolation error coefficient J_1 the two-dimensional Fourier transform and we obtain,

$$\left| DFT(w_{k,l,\acute{k},\acute{l}} - w_{P(k,l,\acute{k},\acute{l})}) \right|^2 = \left| DFT(w_{k,l,\acute{k},\acute{l}}) - u_1 g_2 \right|^2 \qquad (14)$$

In the case where the error made outside of the useful bandwidth is not considered, (14) can be approximated by applying the Taylor development [8] to the following expression,

$$\begin{aligned} Taylor\left[|DFT(w_{k,l,\acute{k},\acute{l}} - w_{P(k,l,\acute{k},\acute{l})})|^2\right] &= \frac{(u_1^2 - 1)^2}{144 \, dp^2} \frac{1}{2\pi} \int_{-\pi}^{\pi} f_1^{\,4} S_{H_k} \, df_1 \\ &+ \frac{(g_2 - 1)^2}{144 \, dp^2} \int_{-\pi}^{\pi} f_2^{\,4} S_{H_l} \, df_2 \\ &+ \frac{1}{2\pi} \int_{-\pi}^{\pi} f_1^2 S_{H_k} \frac{1}{2\pi} \int_{-\pi}^{\pi} f_2^2 S_{H_l} \, df_2 \\ &+ \frac{(u^2 - 1)(g_2^2 - 1)}{72 \, dp^2} \end{aligned} \qquad (15)$$

the functions S_{H_k} and S_{H_l} obtained in (15) are the spectral density function of the channel in the frequency and time domain respectively. In fact the J_2 term is independent of the pilot position. Therefore, a more appropriate pilot position must be chosen in such way that the value of J_1 becomes much reduced. Assuming that u_1 and g_2 are continues variables, the optimum pilot spacing that minimize the term J_1 is obtained by looking for the zero on its derives,

$$\frac{\partial J_1}{\partial u_1}\Big|_{u_1 = \tilde{u}_1} = 0, \quad \frac{\partial J_1}{\partial g_2}\Big|_{g_2 = \tilde{g}_2} = 0 \qquad (16)$$

the calculus of both functions in (16), arise the following equation

$$\widetilde{u}_1 = \frac{1}{\sqrt{dp}} \left(\frac{\frac{1}{2\pi} \int_{-\pi}^{\pi} f_2^4 S_{H_l}(f_2) \, df_2}{\frac{1}{2\pi} \int_{-\pi}^{\pi} f_1^4 S_{H_k}(f_1) \, df_1} \right)^{\frac{1}{8}}$$

$$\widetilde{g}_2 = \frac{1}{\sqrt{dp}} \left(\frac{\frac{1}{2\pi} \int_{-\pi}^{\pi} f_2^4 S_{H_l}(f_2) \, df_2}{\frac{1}{2\pi} \int_{-\pi}^{\pi} f_1^4 S_{H_k}(f_1) \, df_1} \right)^{-\frac{1}{8}} \quad (17)$$

where the values of the nominators in (17) are the 4-th order moment of the Doppler spectrum and the power delay profiles respectively. It can be noted from (17) that the variables are directly linked with dp. They depend essentially on the frequency Doppler, and the power delay profiles of the channel. While the variable $\widetilde{u}_1$ depends directly from the temporal axis, and indirectly from the frequency axis, the variable $\widetilde{g}_2$ depends indirectly from the temporal axis and directly from the frequency axis. Whether the mobility increase, the value of $\widetilde{u}_1$ raise, while $\widetilde{g}_2$ diminish. Therefore, when $\widetilde{u}_1$ goes to infinity, $\widetilde{g}_2$ goes to zero, and the pilot location strategy resulting a block type of pilots as that usually used in WLAN systems for indoor environment. This system is still not complete, because of both vectors **u** and **g** have each one two components (see (3)), and those used still now are three values $(\widetilde{u}_1, \widetilde{u}_2 = 0, and\, \widetilde{g}_2)$. The value of g_1 has not been yet determined. It can be reviewed in previous investigations [1], [6], and [8] that with choosing g_1 equal to the half distance of $\widetilde{u}_1$, we get an hexagonal pilot pattern structure,

$$\mathbf{u} = \left(\widetilde{u}, \frac{\widetilde{g}_2}{2} \right) \; and \; \mathbf{g} = (0, \widetilde{g}_2) \quad (18)$$

4. Simulation results and discussion

In our simulations, we have considered an OFDM system with a total bandwidth of 20 MHz, and $N_c = 256$ carriers, a cyclic prefix of 2 μs, and a frequency carrier of $f_c = 5.8$ GHz. The signal is transmitted over a rayleigh channel with $P = 8$, $\tau_{max} = \{50, 100\}$ ns equi-spaced and uniform. The pilot pattern analyzed are the rectangular and the hexagonal. A range of the signal to noise ratio values from 5 up to 14 dB have been used for the analysis. Each legend's figure depicts the used pilot pattern (*and whether the optimum spacing is used or not*), and the values $\langle \widetilde{u}_1, \widetilde{g}_2 \rangle$. The Fig. 1-(a,b), and Fig. 2-(c,d) are devoted to show the bit error rate (*BER*) performances when different pilot's geometries are used. It can be observed that in 1-(a), both the rectangular, and the hexagonal patterns perform similar BER when the optimum time-frequency spacing algorithm according to (17). However,

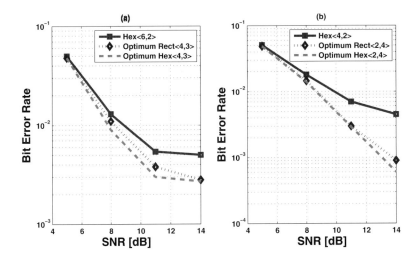

Figure 1. OFDM system performances BER vs. SNR. (a) $v = 60\,Km/h$, $\tau_{max} = 50\,ns$, and $dp = 8.33\%$, (b) $v = 60\,Km/h$, $\tau_{max} = 100\,ns$, and $dp = 12.5\%$

the conventional hexagonal pattern without use of the optimum insertion provides worst performance as the SNR increase up to 14 dB. The same behavior can be observed on Fig. 1-(b), and Fig. 2-(c,d) even when different values of v, τ_{max} and dp are used. Similar performances can be observed in Fig. 2-(d) with the three schemes for SNR values lower than 8 dB. Above 8 dB, the optimum-hexagonal scheme outperforms the rest of the schemes when the $SNR = 14$ dB. Note that in almost the figures that use the optimum time-frequency spacing according to (17) outperform the simply hexagonal spacing, except in Fig. 2-(d). This could be explained by the fact that whether we increase the pilot density, the optimum pilot time-frequency spacing according to (17) loses its interest. Since the main objective behind using different pilot pattern geometries is to be able to use the lowest possible density of pilot, and at the same time locating the PACES using the most appropriate time frequency spacing in order to achieve the best CSI acquisition at the receiver.

5. Conclusions

An adaptive algorithm for the pilot location according to (17) has been analytically designed in this proposal. The algorithm adapts the time-frequency pilot location within the frame according to the changes experienced over the parameters τ_{max} and f_{Dmax}. It is pointed out that the

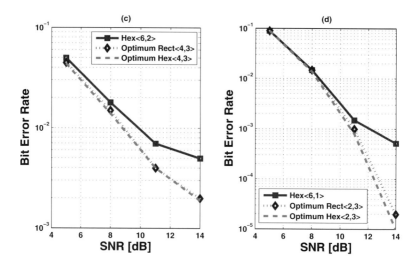

Figure 2. OFDM system performances BER vs. SNR. (c) $v = 100\,Km/h$, $\tau_{max} = 50\,ns$, and $dp = 8.33\%$, (d) $v = 100\,Km/h$, $\tau_{max} = 100\,ns$, and $dp = 16.6\%$

performances of the hexagonal pattern geometry can be improved when the optimum time-frequency spacing of the pilot is implemented.

References

[1] R. Van Nee and R. Prasad, *OFDM for Wireless Multimedia Communications.* Artech House Publisher, 2000.

[2] Tufvensson and T. Maseng,"Pilot Assists Channel Estimation for OFDM in Mobile Cellular Systems", in *Proceedings of the IEEE Vehicular Technology Conference (VTC'97)*, pp.16 Dresden, Germany, June, 1997.

[3] P. Höher, S. Kaiser, and P. Robertson, "Pilot Symbol Aided Channel Estimation in Time and Frequency", in *Proceedings of the IEEE, Global Telecommunications Conference (GLOBECOM'97)*, Phoenix, USA, Nov, 1997.

[4] P. Höher, S. Kaiser, "Two Dimensional Pilot Aided Channel Estimation by Wiener Filtering", *International Conference on Acoustic, Speech and Signal Processing (ICASSP'97)*, Munich, Germany, April 1997.

[5] M. J. Fernandez-Getino Garcia, J. M. Páez Borrallo, S. Zazo, "Pilot Pattern for Channel Estimation in OFDM", in *Proceedings of the IEE Electronic Letters, Vol. 36, Nr 12,* pp. 1049-1050, Jun 2000.

[6] D. E. Dudgon, and M. Mersereau, *Multi-Dimensional Digital Signal Processing.* Prentice-Hall, NJ 1984.

[7] A. V. Oppenheim Ronald W. Schafer, *Discrete-Time Signal Processing*, Prentice Signal Processing Series, 1989.

[8] Erwin Kreyszig, *Advanced Engineering Mathematics*, 8-th edition. John Wiley & Sons. 1999.

Section V

MIMO AND ADAPTIVITY

REDUCED FEEDBACK CLOSED-LOOP SPATIAL MULTIPLEXING FOR B3G SYSTEMS

Rahul Malik and Tan Pek Yew
Panasonic Singapore Laboratories, Singapore

Abstract: Closed-loop MIMO schemes are key to achieving the high spectral efficiencies required of next generation wireless communications systems. While eigen-mode spatial multiplexing is known to realize the optimal diversity and multiplexing gains of the channel, it requires channel knowledge at the transmitter. CSI feedback detracts from the payload carrying capacity of the system and must be minimized. This paper describes receiver techniques that can enable reduced feedback eigen-mode spatial multiplexing – a method particularly applicable to improving capacity in no/low mobility scenarios.

Key words: MIMO, Spatial Multiplexing, Eigen-mode, Reduced Feedback

1. INTRODUCTION

Of late, multi-antenna techniques have received considerable interest in their application to next generation wireless communications systems. While the traditional use of multiple antennas at the transmitter and receiver has been limited to improving the diversity gain of the system, the application of multiple antennas to improving data rates is a relatively new one. The first commercial use of Multiple Input Multiple Output (MIMO) technology is expected to be realized through the use of the much awaited IEEE 802.11n WLAN standard which expects to realize throughputs of in excess of 100Mbps in a 20MHz spectrum. MIMO has also been a topic of interest within 3GPP, especially with the initiation of requirements for the long term evolution of the RAN, which details spectral efficiency targets of upto 5bps/Hz in the DL [1].

There are fundamentally two distinct MIMO modes of operation – (i) spatial-multiplexing; and (ii) diversity. Based on results from information theory [2], the maximum capacity of a MIMO system is min(N_{Tx}, N_{Rx}) times that of an equivalent single-input single-output (SISO) system, N_{Tx} and N_{Rx} being the number of transmit and receive antennas, respectively. Spatial multiplexing refers to a set of techniques the fundamental basis of which involves the transmission of independent data streams from different antennas, in order to realize the MIMO channel capacity.

Although methods such as spatial multiplexing attempt to bridge the capacity gap, a major obstacle to achieving this goal is fading, which impacts link reliability. The use of multiple antennas at the transmitter and receiver result in $N_{Tx}xN_{Rx}$ paths between transmitter and receiver. While the use of receive diversity and combining techniques are well studied, the use of multiple transmit antennas now allows us to better exploit the inherent diversity in the channel, through means such as space-time coding. In general, there is a tradeoff between achieving both multiplexing (increased spectral efficiency) and diversity (increased reliability) gains in a system.

Feedback to the transmitter, allowing it to exploit prevailing channel conditions is an important aspect to improving performance of a communications system. For instance, it can be used to facilitate link adaptation (eg: rate-selection and/or mode-selection – i.e. multiplexing/diversity). Another example of the use of feedback is in eigen-mode spatial multiplexing where knowledge of channel state information (CSI) is used to determine a set of transmit and receive matrices that when applied to a transmission, negate the tradeoff mentioned previously, realizing full diversity and multiplexing gains of the channel [3].

One of the key problems associated with the application of eigen-mode spatial multiplexing to practical systems, is the large amount of feedback information exchange required between receiver and transmitter. While TDD systems can limit feedback to initial transceiver calibration information, subsequently relying on channel reciprocality; they are limited to a specific set of assumptions [3]. In general, FDD (which is more prevalent in cellular systems) requires explicit feedback of channel-state. Previous work describes methods by which feedback is limited to an index parameter, which facilitates the choice of transmit pre-filter from a pre-determined set. Other methods include differential signaling and quantization schemes by which the receiver tracks channel variation and feeds-back only the difference between current and past channel-state (see [4] for references).

As is evinced by the discussion above, achieving high spectral efficiency and reliability with minimal feedback and complexity, are important design goals for future systems. This paper describes receiver techniques that achieve these goals. Section 2 outlines general system assumptions. Section 3 briefly introduces the concept of eigen-mode spatial multiplexing and subsequently describes different receiver approaches based on the assumption of infrequent feedback to the transmitter. In Section 4 we describe simulation parameters and results; and in Section 5, we conclude, with the implications of the results on the design of future wireless systems.

2. SYSTEM ASSUMPTIONS

The use of wideband signaling schemes needed to support higher data rates leads to more dominant multipath effects and consequently frequency selectivity. Of late, orthogonal frequency division multiplexing (OFDM) – a technique that uses frequency domain processing to transform a wideband selective fading channel into a group of narrowband flat fading subcarriers has gained popularity; primarily due to its ability to tackle multipath. Based on its many advantages, there is broad consensus within 3GPP on the adoption of OFDM for the downlink in the evolved UTRAN.

As each subcarrier of an OFDM system can be equalized separately, the receiver techniques described in the remainder of this paper shall be based on a single OFDM subcarrier. Also, as we are considering the application the proposed methods to the downlink of the system, the terms 'receiver' and 'transmitter' are often interchanged with 'UE' and 'BS', respectively.

3. RECEIVERS FOR REDUCED FEEDBACK

As mentioned previously, eigen-mode spatial multiplexing is an optimal space-time processing technique in that it achieves the full multiplexing and diversity gains of the channel. The tradeoff in realizing the benefits of eigen-mode spatial multiplexing lies in the need for CSI feedback. In an eigen-mode spatially multiplexed system the receiver estimates the channel, [H], and performs a transform such as a singular value decomposition (SVD) to determine the matrices of left and right-handed singular vectors – [U] and [V], respectively, as shown in (1).

$$[H] = [U] \cdot [D] \cdot [V]^H \tag{1}$$

The receiver feeds-back CSI to the transmitter, which uses the right-handed singular vectors of the channel, [V], to pre-filter its data [x]. The received signal [y] in (2) also contains a Gaussian noise component, [n].

Corresponding to the transmit filter, the receiver applies the unitary filter $[U]^H$ to estimate the transmitted data. From (3), it can be seen that the application of eigen-mode spatial multiplexing results in an SNR gain proportional to the square of the singular values of the channel – [D].

$$[y] = [H] \cdot [V] \cdot [x] + [n] \qquad (2)$$

$$[\hat{x}] = [U]^H \cdot [y] = [D] \cdot [x] + [U]^H \cdot [n] \qquad (3)$$

Real-world channels are time-varying and thus it is often the case that the transmit filter in use is not perfectly matched to the current channel state. Subscripting individual terms with 'time of feedback' (t_0) and 'current time' (t_n), the received signal in (2) can be re-written as (4). The receiver having knowledge of both – fed-back and current channel state, may derive a receive filter to estimate the transmitted data, as in (5).

$$[y_{t_n}] = [H_{t_n}] \cdot [V_{t_0}] \cdot [x_{t_n}] + [n] \qquad (4)$$

$$[\hat{x}_{t_n}] = [Rxf_{t_n}] \cdot [y_{t_n}] \qquad (5)$$

In the following, we establish different approaches of determining the receive filter. Linear detectors are considered owing to their lower complexity and ease of implementation.

3.1 Ignoring Channel Mismatch

The receive filter may be computed by ignoring the effects of channel mismatch as in (6), where, the receive filter is determined from the current channel state. The signal estimate in (7) suffers from cross-talk between spatial modes as the product $[V_{tn}]^H[V_{t0}]$ is not a perfectly diagonal matrix. However, due to the unitary nature of the receive filter, there is no change to the statistical properties of the noise term.

$$[Rxf_{t_n}] = [U_{t_n}]^H \qquad (6)$$

$$[\hat{x}_{t_n}] = [D_{t_n}] \cdot [V_{t_n}]^H \cdot [V_{t_0}] \cdot [x_{t_n}] + [U_{t_n}]^H \cdot [n] \qquad (7)$$

3.2 Inverse Cancellation Detector

The receive filter may be computed using the zero-forcing approach, as in (8). From the signal estimate in (9), it is apparent that this approach does not result in cross-talk between spatial-modes. However, the receive filter no longer being unitary, results in a coloring of the noise term.

$$[Rxf_{t_n}] = [D_{t_n}] \cdot [V_{t_0}]^H \cdot [V_{t_n}] \cdot [D_{t_n}]^{-1} \cdot [U_{t_n}]^H \qquad (8)$$

$$[\hat{x}_{t_n}] = [D_{t_n}] \cdot [x_{t_n}] + [D_{t_n}] \cdot [V_{t_0}]^H \cdot [V_{t_n}] \cdot [D_{t_n}]^{-1} \cdot [U_{t_n}]^H \cdot [n] \qquad (9)$$

3.3 Wiener Detector

The receive filters in (6) and (8) result in signal distortion and noise-coloring, respectively. An alternate approach to synthesizing the receive filter is based on using both CSI and statistical knowledge of the noise characteristics, such that the resultant cross-talk and noise terms negate one another. Equation (10) depicts such a filter based on the Wiener criterion of minimizing the mean error energy. In (10), [P$_{tn}$] represents the transmit power distribution across spatial modes; and σ^2, the noise variance at the receiver.

$$[Rxf_{t_n}] = \{[D_{t_n}] \cdot [P_{t_n}] \cdot ([H_{t_n}] \cdot [V_{t_0}])^H\} \cdot$$

$$\{[H_{t_n}] \cdot [V_{t_0}] \cdot [P_{t_n}] \cdot ([H_{t_n}] \cdot [V_{t_0}])^H + \sigma^2 [I]\}^{-1} \qquad (10)$$

4. PERFORMANCE EVALUATION

4.1 Simulation Results

Given cost and form-factor constraints of handheld cellular equipment, we assume a UE with two $\lambda/2$ spaced antennas. At the BS, we assume two transmit antennas, spaced at 10λ. We consider operation over 'MIMO Channel F', as specified in [5,6]. Model F was designed to simulate the effects of a hotspot micro-cell environment, typical of an outdoor open space, such as a city square. Although the primary Doppler component of the scattering environment is 6Hz, the third tap of the Doppler spectrum has a spread of 100Hz, in order to capture the effects of nearby vehicular traffic.

Figs. 1 & 2 respectively plot the mean signal to interference and noise ratio (SINR) at the receiver output for the two spatial-modes of the 2x2

MIMO system. The figures plot the SINRs at the detector output vs. the update interval – i.e. the interval at which the CSI is fed-back to the BS; for average input SNRs of 3dB and 15dB. Based on this model, the CSI at the BS is updated periodically, while the UE always has current CSI, resulting in a 'mismatch' between transmitter and receiver. The output SINR is computed as an average taken over the entire update interval.

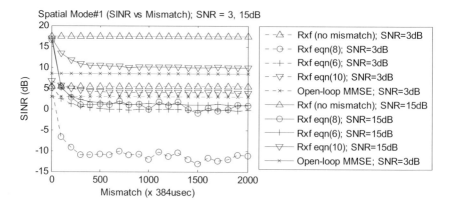

Figure 1. Output SINR vs Channel Mismatch - Spatial Mode#1

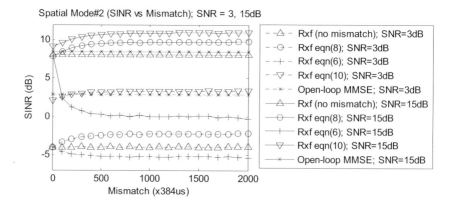

Figure 2. Output SINR vs Channel Mismatch - Spatial Mode#2

As can be seen from the results, the behavior of the output SINRs differs for different input SNRs and also for the two spatial modes. For both spatial modes, the Wiener detector is demonstrated to be the optimal linear receiver.

In the case of the first (stronger) spatial-mode, the performance of the detectors is found to deteriorate with increasing channel mismatch, with the detector in (6) performing significantly better than that in (8) for low SNRs (3dB), both having comparable performance at higher SNRs (15dB). As observed from Fig. 2 for the second spatial mode, the output SINR of the ICD and Wiener detectors are found to improve with increasing mismatch. The performance of the open-loop MMSE is also plotted for comparison.

Fig. 3 depicts the cumulative spectral efficiency of the different detectors vs. channel mismatch of the 2x2 MIMO system at the representative SNRs of 3dB and 15dB. As expected, the Wiener detector consistently outperforms other detectors. Also, at 3dB, the Wiener detector performs better than an 'ideal' eigen-mode spatially multiplexed system due to its inherent ability to compensate for noise.

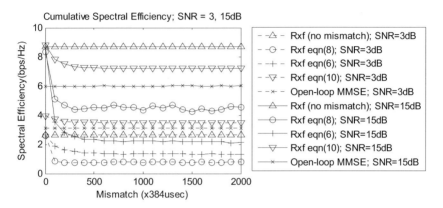

Figure 3. Cumulative Spectral Efficiency vs Channel Mismatch

4.2 Observations

As observed from Figs. 1 & 2, the SINRs (for detectors in (8) and (10)) of both spatial modes behave differently with respect to mismatch. While degradation in SINR wrt mismatch is expected (as for spatial-mode#1), the results of spatial mode#2 are counter-intuitive. For a typical 2x2 channel, the first singular value has a magnitude exceeding 1, while the second, less than 1. For the detectors in (8) and (10) it can be shown that the relative magnitudes of the singular values causes a noise suppression effect resulting in an overall increase in SINR. The performance of the detector in (6) is found to degrade with increasing mismatch, independent of spatial mode and SNR, owing to the increased inter-spatial mode cross-talk.

Although Fig. 3 gives us an indication of the potential capacity, in a real system this is determined by matching a defined modulation and coding scheme (MCS) to the SINR. While the results in this paper assume a uniform power distribution across spatial modes, based on the trend in Figs. 1 & 2, it can be concluded that power-loading may be performed to match the respective spatial modes to suitable MCSs, while operating at an increased mismatch. Operation at an increased mismatch directly translates to a reduction in feedback.

Based on Fig. 3, it can be seen that the after an initial degradation, cumulative spectral efficiency becomes relatively independent of mismatch. At saturation, these are approximately 20% (15dB) and 10% (3dB) better than the optimal open-loop linear receiver (MMSE). This observation suggests the feasibility of a 'pseudo open-loop' mode of operation which relies on relatively infrequent feedback, but, making use of improved receive processing, is able to realize improvements in spectral efficiency over its open-loop counterparts.

5. CONCLUSIONS

Closed-loop modes of operation are important for achieving high spectral efficiencies at reasonable implementation complexity (particularly in low-mobility scenarios). Eigen-mode spatial multiplexing – a linear processing approach, has been shown to realize maximal diversity and multiplexing gains of the MIMO channel. A key requirement for realizing these gains is the knowledge of CSI at the transmitter.

This paper addresses the issue of feedback, which is an expensive overhead that must be minimized. Through the use of receive processing techniques, we have demonstrated how CSI feedback requirements can be reduced, with minimal compromise on link-performance. Based on simulation results, we further describe a 'pseudo open-loop' mode of operation which realizes moderate capacity improvements over its open-loop counterparts. It is expected that further spectral efficiency improvements can be realized through the use of power-loading schemes.

6. REFERENCES

[1] 3GPP TR25.913v2.1.0
[2] J.R. Foschini & M.J. Gans, "On limits of wireless communications in fading environments when using multiple antennas," Wireless Personal Communications, pp. 36-54, Mar 1998.
[3] J.H. Sung, "Transmitter Strategies for Closed-loop MIMO-OFDM," Ph.D. thesis, Georgia Institute of Technology, Jul 2004.
[4] D.J. Love, et al, "What is the Value of Limited Feedback for MIMO Channels?", IEEE Communications Magazine, pp. 54-59, Oct 2004.
[5] J.P. Kermoal, et al, "A Stochastic MIMO Radio Channel Model with Experimental Validation", IEEE JSAC vol. 20 no. 6, Aug 2002.
[6] V. Erceg, et al, "TGn Channel Models", doc: IEEE 802.11-03/940r4, May 2004.

A MIMO-OFDM TRANSMISSION SCHEME EMPLOYING SUBCARRIER PHASE HOPPING

Satoshi Suyama, Kaito Tochihara, Hiroshi Suzuki, and Kazuhiko Fukawa
Tokyo Institute of Technology
2-12-1, O-okayama, Meguro-ku, Tokyo, 152-8550 Japan
{ssuyama, kite, suzuki, fukawa}@radio.ss.titech.ac.jp

Abstract This paper proposes a subcarrier phase hopping scheme for space division multiplexing (SPH-SDM) and applies it to MIMO-OFDM mobile communications. The SPH-SDM scheme is an extended version of the conventional SPH scheme for transmit diversity that enables SDM transmission. The scheme introduces a phase shift into the reference phase of each subcarrier for the transmitted data stream. A unitary random phase matrix carries out the phase shift for a one-packet duration, and the phase shift depends on the indices of streams, subcarriers, and transmit antennas. Such a phase shift can randomize channel frequency responses and reduce the channel correlation between subcarriers. A computer simulation demonstrates that SPH-SDM can improve transmission performance by exploiting the inherent frequency diversity effect, and that it can easily adjust the tradeoff between the diversity gain and the transmission rate.

1. Introduction

MIMO-OFDM transmission has attracted much attention in wireless communications because it can realize more reliable and higher bit-rate transmission systems [1], [2]. A combination of MIMO-OFDM and channel coding achieves good performance even in frequency selective fading channels by exploiting the frequency diversity effect. However, it cannot completely exploit an ability of the channel coding if the delay spread of the channel is small and the interleaver size for the coding is insufficient. To overcome this impairment, the phase diversity and the cyclic delay diversity, which are classified as the transmit diversity (TD) techniques, have been proposed [3]-[6]. Subcarrier phase hopping for the transmit diversity (SPH-TD) is the phase diversity which transmits OFDM signals with the subcarriers reference phases that depend on indices of streams, subcarriers and transmit antennas. Although SPH-TD is effective for improving the frequency diversity effect, it cannot be directly applied to space division multiplexing (SDM) in MIMO-OFDM because it increases the correlation among MIMO channels by SPH.

This paper proposes a new type of subcarrier phase hopping for SDM (SPH-SDM). SPH-SDM is an extended version of SPH-TD for SDM and employs a unitary phase matrix to suppress the channel correlation due to SPH. It is shown that SPH-SDM can improve transmission performance by exploiting the inherent frequency diversity effect.

2. MIMO-OFDM Employing SPH-SDM

2.1 System Model

A block diagram of the MIMO-OFDM transmitter with SPH-SDM using L_T antennas is shown in **Fig. 1**. First, information bits are passed into the cyclic redundancy check (CRC) encoder and are then divided into M data streams where $1 \leq M \leq L_T$. Next, each stream is encoded by a turbo code, and after interleaving the coded bits, the transmitter maps them into a modulation signal at each subcarrier.

Furthermore, the transmitter carries out the subcarrier phase hopping (SPH) for the modulation signals of all streams. SPH shifts the reference carrier phases of the modulation signals for a one-packet duration based on the indices of streams, subcarriers, and transmit antennas. This operation is equivalent to multiplying a modulation signal vector by a random phase matrix, as described in section 2.2. The modulation signals with SPH are converted into a time-domain signal by using IFFT followed by guard interval (GI) insertion.

A block diagram of the MIMO-OFDM receiver with SPH-SDM using L_R antennas is shown in **Fig. 2**. The configuration is the same as that

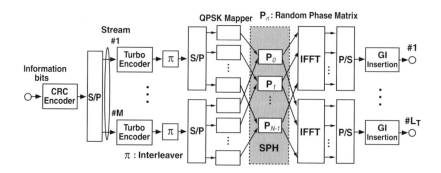

Figure 1. MIMO-OFDM transmitter with SPH-SDM

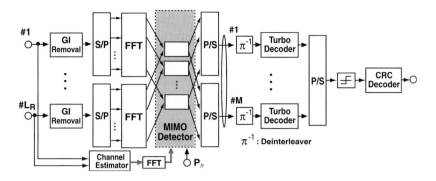

Figure 2. MIMO-OFDM receiver with SPH-SDM

of the conventional MIMO-OFDM receiver except for processing related to the random phase matrix. The receiver first carries out the channel estimation by using the preamble of the packet. Next, the conventional MIMO signal detector, such as a maximum likelihood detector (MLD) and MMSE detector (MMSED), extracts the M data streams by using an equivalent channel matrix. The derivation of the equivalent channel matrix is described in section 2.2. After deinterleaving all soft decisions, the turbo decoder regenerates the information bits and the CRC decoder detects a decision error.

2.2 Signal Model

An L_T-by-1 modulation signal vector $\mathbf{s}_{n,i}$ of the i-th OFDM symbol $(i = 1, 2, \ldots, N_i)$ at the n-th subcarrier $(n = 1, 2, \ldots, N)$ is defined as

$$\mathbf{s}_{n,i}^{\mathrm{T}} = (s_{1,n,i} \ s_{2,n,i} \ \cdots \ s_{M,n,i} \ 0 \ \cdots 0) \tag{1}$$

where the superscript ^{T} denotes transposition and $s_{m,n,i}$ is the modulation signal for the m-th stream $(m = 1, 2, \ldots, M)$, with $M < L_T$. Using an L_T-by-L_T random phase matrix $\mathbf{P}_n$ at the n-th subcarrier, an L_T-by-1 modulation signal vector with SPH $\mathbf{z}_{n,i}$ is given by

$$\mathbf{z}_{n,i} = \mathbf{P}_n \, \mathbf{s}_{n,i} \tag{2}$$

$\mathbf{P}_n$ becomes

$$\mathbf{P}_n = \frac{1}{\sqrt{L_T}} \begin{pmatrix} e^{j\phi_{11,n}} & \cdots & e^{j\phi_{1L_T,n}} \\ \vdots & \ddots & \vdots \\ e^{j\phi_{L_T 1,n}} & \cdots & e^{j\phi_{L_T L_T,n}} \end{pmatrix} = \begin{pmatrix} \mathbf{p}_{1,n} & \cdots & \mathbf{p}_{L_T,n} \end{pmatrix} \tag{3}$$

where $\mathbf{p}_{l_T,n}$ ($l_T = 1, 2, \ldots, L_T$) is an L_T-by-1 phase vector and $\phi_{pq,n}$ ($p, q = 1, 2, \ldots, L_T$) is a random variable that is uniformly distributed from 0 to 2π. The values of the phases change depending on the three indices of the streams, subcarriers, and transmit antennas, and remain constant during one packet.

Next, let $\mathbf{r}_{n,i}$ denote an L_R-by-1 received signal vector at the n-th subcarrier that consists of the received signals after FFT, and let $H_{l_R,l_T,n}$ ($l_R = 1, 2, \ldots, L_R$) denote a channel frequency response from the l_T-th transmit antenna to the l_R-th receive one. In addition, the channel is assumed to be time-invariant during one packet. Using an L_R-by-L_T channel matrix consisting of $H_{l_R,l_T,n}$

$$\mathbf{H}_n = \begin{pmatrix} H_{1,1,n} & \cdots & H_{1,L_T,n} \\ \vdots & \ddots & \vdots \\ H_{L_R,1,n} & \cdots & H_{L_R,L_T,n} \end{pmatrix} = \begin{pmatrix} \mathbf{H}_{1,n}^T \\ \vdots \\ \mathbf{H}_{L_R,n}^T \end{pmatrix} \quad (4)$$

$\mathbf{r}_{n,i}$ can be expressed as

$$\mathbf{r}_{n,i} = \mathbf{H}_n \, \mathbf{z}_{n,i} + \mathbf{n}_{n,i} = \mathbf{H}_{e,n} \, \mathbf{s}_{n,i} + \mathbf{n}_{n,i} \quad (5)$$

where $\mathbf{n}_{n,i}$ is an L_R-by-1 noise vector and $\mathbf{H}_{e,n}$ is the equivalent channel matrix, which is given by

$$\mathbf{H}_{e,n} = \mathbf{H}_n \mathbf{P}_n = \begin{pmatrix} \mathbf{H}_{1,n}^T \mathbf{P}_{1,n} & \cdots & \mathbf{H}_{1,n}^T \mathbf{P}_{L_T,n} \\ \vdots & \ddots & \vdots \\ \mathbf{H}_{L_R,n}^T \mathbf{P}_{1,n} & \cdots & \mathbf{H}_{L_R,n}^T \mathbf{P}_{L_T,n} \end{pmatrix} \quad (6)$$

2.3 Random Phase Matrix

The SPH-SDM receiver extracts the M data streams by using $\mathbf{H}_{e,n}$. Correlation between the elements of $\mathbf{H}_{e,n}$ due to $\mathbf{P}_n$ causes performance degradation of the signal detection. However, SPH-SDM can avoid this degradation by using $\mathbf{P}_n$ of a unitary matrix. This is because transformation by the unitary matrix does not change the statistical properties, and the autocorrelation matrix of $\mathbf{H}_{e,n}$ is equal to that of $\mathbf{H}_n$ as

$$\langle \mathbf{H}_{e,n} \mathbf{H}_{e,n}^H \rangle = \langle \mathbf{H}_n \mathbf{P}_n \mathbf{P}_n^H \mathbf{H}_n^H \rangle = \langle \mathbf{H}_n \mathbf{H}_n^H \rangle \quad (7)$$

where $\langle \ \rangle$ and the superscript ^{H} denote an ensemble average and a Hermitian conjugate, respectively. Furthermore, to conduct the phase shift by $\mathbf{P}_n$, the proposed scheme employs the unitary matrix $\mathbf{P}_n$ of which row vectors are equal to those of the Walsh-Hadamard matrix multiplied

A MIMO-OFDM Transmission Scheme Employing SPH

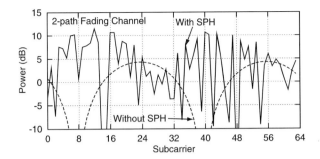

Figure 3. Channel frequency response randomized by SPH

by the phases. With $L_T = 4$, $\mathbf{P}_n$ becomes

$$\mathbf{P}_n = \begin{pmatrix} \mathbf{p}_{1,n} & \mathbf{p}_{2,n} & \mathbf{p}_{3,n} & \mathbf{p}_{4,n} \end{pmatrix}$$

$$= \frac{1}{2} \begin{pmatrix} e^{j\phi_{1,n}} & e^{j\phi_{1,n}} & e^{j\phi_{1,n}} & e^{j\phi_{1,n}} \\ e^{j\phi_{2,n}} & -e^{j\phi_{2,n}} & e^{j\phi_{2,n}} & -e^{j\phi_{2,n}} \\ e^{j\phi_{3,n}} & e^{j\phi_{3,n}} & -e^{j\phi_{3,n}} & -e^{j\phi_{3,n}} \\ e^{j\phi_{4,n}} & -e^{j\phi_{4,n}} & -e^{j\phi_{4,n}} & e^{j\phi_{4,n}} \end{pmatrix} \quad (8)$$

where the phase $\phi_{l_T,n}$ depends on the indices of the transmit antenna l_T and the subcarrier n, and $\mathbf{p}_{l_1,n}^H \mathbf{p}_{l_2,n} = \delta_{l_1,l_2}$ ($l_1, l_2 = 1, 2, 3, 4$) where δ_{l_1,l_2} is the Kronecker's δ. Therefore, the streams become orthogonal to each other. The value of L_T is equal to the power of 2 because $\mathbf{P}_n$ is based on the Walsh-Hadamard matrix.

One squared element of $\mathbf{H}_{e,n}$, $|\mathbf{H}_{1,n}^T \mathbf{p}_{1,n}|^2$ is shown in **Fig. 3**. It is found that the channel frequency response in a 2-path fading channel is randomized by SPH, and the correlation of the channel between the subcarriers decreased if the delay spread of the channel was small.

2.4 Transmit Diversity Effect of SPH-SDM

In the case of $M = 1$, SPH-SDM works as the SPH-TD scheme with the L_T transmit antennas. Then, $\mathbf{s}_{n,i}$ of (1) is given by $\mathbf{s}_{n,i}^T = (s_{1,n,i}\ 0\ \cdots\ 0)$, and thus $\mathbf{z}_{n,i}$ is given by

$$\mathbf{z}_{n,i} = \mathbf{P}_n \mathbf{s}_{n,i} = \mathbf{p}_{1,n} s_{1,n,i} \quad (9)$$

Thus, the above equation is equivalent to SPH-TD, that is, the phase diversity [3].

Furthermore, SPH-SDM can obtain a diversity gain corresponding to the $(L_T - M)$ antennas by transmitting the M streams with SPH from the L_T antennas. It also can easily adjust the tradeoff between the diversity gain and the transmission rate by changing M.

Table 1 SIMULATION CONDITIONS

Number of antennas	L_T: 4, L_R: 4
Number of streams M	Variable (from 1 to 4)
Transmission rate	($M \times 12$) Mbit/s
FFT point	64
Symbol duration	4.0 μs (GI: 0.8 μs)
Modulation	QPSK
Channel code	R=1/2 (punctured), K=4, turbo code
Turbo decoding	Max-Log-MAP algorithm (6 iter.)
Detection scheme	MLD, MMSED
Channel estimation	Ideal
Channel model	16-path with exponential decay
RMS dealy spread	147 ns
Maximum Doppler freq. f_D	0 Hz

These processes for SPH-SDM are performed without channel state information on the transmitter side. Therefore, the SPH-SDM scheme can be realized by a simple structure of the MIMO-OFDM transceiver and is suitable for hardware implementation. An FPGA-board simulator for MIMO-OFDM with SPH-SDM has been implemented [7].

3. Computer Simulation

3.1 Simulation Conditions

Computer simulation to verify the transmission performances of the SPH-SDM transceiver was conducted. The simulation conditions listed in **Table 1** follow the extended versions of 5-GHz wireless LAN [8]. The numbers of antennas, L_T and L_R, were set to 4, and M was changed from 1 to 4. The modulation scheme was QPSK, and the turbo code with the constrained length $K = 4$ and the punctured coding rate $R = 1/2$ was used as the channel coding. The transmission rate was ($M \times 12$) Mbit/s in this modulation and coding. The frequency block interleaving was carried out symbol by symbol along the bit sequence for the subcarriers in one symbol. MLD or MMSED extracted the M streams by using $\mathbf{P}_n$, which was assumed to be known in the receiver. The channel estimation was assumed to be ideal. The channel model was 16-path Rayleigh fading with the average power of each path decaying exponentially and the maximum Doppler frequency f_D was set to 0 Hz.

3.2 Simulation Results

The packet error rate (PER) performance for both MMSED and MLD using SPH-SDM with $M = 4$ is shown in **Fig. 4**. For comparison, curves

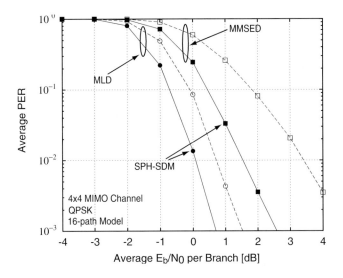

Figure 4. Average PER performance

of the conventional MMSED and MLD were also plotted. Applying SPH-SDM to MMSED and MLD, the receiver can gain 2.0 dB and 0.6 dB of the average E_b/N_0 to achieve PER of 10^{-2} in comparison with MMSED and MLD, respectively. This is because SPH-SDM can randomize channel frequency responses and thus can exploit the frequency diversity effect of channel coding. Moreover, this demonstrates that MMSED employing SPH-SDM can have good performance close to that of the conventional MLD despite the lower complexity, and that the result is effective for the hardware implementation of the MIMO detector.

The throughput performance of MMSED employing SPH-SDM with M as a parameter is shown in **Fig. 5**. It is found that SPH-SDM can realize the most reliable transmission at $M = 1$, while it can achieve the highest transmission rate, 48 Mbit/s, at $M = 4$. In addition, it can easily optimize the tradeoff between the diversity gain and the transmission rate by changing M according to the system conditions.

4. Conclusions

This paper has proposed a MIMO-OFDM transceiver with SPH-SDM that can exploit the inherent frequency diversity effect of channel coding. The SPH-SDM transmitter shifts the phases for the transmitted data streams by multiplying the unitary random phase matrix, and the SPH-SDM receiver extracts the streams by using the equivalent chan-

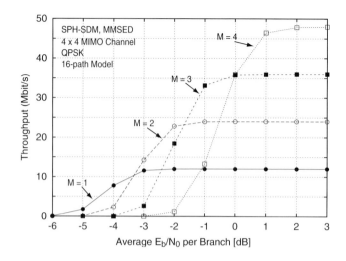

Figure 5. Throughput performance

nel matrix. Computer simulation has demonstrated that the proposed SPH-SDM transceiver with QPSK can achieve 2.0 dB and 0.6 dB gains in average E_b/N_0 at PER of 10^{-2} compared to the conventional MMSED and MLD, respectively, in 4×4 MIMO channel. In addition, it can easily optimize the tradeoff between the diversity gain and the transmission rate.

References

[1] H. Sampath, S. Talwar, J. Tellado, V. Erceg, and A. Paulraj, "A fourth-generation MIMO-OFDM broadband wireless system: design, performance, and field trial results," *IEEE Comm. Mag.*, vol. 40, no. 9, pp. 143-149, Sept. 2002.

[2] A. van Zelst, R. van Nee, and G. A. Awater, "Space division multiplexing (SDM) for OFDM systems," *IEEE VTC 2000-Spring*, pp. 1070-1074, May 2000.

[3] S. Kaiser, "Spatial transmit diversity techniques for broadband OFDM systems," *IEEE GLOBECOM '00*, vol. 3, pp. 1824-1828, Nov.-Dec. 2000.

[4] D. Gore, S. Sandhu, and A. Paulraj, "Delay diversity codes for frequency selective channels," *IEEE ICC 2002*, pp. 1949-1953, April 2002.

[5] H. Suzuki, "Performance of a new adaptive diversity-equalization for digital mobile radio," *IEE Electronics Letters*, vol. 26, no. 10, pp. 626-627, May. 1990.

[6] G. Bauch, "Capacity optimization of cyclic delay diversity," *IEEE VTC 2004-Fall*, vol. 3, pp. 1820-1824, Sept. 2004.

[7] S. Suyama, Y. Sagae, S. Suzuki, and K. Fukawa, "An FPGA-board simulator for MIMO-OFDM transmission," *Intern. OFDM workshop*, pp. 293-297, Aug. 2005.

[8] IEEE Std 802.11a, *High-speed Physical Layer in the 5 GHz Band*, 1999.

MULTICARRIER SDMA SYSTEM WITH REDUCED INTRA-USER CROSS-CORRELATIONS

Nadem H. Dawod, Roshdy Hafez Ph.D, Ian Marsland Ph.D
Dept. of Systems and Computer Eng., Carleton University, Ottawa, Canada

Abstract: The ability of the basestation to separate several users spatially depends on the pair wise cross-correlations between the channel matrices of the users. In this paper we propose an improved null steering downlink MIMO-OFDM system that reduces the intra-user correlations resulting in significant enhancement in system performance. In this system, several basestation multi-antenna arrays are distributed in a given area. Each array is communicating with the basestation via optical fiber links. All signal processing is performed at the basestation. Multi-antenna users are spatially separated such that in each tone of the OFDM symbol only a certain number of users (out of the total number of users) are selected. The selection is based on an algorithm that reduces the pair-wise intra-user correlations. The channel matrix of each user is assumed correlated (generated based on certain angular spectrum and random angular spread) and Ricean distributed (each array has a random K-factor). Several data symbols can be spatially multiplexed to each user over each OFDM tone with high reliability and with very good total system capacity.

Key words: MIMO, OFDM, SDMA, Distributed Antennas

1. INTRODUCTION

Multiple input multiple output (MIMO) systems, orthogonal frequency division multiplexing (OFDM), space division multiple access (SDMA) and radio over fiber (RoF) [1,2] are four techniques that, when combined together, can provide very spectrally efficient data transmission and thereby meet the high speed requirements of future generations of wireless systems.

The correlations between the entries of a user channel matrix and the pair wise cross-correlations between the channel matrices of the mobile users can severely degrade the system performance[3,4]. Many references, like [5,6,7,8], investigated several SDMA methods, but non of them considered the correlations issue. This paper addresses these problems and proposes a system that deals with these correlation issues.

The organization of this paper is as follows. In section 2, a Ricean-correlated channel model of a mobile user is defined. Section 3 presents an improved downlink SDMA system for multi antenna users. Section 4 provides simulation results and discussion. Finally, the conclusions are given in section 5.

2. CHANNEL MODEL

A number of antenna arrays, n_P, are distributed around a microcell. Each one of them is composed of n_L antennas. The arrays are placed at low heights (5~20 meters) above the ground. All of the arrays are connected via optical fibers to a central base station where all signal processing is performed. The mobile station (user) has n_R antennas. The MIMO system model, seen by each user, is now composed of $n_P n_L$ antennas at the base station side and n_R antennas at the user side. The channel is assumed to be Ricean, frequency selective and quasi-static and with known CSI at the basestation. The correlation properties between the antennas within the array depend on the scattering environment and on the antenna configuration.

The lth tap of the $n_R \times n_P n_L$ time domain user's channel matrix is represented as:

$$\mathbf{G}_l = \begin{bmatrix} \mathbf{G}_{l,1}, \mathbf{G}_{l,2}, \ldots, \mathbf{G}_{l,n_P} \end{bmatrix}$$

$$\mathbf{G}_{l,p} = \begin{cases} PDP_l \, PL_p \left(\sqrt{\dfrac{K_p}{K_p+1}} \beta_{LOS,p} \mathbf{G}_{LOS,p} + \sqrt{\dfrac{1}{K_p+1}} \beta_{Ray,l,p} \mathbf{R}_{R,p}^{1/2} \mathbf{G}_{Ray,l,p} \mathbf{R}_{T,p}^{1/2} \right) & \text{if } l=0 \\ PDP_l \, PL_p \beta_{Ray,l,p} \mathbf{R}_{R,p}^{1/2} \mathbf{G}_{Ray,l,p} \mathbf{R}_{T,p}^{1/2} & \text{if } l \neq 0 \end{cases}$$

where $\mathbf{G}_{l,p}$ is the $n_R \times n_L$ channel matrix between the Pth array and the user antennas, $\mathbf{R}_{R,p}$ is the $n_R \times n_R$ correlation matrix at the user's side, $\mathbf{R}_{T,p}$ is the

$n_L \times n_L$ correlation matrix at the Pth array, $\mathbf{G}_{LOS,l,p}$ and $\mathbf{G}_{Ray,l,p}$ are the $n_R \times n_L$ channel matrices between the Pth array and the user for the LOS and the scattering components of the channel, respectively, $\beta_{LOS,p}$ and $\beta_{Ray,l,p}$ are lognormal random variables that represent the shadowing for the LOS and the scattering components of the channel, respectively, PL_p is the normalized pathloss for the Pth array, PDP_l is the normalized gain of the power delay profile of the lth tap, K_p is the Ricean K-factor for the Pth array. The entries of $\mathbf{R}_{T,p}$ and $\mathbf{R}_{R,p}$ are found according to [9].

3. MULTI-USER DOWNLINK NULL STEERING IN A MIMO-OFDM SYSTEM

Two types of correlations has to be taken into account: the correlation between the entries of the channel matrix of each user and the correlations between the channel matrix of a user to that of other users.

In order to achieve spatial separability between different users, we intend to decrease the cross-correlation between the channel matrices of the users using a cross-correlation reduction algorithm.

3.1 Cross-Correlation Reduction Algorithm

- Find all possible sub-groups of n_S users (out of n_U users).
- Loop over all OFDM tones
- Find the pair-wise cross correlation coefficient between the channel matrices of each pair in the sub-group, according to Pearson product-moment correlation:

$$\rho = \frac{\sum_m \sum_n \left(H^p_{mn} - \overline{H^p}\right)\left(H^q_{mn} - \overline{H^q}\right)}{\sqrt{\left(\sum_m \sum_n \left(H^p_{mn} - \overline{H^p}\right)^2\right)\left(\sum_m \sum_n \left(H^q_{mn} - \overline{H^q}\right)^2\right)}}$$

- Find the minimum mean correlation coefficient of each sub-group.
- Choose the sub-group of users with the lowest minimum mean.
- Repeat this over all other tones of the OFDM symbol.

3.2　Signal Model

The channel seen by each tone (of the OFDM symbol) for the kth user is:

$$\mathbf{H}_k = [\mathbf{H}_1^k, \mathbf{H}_2^k, ..., \mathbf{H}_{n_p}^k]$$

$$\mathbf{H}_i^k = \begin{pmatrix} h_{11}^{ik} & h_{12}^{ik} & \cdots & h_{1n_L}^{ik} \\ h_{21}^{ik} & h_{22}^{ik} & \cdots & \vdots \\ \vdots & \vdots & . & \vdots \\ h_{n_R 1}^{ik} & \vdots & . & h_{n_R n_L}^{ik} \end{pmatrix}$$

where $\mathbf{H}_i^k$ is the channel matrix from the kth user antennas to the ith array which is derived from the time domain channel described in section two.

Now the channel seen by the basestation to all users' antennas is

$$\mathbf{H} = \begin{bmatrix} \mathbf{H}_1^T & \mathbf{H}_2^T & \cdots & \mathbf{H}_{n_U}^T \end{bmatrix}^T$$

Our goal is to spatially multiplex to each user as much symbols as the rank of his effective channel matrix, which can be done by eigen-beamforming.

The vector of signals received by all users is:

$$\mathbf{r} = \mathbf{HWd} + \mathbf{n}$$

where:

$$\mathbf{d} = [\mathbf{d}_1, \mathbf{d}_2, \cdots, \mathbf{d}_{n_u}]^T$$
$$\mathbf{d}_k = [d_k^1, d_k^2, \cdots, d_k^{size(\mathbf{W},2)}]$$
$$\mathbf{r} = [\mathbf{r}_1, \mathbf{r}_2, \cdots, \mathbf{r}_{n_u}]^T$$
$$\mathbf{r}_k = [r_1, r_2, \cdots, r_{n_R}]$$

The row vector $\mathbf{d}_k$ is a modulated data vector transmitted to the kth user, $\mathbf{r}_k$ is the received data vector at the kth user antennas, $\mathbf{W}$ is the null steering matrix used for all users and $\mathbf{n}$ is an additive zero mean complex Gaussian noise vector at the antennas of the users.

In order for the kth user vector, $\mathbf{d}_k$, not to be seen by the other users, it should be rotated towards the null space of the total channel matrix, $\mathbf{H}_k^C$, of the other users:

$$\mathbf{H}_k^C = \left[\mathbf{H}_1^T, \cdots, \mathbf{H}_{k-1}^T, \mathbf{H}_{k+1}^T, \cdots, \mathbf{H}_{n_u}^T \right]$$

$$\mathbf{H}_k^C = \mathbf{U}_{\mathbf{H}_k^C} \Lambda_{\mathbf{H}_k^C} (\mathbf{V}_{\mathbf{H}_k^C})^H$$

If $\mathbf{W}_k$ is the matrix whose columns are the last ($n_p.n_L - n_R (n_U-1)$) columns of $\mathbf{V}_{\mathbf{H}_k^C}$ and

$$\mathbf{W} = [\mathbf{W}_1, \mathbf{W}_2, \cdots, \mathbf{W}_{n_u}]$$

then (for all users),

$$\mathbf{H}_k^C \mathbf{W}_k = 0$$

The last equation implements the zero interference constraint which means that a signal transmitted to a user is nulled to all other users. The efficiency of this null steering depends mainly on the intra-user cross correlations. For this null steering to work, two conditions has to be fulfilled, we should have at least one null singular vector for each of the (n_U-1) users and the matrices in $\mathbf{H}_k^C$ has to be pair-wise less correlated.

When the basestation antennas are distributed then the users instead of communicating with one point, they will communicate with an enlarged space dimension (multi-point basestation) and when exploiting the frequency dimension, represented by the OFDM tones, the spatial separation ability of users can be much improved using the cross-correlation reduction algorithm mentioned above. In this case n_U is replaced by n_S in the previous signal model, where n_S is the number of selected users (out of n_U) for each tone of the OFDM symbol and the total channel matrix,, will be reduced to

$$\mathbf{H} = \left[\mathbf{H}_1^T, \mathbf{H}_2^T, \cdots, \mathbf{H}_{n_s}^T \right]^T$$

4. SIMULATION RESULTS AND DISCUSSION

We used a 64 tone OFDM symbol each loaded with a QPSK signal. A Ricean-correlated channel matrix is generated for each array assuming a Gaussian angular spectrum with angular spread ranging randomly between 5-30 degrees. The angles of departure are also randomly generated. A Ricean K-factor is randomly selected between 0-10 dB for each array matrix. When the antennas of the basestation are assumed co-located, a Laplacian angular spectrum is assumed with 2 degrees angular spread.

Fig. 1 depicts the case when the basestation antenna arrays are distributed around the cell ($n_P=8$) with two antennas each ($n_L=2$) and we intentionally correlated the pair-wise channel matrices of users 1&2 (out of 5 users). No cross-correlation reduction algorithm is used. Very bad BER curves for users 1&2 is recognized, which necessitates dropping users 1&2 or the collapse of the SDMA system.

Fig. 2 examines the performance of the system when the number of correlated users is increased. Curves 1&2 show a bad performance when no user selection is used. Curve 3 shows the case when users 1&2 and 3&4 are intra-correlated, but using the cross-correlation reduction algorithm, the

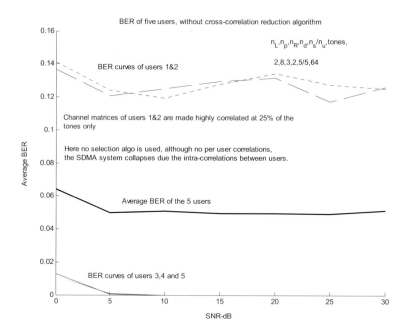

Figure 1. Average BER, without using cross-correlation reduction algorithm

SDMA MIMO-OFDM system works very good. Same good performance is seen in curve 4 when 3 pairs of the users (1&2, 3&4 and 5&6) are intra-correlated. The more correlated users the less number of total gained tones by each user.

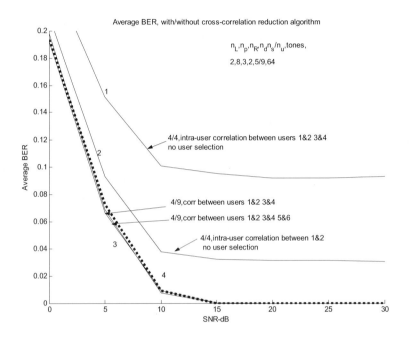

Figure 2. Average BER, with/without using cross-correlation reduction algorithm

5. CONCLUSION

We have proposed an SDMA MIMO-OFDM scheme that, based on null steering, spatially multiplexes to each user several data symbols in each channel use, even if the users' channel matrices seen by several OFDM tones are cross-correlated. The system exploits the enlarged space and the frequency domain to guarantee uncorrelation between the channel matrices of the users who share the same OFDM tone. A cross-correlation reduction algorithm was used for this target. Computer simulations show significant improvement in the performance of all mobile users without dropping any one of them.

References

[1] Nadem H. Dawod, R. Hafez, and I. Marsland, "Performance of MIMO-OFDM in widely spaced antenna arrays", Wireless 2004, Calgary, Canada, July, pp. 42–46.

[2] I. Haroun, F. Gouin and R.H.M. Hafez, "Experimental study of 802.11a WLAN over optical fiber", Microwave Magazine, May, 2003.

[3] Nadem H. Dawod, R. Hafez, and I. Marsland, "Performance of Widely Spaced Antenna Arrays in Correlated-Ricean MIMO-OFDM Channels", WWC'2005 May 2005. San Francisco USA.

[4] H. Bolcskei, D. Gesbert and A. Paulraj. "On the capacity of OFDM-based spatial multiplexing systems", IEEE Transactions on Communications, vol. 50, pp. 225–234, February 2002.

[5] Nishimura. T, Ohgane. T, Ogawa. Y, Doi. Y, Kitakado. J, "Downlink beamforming performance for an SDMA terminal with joint detection", Vehicular Technology Conference, 2001. VTC 2001 Fall. IEEE VTS 54th, Volume: 3, 7-11 Oct. 2001 Pages: 1538–1542.

[6] Minjoong Rim, "Multi-user downlink beamforming with multiple transmit and receive antennas", Electronics Letters, Volume: 38, Issue: 25, 5 Dec. 2002 Pages: 1725–1726.

[7] Lai-U. Choi, Murch. R.D. "A transmit preprocessing technique for multiuser MIMO systems using a decomposition approach", Wireless Communications, IEEE Transactions on, Volume: 3, Issue: 1, Jan. 2004 Pages: 20–24.

[8] Spencer. Q.H., Swindlehurst. A.L., Haardt. M, "Zero-forcing methods for downlink spatial multiplexing in multiuser MIMO channels", Signal Processing, IEEE Transactions on, Volume: 52, Issue: 2, Feb. 2004 Pages: 461–471.

[9] B. Vucetic and J. Yuan, *Space-Time Coding*, May 2003, ISBN: 0-470-84757-3.

EFFECT OF ADAPTIVE MODULATION AND ERROR CORRECTION ON MULTI-BAND OFDM-MIMO SYSTEM

Mitsugu Ohkawa
Next-generation Laser Communication Satellite Technology Research Center,
National Institute of Information and Communications Technology(NICT)
1-33-16 Hakusan, Bunkyo-ku, Tokyo, 113-0001, Japan
okawa@nict.go.jp

Ryuji Kohno
Division of Physics, Electrical and Computer Engineering, Yokohama National University
79-5 Tokiwadai, Hodogaya-ku, Yokohama, 240-8501, Japan
kohno@ynu.ac.jp

Abstract We have devised a new system for forming the antenna beam in a combined OFDM and MIMO scheme. In this scheme, the OFDM signal is divided among multiple bands, and the divided signals are transmitted on an appropriate beam of each sub-band. In this paper, We have described the system configuration of the proposed multi-band OFDM-MIMO scheme for use over multipath wireless channels. And, we have discussed the performance of adaptive modulation and error correction of the system.

Keywords: MIMO, OFDM, adaptive modulation, maximal ratio combining

1. Introduction

As multimedia services continue to grow in popularity, there will be increasing pressure regarding frequency band availability. To help alleviate this pressure, we need to establish highly efficient signal transmission technologies which will enable improved frequency and power efficiency for radio propagation and signal transmission.

In this paper, we introduce the effect of adaptive modulation and error correction on a multi-band OFDM-MIMO system in which multi-band orthogonal frequency division multiplexing (OFDM) is combined with the multiple input, multiple output (MIMO) concept[1,2,3]. The MIMO method separates and recomposes signals on the reception side using many independent radio propagation paths so that different multiple signals can be transmitted on the same frequency band. We have developed a method that allows the spectrum of multi-band OFDM signals to be adaptively shared to improve the path gain for MIMO transmission using adaptive modulation and error correction. In this system, the OFDM signal is divided into multi-bands, and the divided signals are transmitted on the appropriate beam in each multi-band.

The MIMO paths consist of parallel beams because of the beam-forming antenna. Multi-band OFDM is used as the modulation scheme so the carriers are orthogonal to each other. In the receiver, the OFDM signals of each MIMO path are synthesized. In radio propagation where obstacles might be a problem, the proposed system can construct signal paths to improve the utilization efficiency with respect to frequency and power.

The multi-path delay can be absorbed by the OFDM guard interval, and the effect of the time delay equalizer is expected in the multi carrier of OFDM (guard interval). The proposed method enables construction of an adaptive modulation and error correction controlling system over the frequency selected propagation environment where the H matrix changes over time. An adjustment system can be used which changes the weights through an adjustment algorithm. This adjustment system changes the MIMO weights and path gain weights by changing the H matrix while controlling the beams to obtain the best possible gain.

2. Multi-Band OFDM-MIMO System[4]

The configuration of the proposed multi-band OFDM-MIMO scheme is shown in Figure 1. Multi-band OFDM can use various modulation methods (BPSK, QPSK, 8PSK, etc.) in which the carriers are orthogonal to each other. The advantageous sub-band paths and disadvantageous sub-band paths are classified into groups depending on the power capacity. Also, the OFDM power distribution is controlled over the multiple bands because the frequency-selective characteristic is compensated.

After that, the MIMO signals of each group are transmitted through overlapping frequency bands. Each group makes up a MIMO transmission system for an advantageous power environment given the wireless link situation, and the MIMO weights for each sub-carrier are set.

Beam sets which adjust to the broadband frequency characteristic cooperatively with the beams of other sub-bands are prepared, because

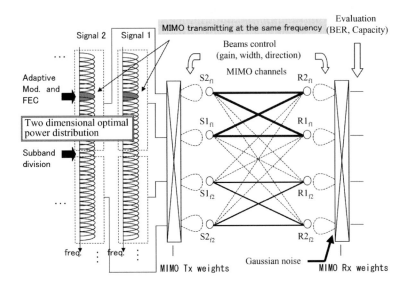

Figure 1. Institute of Communications and Navigation

multiple beams are needed to provide many independent paths in MIMO transmission.

When multiple beams are formed, OFDM signals should not use any frequency band where the receiving power is significantly attenuated. A beam will be formed in the frequency band which can be used given the band's power capacity.

This multi-band OFDM-MIMO scheme controls the link parameters of two stages related to the frequency domain and space domain. One of these OFDM power distribution control stages forms beam sets adjusted in the broad frequency band, and the other performs the S/N maximal ratio combining (MRC) for the MIMO multiple transmission. After a beam is formed and the state of the OFDM power distribution decided, the S/N MRC for each MIMO eigenmode is performed at the same frequency for the signals of each sub-carrier. The OFDM signals for each MIMO path are synthesized for the sub-carriers in the receiver.

Adaptive control is applied for the symbol time according to changes in the channels. In adaptive control, the channel estimation is done using pilot symbols at the OFDM frequency and in the MIMO eigenvalue domain, and the H matrix is estimated. The eigenvalues and the receiving power per sub-band are derived as average values over a short time period. The selection of modulation and forward error correction

(FEC) are adaptively performed for each sub-band and eigenvalue using the water-filling algorithm as the optimum power distribution.

In the case of radio propagation in the presence of obstacles, this system can create appropriate signal paths at high frequency and power utilization efficiency. In the adaptive transmission control, the processing of calculation can be decreased only the ratio of the number of subband and subcarrier.

3. Performance analysis of Multi-band OFDM-MIMO scheme

3.1 Multi-band OFDM-MIMO signal

In the multi-band OFDM-MIMO signal, the multi-carrier complex transmission signal, $s(t)$, is expressed as

$$s(t) = \sum_{i=-\infty}^{\infty} \sum_{t_f=1}^{T_f} \sum_{k=B_{t_f}}^{E_{t_f}-1} \text{Real}[C_{ki} e^{j2\pi f_k(t-iT_s)}]$$

$$\cdot f(t - iT_s)$$

$$f(t) = \begin{cases} 1 & 0 \leq t \leq T_s \\ 0 & t < 0 : t > T_s \end{cases}$$

$$f_k = f_0 + \frac{k}{T_s}, \tag{1}$$

where T_s is the symbol time, f_k is the carrier frequency, and C_{ki} is the complex amplitude for the $i-th$ symbol of carrier k. Also, t_f is the sub-band number, T_f is the number of sub-bands, B_{t_f} is the first carrier number for the $t_f - th$ sub-band, and E_{t_f} is the final carrier number.

In addition, the signals of different sequences are transmitted at the same frequency band in the MIMO system. The number of eigenvalues is assumed to be M, and eigenmode multiple signals of $s_1(t), s_2(t), \ldots,$ and $s_M(t)$ are transmitted.

The reception signal vector, $\vec{R}$, of the MIMO eigenmode transmission is given by

$$\vec{R}(t) = V^H(t)(H(t)W(t)\vec{s}(t) + \vec{Z}(t)) \tag{2}$$

where $H(t)$ is the channel matrix, $W(t)$ is the transmission weights matrix, $V(t)$ is the reception weights matrix, and $\vec{Z}(t)$ is a noise vector.

The channel matrix H for multi-band OFDM can be expressed as H' because the elements of $\vec{s}(t)$ are the signals of each sub-band into which the frequency band is divided and taking into consideration the orthogonal characteristic on the frequency axis.

$$H' = \begin{bmatrix} H_{band1} & 0 & \cdots & 0 \\ 0 & H_{band2} & \cdots & 0 \\ \vdots & \vdots & \ddots & \vdots \\ 0 & 0 & \cdots & H_{bandT_f} \end{bmatrix} \quad (3)$$

Therefore, the MIMO eigenmode transmission of each sub-band is possible. The diagonal elements of H' have the frequency selected characteristic of the propagation. This frequency characteristic is compensated by the OFDM power distribution control.

3.2 Maximal ratio combining (MRC) of an OFDM-MIMO scheme

The S/N MRC method in the OFDM-MIMO scheme for a narrow-band signal has already been studied and the best MIMO weight coefficient has been derived[5]. Transmission weight $\vec{W}$ is assumed in the transmitter, and reception weight $\vec{V}$ is assumed in the receiver. In each band, m transmission beams and n receiving beams are formed. The signal $\vec{s}(t)$ is transmitted with the average transmission power $\vec{P}_{signal}$.

The transmission power $\vec{P}_{signal}$ and noise power $\vec{P}_{noise}$ are defined by

$$\vec{P}_{signal} = (\langle s_1^* s_1 \rangle, \langle s_2^* s_2 \rangle, \ldots \langle s_M^* s_M \rangle) \quad (4)$$

$$\vec{P}_{noise} = (\langle Z_1^* Z_1 \rangle, \langle Z_2^* Z_2 \rangle, \ldots \langle Z_M^* Z_M \rangle) \quad (5)$$

The received signals for the $M-th$ eigenvalue is expressed by

$$R_M = \vec{V}_M^H (H \vec{W}_M s_M + Z_M). \quad (6)$$

The receiving power P_{RM} for the $M-th$ eigenvalue is given by

$$P_{RM} = |\vec{V}_M^H H \vec{W}_M|^2 P_{signal_M} + \vec{V}_M^H \vec{V}_M^* P_{noise_M}. \quad (7)$$

The MIMO weights, $\vec{W}_M$, and $\vec{V}_M$ which satisfy the S/N MRC in each eigenmode are derived through singular value decomposition (SVD) as singular value vectors (eigenvalue vectors in the case of a square matrix).

The transmission weight W is given by

$$SVD(H^T H^*) = W \Lambda W^H. \quad (8)$$

The reception weight V is given by

$$SVD(H H^H) = V \Lambda V^H. \quad (9)$$

The matrix Λ is the eigen matrix, and the diagonal Λ is ($\lambda_1(t), \lambda_2(t), \ldots, \lambda_M(t)$). $()^T$, $()^*$, and $()^H$ denote transpose, complex conjugate, and Hermitian operations, respectly.

3.3 Bit error rate of a multi-band OFDM-MIMO scheme

The average BER characteristics of the multiband OFDM-MIMO communications scheme are derivable from an expression including the probability density function of the eigenvalues and the error function of the multilevel modulation schemes. The average BER characteristics for the first eigenvalue and the other order eigenvalues in Rayleigh fading are given below.

The probability density function of the first eigenvalue of Rayleigh fading can be approximated by a gamma distribution[5].

If the numbers of input terminal, m, and output terminal, n, are set in the MIMO system, the probability density function agree with one of space diversity for $(1,mn)$. Therefore, the probability density function can be use to derive the BER of MIMO communication systems.

The average BER in Rayleigh fading to the first eigenvalue is thus

$$P_{\lambda 1} = \int_0^\infty g(\gamma) \cdot \frac{1}{(mn-1)!} \frac{\gamma^{mn-1}}{(\sigma^2 \overline{\lambda_1}/mn)^{mn}} \cdot exp(-\frac{\gamma}{\sigma^2 \overline{\lambda_1}/mn}) d\gamma \tag{10}$$

where $g(x)$ is a function of BER of the multilevel modulation (QPSK,8PSK,16QAM,etc.). $\overline{\lambda_1}$ is the first average eigenvalue normalized by the total value of the eigenvalues.

Averrage BER to the middle order eigenvalue of Rayleigh fading can be calculated for each eigenvalues using the probability density functions of a gamma distribution as same as Eq.(10).

The probability density functions of Rayleigh fading in the last order eigenvalue are expressed by exponential distributions. Average BER to the last order eigenvalue is given as

$$P_{\lambda M} = \int_0^\infty g(\gamma) \cdot \frac{1}{(\sigma^2 \overline{\lambda_M}/mn)} \cdot exp(-\frac{\gamma}{\sigma^2 \overline{\lambda_M}/mn}) d\gamma \tag{11}$$

where $\overline{\lambda_M}$ is the last order normalized eigenvalue.

Then σ^2 at the maximum Doppler frequency f_d with j OFDM carriers and u sub-carriers becomes

$$\sigma^2 = \{(M(S/N)_i)^{-1} + (\frac{b_0'}{\sigma_c^2})^{-1}\}^{-1}, \tag{12}$$

$$b_0' = 1 - \frac{(\pi f_d T_s)^2}{6}, \tag{13}$$

$$\sigma_c^2 = \sum_{k=0, k\neq j}^{u-1} \frac{(f_d T_s)^2}{2(k-j)^2}. \tag{14}$$

where b_0' are desired signal powers, and σ_c^2 is channel interference.

The BER used block code like Reed-Solomon (RS) code is given by

$$P_{FEC} \approx \frac{q+1}{2q^2} \sum_{j=r+1}^{q} j \cdot \frac{q!}{(q-j)! j!} \cdot P_W^j (1-P_W)^{q-j}, \tag{15}$$

where $q = 2^b - 1$ is the block length (b is the number of bits in a symbol), p is information data length, and r is the number of correctable symbols $(q - p = 2r)$[6]. The symbol error rate P_W is written as

$$P_W = (2q)/(q+1) \cdot P_{\lambda_i}. \tag{16}$$

The BER can be calculated from equations (10)-(16) for the MIMO of each frequency divided path. Because the delay amount τ is assumed to be absorbed by the guard interval in the OFDM method, the symbol interference is not considered.

4. Adaptive Modulation and Error Correction

Figure 2 shows the flowchart for adaptive modulation for the MIMO eigenvalues and the OFDM sub-bands. Adaptive selection of the modulation and FEC effectively distributes the power. The eigenvalues are derived for each eigenmode by changing the propagation matrix. Then, the power distribution is determined for the MIMO eigenmodes and OFDM frequency sub-bands from the eigenvalues and the receiving S/N. Ideally, the power distribution accords to the water-filling theorem. The appropriate modulation and FEC is set for each path so that the power distribution is approximately ideal. The appropriate modulation and FEC mean the scheme to achieve the maximum frequency utilization efficiency in the power.

The mode of modulation and FEC are selected from candidates in a look-up table (LUT) for which the required S/N satisfies the power distribution for a certain BER. The candidates of high frequency utilization efficiency will be from an appropriate interval on which the required S/N is large.

Figure 3 shows the timing of adaptive transmission control. In this case, transmitting power is distributed per sub-band as the average power over a short duration in the frequency domain and time domain. After the average eigenvalues and the receiving S/N of a short duration in each subband have been presumed from the channel estimation, the

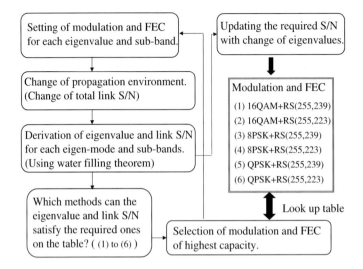

Figure 2. Flowchart of adaptive modulation for MIMO eigenvalue and OFDM subband

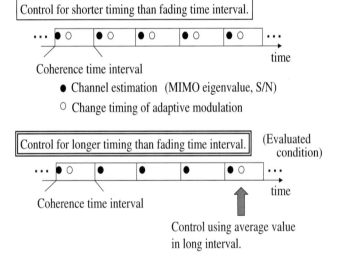

Figure 3. Timing of adaptive transmission control

eigenvalues are used to update the LUT with the required S/N of each transmission scheme to satisfy the required BER. The scheme that satisfies the required S/N is allocated to the mode with the large receiving S/N. The eigenmodes of the subband are judged to be unusable if the transmission scheme is not able to satisfy the required S/N in the LUT. Because the transmitting power in this part becomes useless, the power of the mode to which S/N greatly deteriorates is evenly allotted to all other modes. This process of selection and allocation is continuously repeated.

5. Evaluation of the system performance

The evaluations were done with QPSK, 8PSK, and 16QAM comprising the multilevel modulation scheme and RS (255,239) and (255,223) (Reed Solomon codes) as the error correcting code. The communication capacity of adaptive transmission was numerically evaluated. The element communication capacity, $C_i(f)$, of each eigenmode is decided from the modulation and FEC of the adaptive transmission. The received $(S/N)_i(f)$ should match the required S/N to satisfy the chosen BER.

Figure 4 shows an example of the frequency response of a two-wave propagation model ($\tau = 0.1\mu sec$) on the frequency selective fading channel. As for the divided subband, flat fading is assumed, and the performance of the adaptive control is evaluated.

Figure 5 shows the improvement of communication capacity relative to the number of subbands at 20dB of the setting S/N. The upper bound can be reached by increasing the number of partitions, and a value of 100 close to the saturation value be achieved ($\tau = 0.1\mu sec$: distance 30 m). It is possible to compensate for the power loss due to the channel frequency response by using the power distribution of the signal to the multiband. Although this evluation shows that the frequency response can be compensated with a lot of subbands, the suitable number of partitions for the propagation channel is determined by the trade-off of the effect of the partition and the complexity degree. When there are few partitions, the number of switchings of the adaptive transmission of each subband can be reduced.

The evaluation in the Rayleigh fading environment for no line of the sight shows the effect of the second eigenvalue of MIMO. Figure 6 shows S/N vs. BER characteristics of various modulation schemes of MIMO(2,2) in the Rayleigh fading environment. The frequency response of the two wave model assumed delay waves with equal delays of $0.1\mu sec$. The required S/Ns of various modulation schemes (QPSK,8PSK,16QAM + RS(255,239)or(255,223)) corresponding to the first eigenvalue and the second eigenvalue for a BER of 10^{-5} are listed in the look up table (LUT), and these are used in the evaluation of the selection using adaptive transmission.

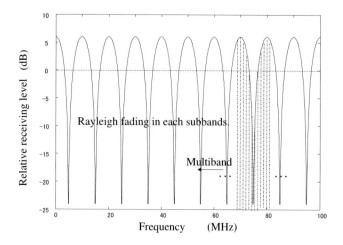

Figure 4. Frequency characteristics for two wave model of $\tau = 0.1 \mu sec$

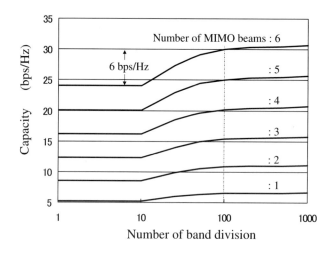

Figure 5. Upper limited capacity relative to the number of subbands

Figure 7 shows the capacity of our multi-band OFDM-MIMO system using adaptive modulation and error correction for a frequency selected Rayleigh fading channels when the received S/N changes. The number of subbands is assumed to be 100, and the required BER is assumed to be 10^{-5}. The figure confirms the effect of the second eigenmode in the point

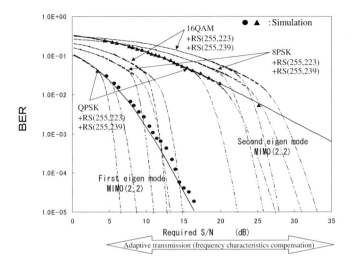

Figure 6. S/N vs. BER characteristics for Rayleigh fading

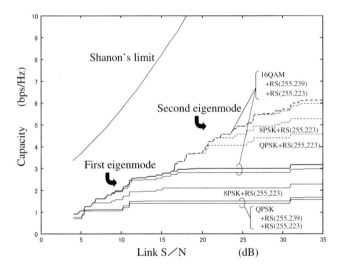

Figure 7. Capacity achieved by using adaptive modulation and error correction

of large receiving S/N and that the communication capacity increases as a result of using the multilevel modulation scheme. The multi-band effect raises the capacity due to the degradation of frequency selective channels by using adaptive modulation and error correction selecting.

6. Conclusions

We have developed a multi-band OFDM-MIMO scheme in which the spectra of OFDM signals are shared to improve the path gain for MIMO transmission. In this system, OFDM signals whose characteristics are opposite to the frequency characteristics are formed through OFDM power distribution control.

The effect of the adaptive transmission of the multiband OFDM-MIMO communication scheme was verified to improve communication efficiency in a multipath propagation environment with frequency selective fading. A study of the Rayleigh fading environment showed that adaptive transmission compensates the power loss of the wideband frequency response and the second eigenmode of MIMO in the case of large receiving S/N increases the communication capacity.

This demonstrates the beneficial effect of the adaptive modulation and error correction in the multiband OFDM-MIMO communications scheme based on two-dimensional power distribution to the spatial eigenmode of MIMO and frequency subbands of OFDM.

References

[1] A. Goldsmith, S. A. Jafar, N. Jindal, and S. Vishwanath. Capacity limits of MIMO channels. *IEEE Journal on Selected Areas in Communications*, 21:684–702, June 2003.

[2] H. Sampath, S. Talwar, J. Tellado, V. Erceg, and A. Paulraj. A fourth-generation MIMO OFDM broadband wireless system. *IEEE Communication Magazine*, 40:143–149, Sep. 2002.

[3] Ye. Li, J.H. Winters, and N.R. Sollenberger. MIMO-OFDM for wireless communications: signal detection with enhanced channel estimation. *IEEE Transactions on Communications*, 50:1471–1477, Sep. 2002.

[4] M. Ohkawa and R. Kohno. Effect of a multi-band OFDM-MIMO system on frequency-selective propagation characteristics. *IEICE Transactions on Communications*, E88-B:19–27, Jan. 2005.

[5] Y. Karasawa. MIMO propagation channel modeling. *IEICE Transactions on Communications*, J86-B:1706–1720, Sep. 2003.

[6] G. C. Clark and J. B. Cain. *Error Correction Coding for Digital Communications*. Plenum Press, New York, 1981.

NEAR OPTIMAL PERFORMANCE FOR HIGH DATA RATE MIMO MC-CDMA SCHEME

P.-J. Bouvet and M. Hélard
France Telecom R&D Broadband Wireless Access Laboratory
4 rue du Clos Courtel, 35512 Cesson-Sévigné, France
maryline.helard@francetelecom.com

Abstract   MC-CDMA is studied in a multiple antenna context in order to efficiently exploit robustness against multipath effects and multiuser flexibility of MC-CDMA and channel diversity offered by MIMO systems for radio mobile channels. The MIMO MC-CDMA transmission scheme is detailed for a 4×4 and two 4×2 MIMO systems as well as their associated receivers performed in an iterative way. The simulation results show the efficiency of the proposed systems that can provide high capacity for a synchronous downlink while offering a good trade-off between performance and complexity. Results also take into account performance degradation due to non perfect channel estimation.

1. Introduction

In this paper MC-CDMA and MIMO techniques are combined in order to increase the bandwidth efficiency while keeping robustness against channel selectivity. On the one hand MC-CDMA efficiently combines OFDM and CDMA techniques providing high spectral efficiency, good robustness against multipath effects, multiple access flexibility and low multi-user interference [1]. On the other hand, MIMO techniques offer efficient schemes to improve capacity thanks to spatial diversity. Space-time block coding (STBC) associated with MC-CDMA was already demonstrated to achieve high performance while keeping reduced complexity receiver [2]. By considering spatial multiplexing with MC-CDMA, data rate can be substantially enhanced but residual interference terms have to be treated by using more complex receiver. More precisely, in [3] and [4], an iterative receiver issued from [5] is proposed in order to cancel co-antenna interference (CAI) coming from the MIMO architecture. In this paper we investigate an iterative receiver that combines CAI and multiple access interference (MAI) cancellation providing near optimal performance whatever the load. Simulation results including channel estimation are presented for a 4×4 and two 4×2 MIMO systems for a synchronous downlink of a radio mobile application.

2. MIMO MC-CDMA system model

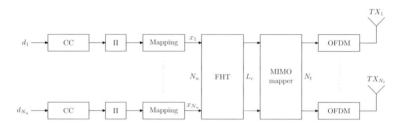

Figure 1. MIMO MC-CDMA transmitter

The transmitter scheme is shown in Fig. 1. Each user's data d_j is first individually channel encoded, bit interleaved and mapped into complex symbols belonging to the constellation $\mathcal{A}$. Let $\mathbf{x}$ be a coded symbol vector collecting the N_u active users: $\mathbf{x} = [x_1, \ldots, x_{N_u}]^\mathrm{T}$. This sequence is spread using for instance a fast Hadamard transform (FHT) well known to provide good performance for a synchronous downlink scenario. Let L_c be the length of the spreading sequence and $\mathbf{C} = [\mathbf{c}_1, \ldots, \mathbf{c}_{N_u}]$ the $L_c \times N_u$ matrix of user code, the spread vector is simply expressed as:

$$\mathbf{y} = \mathbf{C}\mathbf{x} \qquad (1)$$

The spread signal is then divided into blocks $\mathbf{s}$ of length Q, submultiple of L_c, and mapped into N_t sub-streams using a space-time coding scheme represented by the coding matrix $\mathbf{S} \in \mathbb{C}^{T \times N_t}$ where T is the latency code i.e. the number of symbols for which the propagation channel is assumed to be constant. Finally, sub-streams are individually Orthogonal Frequency Division Multiplex (OFDM) modulated before transmission on one antenna. The MIMO channel connects N_t transmit and N_r receive antennas where each link is considered time and frequency selective. At the receive side an OFDM demodulation is performed on each antenna. Let h_{ij} be the flat fading coefficient provided by the OFDM modulation representing the link between antenna i and antenna j. These coefficients can be rearranged in matrix $\mathbf{H} \in \mathbb{C}^{TN_r \times Q}$ representing both space-time coding scheme and OFDM MIMO channel such as:

$$\mathbf{r} = \mathbf{H}\mathbf{s} + \mathbf{n} \qquad (2)$$

where $\mathbf{r} \in \mathbb{C}^{TN_r \times 1}$ is the equivalent receive vector, $\mathbf{s} \in \mathbb{C}^{Q \times 1}$ the incoming signal, and $\mathbf{n} \in \mathbb{C}^{TN_r \times 1}$ an equivalent noise vector [6]. Let $\mathbf{H}_g$ be a diagonal per block matrix of size $\frac{TN_rL_c}{Q} \times N_u$ representing L_c/Q different observations of the equivalent MIMO matrix $\mathbf{H}$:

$$\mathbf{H}_g = \text{diag}(\underbrace{\ldots, \mathbf{H}, \ldots}_{L_c/N_t}) \quad (3)$$

Let $\mathbf{z}$ be a global receive vector of size $\frac{TN_rL_c}{Q}$ obtained by stacking L_c/Q consecutive vectors $\mathbf{r}$, then the global system can be expressed as:

$$\mathbf{z} = \mathbf{H}_g\mathbf{C}\mathbf{x} + \mathbf{n} = \underbrace{\mathbf{H}_g\mathbf{c}_k x_k}_{\text{desired signal}} + \underbrace{\sum_{n=1, n \neq k}^{N_u} \mathbf{H}_g\mathbf{c}_n x_n}_{\text{interference terms}} + \underbrace{\mathbf{w}}_{\text{noise}} \quad (4)$$

where $\mathbf{w}$ is a noise vector with zero mean and variance σ_w^2.

3. Proposed iterative receiver

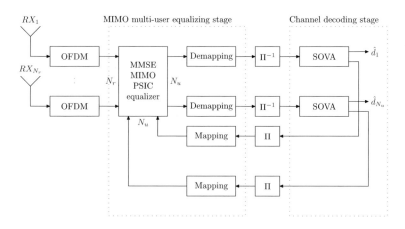

Figure 2. MIMO MC-CDMA proposed receiver

As shown in (4), the receive signal is corrupted by interference terms coming from both orthogonality loss between users (MAI) due to channel frequency selectivity as in SISO (Single Input Single Output) case [7] and the MIMO scheme (CAI). In the proposed iterative receiver represented in Fig. 2, the MAI and CAI terms are globally removed in an iterative loop by benefiting from the channel decoding of all user data. The multi-user MIMO equalization stage consists of an interference canceller inspired from turbo-equalization that can be expressed with the following equation [8, 9]:

$$\tilde{x}_k = \mathbf{p}_k^H \mathbf{z} - \mathbf{q}_k^H \begin{bmatrix} \hat{x}_1 & \ldots & \hat{x}_{k-1} & 0 & \hat{x}_{k+1} & \ldots \hat{x}_{N_u} \end{bmatrix}^T \quad (5)$$

where $\hat{x}_k$ is an estimate of x_k given by the previous iteration. The two weight vectors $\mathbf{p}_k^H$ and $\mathbf{q}_k^H$ are optimized under the MMSE criterion [9]:

$$\mathbf{p}_k^H = \lambda_k \bar{\mathbf{p}}_k^H \qquad (6)$$
$$\mathbf{q}_k^H = \mathbf{p}_k^H \mathbf{H}_g \mathbf{C} \qquad (7)$$

with

$$\lambda_k = \frac{\sigma_x^2}{\sigma_x^2 + \sigma_{\hat{x}}^2 \mathbf{e}_k^T \mathbf{H}_g^H \bar{\mathbf{p}}_k} \qquad (8)$$

and

$$\bar{\mathbf{p}}_k = \sigma_x^2 \mathbf{e}_k^T \left(\mathbf{C}^H \mathbf{H}_g^H \mathbf{H}_g \mathbf{C}(\sigma_x^2 - \sigma_{\hat{x}}^2) + \sigma_w^2 \frac{L_c}{N_u} \mathbf{I} \right)^{-1} \mathbf{C}^H \mathbf{H}_g^H \qquad (9)$$

where $\mathbf{e}_k$ denotes a vector whose elements are 0 except the k-th element which is equal to 1 whereas $\sigma_{\hat{x}}^2$ is the variance of the estimated symbol. At first iteration, $\sigma_{\hat{x}}^2$ is set to 0 leading to $\lambda_k = 1$, thus the expression of the equalized symbol becomes:

$$\tilde{x}_k = \sigma_x^2 \mathbf{c}_k^H \left(\mathbf{H}_g^H \mathbf{H}_g \sigma_x^2 + \sigma_w^2 \frac{L_c}{N_u} \mathbf{I} \right)^{-1} \mathbf{H}_g^H \mathbf{z} \qquad (10)$$

At full load, (9) can be easily simplified as:

$$\bar{\mathbf{p}}_k = \sigma_x^2 \mathbf{c}_k^T \left(\mathbf{H}_g^H \mathbf{H}_g (\sigma_x^2 - \sigma_{\hat{x}}^2) + \sigma_w^2 \frac{L_c}{N_u} \mathbf{I} \right)^{-1} \mathbf{H}_g^H \qquad (11)$$

avoiding a $N_u \times N_u$ inversion matrix. Equation (11) will be further used whatever the load even if it is slightly different from (9) at non full load. Such approximation is motivated by the fact that at the first iteration (11) is equal to a single user MMSE equalization well known to provide good performance. At other iterations since the term $\sigma_x^2 - \sigma_{\hat{x}}^2$ tends towards 0, the difference between (9) and (11) becomes very small. While using (11), the multi-user spatial equalizer involves only matrix multiplications and inversions of $Q \times Q$ block diagonal matrix whatever the load. Thus the overall complexity of the receiver is essentially dominated by the channel decoding of each users. Moreover its complexity does not depend on the constellation size but mainly on the number of antennas.

4. Performance results

Simulation results are obtained with 1/2-rate convolutional encoder with generator $(133, 177)^o$, random bit interleaving, QPSK modulation with Gray mapping and Hadamard spreading matrix of size $L_c = 64$. The proposed iterative receiver is implemented for a 4×4 and two 4×2 MIMO systems. BER performance are provided over Rayleigh i.i.d. channels and over a more realistic channel including channel estimation.

4.1   4 × 4 MIMO system

Performance results are provided over Rayleigh i.i.d. MIMO channel leading to the best achievable performance of OFDM transmission over time and frequency selective channel. We considered a spatial multiplexed scheme to optimally exploit the capacity of the 4×4 system [10]. The spatial scheme is therefore represented by:

$$\mathbf{S} = \begin{bmatrix} s_1 & s_2 & s_3 & s_4 \end{bmatrix} \text{ and } \mathbf{H} = \begin{bmatrix} h_{11} & h_{21} & h_{31} & h_{41} \\ h_{12} & h_{22} & h_{32} & h_{42} \\ h_{13} & h_{23} & h_{33} & h_{43} \\ h_{14} & h_{24} & h_{34} & h_{44} \end{bmatrix}$$

For this scheme $Q = 4$ and $T = 1$, thus the interference cancellation process requires 4×4 matrix inversions. In order to highlight the performance of the receiver described in part 3, simulation results are provided for two alternative receivers. The first one is a non iterative single user (SU) MMSE receiver that is exactly equal to the first iteration of the proposed iterative receiver given by (10) and the second one is an iterative receiver issued from [4] where only the CAI is treated inside the iterative process. Please note that the two iterative receivers have similar complexity. In Figure 3(a), the BER performance results of these three receivers are presented as a function of E_b/N_0 for a 4×4 system. By treating interference terms in a global manner, the proposed receiver overcomes the receiver of [4] about 0.8 dB at 10^{-4} whereas more than a 6 dB gain is noted compared to the non iterative SU MMSE receiver. The proposed iterative receiver tends towards the AWGN curve meaning that all interference terms are perfectly cancelled with only 4 iterations. Figure 3(b) provides BER performance versus the load. The performance of the SU MMSE obviously improves when the number of users decreases whereas performance results of iterative receivers hardly depend on the load since near optimal performance is already reached at full load.

4.2   Comparison of two 4 × 2 systems

Following the IST 4MORE project specifications for downlink transmission [11], we now consider a 4×2 antenna system, where each subchannel corresponds to a 17-taps BRAN E model and the mobile speed is set to 16.6 m/s. We assume no spatial correlation between antennas. The carrier frequency is set to 5 GHz and the sampling frequency is 61.44 MHz. The FFT size is 1024 and 695 subcarriers are modulated leading to an occupied bandwidth of 41.2 MHz. The cyclic prefix is equal to 256 samples. All these parameters have been chosen according

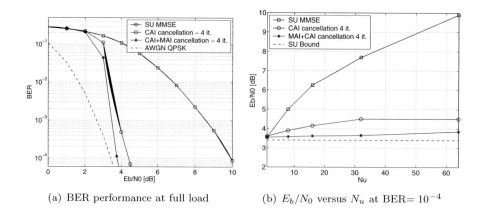

(a) BER performance at full load (b) E_b/N_0 versus N_u at BER= 10^{-4}

Figure 3. 4×4 Rayleigh i.i.d. channel, 4 bps/Hz

to the time and frequency coherence of the channel in order to reduce the inter-carrier and the inter-symbol interferences. Bit interleaving depth is chosen in order to respect the 4MORE frame leading to a size per user of 504 bits for a QPSK and 1008 bits for 16-QAM. In order to maximize the data rate, we choose the following MIMO scheme issued from linear dispersion coding theory [12]:

$$\mathbf{S} = \begin{bmatrix} s_1 & s_2 & s_3 & s_4 \\ -s_2^* & s_1^* & -s_4^* & s_3^* \end{bmatrix} \text{ and } \mathbf{H} = \begin{bmatrix} h_{11} & h_{21} & h_{31} & h_{41} \\ h_{21} & h_{11} & h_{32} & h_{42} \\ h_{21}^* & -h_{11}^* & h_{41}^* & -h_{31}^* \\ h_{22}^* & -h_{12}^* & h_{42}^* & -h_{32}^* \end{bmatrix}$$

For this scheme, $Q = 4$ and $T = 2$ thus the interference cancellation requires inversions of 4×4 size matrices. We propose to compare this system with the orthogonal Alamouti antenna switched (AS) scheme as space-time code [6]:

$$\mathbf{S} = \begin{bmatrix} s_1 & s_2 & 0 & 0 \\ -s_2^* & s_1^* & 0 & 0 \\ 0 & 0 & s_3 & s_4 \\ 0 & 0 & -s_4^* & s_3^* \end{bmatrix} \quad (12)$$

Thanks to the orthogonality of the Alamouti scheme, a SU MMSE equalizer can be used while keeping good performance [2]. However, in order to provide same spectral efficiency than the LD code, a 16-QAM modulation has been considered. Fig. 4(a) shows BER performance of the two 4×2 systems at full load. First, we notice that for both schemes, the degradation with respect to the Rayleigh i.i.d. channel is relatively small. The LD coding scheme associated with the proposed receiver outperforms the Alamouti AS scheme with SU MMSE equalization. In fact, for a 4×2 channel, the maximal mutual information attainable with the LD code is greater than those obtained with the Alamouti scheme [12]. Moreover, for Alamouti AS scheme residual MAI terms remain at the output of the SU MMSE equalizer leading to performance degradation. An iterative process could also be used but performance would not outperform the 16-QAM AWGN curve giving an advantage to the LD scheme. Fig. 4(b) provides BER performance of the two same systems by including pilots based channel estimation [11]. Simulation results show that the channel estimation has a similar impact on the SU MMSE equalizer than on the proposed iterative receiver. Only 1.8dB degradation is observed at 16.6 m/s at BER= 10^{-4}.

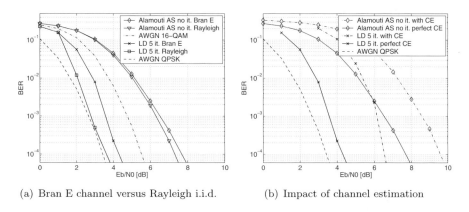

(a) Bran E channel versus Rayleigh i.i.d.

(b) Impact of channel estimation

Figure 4. 4×2 MIMO channel, full load, 2 bps/Hz

5. Conclusion

In this paper, MC-CDMA has been combined with different space-time coding techniques in order to exploit both MIMO capacity and antenna diversity. We have proposed an efficient iterative receiver based on linear equalization providing near single user bound performance for all load configurations. By iteratively removing MAI and CAI, the proposed receiver can be associated with high rate space-time coding scheme that optimally exploits MIMO capacity while keeping advantages of MC-CDMA. BER performance on realistic channel including channel estimation shows only small degradation compared to the theoretical channel. The proposed iterative receiver needs to decode the data of all active users. Nevertheless, the interference cancellation stage does not depend on the number of users but only requires inversions of 4×4 matrices for the considered 4×4 and 4×2 systems. Associated with the proposed iterative receiver, the 4×2 LD scheme was proved to provide good BER performance and quite high capacity with only 4 iterations.

Acknowledgment

Part of this work has been carried out within the framework of the IST 4MORE project [11].

References

[1] N. Yee, J. Linnartz, and G. Fettweis. Multi-carrier CDMA in indoor wireless radio networks. In *Proceedings of PIMRC'93*, Yokohama, Japan, September 1993.

[2] V. Le Nir, M. Hélard, and R. Le Gouable. Space-time block coding applied to turbo coded multicarrier CDMA. In *Proceedings of VTC Spring'03*, pages 577–581, Jeju, Korea, May 2003.

[3] M. Vehkapera, D. Tujkovic, Zexian Li, and M. Juntti. Layered space-frequency coding and receiver design for MIMO MC-CDMA. In *Proceedings of ICC'05*, pages 3005–3009, Seoul, Korea, May 2005.

[4] P-J. Bouvet, V. Le Nir, M. Hélard, V. Le Nir, and R. Le Gouable. Spatial multiplexed coded MC-CDMA with iterative receiver. In *Proceedings of PIMRC'04*, Barcelona, Spain, September 2004.

[5] M. Sellathurai and S. Haykin. TURBO-BLAST for wireless communications: theory and experiments. *IEEE Trans. Signal Processing*, 50(10):2538–2546, October 2002.

[6] P-J. Bouvet, M. Hélard, and V. Le Nir. Low complexity iterative receiver for non-orthogonal space-time block code with channel coding. In *Proceedings of VTC Fall'04*, Los Angeles, USA, September 2004.

[7] M. Hélard, R. Le Gouable, J. F. Helard, and J. Y. Baudais. Multicarrier CDMA techniques for future wideband wireless networks. *Annal. Télécom*, 56:260–274, 2001.

[8] X. Wang and H.V. Poor. Iterative (turbo) soft interference cancellation and decoding for coded CDMA. *IEEE Trans. Commun.*, 47(7):1046–1061, July 1999.

[9] C. Laot, R. Le Bidan, and D. Leroux. Low complexity linear turbo equalization: A possible solution for EDGE. *IEEE Trans. Wireless. Commun.*, 2005. "to be published".

[10] G. J. Foschini. Layered space-time architecture for wireless communication in a fading environment when using multielement antennas. *Bell Syst. Tech. Journal*, 1:41–59, October 1996.

[11] IST 4MORE project. 4G MC-CDMA multiple antenna system on Chip for Radio Enhancements. http://www.ist-4more.org.

[12] B. Hassibi and B. M. Hochwald. High-rate codes that are linear in space and time. *IEEE Trans. Inform. Theory*, 48(7):1804–1824, July 2002.

STBC-TCM FOR MC-CDMA SYSTEMS WITH SOVA-BASED DECODING AND SOFT-INTERFERENCE CANCELLATION

Luis Alberto Paredes Hernández and Mariano García Otero
Dept. Señales, Sistemas y Radiocomunicación
ETSI de Telecomunicación
Universidad Politécnica de Madrid. Spain
{paredes, mariano}@gaps.ssr.upm.es

Abstract   This paper analyzes several interference cancellation schemes applied to STBC-TCM for MC-CDMA. In these systems, a linear MMSE detector is conventionally used to reduce interference generated by multipath, multiuser, and multiple antennas propagation. To obtain further performance improvements, a more efficient SOVA-based iterative MMSE scheme is considered. This receiver performs soft interference cancellation for every user based on a combination of the MMSE criterion and the turbo processing principle. It is shown that these block space-time concatenated with coding channel detectors can potentially provide significant capacity enhancements over the conventional matched filter receiver.

1. INTRODUCTION

In mobile communications systems, diversity techniques have proved to be very useful to combat fading and to increase the capacity of the channel. Recently, new space diversity schemes have been developed in combination with channel coding and equalization, resulting in very efficient schemes for single carrier systems on frequency selective fading channels [1-3]. This paper discusses the performance of a STB-TCM coded MC-CDMA system, over frequency selective MIMO channels with multiuser detection based MMSE and soft-interference cancellation by SOVA-based decoder.

2. SYSTEM DESCRIPTION

The system block diagram is shown in Figure 1, for the case of a downlink transmission. The binary information data, $\{b^k\}$, for user k are 8-PSK TCM-STB encoded. At the STB encoder output, each symbol is multiplied by the spreading code of the specific user, $\mathbf{c}^k = [c_1^k \ldots c_f^k \ldots c_{L_c}^k]^T$, where c_f^k is the fth subcarrier of the spreading code of the kth user. And the coded symbols are transmitted by $N_t = 2$ antennas. Since, we use the Alamouti's scheme [4]. The length L_c of the spreading sequence is equal to the number N_c of subcarriers. So, STBC is carried out on two adjacent OFDM symbols, and the receiver has to process two successive symbols as a whole block. The received signal at rth receive antenna is equal to:

$$\begin{bmatrix} \mathbf{s}_r(t) \\ \mathbf{s}_r(t+1) \end{bmatrix} = \begin{bmatrix} \mathbf{H}_{1r} & \mathbf{H}_{2r} \\ \mathbf{H}_{2r}^* & -\mathbf{H}_{1r}^* \end{bmatrix} \begin{bmatrix} \mathbf{C} & 0 \\ 0 & \mathbf{C} \end{bmatrix} \begin{bmatrix} \mathbf{x}_1 \\ \mathbf{x}_2 \end{bmatrix} + \begin{bmatrix} \mathbf{n}_r(t) \\ \mathbf{n}_r(t+1) \end{bmatrix} \quad (1)$$

Where, $\mathbf{s}_r(t) = [s_{r,1}(t) \ldots s_{r,f}(t) \ldots s_{r,N_c}(t)]^T$ is the vector of N_c received signals, $\mathbf{H}_{pr}$ ($p \in \{1,2\}, r=1$) is a diagonal matrix of channel impulsive response, $\mathbf{C} = [\mathbf{c}^1 \mathbf{c}^2 \ldots \mathbf{c}^K]$ is an L_c order square matrix of users spreading codes, $\mathbf{x}_1$ and $\mathbf{x}_2$ are vectors of the data symbols transmitted in a block by the K active users, and $\mathbf{n}_r(t)$ is the complex AWGN vector (variance $E|n_{r,f}|^2 = N_0 \ \forall f, r$).

The resulting received signal $y = [\mathbf{y}_1^T \ \mathbf{y}_2^T]^T$ is the addition of the combined signals from all receives antennas:

$$y = \sum_{r=1}^{Nr} \mathcal{G}_r s_r \text{ with } \mathcal{G}_r = \begin{bmatrix} \mathbf{G}_{1r} & \mathbf{G}_{2r}^* \\ \mathbf{G}_{2r} & -\mathbf{G}_{1r}^* \end{bmatrix} \quad (2)$$

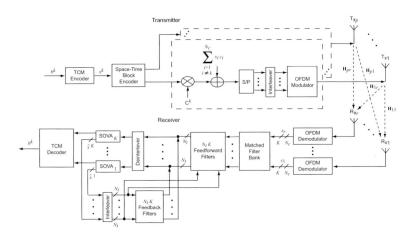

Figure 1. Block diagram of the STB-TCM coded MC-CDMA system for the kth user.

After despreading and threshold detection, the detected symbols correspond to the sign of the scalar product of the received signals $y_{1,f}$, $y_{2,f}$ and the specific spreading code $\mathbf{c}^k$:

$$\hat{x}_{1,2}^k = sign[\langle \mathbf{y}_{1,2}, \mathbf{c}^k \rangle] = sign\left[\sum_{f=1}^{N_c} c_f^k y_{1,2,f}\right] \qquad (3)$$

In order to improve the detection robustness, the interference terms of the MAI and ISI have to be suppressed without increasing the noise term. Hence, to combat the channel fading and the MAI, the detection schemes have to be optimized to extract the desired user's signal while reducing other user's interferences.

3. DETECTION MMSE

In [1], Auffray *et al.* concluding that the linear MMSE detector gives better behaviour (measured in terms of BER versus E_b/N_0). We applied the iterative MMSE detector derived by Vucetic *et. al.* [5], but it will concatenated with SOVA-based soft-interference cancellation.

3.1 Iterative MMSE Detector

There are two known Soft-Input Soft-Output decoding methods: MAP decoder, and SOVA. While the former provides the best performance in terms of minimizing the decoding errors, the latter has significantly lower complexity with only slight degradation in the decoding performance. Hence, SOVA is more suitable for hardware implementation. For the iterative detector, there are two different proposed SOVA implementations; one proposed by C. Battail [6] and the other by Hagenauer [7]. The key difference between these two algorithms is the updating rule. Besides, in [8] it shows that the performance of the SOVA proposed by Battail is 0.5 dB better than that of Hagenauer. However, in this schemes it can be easily seen that they have the same structure but the latter has a simplified version of the updating rule, which from an implementation point of view is preferable since it does not need any mathematical manipulation, only a single comparator. For these reasons we selected Hagenauer-SOVA for our iterative receiver.

In an iterative detector, the interference estimate for the pth transmit antenna of the kth user is formed by adding the regenerated signals of all users and all transmit antennas, except the one for the desired user k and antenna p. After each decoding iteration, the soft decoder outputs, calculated by the tools given by Hagenauer in [9], are used to update the a priori probabilities of the transmitted symbols. These updated probabilities are used in the calculation of the MMSE filter feedforward and feedback coefficients. Assuming that $z_p^k(t)$ is the input to the kth user decoder corresponding to the pth transmit antenna at time t, it is represented by $z_p^k(t) = \left(\mathbf{w}_{fp}^k(t)\right)^H s(t) + \left(\mathbf{w}_{bp}^k(t)\right)^H \hat{\underline{\mathbf{x}}}_p^k(t)$. Where, $\mathbf{w}_{fp}^k(t)$ is the $N_r N_c \times 1$ optimized feedforward coefficients vector, $\mathbf{w}_{bp}^k(t)$ is the $(N_t K - 1) \times 1$ feedback coefficients vector, $\hat{\underline{\mathbf{x}}}_p^k(t)$ is the $(N_t K - 1) \times 1$ vector representing the feedback soft decisions for all users and all transmit antennas except the decision corresponding to the pth transmit antenna of user k. Note that the feedback coefficients appear only through their sum in $z_p^k(t)$. Hence, we can assume, without loss of generality, that $_{bp}^k(t) = \left(\mathbf{w}_{bp}^k(t)\right)^H \hat{\underline{\mathbf{x}}}_p^k(t)$. Where, $_{bp}^k(t)$ is a single coefficient that represents the sum of the feedback terms.

The coefficients $\mathbf{w}_{fp}^k(t)$ and $_{bp}^k(t)$ are obtained by minimizing the mean square error value ε between the data symbols and their estimates, given by:

$$\begin{aligned}\varepsilon &= E\left[\left|z_p^k(t) - x_p^k(t)\right|^2\right] \\ &= E\left[\left|\left(\mathbf{w}_{fp}^k(t)\right)^{\mathrm{H}}\left\{\mathbf{h}_p^k x_p^k(t) + \underline{\mathbf{H}}_p^k \underline{\mathbf{x}}_p^k(t) + n(t)\right\}\right.\right. \\ &\quad \left.\left. + {}_{bp}^k(t) - x_p^k(t)\right|^2\right]\end{aligned} \qquad (4)$$

Where $\mathbf{h}_p^k$ is the $N_r N_c \times 1$ signature vector for the pth transmit antenna of the kth user, $\underline{\mathbf{H}}_p^k$ is an $N_r N_c \times (N_t K - 1)$ matrix composed of the signature vectors of all users and transmit antennas except the pth antenna of the kth user, and $\underline{\mathbf{x}}_p^k(t)$ is the $(N_t K - 1) \times 1$ transmitted data vector from all users and transmit antennas except the pth antenna of the kth user. The optimum feedforward and feedback coefficients, $\mathbf{w}_{fp}^k(t)$ and $_{bp}^k(t)$, respectively, can be represented by the following expressions:

$$\begin{aligned}\mathbf{w}_{fp}^k(t) &= \left(A + B + R_n - FF^{\mathrm{H}}\right)^{-1} \mathbf{h}_p^k \\ {}_{bp}^k(t) &= -F^{\mathrm{H}}\left(\mathbf{w}_{fp}^k(t)\right)\end{aligned} \qquad (5)$$

Where:

$$\begin{aligned}A &= \mathbf{h}_p^k \left(\mathbf{h}_p^k\right)^{\mathrm{H}} \\ B &= \underline{\mathbf{H}}_p^k \left[\mathbf{I}_{N_t K - 1} - Diag\left(\underline{\mathbf{x}}_{E_p}^k \left(\underline{\mathbf{x}}_{E_p}^k\right)^{\mathrm{H}}\right)\right. \\ &\quad \left. + \underline{\mathbf{x}}_{E_p}^k \left(\underline{\mathbf{x}}_{E_p}^k\right)^{\mathrm{H}}\right]\left(\underline{\mathbf{H}}_p^k\right)^{\mathrm{H}} \\ F &= \underline{\mathbf{H}}_p^k \underline{\mathbf{x}}_{E_p}^k \\ R_n &= \tfrac{2}{n}\mathbf{I}_{N_r N_c}\end{aligned} \qquad (6)$$

And $\mathbf{I}_N$ denotes the identity matrix of size N, $\underline{\mathbf{x}}_{E_p}^k$ is the $(N_t K - 1) \times 1$ vector of the expected values of the transmitted symbols from the other $N_t K - 1$ users and their transmit antennas. In iterative detector, during first decoding iteration, we assume that the *a priori* probabilities for transmitting all symbols are equal, and hence, $\underline{\mathbf{x}}_{E_p}^k = 0$.

After each iteration, $\underline{x}_{E_p}^k$ is updated from the soft outputs of the decoders, as derived in [9] by Hagenauer *et al.*, and then used to generate the new set of filter coefficients.

4. SIMULATION RESULTS

We have evaluated, by simulations, the system performance, measured in terms of BER and FER versus E_b/N_0 and active users. In this system, both linear and iterative MMSE detectors have been used. The different subcarriers are affected by independent frequency selective Rayleigh fading. We further assumed that the signals from all different users are received with the same power, and that the receiver perfectly knows the channel response. Each frame is composed by 64 symbols that are transmitted from $N_t = 2$ transmit antennas, thus forming 32 transmission symbol blocks on each frame. The channel remains constant during the transmission of a whole symbol block. The codes used by the 4-states TCM encoder with two transmit antennas are those obtained by Ungerböeck in [10] for the case of an 8-PSK modulation scheme: $H^0 = 5_8$, $H^1 = 2_8$, $H^2 = 0$, with a constraint length $v = 2$. We have assumed full system load, that is, there are $K = 8$ active users simultaneously, with $L_c = 8$ spreading codes present in the system. The Figure 2 and Figure 3 depict BER and FER performance for several MMSE receivers on frequency selective Rayleigh fading channels, respectively. We can see that iterative schemes perform significantly better when compared to the linear MMSE detector. Besides we found negligible differences between iterative schemes with varying number of iterations for low SNRs values, so we suggest to always using a single iteration equalizer for this values. Also, the STB concatenated with TCM coded system improves in more than 4 dB to the MC-CDMA system without TCM coding.

STBC-TCM for MC-CDMA systems

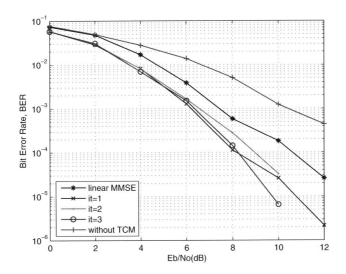

Figure 2. BER performance of a STB-TCM coded MC-CDMA system with the iterative MMSE receiver.

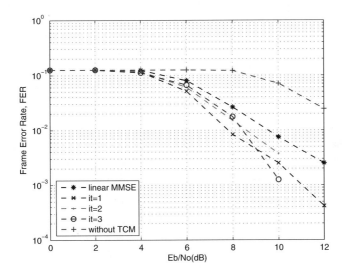

Figure 3. FER performance of a STB-TCM coded MC-CDMA system with the iterative MMSE receiver.

5. CONCLUSIONS

In this paper, we have evaluated the performance of a synchronous downlink in a multiuser STB-TCM coded MC-CDMA system, operating over frequency selective Rayleigh channels. We have obtained significant BER and FER improvements when an iterative MMSE detector is used, and the STBC coupled with TCM outperform the MC-CDMA performance. On the other hand, we found negligible differences between iterative schemes with varying number of iterations for the low SNRs, so we suggest to always using a single iteration equalizer for this values.

6. REFERENCES

[1] J. M. Auffray, J. F. Hélard, "Performance of Multicarrier Technique Combined with Space-Time Block Coding over Rayleigh Channel," *IEEE 7th Int. Symp. On Spread-Spectrum Tech. & Appl.*, Prague. Sept., 2002.

[2] H. Yang, J. Yuan, B. Vucetic, "Interference Suppression Schemes for Space-Time Trellis Coded CDMA Systems," *IEEE 57th Semiannual VTC 2003*, vol. 1, pp. 717-721, April, 2003.

[3] E. Biglieri, A. Nordio, G. Taricco, "Suboptimum Receiver Interfaces and Space-Time Codes," *IEEE Trans. on Signal Processing*, vol. 51. No. 11, Nov., 2003.

[4] S. Alamouti, "A Simple Transmit Diversity Technique for Wireless Communications," *IEEE J. Select. Areas Commun*, vol. 16, pp. 1451-1458, Oct., 1998.

[5] B. Vucetic, J. Yuan, *Space-Time Coding*. New York: Willey, 2003.

[6] G. Battail, "Pondération des Symboles Décodés par l'Algorithme de Viterbi," *Ann. Télécommun., Fr.*, vol. 42, No. 1-2, pp. 31-38, Jan., 1987.

[7] J. Hagenauer, P. Robertson, "Iterative (TURBO) Decoding of Systematic Convolutional Codes with the MAP and SOVA Algorithms," *ITG-Fachberichte*, vol. 130, pp. 21-29, 1995.

[8] L. Lin, R. S. Cheng, "Improvements in SOVA-based decoding for turbo codes," *Communications, 1997. ICC 97 Montreal, 'Towards the Knowledge Millennium'. 1997 IEEE International Conference*, vol. 3, pp. 1473-1478, 8-12 June, 1997.

[9] J. Hagenauer, E. Offer, L. Papke, "Iterative Decoding of Binary Block and Convolutional Codes," *IEEE Trans. Inform. Theory*, vol. 42. No. 2, pp. 429-445, March, 1996.

[10] G. Ungerböeck, "Channel Coding with Multilevel/Phase Signals," *IEEE Trans. Inform. Theory*, vol. 28, Issue 1, pp. 55-67, Jan., 1982.

OFDMA WITH SUBCARRIER SHARING

Stephan Pfletschinger and Faouzi Bader
Centre Tecnològic de Telecomunicacions de Catalunya (CTTC)
Parc Mediterrani de la Tecnologia
Av. del Canal Olmpic s/n, 08860 Castelldefels (Barcelona), Spain
{stephan.pfletschinger,faouzi.bader}@cttc.es

Abstract In this paper we describe an adaptive method for maximizing the sum rate in the uplink of a multi-user multi-carrier system. Based on the optimum power allocation, which is computed by the capacity-achieving iterative water-filling algorithm, we accordingly assign the users to the corresponding subcarriers. In some cases, the optimum solution requires the concurrent allocation of more than one user to the same subcarrier. In such a case, the multiple access *per subcarrier* is resolved by a CDMA component, which is built on top of an OFDMA system. We evaluate the possible gains in spectral efficiency of this OFDMA system with subcarrier sharing.

1. Introduction

The uplink of a multi-user system can be described as a multiple-access channel (MAC). For the practically important case that the channel is frequency-selective and the noise is Gaussian, the capacity region and the sum capacity have been described by Cheng and Verdú [1]. The sum capacity is achieved with the multi-user water-filling solution (see Figure 1), which allocates to each user a part of the spectrum and requires that the power spectral density (PSD) of each user follow a water-filling distribution. The important conclusion from this generalization of the well-known water-pouring principle [2] is that for achieving maximum throughput, the multiple-access scheme is FDMA. However, this only holds for arbitrarily fine granularity in the frequency domain and is approximated by an OFDMA system with a large number of subcarriers in relation to the number of users, i.e. $N_c \gg U$, where N_c denotes the number of subcarriers and U the number of users.

Alternatively and without the need for approximations, the OFDMA uplink can be described as a Gaussian multiple-access channel with

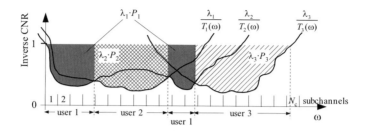

Figure 1. Multi-user water-filling diagram: A separate frequency band is allocated to each user. The channel to noise ratios (CNR) $T_u(\omega) = |H_u(\omega)|^2/\tilde{N}_0$ are scaled by the multipliers λ_u such that the "water-level" is unity.

vector inputs and vector output:

$$\mathbf{y} = \sum_{u=1}^{U} \mathbf{H}_u \mathbf{x}_u + \mathbf{w} \qquad (1)$$

For this channel, Yu et al. [3] describe a numerical method, called *iterative water-filling* (IWF), which maximizes the sum rate under user-individual power constraints. The salient feature of the IWF algorithm is its fast convergence which has been mathematically proven [3].

In the following, we will apply this channel model and the corresponding capacity-achieving solution to an OFDMA system with and without subcarrier sharing and discuss their relative merits. We will focus on the uplink and the sum rate maximization problem only. Analogous results can be expected for the downlink due to the duality between the multiple-access and the broadcast channel [4, 5].

2. System Model

We consider two system models: Pure OFDMA, where non-overlapping sets of subcarriers are allocated to the users, and an OFDMA-CDMA system, which allows several users to transmit over the same subcarrier. In the latter system, the users which share a subcarrier, are separated with an additional CDMA component. Since the former system is a special case of the latter, we describe both variants with the same system model, which is depicted in Figure 2. For pure OFDMA, the $L \times L$ spreading matrix $\mathbf{C}$ reduces to the scalar $C = 1$.

In our system model as illustrated in Figure 2, the data bits of user u are first adaptively encoded and modulated (ACM) and then transmitted over a set of dedicated subchannels. Provided that the cyclic prefix is longer than the channel impulse response, the received signal

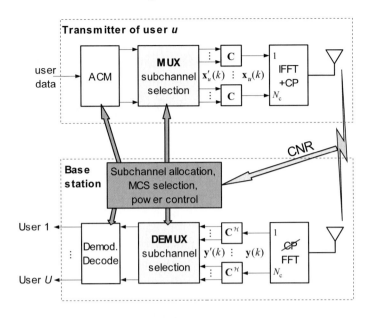

Figure 2. Uplink of an OFDMA system with subcarrier sharing. A controller located in the base station selects the appropriate subchannel allocation, the modulation and coding scheme (MCS) and the transmit powers for all users in the uplink.

on subcarrier n at time slot k can be expressed as

$$y_n(k) = \sum_{u=1}^{U} H_{n,u} \cdot x_{n,u}(k) + w_n(k), \quad n = 1, \ldots, N_c \quad (2)$$

In order to allow up to L users to share one OFDM subcarrier, we introduce a CDMA component, which is depicted as the block with the spreading matrix $\mathbf{C} = (\mathbf{c}_1 \cdots \mathbf{c}_L)$. This subcarrier sharing is implemented with L orthogonal spreading sequences $\{\mathbf{c}_i\}_{i=1}^{L}$ of length L, e.g. with Walsh-Hadamard spreading codes, which fulfill $\mathbf{CC}^{\mathcal{H}} = \mathbf{I}_L$. Note that in this formulation, TDMA and FDMA are included as the special cases $\mathbf{C} = \mathbf{I}_L$ and $\mathbf{C} = 1/\sqrt{L}\left[\exp(j\frac{2\pi}{L}\nu\mu)\right]_{\nu,\mu=0,\ldots,L-1}$. This subcarrier sharing with orthogonal spreading codes actually creates L parallel subchannels on each subcarrier, yielding $L \cdot N_c$ subchannels in total. We can define the equivalent channel matrix $\mathbf{H}'_u$ with diagonal entries

$$H'_{n,u} = H_{\lfloor (n-1)/L \rfloor + 1, u}, \quad n = 1, \ldots, LN_c \quad (3)$$

and write in analogy to (2)

$$y'_n(k) = \sum_{u=1}^{U} H'_{n,u} \cdot x'_{n,u}(k) + w'_n(k), \quad n = 1, \ldots, LN_c \quad (4)$$

With $N = LN_c$ and by defining the vectorial transmit and receive signals as the vectors $\mathbf{x}'_u = (x_{1,u}, \ldots, x_{N,u})^T$, $\mathbf{y}' = (y_1, \ldots, y_N)^T$, the channel matrix $\mathbf{H}'_u = \mathrm{diag}(H'_{1,u}, \ldots, H'_{N,u})$ and the noise as $\mathbf{w}' = (w_1, \ldots, w_N)^T$, we can write both (2) and (4) in the form (1). The power on subcarrier n of user u is given by $p_{n,u} = \mathrm{E}[|x'_{n,u}|^2]$ and the total noise power is $N_0 = N \cdot \mathrm{E}[|w'_n|^2]$. The channel gain to noise ratio (CNR) is hence defined by

$$T_{n,u} = \frac{|H'_{n,u}|^2}{\Gamma_u \cdot N_0/N}, \quad n = 1, \ldots, N;\ u = 1, \ldots, U \qquad (5)$$

where Γ_u is the SNR gap which will be introduced later. For the time being, we set $\Gamma_u = 1$.

The introduction of subcarrier sharing thus increases the degrees of freedom for the resource allocation. In the next sections, we will evaluate the possible increase of the sum rate due to the CDMA component.

3. Multi-User Water-Filling

The maximum information bitrate of user u, i.e. the channel capacity of this user, is given by the mutual information between its transmit signal $\mathbf{X}'_u$ and the receive signal $\mathbf{Y}'$ as $R_u = I(\mathbf{X}'_u; \mathbf{Y}')$. The sum rate, expressed in units of bits per channel use, which in our case is equivalent to bps/Hz, is given by

$$R = \sum_{u=1}^{U} R_u = \frac{1}{N} \sum_{n=1}^{N} \log_2 \left(1 + \sum_{u=1}^{U} p_{n,u} T_{n,u}\right) \qquad (6)$$

With the formulation according to (1),(4), (5) and $N = LN_c$, both system variants are covered: $L = 1$ for pure OFDMA and $L > 1$ for OFDMA-CDMA. In the following, we consider the sum-rate maximization problem:

$$\text{maximize} \quad R \qquad (7)$$

$$\text{subject to} \quad \sum_{n=1}^{N} p_{n,u} \leq P_u \quad \forall u = 1, \ldots, U \qquad (8)$$

$$p_{n,u} \geq 0 \quad \forall u = 1, \ldots, U;\ n = 1, \ldots, N \qquad (9)$$

The main difference of this formulation to the corresponding problem for the downlink is the power constraint (8), which has to hold individually for each user, while in the downlink there is only one constraint for the sum power. Nevertheless, there exists a deep relationship between both cases and many results for one case can be applied to the other by making use of duality [4, 5].

Since the uplink of both systems can be described as a Gaussian vector multiple-access channel (1), the iterative water-filling algorithm of Yu et al. [3] can be applied to find the maximum sum rate (see also [6]). If the channel can be decomposed into N parallel subchannels like in (4), the IWF algorithm reduces to the simple form:

```
p_{n,u} = 0  ∀n, u
repeat
  for u = 1 to U
```
$$z_n = \sum_{j=1, j \neq u}^{U} |H_{n,j}|^2 p_{n,j} + N_0$$
$$\mathbf{p}_u = \underset{\mathbf{q}=(q_1,\ldots,q_N)}{\arg\max} \left\{ \sum_{n=1}^{N} \log_2 \left(|H_{n,u}|^2 q_n + \Gamma_u z_n \right) : \sum_{n=1}^{N} q_n \leq P_u, \, q_n \geq 0 \right\}$$
```
  end
until the desired precision is reached
```

This algorithm delivers the optimum transmit power vector $\mathbf{p}_u = (p_{1,u}, \ldots, p_{N,u})^\mathrm{T}$. Given these powers, the sum rate can be calculated easily by (6). This sum rate denotes the Shannon capacity, which can serve as an upper bound for practically implementable transmission schemes, but it does not indicate how to construct the optimum scheme. A more realistic upper bound can be found by using the gap approximation [7, 8]: The SNR gap Γ_u describes the additional SNR required by a transmission scheme to achieve a rate equal to the Shannon capacity. This SNR gap can be introduced easily in the definition of the CNR by setting $\Gamma_u > 1$ and can also be considered in the IWF algorithm. For uncoded M-QAM with symbol error probability $\mathcal{P}_\mathrm{S}$, the SNR gap is approximated by $\Gamma = \frac{1}{3} \left(Q^{-1}(\mathcal{P}_\mathrm{S}/4) \right)^2$. The sum rate with an SNR gap $\Gamma_u > 1$ can also be calculated by (6), provided that the same Γ_u has been considered in the IWF algorithm to compute the power allocation $p_{n,u}$.

The power allocation obtained with the IWF algorithm does not consider the "OFDMA constraint" that subcarriers are exclusively allocated. Nevertheless, the required subcarrier allocation for OFDMA can be derived directly from the power allocation by assigning to each subcarrier the user with the highest allocated power:

$$a_n = \arg\max_u (p_{n,u}), \quad n = 1, \ldots, N \tag{10}$$

Here, $\mathbf{a} = (a_1, \ldots, a_N)$ denotes the subcarrier allocation vector, which contains in each position the assigned user. Now, we have for each user a set of dedicated subcarriers $\mathcal{N}_u = \{n : a_n = u\}$ and thus the multi-user problem is decoupled into U single-user problems. In order not to

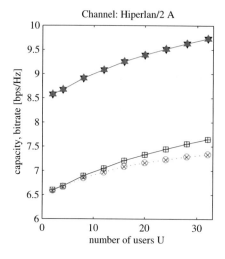 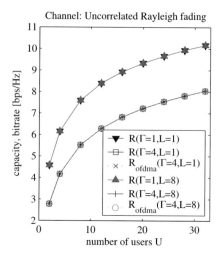

Figure 3. Cell capacity and achievable bit rates for two channel models: The Hiperlan/2 A channel is a multipath model and considers the path loss of the uniformly distributed users. The second model assumes for each subcarrier uncorrelated Rayleigh fading, *i.e.* $H_{n,u} \sim \mathcal{CN}(0,1)$.

waste power by the quantization step (10), the power allocation has to be recomputed:

$$\mathbf{p}_u = \arg\max_{\mathbf{q}} \left\{ \sum_{n \in \mathcal{N}_u} \log_2\left(1 + q_n T_{n,u}\right) : \sum_{n \in \mathcal{N}_u} q_n \leq P_u,\ q_n \geq 0 \right\} \quad (11)$$

From this power allocation, which is nothing else than the well-known single-user water-filling distribution, the rate sum can be calculated by

$$R_{\text{ofdma}} = \sum_{u=1}^{U} R_u = \frac{1}{N} \sum_{u=1}^{U} \sum_{n=1}^{N} \log_2\left(1 + p_{n,u} T_{n,u}\right) \quad (12)$$

For the case that the IWF algorithm allocates power to only one user per subchannel, *i.e.* $p_{n,u} = \delta[u - a_n] p_{n,a_n}$, the single-user water-filling (11) does not alter the power allocation and thus both (6) as well as (12) reduce to $R = \frac{1}{N} \sum_{n=1}^{N} \log_2(1 + p_{n,a_n} T_{n,a_n})$.

4. Simulation Results and Discussion

Figure 3 illustrates the maximum sum rates according to (6) and (12) for two channel models, SNR gaps $\Gamma_u \in \{1,4\}$ and spreading factors $L \in \{1,8\}$. It can be seen that the sum rate $R(\Gamma, L)$ according to (6) does actually not depend on the spreading factor L, *i.e.* the introduction of the CDMA component does not change the maximum rate. This result, which is only surprising at the first glance, can be explained most easily if the subcarrier sharing is implemented with TDMA: If the optimum power allocation is recomputed for each time slot, the result will not change since the channel is assumed to be constant. A similar argument holds for FDMA: Since the subcarriers are assumed to be flat, no gains can be expected by a further frequency division of each subcarrier.

The optimum multiple-access scheme *per subcarrier* would allocate the powers as computed by the IWF algorithm to the users and apply superposition coding with successive decoding. Since the capacity region of an orthogonal multiple-access scheme is a subset of the MAC capacity region and the optimum power allocation does not necessarily correspond to the maximum sum rate *per subcarrier*, an orthogonal multiple-access scheme incurs a (small) performance loss.

For any Γ, L_1, L_2, it must hold $R(\Gamma, L_1) \geq R_{\text{ofdma}}(\Gamma, L_2)$. This is confirmed by the simulations, although the differences are rather small: For the Hiperlan/2 channel, a small gap is visible which increases with U, but for uncorrelated Rayleigh fading we have $R(\Gamma, L_1) \approx R_{\text{ofdma}}(\Gamma, L_2)$. The comparison of $R_{\text{ofdma}}(\Gamma = 4, L = 1)$ and $R_{\text{ofdma}}(\Gamma = 4, L = 8)$ reveals that the sum rate for $L = 8$ is indeed slightly higher for the Hiperlan/2 channel (although too small to be visible in the plot).

These results thus show that the gains obtained by the introduction of the CDMA component are in the best case very marginal or not even detectable by numerical simulations. This result is surprising in the sense that in [9], a gain of up to 5 dB has been reported by the introduction of a CDMA component. A probable explanation for this considerable difference might be that – although a certain duality exists between uplink and downlink as well as between rate maximization and power minimization – the optimization problems are not exactly dual due to different constraints. This interesting difference also suggests that we might arrive at different conclusions when we formulate our problem with rate constraints per user or when we aim for other points in the capacity region.

Another more practical drawback of subcarrier sharing is the requirement of a longer coherence time: The introduction of the CDMA component requires the channel to be constant for L OFDM symbols, *i.e.* the coherence time must be L times longer with subcarrier sharing. However, the coherence time of the channel also limits the number of subcarriers in an OFDM system and hence in this aspect the CDMA component has no advantage over an OFDM system with LN_c subcarriers.

5. Conclusions

We have considered the maximization of the sum rate in the uplink of an OFDMA system with a *per subcarrier* CDMA component which allows to share each subcarrier by L users. The introduction of this secondary multiple-access component creates L times more subchannels and thus increases significantly the degrees of freedom for the resource allocation. Interestingly, this has a rather marginal effect on the achievable sum rate. Hence, the performance of an uplink OFDMA system in terms of sum rate does benefit only very marginally from the introduction of subcarrier sharing.

Acknowledgement

This work has been performed in the framework of the IST project IST-2003-507581 WINNER, which is partly funded by the European Union. The authors would like to acknowledge the contributions of their colleagues. In addition, the authors wish to express their gratitude to Dr. C. Ibars and M. Realp for helpful discussions.

References

[1] R. S. Cheng and S. Verdú, "Gaussian multiaccess channels with ISI: capacity region and multiuser water-filling," *IEEE Trans. Inform. Theory*, vol. 39, no. 3, pp. 773-785, May 1993.

[2] R. G. Gallager, *Information Theory and Reliable Communication*, Wiley 1968.

[3] W. Yu, W. Rhee, S. Boyd and J. Cioffi, "Iterative water-filling for Gaussian vector multiple-access channels", *IEEE Trans. Inform. Theory*, vol. 50, no. 1, pp. 145-152, Jan. 2004.

[4] P. Viswanath and D. Tse, "Sum capacity of the vector Gaussian broadcast channel and uplink-downlink duality", *IEEE Trans. Inform. Theory*, vol. 49, no. 8, pp. 1912-1921, Aug. 2003.

[5] N. Jindal, S. Vishwanath and A. Goldsmith, "On the duality of Gaussian multiple-access and broadcast channels", *IEEE Trans. Inform. Theory*, vol. 50, no. 5, pp. 768-783, May 2004.

[6] S. Pfletschinger, "From cell capacity to subcarrier allocation in multi-user OFDM", *IST Mobile & Wireless Communications Summit*, Dresden, June 2005.

[7] G. D. Forney and M. V. Eyuboglu, "Combined equalization and coding using precoding", *IEEE Commun. Mag.*, pp. 25-34, Dec. 1991.

[8] J. M. Cioffi, "A multicarrier primer", *ANSI T1E1.4 Committee Contribution*, Nov. 1991.

[9] P. Trifonov, E. Costa and E. Schulz, "Adaptive user allocation, bit and power loading in multi-carrier systems", *9th Int. OFDM-Workshop*, Dresden, Sept. 2004.

NEW LOADING ALGORITHMS FOR ADAPTIVE SS-MC-MA SYSTEMS OVER POWER LINE CHANNELS: COMPARISON WITH DMT

Matthieu Crussière, Jean-Yves Baudais, and Jean-François Hélard
Electronics and Telecommunications Institute of Rennes (IETR)
INSA, 20 avenue des Buttes-de-Coësmes, CS 14215, 35043 Rennes Cedex, France
{matthieu.crussiere,jean-yves.baudais,jean-francois.helard}@insa-rennes.fr

Abstract In this paper, we propose to combine adaptive loading principles with the spread-spectrum multicarrier multiple access (SS-MC-MA) scheme. Such an approach has a particular interest in the context of powerline communications (PLC), where the transmitters have not only to exploit robust transmission techniques, but also to adapt the modulations to the channel response. We introduce a finite-granularity loading algorithm that dynamically handles the configuration of the system under power spectral density constraints. The presented algorithms assign subcarriers, spreading codes, bits and energy to each user in order to maximize either the data rate or the noise margin at a given target symbol error rate (SER). Simulation results of the new scheme are presented for different measured PLC channels and are compared with those of the classical discrete multitone modulation (DMT) approach. It is shown that the adaptive SS-MC-MA scheme performs significantly better than DMT, due to its natural energy gathering capability which leads to a more efficient bits and energies distribution.

1. Introduction

In the past few years, the intense demand for high-speed home Internet has given rise to permanent necessity for additional transmission capacities on the access network. High data rate communications over the so-called last-mile have become a challenging task and have motivated the study of new telecommunication networks and new transmission technologies. A promising possibility is then offered by powerline communications (PLC), which are today considered as a convenient and cheap alternative to already available digital subscriber line (DSL), cable or wireless technologies. However, power distribution networks have not been designed for communication purposes and do not present a favorable transmission medium. The power line channel exhibits multipaths caused by reflections on the discontinuities of the tree-like-structured network, and is characterized by strong frequency-dependent cable losses [1]. Unfavorable noise conditions have also to be considered among which the most unpopular impulsive noise [2].

To cope with the impairments of such a hostile channel, PLC systems have to apply robust and efficient modulation techniques such as spread spectrum (SS) and orthogonal frequency division multiplexing (OFDM) schemes [3]. As we dedicate our study to outdoor PLC, multiuser and multicellular needs are moreover to consider. Like in the wireless context, multicarrier spread spectrum (MC-SS) schemes, which combine both above-mentioned techniques, then represent suitable schemes to investigate. They have shown very good performances in the case of multiuser communications in difficult environment and are today proposed for beyond 3G mobile systems.

On the other hand, it turns out that the PLC channel is invariant for periods of time that are dramatically long compared to classical symbol durations. This quasi-static behavior encourages the use of adaptive modulation schemes like the well-known digital multitone (DMT) modulation, which has been selected as a standard for transmission over DSL. DMT actually uses a so-called loading algorithm based on the classical waterfilling solution, known as optimal. Combined with FDMA (frequency division multiple access), DMT can easily be adapted to PLC multiuser communications.

In this article, we propose to combine adaptive loading principles with MC-SS systems. The basic idea is to take advantage of both approaches in order to obtain a robust, efficient and suitable scheme for the PLC context. The proposed scheme consists in a spread spectrum multicarrier multiple access (SS-MC-MA) system [4] combined with a subcarrier, code, bit and energy allocation algorithm. The algorithm handles the system configuration in order to maximize either the throughput or the noise margin of each user. As in practical PLC systems, finite order modulations and power spectral density (PSD) limitations will be considered to derive the loading algorithms. Under such constraints, the DMT performance reduces and we will show that the introduced system clearly gives the best results.

This paper is organized as follows. Section 2 describes the used SS-MC-MA system. The main mathematical results of maximization problems are developed in section 3. Section 4 gives rate and noise margin results in the case of a 4-user multiple access communication over PLC channels. Finally, section 5 concludes this paper.

2. System description

The generated multicarrier signal is composed of $N = 2048$ subcarriers transmitted in the band 0–20 MHz, which is the suggested band for outdoor PLC. The intercarrier spacing is 9,765 kHz and a guard interval of 5 μs is employed to avoid inter-symbol interference. Moreover, we suppose that synchronization and channel estimation tasks have been processed successfully. Furthermore, we consider that perfect channel state information is available at the transmitter, which can fairly be assumed from the quasi-static behavior of PLC channels. In simulations, we use powerline channel responses that have been measured in an outdoor network by the French power supply company (EDF) (see Fig. 1).

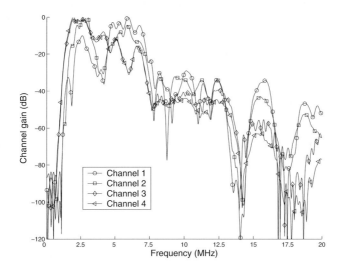

Figure 1. Examples of measured PLC channel responses

Let us recall that SS-MC-MA is a multicarrier modulation that combines CDMA and FDMA (frequency division multiple access). The FDMA component consists in the transmission of several subsets of subcarriers in parallel, each subset being exclusively assigned to a single user. The CDMA component allows each user to simultaneously transmit several symbols on the same subset by spreading them in the frequency domain. As with SS-MC-MA, each subcarrier is only used by a single user and then corrupted by a single channel, the subcarrier, bit and energy allocation process is made easier and particularly efficient in a multiuser scenario, as we will se later on. The subcarrier subsets are denoted S_b, $b \in [1; \lfloor \frac{N}{L} \rfloor]$, where L is the spreading factor. $\mathcal{B}_u$ are the sets containing the indices b of the subsets S_b collected by user u, $u \in [1; U]$. In a general approach, each subset S_b receives K_b spreading sequences associated to different modulation orders $R_{k,b}$ and different transmit energy levels $E_{k,b}$, $k \in [1; K_b]$. Moreover, the subcarriers are adaptively distributed among users. The proposed loading algorithm actually handles the optimal configuration of the system depending on the optimization policy. Recall that we assume finite granularity of modulations, i.e. $R_{k,b} \in \mathbb{N}$, and PSD constraint, i.e. $\sum_k E_{k,b} \leq E$. Note that the proposed system can be applied either to the downlink or the uplink of an indoor or an outdoor PLC network.

3. Main results

In this section, we introduce the main theoretical results of the paper. In order to derive the loading algorithms, we establish specific propositions that give the optimal loading strategy to use in each subcarrier subset. Considering infinite granularity, the throughput achieved by user u over its set of subsets S_b, $b \in \mathcal{B}_u$, is given by,

$$R_u = \sum_{b \in \mathcal{B}_u} R_b = \sum_{b \in \mathcal{B}_u} \sum_{k=1}^{K_b} R_{k,b} = \sum_{b \in \mathcal{B}_u} \sum_{k=1}^{K_b} \log_2 \left(1 + \frac{1}{\gamma_u \Gamma} \frac{L^2}{\sum_{l \in S_b} \frac{1}{|h_{l,u}|^2}} \frac{E_{k,b}}{N_0} \right), \tag{1}$$

where Γ is the SNR gap and γ_u is an additional margin for users u, called the noise margin gap, as defined in [5]. $h_{l,u}$ denotes the channel gain of user u's subcarrier l. Both of the approaches, thoughput and noise margin maximization, are considered in this paper. Optimization is first led assuming a fixed subcarrier sharing among users.

3.1 Throughput maximization

In this section, as throughput maximization is considered, we set $\gamma_u = 1$. Given a subset partionning of the spectrum, the Lagrange optimization procedure applied to the throughput maximization of R_u leads to a fairly simple solution which consists in achieving a uniform distribution of bits and energy between the L available codes in each subset S_b. In that case, we have $E^*_{k,b} = E/L$ and $R^*_{k,b} = R^*_b/L$, where superscript * denotes optimality. Assuming $E = 1$ without loss of generality, R_b^* is obtained substituting $E_{k,b}$ with $E_{k,b}^*$ in (1),

$$R_b^* = L \, \log_2 \left(1 + \frac{1}{\Gamma} \frac{L}{N_0 \sum_{l \in \mathcal{S}_b} \frac{1}{|h_{l,u}|^2}} \right) \qquad (2)$$

When finite order modulations are used, the Lagrange solution can not be applied and the problem gets a bit trickier. We then need to introduce the optimal bits and energy distribution given in Prop. 1.

PROPOSITION 1 *Let denote $\mathcal{R}$ the bit allocation policy. For a given subset allocation among users, the throughput of the system is maximized if, in each subset $\mathcal{S}_b$, $\mathcal{R}$ assigns $(\lfloor R_b^*/L \rfloor + 1)$ bits to n_b codes and $\lfloor R_b^*/L \rfloor$ bits to $(L - n_b)$ codes, where $n_b = \lfloor L(2^{R_b^*/L - \lfloor R_b^*/L \rfloor} - 1) \rfloor$. The achieved rates are denoted $\bar{R}_{k,b}^*$ and the related energies express*

$$\bar{E}_{k,b}^* = \frac{\Gamma}{L^2} \sum_{l \in \mathcal{S}_b} \frac{N_0}{|h_{l,u}|^2} \left(2^{\bar{R}_{k,b}^*} - 1 \right) \qquad (3)$$

The rate on each subset $\mathcal{S}_b$ reaches the value $\bar{R}_b^* = \sum_k \bar{R}_{k,b}^*$, and the maximal rate of user u writes $\bar{R}_u^* = \sum_b \bar{R}_b^*$. Hence, the throughput maximization is fairly simple and requires the computation of R_b^*, n_b and of only two values of rates $R_{k,b}^*$ and energies $E_{k,b}^*$.

3.2 Noise margin maximization

As evident from equation (1), no simple closed form solution can be derived to obtain the maximal noise margin γ_u for a target throughput R_u. In each individual subset however, a Lagrange optimization applied to the maximization of the noise margin for a target rate R_b and infinite granularity yields the solution: $R_{k,b}^* = R_b/L$ and $E_{k,b}^* = E/L$, which is very close to the throughput maximization solution. Let us denote γ_b the noise margin of subset $\mathcal{S}_b$. In the case of finite granularity, a modified loading procedure has to be established as proposed in Prop. 2.

PROPOSITION 2 *Let denote $\mathcal{R}$ the bit allocation policy. For a given subset allocation among users and given target rates R_b, the noise margin of the system is maximized if, in each subset $\mathcal{S}_b$, $\mathcal{R}$ assigns $(\lfloor R_b/L \rfloor + 1)$ bits to $(R - \lfloor R_b/L \rfloor L)$ codes and $\lfloor R_b/L \rfloor$ bits to $L - (R - \lfloor R_b/L \rfloor L)$ codes.*

The achieved rates are denoted $\bar{R}_{k,b}^*$ and the related energies express

$$\bar{E}_{k,b}^* = \frac{2^{\bar{R}_{k,b}^*} - 1}{\sum_{k=1}^{L}(2^{\bar{R}_{k,b}^*} - 1)} E \qquad (4)$$

Using (1) the optimal achieved noise margin on subset $\mathcal{S}_b$ then writes

$$\bar{\gamma}_b^* = \frac{1}{\sum_k 2^{\bar{R}_{k,b}^*} - 1} \frac{1}{\Gamma} \frac{L^2}{\sum_{l \in \mathcal{S}_b} \frac{1}{|h_{l,u}|^2}} \frac{E}{N_0}. \qquad (5)$$

Prop. 2 assumes that the individual target rates R_b for each subset $\mathcal{S}_b$ are known. As target rates R_u rather than R_b are specified, the loading algorithm must handle the rate distribution among the subsets while maximizing the noise margin in each subset $\mathcal{S}_b$. Denoting $\bar{\gamma}_u^*$ the optimal noise margin of user u, we have to solve $\bar{\gamma}_u^* = \max_{R_b} \min_b \bar{\gamma}_b^*$. Since the noise margin $\bar{\gamma}_b^*$ resulting from the assignment of a certain number of bits in a subset is independent of the numbers of bits assigned to other subsets, it turns out that a greedy approach is optimal [6]. The basic idea is to iteratively assign bits to the subsets, one bit at a time, and at each iteration, an additional bit is assigned to the subset that will exhibit the highest noise margin γ_b after receiving one more bit. The loading procedure can then be described as follows

```
1. Initialization. Set ∀b R_b = 0. Compute ∀b γ̄*_b considering R_b + 1
3. While ∑_b R_b < R_u
    a. Find b = arg min_b (γ̄*_b). Assign one more bit to the selected subset
       S_b,
    c. Update R_b = R_b + 1 for the found b and compute new γ̄*_b for R_b + 1
```

It is important to note that each additionnal bit assignment must be processed with respect to Prop. 2. The presented algorithm is a bit-additive algorithm, with an initialization state $\forall b\ R_b = 0$. A bit-removal approach could also be led starting from the state $\forall b\ R_b = \bar{R}_b^*$, without changing the final allotment result. It arises that the absence of closed form solution to the noise margin maximization problem leads to a more complex procedure than that of the rate maximization task.

3.3 Adaptive subcarrier distribution algorithms

Both optimization procedures of the previous sections consider that the sharing of the spectrum between the users has already been processed. The subcarrier distribution policy has to be studied in order to optimize the spectrum resource. The aim is not to maximize the total throughput (resp. the total noise margin) but instead to maximize each individual throughput (resp. each individual margin). The maximization problems can then be stated

$$(P1): \max_{\{S_b, B_u\}} \left(\min_u \bar{R}_u^* \right) \qquad (P2): \max_{\{S_b, B_u\}} \left(\min_u \bar{\gamma}_u^* \right) \qquad (6)$$

These are classical max-min problems and the maximization task essentially consists in finding the different subsets so that each user can maximize its specific metric. Let us denote η_u this metric. For throughput maximization problem $(P1)$, η_u is the rate $\bar{R}_u^*$ assigned to user u, and for margin maximization $(P2)$, η_u is the margin $\bar{\gamma}_u^*$ of user u. To find the optimal solution, each problem should be formulated into a standard convex optimization problem. However, as the resulting algorithm would require a prohibitive intensive computation, a sub-optimal but fairly simple solution is proposed as in [7]. The solution consists in iteratively assigning one block at a time to the user u that benefits from the lowest metric η_u. In order to increase η_u as more as possible, each new block should be made up of the best available subcarriers. After each assignment, the metric value is updated with respect to the previously stated propositions, namely Proposition 1 for problem $(P1)$ or Proposition 2 for problem $(P2)$. A priority order among user is required to allocate a first block to each user and provide initial individual metric values. A convenient solution is to sort the users by ascending order of channel response amplitude. The detailed algorithm writes

---

1. Initialization. Set $\forall u\ \eta_u = 0,\ B_u = \emptyset,\ \mathcal{H} = \{1, \ldots, N\}$.
   Compute $\forall u\ \mathfrak{H}_u = \sum_{i \in \mathcal{H}} |h_{i,u}|$
2. While $\exists u\ B_u = \emptyset$
   a. Find $u = \arg\min_u (\mathfrak{H}_u)$ with $\text{card}(B_u) = 0$
   b. Select the best L_c subcarriers of user u defining the subset S_b
   c. Compute η_u. Update $\mathcal{H} = \mathcal{H} - S_b$, $B_u = \{b\}$
3. While $\mathcal{H} \neq \emptyset$
   a. Find $u = \arg\min_u (\eta_u)$
   b. Select the best L_c subcarriers of user u defining a new subset S_b
   c. Compute new η_u. Update $\mathcal{H} = \mathcal{H} - S_b$, $B_u = B_u + \{b\}$

The proposed iterative algorithm actually consists in a greedy procedure applied to the subcarrier distribution. As an FDMA approach is carried out, assigning a subcarrier to a particular user prevents other users from using that subcarrier. This dependency makes *any* greedy

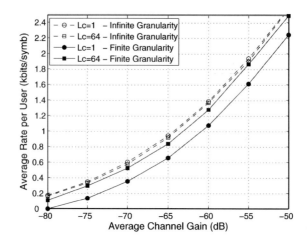

Figure 2. Achieved throughputs for the proposed algorithm

algorithm sub-optimal [6]. Nevertheless, we will see in simulations that the proposed scheme offers very satisfying results.

4. Simulation results

We apply the proposed algorithms to the case of a 4-user multiple access communication over the channels presented Fig. 1. The performance of the new system are compared to those obtained with a DMT/FDMA approach. Note that DMT is equivalent to the proposed SS-MC-MA system with $L = 1$. We assume a background noise level at -110 dBm/Hz and a flat transmission PSD of -30 dBm/Hz. We use the 2^q-ary QAM constellations specified for DSL. Results are given for a target SER of 10^{-3} without channel coding, corresponding to an SNR gap $\Gamma = 6$ dB.

Fig. 2 shows the average rates per user achieved when the system is configured to maximize the throughput. Results are given versus the average channel gain $G = \frac{1}{N}\sum_n |h_n|^2$ which conveys the attenuation experienced by the signal through the channel. The corresponding SNR is then given by $\text{SNR}_{dB} = -30 + G_{dB} + 110$. The system performance is presented in the case of infinite and finite granularity and for a spreading factor $L = 64$ with SS-MC-MA and $L = 1$ corresponding to DMT. Achieved rates are slightly higher with DMT when granularity is infinite, since DMT represents the optimal solution. In the finite granularity approach, both schemes achieve lower throughputs, which is due to the

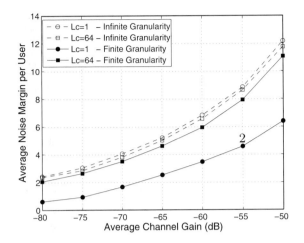

Figure 3. Noise margin for the proposed algorithm

combined effect of the PSD constraint and discrete modulation requirements. Nevertheless, SS-MC-MA exhibits the highest throughput and performs closer to the infinite granularity upper-bound than DMT. For example, for a channel gain of -60 dB, corresponding to an SNR of 20 dB, SS-MC-MA can transmit 1280 bits per symbol and for each of the 4 users, while DMT only offers 1080 bits per symbol and per user. The corresponding total throughputs are around 48 Mbps for SS-MC-MA and only 40 Mbps for DMT. Moreover, note that the worst the SNR, the better the relative throughput gain. It is then clear that the spreading component provides throughput gain. This behavior can be explained by the energy gathering capability of SS-MC-MA within each subcarrier block. Contrary to DMT, the proposed system can advantageously collect and exploit the residual energies, lost on each subcarrier of the DMT system because of the finite granularity of the QAM modulations.

If we now run the noise margin algorithm, we obtain the results presented Fig. 3 which depicts the computed average noise margins per user versus the SNR. For each SNR, the specified target rate corresponds to the half of the maximal achievable rate, i.e. the half of the thoughput previously achieved with infinite granularity when $L = 1$ (see fig 2). As evident from the plotted curves, the proposed scheme is able to perform very close to the infinite granularity upper limit and straightforwardly outperforms the DMT system. For instance, for a channel gain of -60 dB, the exhibited noise margin with SS-MC-MA is almost twice as high as with DMT. These results indicate that the transmitted bits are better distributed across the spectrum when spreading is processed. Recall that any noise margin increase improves the robustness of the system, which is worth of interest in a transmission context as noisy as the PLC's.

5. Conclusion

Loading procedures have been proposed as a solution to the throughput and the noise margin maximization tasks within each subcarrier subset of an SS-MC-MA system under PSD constraints. We also proposed a subcarrier sharing algorithm that either maximizes the minimum data rate or the minimum noise margin among users. It was shown that the energy gathering capability of the spreading function leads to a better distribution of bits and energies across the subcarriers with SS-MC-MA than with DMT. Equivalently, SS-MC-MA better exploits the energy resource than DMT does, so that the spared energies can be efficiently used to increase either the throughput or the noise margin of each active user. This interesting feature eventually makes SS-MC-MA outperform DMT in terms of data rates and noise margins. Typically, it was shown that a relative throughput gain superior to 20% is obtained for average SNRs less than 20 dB, and that the obtained noise margin is increased twofold at such SNR values. We finally conclude that the proposed adaptive SS-MC-MA scheme can advantageously be exploited in quasi-static multiuser environments such as PLC's, allowing an increase of either the data rate or the robustness of the systems.

References

[1] M. Zimmermann and Klaus Dostert. A multipah model for the powerline channel. *IEEE Trans. Commun.*, 50(4):553–559, April 2002.

[2] M. Zimmermann and Klaus Dostert. Analysis and modeling of impulsive noise in broad-band powerline communications. *IEEE Trans. Electromagn. Compat.*, 44(1):249–258, February 2002.

[3] E. Biglieri. Coding and modulation for a horrible channel. *IEEE Trans. Commun.*, 41(5):92–98, May 2003.

[4] S. Kaiser and W-A. Krzymien. Performance effects of the uplink synchronism in SS-MC-MA system. *European Trans. Commun.*, 10, July 1999.

[5] J.M. Cioffi. A multicarrier primer. Technical report, ANSI T1E1.4/91–157, Committee contribution, 1991.

[6] A. Federgruen and H. Groenevelt. The greedy procedure for resource allocation problems: necessary and sufficient conditions for optimality. *Operations research*, 34(6), Nov./Dec. 1986.

[7] W. Rhee and J.M. Cioffi. Increase in capacity of multiuser ofdm system using dynamic subchannel allocation. In *Proc. IEEE Vehicular Technology Conference (VTC-Spring)*, volume 2, pages 1085–1089, May 2000.

AN INVESTIGATION OF OPTIMAL SOLUTION FOR MULTIUSER SUB-CARRIER ALLOCATION IN OFDMA SYSTEMS

Ying Peng, Simon Armour, Angela Doufexi and Joe McGeehan
Electrical and Electronic Engineering Department, University of Bristol, Woodland Road, Bristol, U.K., BS8 1UB

Abstract: In this paper, the problem of multi-user sub-carrier allocation by metric of channel gain in an OFDMA system is formulated as an assignment problem in order to investigate an optimal solution. An optimal channel sub-carrier allocation (OCSA) strategy (extension of Hungarian Method [1]) is proposed for fair bandwidth allocation and highest total perceived channel gain. Additionally, two sub-optimal algorithms are considered: One is an extension of the DSA algorithm in [2], the other is maximum gain sort-swap (MGSS) which is an extension of the algorithm in [3]. The DSA algorithm is of particular interest since it has been shown in [5] to combine well with MIMO. The algorithms can outperform OFDMA with pseudo-random sub-carrier allocation by up to 6dB. Simulation results show that the DSA algorithm achieves perceived channel gains within a fraction of a dB of the optimal solution.

Key words: 4G systems, Wireless access, OFDMA, Sub-carrier allocoation

1. INTRODUCTION

Orthogonal Frequency Division Multiplexing Access (OFDMA) is a subject of considerable recent interest for broadband wireless communications. OFDMA matches well to the multi-user scenario. It makes use of OFDM modulation whilst allowing multiple access by separating symbols in both time and frequency.

As RF bandwidth and transmission power in the OFDMA system are limited, the key issue is how the utilization of the channel can be maximized, whilst maintaining a fair share of channel bandwidth for all users. Previous work has already proven that in a frequency selective channel, sub-carriers will perceive a large variation in channel gain and the perceived channel will be different for each user. If a deterministic allocation of sub-carriers is employed, this multi-user diversity can be exploited. In this way, the majority of sub-carriers are allocated to each user so as to ensure a high perceived gain (relative to the mean) rather than attenuation.

A low complexity sub-carrier allocation algorithm for adaptive modulation OFDMA systems was proposed in [2]. It used channel gain as the metric to achieve high perceived channel gain and fair bandwidth allocation for each user, whilst simultaneously preventing users from sharing the same sub-carriers (interfering with each other). However, whilst achieving significant gain, this algorithm did not reach the optimal solution (in terms of achieving the total maximum channel gain for users).

In [3], the Sort-Swap algorithm was examined to make this allocation problem a classic assignment problem to minimise the required transmit power. In this paper, we extend this idea to maximizing the total channel gain as an assignment problem. The Hungarian Method considered in [1,3,4] is modified to solve this assignment problem and achieve an optimal solution. The Sort-Swap algorithm in [3] and DSA algorithm in [2] are considered both separately and in combination as sub-optimal solutions. The DSA algorithm is of further interest since it has been shown to offer the ability to mitigate the debilitating effects of channel correlation on MIMO systems [5].

2. SYSTEM MODEL

The OFDMA system considered here includes one Base Station (BS) and multiple Mobile Stations (MSs). In this paper, the downlink is considered for the sake of simplicity. However, the DSA algorithm and the diversity gains which it achieves are equally applicable to the uplink (which will enjoy the further benefits of OFDMA processing gain). The BS is considered to communicate simultaneously with multiple MSs, each of which is allocated a single sub-channel consisting of a number of OFDMA sub-carriers (equal numbers of sub-carriers per sub-channel and a single sub-channel per MS is assumed for simplicity in this paper but this is not essential to the functionality of the DSA algorithm). The baseband process is illustrated in Fig. 1. The channel estimator is not explicitly considered in this paper (ideal channel knowledge at the BS for sub-carrier allocation purposes is assumed), but must be applied in practice.

3. ASSIGNMENT PROBLEM

Assuming in an OFDMA system, provision of QoS and data rate requests have been fixed, the number of sub-carriers is specified by the request of submitted data rate for each user m. The channel gain matrix H of size of $M \times N$ where N is the number of all useable sub-carriers and M is the number of all users (sub-channels as well in this case) is assumed to be known in the BS for sub-carrier allocation. The total perceived channel gain P_{total} for all users is considered in order to provide good sub-channels for all users.

The allocation problem then can now be formulated as:

$$\text{Maximize } P_{total} = \sum_{n}^{N}\sum_{m}^{M} C_{m,n} H_{m,n} \tag{1}$$

$$\text{Subject to } C_{m,n} = \begin{cases} 1, & \text{subcarrier } n \text{ is assigned to user } m, \\ 0, & \text{otherwise,} \end{cases} \tag{2}$$

$$\sum_{m} C_{m,n} = 1 \tag{3}$$

$$\sum_{n} C_{m,n} = K \tag{4}$$

where $H_{m,n}$ is channel gain of sub-channel m (for user m) and sub-carrier n; $C_{m,n}$ is allocation matrix element for user m and sub-carrier n. K is the number of sub-carriers used per user. In this paper the case of equal numbers of sub-carriers per sub-channel is assumed and hence $K=N/M$. The maximum total perceived channel power is indicated by P_{total}^{opt}.

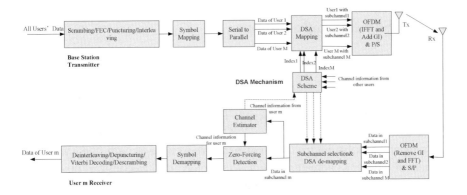

Figure 1. Downlink OFDMA system model

4. SOLUTIONS

4.1 Optimal solution: OCSA – Extended Hungarian Method

The Hungarian algorithm aims to minimize the cost by computation with a cost matrix (as applied in [1] and [4]) for the optimal solution to the allocation problem. Following this idea, the solution for (1) can be found by an inverted form of the Hungarian algorithm due to request of tracking down maximum total channel gain. This inversion can be made as following basic processing rules based on a reformulated channel gain matrix $\hat{H}$ in which each user's entry is duplicated for K times so that the matrix becomes size of $N \times N$ instead of $M \times N$, to extend the Hungarian algorithm:

$$P_{total}^{N \times N} = \sum_{n}^{N} \sum_{km=1}^{N} c_{km,n} |h_{km,n}|^2 \tag{5}$$

1. Always find maximum element in each row and column of channel gain matrix instead of minimum element in cost matrix;

2. Always make locations with maximum element recorded (such as locating zeros to differentiate other elements).

This is similar to Hungarian Method as a kind of exhaustive search which makes use of computation to check all pairs of 'user' and 'sub-carrier' one by one and compare the total gain for maximum value. When the size of the channel gain matrix increases (due to more users and/or sub-carriers) the computational effort increases substantially. Thus, in this paper, the optimal solution is considered as a reference upper bound of the system performance and is not suggested for use in practice.

4.2 Sub-optimal solutions

4.2.1 MGSS

Due to the unfeasible complexity of OCSA, a lower complexity sub-optimal solution must be considered. The algorithm previously proposed in [3] is extended, namely, Maximum Gain Sort-Swap (MGSS) which sorts sub-carrier pairs by metric of total perceived channel gain and swaps sub-carrier allocations between users to exploit the maximum power. The iteration is applied to make the most of the sort-swap process to achieve a near optimal solution. The channel gain matrix which is applied in computation in this solution is H with original and decreased (relative to OCSA) size of $M \times N$. The algorithm can be summarized as:

I. First, the channel gain matrix H is sorted in descending order and sub-carrier index matrix and initial allocation matrix $C_{initial}$ is formed according to the sorted entry of H.

II. Then, there is a swap iteration process to improve the initial allocation. The sub-carriers are swapped between users when it leads to the total channel gain increase. A gain increase factor is found by summing the minimum gain changed value when a certain sub-carrier n changed its allocation from user m_i to m_j and vice versa. The minimum value among all these gain increase factors for all user pairs is chosen and the corresponding sub-carrier is replaced in the chosen user pair. The above process is repeated until the minimum gain increase factor is not negative, i.e. allocation cannot be improved any more.

Due to reasonable and low complexity initial process, the iteration of sort and swap decreases, consequently, resulting in the lower complexity and better practicability relative to OCSA.

4.2.2 DSA and DSA-swap

The Dynamic Sub-carrier Allocation (DSA) algorithm [2] achieves sub-carrier allocation by metric of magnitudes of sub-channel gain for each user. It aims to maximise the average perceived channel gain for each user without minimizing it in other users. Thus a fair distribution of benefits is offered to every user whilst not adversely affecting overall system capacity.

In the following, P_m represents the average received power for user m, $\hat{N}$ is a vector containing the indeces of the useable sub-carriers (i.e. $\hat{N} = \{1,2,3,...,N\}$). 0_N is a vector of zeros of length N. $C_{m,s}$ is the location matrix for user m and sub-carrier s.

I. Initialization
Set $P_m=0$, $C_{m,s}=0$ for all users $m=1,...,M$ and $s=1$
II. Main process
While $\hat{N} \neq 0_N$
{(a) Make a short list according to the users that have less power. Find user m satisfying $P_m \leq P_i$ for all i, $1 \leq i \leq M$

(b) For the user m chosen in (a), Find sub-carrier n satisfying $H_{m,n} \geq H_{m,j}$ for all $j \in \hat{N}$

(c) Update P_m, $\hat{N}$ and $C_{m,s}$ with the n from (b) according to: $P_m = P_m + H_{m,n}$; $\hat{N} = \hat{N} - n$; $C_{m,s} = n$; $s = s+1$

(d) go to the next user in the short list determined in (a) until all users are allocated another sub-carrier.

This algorithm also can be improved by applying the sort-swap method as mentioned in 4.2.1 (II) with which the DSA algorithm is treated as an initial process and similar $C_{initial}$ in 4.2.1(I) can be achieved straightforwardly by $C_{m,s}$. This combination is referred to as DSA-swap. Both DSA and DSA-swap algorithms retain low complexity.

5. SIMULATION RESULTS

In the system which is proposed in section 2, the assignment algorithms proposed in section 4 are used to allocate the sub-carriers (conventional OFDMA with pseudo-random sub-carrier allocation is shown as a reference). The channel is modeled as a tapped delay line with a normalized RMS delay spread of 25. 2000 iid quasi-static channel samples are used in each simulation. $M = 16$ and $N = 768$ are assumed.

In order to simplify the equation (1) and normalize power per user per sub-carrier:

$$P_{norm}(dB) = 10\log_{10}(\frac{1}{16} \times \frac{1}{48} \times \sum_{user=1}^{16} \sum_{sub=1}^{48} H_{user,sub}) \text{ (dB)} \qquad (6)$$

where $H_{user,sub}$ is the channel gain of a certain user at allocated sub-carrier (1~48 is the allocated sub-carrier index per user, not the index in the entire 768 sub-carrier sequence) in a certain simulation time.

The resulting power gains as a function of iteration number are illustrated for one typical channel instance in Fig. 2. Both MGSS and DSA-Swap tend very close to the optimum solution (achieved by OCSA) after 3-5 iterations. The result of DSA-swap is slightly better than that of MGSS. Care should be taken to note the scale on the power gain axis since, whilst DSA and OCSA look far apart on this graph, DSA actually achieves 97.54% of the power gain of OCSA. Hence, the DSA algorithm (without swapping) can be considered a low complexity, near optimal solution. MGSS is also worthy of interest because it also offers a low complexity, near optimal sub-carrier allocation. DSA-swap is of less interest since it offers minimal improvements over DSA and MGSS in return for increased complexity (it essentially requires the sum of the complexity of the other two).

Fig. 3 presents the complementary cumulative distribution function (CCDF) of all 2000 random channels for DSA, DSA-Swap and MGSS (left figure) and 'conventional' OFDMA (right figure). OCSA cannot be shown because of high complexity in simulation. This shows that the similar performance of these algorithms is consistent across the entire statistical sample. DSA outperforms the random allocation strategy by up to 6dB and results in significantly lower variation around the mean.

An Investigation of Optimal Solution

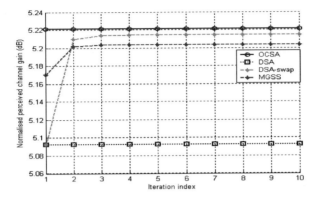

Figure 2. Comparison of all solutions : OCSA, MGSS, DSA and DSA-swap

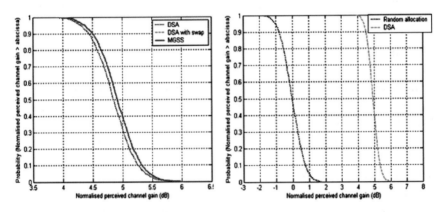

Figure 3. Comparison of CCDF of total channel gain for: left- all sub-optimal solutions; right- conventional OFDMA and DSA solutions

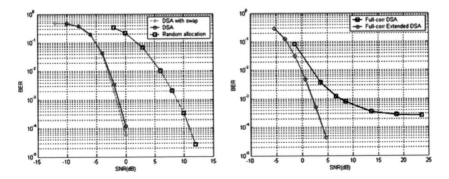

Figure 4. Comparison of BER performance: left: DSA with swap, DSA and Conventional (random allocation) in SISO- OFDMA; right: DSA and Extended DSA for very high correlated channel environment in Spatial Multiplexing MIMO-OFDMA case

Fig. 4 shows a comparison of BER performance. It is obvious (from the left figure) that DSA with swap and DSA solution achieve considerable gain (11~12 dB) relative to conventional (random allocation) OFDMA. The capability of the DSA algorithm to mitigate the effects of correlation is also shown (right figure). For more detail on this capability, the reader is referred to [5].

6. CONCLUSIONS

In this paper, an optimum sub-carrier allocation algorithm is investigated with channel gain as the metric. The optimum solution, OCSA, provides the best performance. The DSA algorithm (a sub-optimal algorithm proposed in [2]) can be improved when iterations are performed, meanwhile the advantages of DSA algorithm can be remained. Another sub-optimal algorithm MGSS makes the allocation process more efficient than DSA-Swap and achieves similar performance.

The results in this paper assume perfect channel estimation and perfect synchronization. As channel estimation in wideband channels is in general not ideal, the effect of non-ideal channel information will be a very important issue.

REFERENCES

1. H. W. Khun, "The Hungarian method for the assignment problem," Naval Research Logistics Quarterly, Q.2, pp. 473-489, Aug. 1955.
2. Y. Peng, A. Doufexi, S. Amour, J. McGeehan "An Investigation of Adaptive Sub-carrier Allocation in OFDMA Systems," VTC Spring 2005.
3. S. Pietrzyk, G. J. M. Janssen, "Multiuser subcarrier allocation for QoS provision in the OFDMA systems", Vehicular Technology Conference Proceedings, IEEE 56th, Vol 2, 24-28 Sept. 2002, pp. 1077-1081, 2003.
4. T. Issariyakul, E. Hossain, "Optimal Radio Channel Allocation for Fair Queuing in Wireless Data Networks", IEEE International Conference on Communications, '03, Vol 1, pp. 142-146, May 2003.
5. Y. Peng, S. Armour, "Application of Adaptive Sub-carrier Allocation to MIMO-OFDMA Systems to Combat Channel Correlation," World Wireless Congress 2005, May 2005.

OPTIMAL SOLUTION TO ADAPTIVE SUBCARRIER-AND-BIT ALLOCATION IN MULTICLASS MULTIUSER OFDM SYSTEM

Kainan Zhou, Yong Huat Chew and Yan Wu
Institute for Infocomm Research,
21 Heng Mui Keng Terrace, Singapore, 119613
Department of Electrical and Computer Engineering,
National University of Singapore, Singapore 117576
Kainan@nus.edu.sg, chewyh@i2r.a-star.edu.sg

Abstract Subcarrier-and-bit allocation (SBA) has been extensively investigated in the literature to improve spectral efficiency of multiuser orthogonal frequency division multiplex (OFDM) systems. However, in previous studies, only suboptimal solutions were given and only single-class case was considered. In this paper, a unified analysis of adaptive SBA for multiclass multiuser OFDM system is presented. The constrained power optimization is formulated as a mixed integer nonlinear programming (MINLP) problem. The optimal solution to this problem is derived, i.e., the instantaneous total transmit power is minimized with the quality-of-service (QoS) (data rate and bit error rate) of each class guaranteed. The optimized system performance is compared with both the fixed subcarrier and bit allocation scheme and the rate-adaptive scheme without multiuser diversity. The framework presented in this paper, based on the optimal solution, can be used as a benchmark for future developed heuristic algorithm.

1. Introduction

Next generation mobile communication (NextG) is featured by providing high rate, high quality data transmission. The system should be able to support multiclass services with the satisfaction of individual quality-of-service (QoS) requirements; this, in turn, demands more and more of the limited spectrum and calls for wise use of system resources. Therefore, adaptive resource allocation has become an essential topic in NextG system design.

OFDM technique is a very attractive candidate for nextG, not only because it can exploit frequency diversity to combat multipath fading, and thus enhance the system capacity, but also because of its capability of flexible frequency access to facilitate adaptive resource allocation. It can assign more bits on those better subcarriers which have higher signal-to-noise ratio (SNR) and less bits on worse subcarriers or simply switch it off, i.e., no data is transmitted on those subcarriers. This concept is generally called "water-filling". For multiuser OFDM, system can even dynamically distribute subcarriers among different users according to their respective channel conditions rather than allocate fixed groups of subcarriers to a certain user. In this way, multiuser diversity is exploited, which is why adaptive subcarrier-and-bit allocation (SBA) is able to achieve higher spectral efficiency in the multiuser environment, compared with the fixed subcarrier and bit allocation scheme or the rate-adaptive scheme without dynamic subcarrier selection.

Recently, many SBA algorithms for multiuser OFDM system were proposed. In [1], Wong aimed to minimize the overall transmit power with given QoS requirements. The integer constraints were relaxed and an assumption of time-sharing subcarriers was made. Then, a Lagrangian-based algorithm is proposed to solve the modified problem to give the lower bound of the minimum transmit power. In order to reduce computational complexity of this algorithm, a heuristic subcarrier allocation scheme was proposed in [2], with the assumption of fixed modulation modes. Later in [3], Zhang proposed another reduced-complexity subcarrier-bit-and-power allocation algorithm to maximize the overall system throughput. Equal power distribution over the subcarriers was assumed so that SBA problems can be decoupled, and therefore the modified problem could be easily solved through linear integer programming.

These algorithms and those reported elsewhere [4] [5], however, only gave the suboptimal solutions to the original problems with certain assumptions or relaxations. For example, some algorithm avoided integer programming by relaxing discrete integer set to real set [1] and allowing subcarrier sharing [1] [4]; some avoided nonlinear programming by converting the nonlinear objective function into a linear one based on some assumptions [2] [3]; still some used a two-step adaptation to decouple the combinatorial problem [5] [8]. Although suboptimal solutions were shown to be efficient to reduce the computational complexity for realtime implementation, we still have no idea about the gap between the suboptimal and optimal performance. This is obviously because the optimal solution has yet been given so far, and it remains a challenging topic in this area. Moreover, variety of service classes was not taken into consideration in the previous literature, which restricted the application of these algorithms in only single-class case. In this paper, we deal with transmit power optimization of adaptive SBA for multiuser OFDM system supporting multiclass services. Each class has its own QoS requirements such as target data rate and bit error rate (BER). The purpose is to conduct a unified analysis of this constrained optimization problem and then derive its exact optimal solution. The results obtained from this theoretical framework will be useful for comparing the accuracy of any developed suboptimal or heuristic algorithm.

2. Problem Definition

Here we consider a rate-adaptive downlink OFDM system supporting two classes of services. Class 1 needs a constant data rate of R_1 bits/OFDM symbol and a target BER of P_{e_1}; while Class 2 needs a minimum data rate of R_2 bits/OFDM symbol and a target BER of P_{e_2}. The number of subcarriers is N, shared by K_1 Class 1 users and K_2 Class 2 users. With careful design of OFDM signal, i.e., if the length of cyclic prefix is longer than the maximum delay of the multipath channel, intersymbol interference is mitigated; hence, each subcarrier experiences only flat fading. Here we let $g_{k,n}$ denote fading gains as seen by the kth user on the nth subcarrier. Also let N_0 denote the power spectral density (PSD) of the white Gaussian noise, and assume it is the same for all subcarriers and all users.

Further, we let $s_{k,n}$ denote the assignment indicator, i.e., if the nth subcarrier is assigned to the kth user, $s_{k,n} = 1$; otherwise $s_{k,n} = 0$. In our system, no subcarrier can be assigned to more than one user. Therefore, if $s_{k,n} = 1$, $s_{k',n} = 0$ for all $k' \neq k$. Let $c_{k,n}$ denote the number of bits of the kth user assigned onto the nth subcarrier. Assume that square signal constellations 4QAM, 16QAM and 64QAM are considered in our system model, so the number of bits within each symbol has three possible values: 2,4 and 6. Hence, the integer sets for $s_{k,n}$ and $c_{k,n}$ are respectively $S = \{0,1\}$ and $C = \{0,2,4,6\}$. $c_{k,n} = 0$ means that the kth user transmits no information bits on the nth subcarrier.

In [1], the function of the required received power at a given BER P_e and constellation of c bits/symbol for QAM signals is presented:

$$f(c) = \frac{N_0}{3} \left[Q^{-1} \left(\frac{P_e}{4} \right) \right]^2 (2^c - 1) \tag{1}$$

where

$$Q(x) = \frac{1}{\sqrt{2\pi}} \int_x^\infty e^{-t^2/2} dt \tag{2}$$

Whereas, in order to maintain QoS requirements of each user, the assigned power at the transmitter for the kth user on the nth subcarrier is

$$P_{k,n} = f(c_{k,n})/g_{k,n}^2$$
$$= \frac{N_0}{3}\left[Q^{-1}\left(\frac{P_e}{4}\right)\right]^2 (2^{c_{k,n}} - 1)/g_{k,n}^2 \quad (3)$$

on condition that the nth subcarrier is assigned to the kth user, i.e., $s_{k,n} = 1$; otherwise, $P_{k,n} = 0$.

In regard to the adaptive SBA problem concerned in this paper, our target is to minimize the overall transmit power while satisfying all the data rate and BER constraints for both Class 1 and Class 2 users. Referring to (2) and its condition, the problem can be numerically formulated as below:

$$\min_{s_{k,n},c_{k,n}} \sum_{k=1}^{K_1}\sum_{n=1}^{N} \rho_1(2^{s_{k,n}c_{k,n}} - 1)/g_{k,n}^2$$
$$+ \sum_{k=K_1+1}^{K_1+K_2}\sum_{n=1}^{N} \rho_2(2^{s_{k,n}c_{k,n}} - 1)/g_{k,n}^2 \quad (4)$$

subject to

$$\sum_{n=1}^{N} s_{k,n}c_{k,n} = R_1, \quad k = 1, 2, \ldots, K_1 \quad (5)$$

$$\sum_{n=1}^{N} s_{k,n}c_{k,n} \geq R_2, \quad k = K_1+1, K_1+2, \ldots, K_1+K_2 \quad (6)$$

$$\sum_{k=1}^{K_1+K_2} s_{k,n} = 1, \quad n = 1, 2, \ldots, N \quad (7)$$

and

$$s_{k,n} \in S \text{ integer}, \quad c_{k,n} \in C \text{ integer} \quad (8)$$

where

$$\rho_1 = \frac{N_0}{3}\left[Q^{-1}\left(\frac{P_{e_1}}{4}\right)\right]^2$$

$$\rho_2 = \frac{N_0}{3}\left[Q^{-1}\left(\frac{P_{e_2}}{4}\right)\right]^2 \quad (9)$$

are constants related to service category.

Table 1. $K_1 = K_2 = 2, N = 4, R_1 = 2$ bits/OFDM symbol, $R_2 = 4$ bits/OFDM symbol, $P_{e_1} = 10^{-5}$, $P_{e_2} = 10^{-3}$, $N_0 = 1$.

$s_{k,n}/c_{k,n}$	$n=1$	$n=2$	$n=3$	$n=4$
$k=1$	1/2	0/0	0/2	0/6
$k=2$	0/0	0/4	1/2	0/0
$k=3$	0/0	0/2	0/0	1/4
$k=4$	0/2	1/4	0/6	0/6

Table 2. Fading gains on each subcarrier for each user.

$g_{k,n}$	$n=1$	$n=2$	$n=3$	$n=4$
$k=1$	1.0332	0.386	0.77132	2.323
$k=2$	0.94289	1.2926	1.5579	0.92533
$k=3$	0.65558	1.1125	0.94374	2.0745
$k=4$	0.30256	2.0287	1.6813	1.1541

This formulation, as defined by (3)-(8), is a mixed-integer nonlinear programming (MINLP) problem with a nonlinear objective function (3) and $2(K_1+K_2)N$ integer optimization variables on discrete set. It has K_1 nonlinear equality constraints in (5), K_2 nonlinear inequality constraints in (6) and N linear constraints in (7).

3. Solution and Results

The gradient matrix and Hessian matrix for both the nonlinear objective function and the constraints are obtained. Nonlinear branch and bound algorithm [6] is used to search for the optimum solution for our problem. No relaxations or approximations are made in this process. The detail of the algorithm can be found in [7].

The optimum solution can be achieved for any given number of users and subcarriers. A simple case study is shown in Table 1-2, where the optimum solution is obtained at a snapshot of channel gains on each subcarrier for each user. The assignment results when the minimum power are achieved is presented in Table 1, where we can see the optimal subcarrier allocation and constellation selection on each subcarrier for each user with every constraints fulfilled. The channel gains on each subcarrier as seen by each user are also tabulated in Table 2 for this case. The minimized total transmit power is 56.9 unit. In order to further explain the result, if we change $c_{3,1} = 4$ and $c_{4,4} = 4$, and accordingly assign $s_{1,3}$, $s_{2,2}$, $s_{3,1}$ and $s_{4,4}$ equal to 1, the resulted total

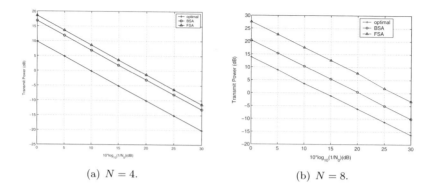

(a) $N = 4$.   (b) $N = 8$.

Figure 1. Performance comparisons between the optimal solution and other schemes. $K_1 = 2$, $K_2 = 1$, $R_1 = 2$ bits/OFDM symbol, $R_2 = 6$ bits/OFDM symbol, $P_{e_1} = 10^{-2}$, $P_{e_2} = 10^{-4}$.

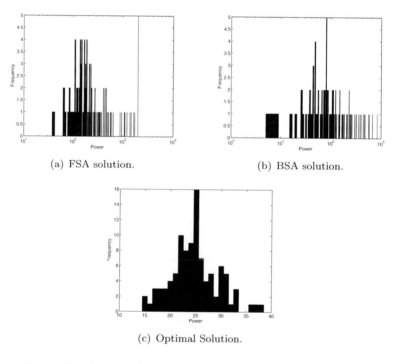

(a) FSA solution.   (b) BSA solution.

(c) Optimal Solution.

Figure 2. Probability density distributions for minimized transmit power of the 3 SBA schemes with a number of channel gain observations. $K_1 = 2$, $K_2 = 1$, $N = 8$, $R_1 = 2$ bits/OFDM symbol, $R_2 = 6$ bits/OFDM symbol, $P_{e_1} = 10^{-2}$, $P_{e_2} = 10^{-4}$, $N_0 = 1$.

transmit power needed for the new assignment is 233.9 unit, higher than the optimal value by 6.14dB.

The performance of our optimal SBA scheme is compared with the rate-adaptive scheme without multiuser diversity as well as a simple subcarrier allocation scheme discussed in [8]. In fixed subcarrier allocation scheme like traditional FDMA scheme, subcarriers are allocated to the users according to certain predetermined mapping rule. In other words, $\{s_{k,n}\}$ are known before the optimization process. The optimal constellations on each subcarrier for each user $\{c_{k,n}\}$ are searched to achieve the maximum revenue. Moreover, a simple subcarrier allocation scheme is also proposed in [8] for multiuser OFDM systems, in which each subcarrier is allocated to the user who sees the highest channel gain. In this way, the subcarrier and bit allocations are decoupled. $\{s_{k,n}\}$ are firstly decided according to the ranking of channel gains, and then $\{c_{k,n}\}$ can be easily obtained to maximize the system revenue while all the constraints are satisfied. Hereafter, we address the two schemes mentioned above FSA and BSA, respectively, for the convenience of future reference.

The transmission power of the multiclass multiuser OFDM system with different number of subcarriers is presented in Fig. 1 for $N = 4$ and $N = 8$, where the performance is also compared with FSA and BSA schemes. The curves are generated based on 100 observations when the channel changes and the average values are plotted against the noise spectral density N_0. The required transmission power of our presented optimal solution is always well under the other two schemes, as we can see from the gap between the optimal curve and the FSA, BSA curves in both (a) and (b). Furthermore, when the number of subcarriers increases from 4 to 8, the power difference between our optimal solution and FSA goes from 8dB to 13dB. The wider gap indicates that the performance degradation of fixed subcarrier allocation becomes worse as the system complexity increases in the multiclass multiuser environment. For BSA scheme, although the gap closes up by less than 1dB when N increases, the required transmit power is still 6dB more than that of the optimal scheme. This manifests large potential for performance enhancement of the current heuristic algorithms and imposes great need to explore more efficient suboptimal subcarrier allocation schemes to close the wide gap. Thus, the presented theoretical framework can be used as the benchmark for current and future developed heuristic SBA schemes for multiclass multiuser OFDM systems.

In order to study the power variations with changing channel conditions, the probability density distributions for the minimum required transmit power of the 3 schemes concerned in this paper is also presented in Fig. 2. The histograms for the minimum power at the 100 snapshots of channel gains are drawn for different SBA schemes, where we can see that not only the values of power by the optimal scheme are generally much smaller than the other two schemes, the variance of the optimal solution at different channel gains is much smaller than the other two schemes. The smaller variance shows the advantage of the optimal scheme as it reduces the dynamic range of the transmit power which leads to an easier and less expensive transmitter design.

4. Conclusions

In this paper, we present the theoretical analysis for adaptive SBA algorithm in multiclass multiuser OFDM systems. The downlink transmission supporting 2 user classes is examined. The transmission power is minimized with fulfillment of the QoS constraints of each class. The exact optimal solution is obtained using MINLP. Some simple case studies are presented and the performance curves are presented and compared with rate-adaptive OFDM with fixed subcarrier allocation and also OFDM with only a simple subcarrier allocation scheme. Our preliminary results gain an insight into the gap between the system performance of the heuristic schemes and that of the optimal solution. The proposed optimal scheme, therefore, can provide the benchmark for the current and future research of SBA algorithms for multiclass rate-adaptive systems.

References

[1] C.Y. Wong, R.S. Cheng, K.B. Letaief, and R.D. Murch. Multiuser OFDM with adaptive subcarrier, bit, and power allocation. *IEEE J. Select. Areas commun.*, vol. 17, no. 10, pp. 1747-1758, Oct. 1999.

[2] C.Y. Wong, C.Y. Tsui, R.S. cheng, and K.B. Letaief. A real-time subcarrier allocation scheme for multiple access downlink OFDM transmission. In *IEEE Proc. VTC'99*, vol. 2, 1999, pp. 1124-1128.

[3] Y.J. Zhang and K.B. Letaief. Multiuser Adaptive subcarrier-and-bit allocation with adaptive cell selection for OFDM systems. *IEEE Transactions on Wireless Communications*, vol. 3, no. 5, pp. 1566-1575, Sep. 2004.

[4] W. Rhee and J.M. Cioffi. Increase in capacity of multiuser OFDM system using dynamic subchannel allocation. *IEEE Proc. VTC'2000*, 2000, pp. 1085-1089.

[5] H. Yin and H. Liu. An efficient multiuser loading algorithm for OFDM-based broadband wireless system. *IEEE Global Telecommunications Conference*, vol. 1, 2000, pp. 103-107.

[6] I.E. Grossmann. Mixed-integer nonlinear programming techniques for process systems engineering. *Lecture notes, Department of Chemical Engineering, Carnegie Mellon University*, Jan. 1999.

[7] K. Zhou, and Y.H. Chew. Exact solution to adaptive subcarrier-and-bit allocation in multiclass multiuser OFDM systems. submitted to *IEEE Trans. Veh. Technol.*

[8] J. Jang and K.B. Lee. Transmit power adaptation for multiuser OFDM systems. *IEEE J. Select. Areas Commun.*, vol. 21, no. 2, pp. 171-178, Feb. 2003.

DYNAMIC AND SCALABLE BANDWIDTH ALLOCATION FOR BEYOND 3G CDMA SYSTEMS

Mo-Han Fong[1,2], Geng Wu[1], Wen Tong[1], Jun Li[1], T. Aaron Gulliver[2], Vijay K. Bhargava[3]
1. Nortel, Ottawa, Ontario, Canada; 2. Department of Electrical and Computer Engineering, University of Victoria, Victoria, BC, Canada; 3. Department of Electrical and Computer Engineering, University of British Columbia, Vancouver, BC, Canada

Abstract: In this paper, we introduce an enhanced Direct Spread Code Division Multiple Access (DS-CDMA) based system, called Multi-Carrier DS-CDMA (MC DS-CDMA), that supports dynamic and scalable bandwidth allocation based on service and QoS requirements. This system can support concurrent transmissions on N DS-CDMA carriers, where N is scalable based on system bandwidth and each carrier transmits an independent DS-CDMA waveform. MC DS-CDMA is a fundamentally different concept than the straightforward bandwidth expansion of current CDMA systems in several aspects. First, it is strictly backward compatible, and thus can overlay with existing single carrier DS-CDMA system without any performance degradation. Moreover, variable and asymmetric radio bandwidth (or number of carriers) can be dynamically scheduled to a Mobile Station (MS) on the Forward Link (FL) and the Reverse Link (RL) on a per-scheduling time-slot basis. This introduces another dimension of fast resource/user multiplexing in the carrier domain, in addition to the time domain as in existing 3rd generation systems, e.g. UMTS HSDPA, 1xEV-DO, 1xEV-DV. We also propose a novel service-driven layer 2/3 protocol design across multiple carriers to efficiently multiplex user traffic across multiple carriers and to exploit both multi-carrier diversity and multi-user diversity. Our performance results show that MC DS-CDMA provides significant improvements over single-carrier systems.

Key words: DS-CDMA, multi-carrier (MC), dynamic and scalable bandwidth allocation, 1xEV-DV

1. INTRODUCTION

As an evolution of 3rd generation (3G) wireless systems, 1xEV-DO (TIA/EIA/IS-856 [1]), 1xEV-DV (TIA/EIA/IS-2000 Release C/D [2][3]) and HSDPA (UMTS Release 5 [4]) technologies introduced in the past few years have significantly increased the spectral efficiency of conventional CDMA networks. Much of the spectral efficiency gain of the above systems is obtained through fast Time-Division-Multiplex (TDM) scheduling, adaptive modulation and coding, and hybrid ARQ techniques. These techniques exploit the channel variation in the time domain in terms of fast fading and long-term shadowing, to assign the optimum modulation/coding to a user in temporally good channel condition. Thus, they are most beneficial for best effort or delay tolerant data services where opportunistic scheduling can be effectively employed on the TDM fat-pipe channel, e.g. the forward packet data channel (F-PDCH) in 1xEV-DV [2][3] and the high speed downlink shared channel (HS-DSCH) in HSDPA [4]. The sector capacity improvement for best effort services can be as high as three times. However, the above techniques are less effective for delay sensitive services such as voice over IP (VoIP) and audio/video streaming since opportunistic scheduling, adaptive modulation/coding assignment and HARQ cannot be effectively employed when packets with relatively static inter-arrival times have to be delivered within a small delay bound.

To improve the system performance for delay sensitive packet data services and to enable the support of packet data services with highly asymmetrical and variable bandwidth demand (envisioned for beyond 3G systems), we propose Multi-Carrier (MC) DS-CDMA. MC DS-CDMA supports concurrent transmissions on N DS-CDMA carriers, where N is scalable based on system bandwidth, and each carrier transmits an independent DS-CDMA waveform. MC DS-CDMA is a fundamentally different concept than the straightforward bandwidth expansion of current CDMA systems in several aspects. First, variable and asymmetrical radio bandwidth (or number of carriers) can be assigned to an MS on the FL and the RL. The number of carriers and the specific carrier(s) assigned to an MS on the forward link (FL) and the reverse link (RL) can be dynamically changed on a per-scheduling time-slot basis. This flexibility introduces another dimension of fast resource/user multiplexing in the carrier or frequency domain, in addition to the time domain multiplexing as in existing 3G systems. Moreover, MC DS-CDMA also allows for more efficient spectrum sharing and user multiplexing based on service and QoS requirements. Second, we propose a novel service-driven layer 2/3 protocol design across multiple carriers to efficiently multiplex user traffic across multiple carriers and to exploit both multi-carrier diversity and multi-user diversity. Third, MC DS-CDMA can provide backward compatibility to existing single carrier DS-CDMA systems, thus allowing for the overlay of existing single carrier DS-CDMA system with MC DS-CDMA while the network deployment migrates towards broadband support.

In Section 2 of this paper, we present the detailed MC DS-CDMA system concept. Section 3 illustrates the performance of the MC DS-CDMA system. This is followed by some concluding remarks in Section 4.

2. MC DS-CDMA SYSTEM DESCRIPTION

In this section, we present a detailed description of the proposed MC DS-CDMA system.

2.1 System Overview

In this paper, for illustration purpose, we choose 1xEV-DV as a reference for the single carrier DS-CDMA system. For simplicity, we call the MC 1xEV-DV system as MC-DV. Note that similar multi-carrier enhancement can be applied to other single carrier DS-CDMA systems such as UMTS.

The baseband transmitter block diagram of MC-DV is shown in Figure 1. Each carrier transmits an independent DS-CDMA waveform at 1.2288Mcps. Adaptive modulation and coding is performed independently on each carrier based on the channel condition experienced by the user and size of the user's packet scheduled on the carrier on a particular time slot. At each scheduling time slot, the layer 2 scheduler/packet multiplexer maps the layer 2 packets of different users onto the N carriers. The mapping aims to optimize the system capacity while meeting the QoS requirement of each user's service. More details of the service driven layer 2/3 protocol will be given in Section 2.2.

Figure 2 shows the overlay of 1xEV-DV/1xRTT operable carriers within MC-DV carriers over the Nx1.25MHz spectrum allocation of the system. The 1xEV-DV/1xRTT operable carriers serve the legacy 1xEV-DV/1xRTT terminals, and also serve as the primary carriers for MC-DV terminals for system access from null or dormant states, because the 1xEV-DV/1xRTT operable carriers contain the full system access channels (e.g. paging, sync, random access). The primary carriers can also support voice, circuit data and packet data services to the MC-DV terminals. The MC-DV-only carriers, called the supplemental carriers, serve only the MC-DV terminals in active state. The supplemental carriers are optimized for high-speed packet data operation whereby all the Walsh codes and power in the FL excluding those used by the pilot channel are used for the high-speed packet data channels.

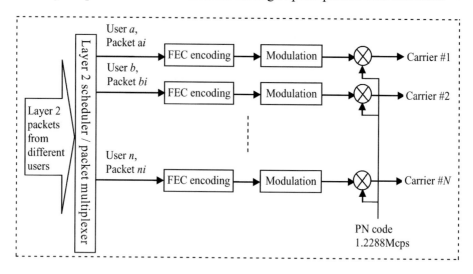

Figure 1. Baseband transmitter block diagram of MC-DV.

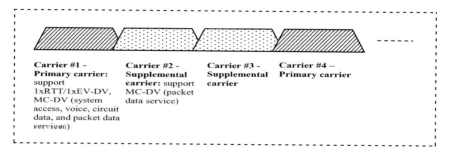

Figure 2. MC-DV carrier configuration (an illustration).

A unique feature of MC-DV is that it supports asymmetric radio bandwidth allocation to a particular MS in the FL and the RL according to the service requirements. This is illustrated in Figure 3. For example, to support FL intensive applications such as real-time movie downloading, three FL carriers and one RL carrier can be assigned to the user. To support applications that require symmetrical high rate FL and RL such as interactive gaming, three carriers can be assigned to the user on both the FL and the RL. On the other hand, conventional voice users can be assigned a single carrier on both links. The modes can be changed dynamically for each MS based on the service requirements.

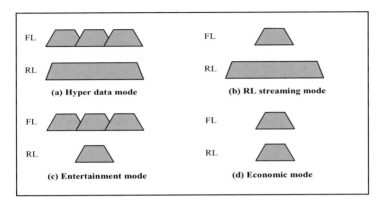

Figure 3. Multi-mode carrier allocation.

2.2 Service Driven Protocol Design

MC-DV supports fast dynamic resource allocation in the time domain in terms of TDM slots and in the carrier domain in terms of the number of carriers and the specific carrier(s) assigned to each MS. The resource allocation scheme at the layer 2 scheduler is driven by two factors: 1) to meet the service requirements of the MS, and 2) to maximize the overall system capacity.

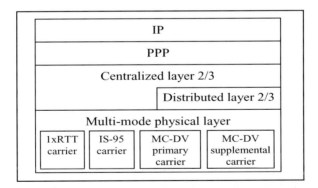

Figure 4. Service-driven protocol structure for MC-DV

We propose a versatile service-driven protocol design consisting of common layer 2 and layer 3 protocol stacks as shown in Figure 4, to support multiple, inhomogeneous carriers and physical layer configurations (e.g. 1xRTT, 1xEV-DV and MC-DV). The layer 2 and 3 protocols provide a common interface with the wireline upper layer protocols such as PPP/IP/TCP. The layer 2 and 3 protocols interface with the multi-mode physical layer by selecting the appropriate physical layer resource in both the frequency domain and the time domain to meet the quality of service required by upper layers applications as well as the subscriber's profile. The layer 2/3 protocols can be split into a centralized portion (e.g. at the Base Station Controller (BSC)) and a distributed portion (e.g. at the Base Transceiver Subsystem (BTS)).

The multi-mode physical layer consists of 1 to N carriers, where each of the carriers can have different physical layer configurations, as described earlier. Each of the carriers can also be configured differently in terms of the QoS it provides to the upper layers. The layer 2 protocol consists of one or more Radio Link Protocols (RLPs) and the Medium Access Control (MAC) sublayer. RLP provides transparent (i.e. no ARQ), or non-transparent (i.e. with ARQ), link layer control. The data plane of the MAC sublayer provides dynamic multiplexing and demultiplexing of layer 2 frames from one or more MSs to/from physical layer frames. The control plane of the MAC sublayer manages the MAC state machine for each MS. The layer 3 protocol defines a set of signaling messages and signaling flows that controls the overall air-interface operations.

2.3 System Operation

A detailed description of the MC-DV system operation is given below.

2.3.1 System Access

When in the power-on or dormant states, an MC-DV MS monitors the primary carrier(s) for paging information from the network. To initiate a call, an MS performs random access on the primary carrier(s). The particular primary carrier(s) to monitor or perform system access can be determined based on hashing on the MS's IMSI or ESN.

2.3.2 High Speed Packet Data Operation

On the primary carrier(s), 1xRTT, 1xEV-DV and MC-DV MSs share the same code and power space in the FL, and share the same Rise-Over-Thermal (ROT) space in the RL. The 1xEV-DV and MC-DV MSs may share the same high-speed packet data channels, i.e. F-PDCH(s) on the primary carrier or they may each have independent F-PDCH(s) on the primary carrier. The layer 2 scheduler performs fast TDM scheduling on the pools of F-PDCHs across multiple carriers to maximize the overall system capacity through multi-user and multi-carrier diversity while meeting the service requirements of all the MSs. To select an MS's packet for transmission on a particular time slot, the scheduler prioritizes each of the MSs' packets according to the channel condition experienced by the MS on each of the associated carriers, and the QoS requirements (e.g. delay bound, minimum data rate, etc.) of the packets. The highest priority packet on each carrier is selected for transmission on that carrier. An MC-DV MS can be scheduled to transmit on one or more primary or supplemental carriers on a particular time slot. The prioritization and scheduling of HARQ retransmission packets is treated in a similar way as new layer 2 packets, whereby the HARQ retransmission packets can be scheduled to transmit on any of the associated primary and supplemental carriers. In addition to fast scheduling, dynamic load balancing is performed across the carriers to ensure optimum loading of different types of MSs, e.g. 1xRTT, 1xEV-DV and MC-DV MSs, on the primary and supplemental carriers.

3. PERFORMANCE OF MC DS-CDMA

In this section, we evaluate the FL performance of MC-DV (with $N = 3$ carriers), versus 1xEV-DV. Full system-level simulation is performed based on the methodology defined in [8]: 19 tri-sector cells, modified Hata path loss model, log-normal shadowing, fast fading with mixed mobile speeds ranging from stationary to 120 km/h. All three MC-DV carriers are configured as primary carriers. Both MC-DV and 1xEV-DV share the same F-PDCH on each carrier. Fast scheduling and rate adaptation per 1.25ms is performed for both 1xEV-DV and MC-DV. The chosen scheduler scheme aims to maximize system capacity while meeting the specific QoS requirements of the applications. Performance is evaluated for both the full-buffer traffic case and the non-full-buffer traffic case, i.e. FTP, HTTP and streaming video. Full TCP/IP protocols stack are modeled for FTP and HTTP. Full MAC states [2][3] of null, dormant, control-hold and active states are modeled for the non-full-buffer traffic cases. The performance metric of interest is the system capacity in terms of sector throughput and number of users supported per sector based on the system outage criteria defined in [8]. The system is considered in outage (and thus capacity is reached), when more than 2% of the system users experience QoS outage. Different QoS outage criteria are defined in [8] for different traffic types.

Table 1. Summary of normalized per-carrier system capacity gain of MC-DV (3x) over 1xEV-DV.

Traffic type	User capacity gain	Sector throughput gain
Full queue	N/A	11%
FTP	16%	10%
HTTP	22%	22%
Streaming video (32kbps with 5 sec delay bound)	42%	42%
Real-time video/gaming (modeled as 32kbps with 100ms delay bound)	2–3 times	2–3 times
Mixed circuit voice and FTP	50%	10%

As expected, the performance gain of MC-DV over 1xEV-DV is more prominent for streaming and real-time services because of the additional dimension for fast resource allocation in the carrier domain. In addition, MC-DV provides trunking efficiency gain, which is important for real-time and streaming services. With dynamic load balancing of services/users across the MC-DV carriers, MC-DV also provides capacity gain for the mixed 1xRTT voice and packet data (e.g. FTP as shown in Table 1) scenario.

4. CONCLUSIONS

In this paper, we provided a detailed description of the proposed MC DS-CDMA system. The performance advantage of MC DS-CDMA over conventional single carrier DS-CDMA was also demonstrated.

5. REFERENCES

[1] TIA/EIA-856, cdma2000 High Rate Packet Data Air Interface Specification
[2] TIA-2000-C, CDMA 2000® Series of Standard for Spread Spectrum Systems, Release C
[3] TIA-2000-D, CDMA 2000® Series of Standard for Spread Spectrum Systems, Release D
[4] 3GPP TR 25.855 V5.0.0, Technical Specification Group Radio Access Network; High Speed Downlink Packet Access; Overall UTRAN Description (Release 5).
[5] TIA/EIA-2000-0, CDMA 2000® Series of Standard for Spread Spectrum Systems, Release 0
[6] TIA/EIA-2000-A, CDMA 2000® Series of Standard for Spread Spectrum Systems, Release A
[7] TIA/EIA/IS-2000-B, CDMA 2000® Series of Standard for Spread Spectrum Systems, Release B
[8] 3GPP2 C.R1002-0, cdma2000 Evaluation Methodology

EFFECTIVE SINR MAPPING FOR AN MC-CDMA SYSTEM

Robert Elliott[1], Alexander Arkhipov[2], Ronald Raulefs[2], and Witold A. Krzymień[1]

[1]*TRLabs/University of Alberta, Edmonton, Alberta, Canada;* [2]*German Aerospace Center (DLR), Oberpfaffenhofen, Germany*

Abstract: One technique employed by modern cellular systems to increase throughput and reduce errors is the use of hybrid automatic repeat request (H-ARQ) schemes to retransmit packets that are in error. Adaptive modulation and coding can also be used between transmissions to adjust to the channel conditions for each transmission. However, multiple transmissions and modulation formats lead to a large number of possible combinations for which we need to determine the performance. Effective SINR mapping is a method to map the various combinations into a smaller group of reference states (e.g. a constant modulation and AWGN channel for all transmissions). Previous work has focused on the single-carrier, constant-modulation case in flat fading. In this work, we examine how to perform SINR mapping in a multi-carrier CDMA system with adaptive modulation and frequency-selective fading.

Key words: SINR mapping, adaptive modulation and coding, multi-carrier CDMA, hybrid ARQ

1. INTRODUCTION

As cellular systems evolve past third generation (3G) designs and into the fourth generation (4G) and beyond, the transfer of high speed packet data is becoming an increasingly important factor in their design. Such data needs to be transmitted not only at high data rates, but also with decreased latency and a reasonable reliability, e.g. packet error rates around 1% at the physical layer. To enable such requirements, a variety of advanced techniques are used. To deal with the growing throughput demands, designs using multiple carriers are becoming common. In particular, this paper deals with a multi-carrier code division multiple access (MC-CDMA) system [1]. In such a system, several users may receive data simultaneously, with each user's data being distinguished by an orthogonal spreading code. The number of simultaneously supportable users is equal to the length L of the spreading code. As opposed to pure orthogonal frequency division multiplexing (OFDM), the sequences are spread and interleaved over all available subcarriers to achieve a frequency diversity gain. The spread sequences are then transmitted using OFDM.

To achieve a target packet error rate with the maximum possible throughput, adaptive modulation and coding can be used, wherein the transmitter adjusts the modulation format and error coding rates based on what the channel can currently support. In addition, physical layer ARQ (automatic repeat request) schemes [2] can be employed to recover packets that are initially in error. In particular, this submission considers a Type II hybrid ARQ scheme using incremental redundancy. Retransmissions for a given packet consist of additional parity bits, which are combined with the previous transmissions to obtain a more reliable copy of the packet.

One problem encountered when using an ARQ scheme with adaptive modulation and coding is deciding what type of modulation to use for retransmissions, given that some data has already been sent. Usually, error rate curves are determined based on the assumption that the channel conditions are constant for the entire packet. With retransmissions this is unlikely to be true. Some systems, such as 1xEV-DO [3], keep the modulation and coding constant between transmissions, which helps to reduce the complexity of the problem. However, in general, the modulation format and number of symbols sent may vary between transmissions. Consider a system with K possible formats for transmitting packets and N allowed transmission attempts per packet, resulting in K^N possible combinations of transmissions, assuming that all N transmission attempts are used. For full knowledge of the supportable rates, a lookup table or error curve would need to be available for each combination, which is clearly infeasible. Hence, a lower complexity approach is desireable.

Effective SINR mapping is a method to reduce the number of reference curves that must be stored. Essentially, the different SINR values and transmission formats are mapped to a single SINR value. That value is then compared with a small number (ideally one) of reference curves to determine the resulting packet error rate. Previous work in this area has been done in the context of single-carrier systems, such as IS-2000 [4]. However, those results are not directly transferable to a multi-carrier system, for two main reasons:

1. The previously defined mappings were used with a constant modulation between retransmissions. In this work, we will be considering adaptive modulation and coding, wherein the number of bits transmitted and the modulation can change between retransmissions.
2. The prior work considers a block fading model, wherein for each frame, the SINR is considered constant for one frame, and then changes for the next frame. With a wideband MC-CDMA system, this situation may not be true.

In particular, different subcarriers will likely experience different fading conditions, and so the SINR can vary within one given frame. The SINR mapping should account for this.

The remainder of this paper is organized as follows. Section 2 will describe the existing methods for SINR mapping, and how those methods can be adapted to a scenario with multiple carriers and adaptive modulation. Section 3 describes the simulation setup and system being considered. Section 4 describes the results of the simulations. Finally, Section 5 provides some final discussions and conclusions on the topic.

2. EFFECTIVE SINR MAPPING METHODS

As mentioned earlier, prior work on SINR mapping tends to assume that the SINR is constant for an entire transmission, but changes between re-transmissions. Of these existing methods, probably the simplest is the linear mapping. An effective SINR value is found by taking the average of the N SINR values for each of the transmissions:

$$\gamma_{eff} = \frac{1}{N}\sum_{i=1}^{N}\gamma_i \tag{1}$$

In [5] and [6], several mapping methods were investigated to convert fading SINR values to equivalent AWGN SINR values in the context of the IS-2000 reverse supplemental channel (R-SCH)[4]. Of the proposed methods, the one that appears to work the best is the capacity or mutual information (MI) mapping. In [5], the mapping is done by calculating the complex AWGN channel capacity for each of the N SINR values, averaging the capacity values, and finding the corresponding SINR for that average:

$$C(\gamma_{eff}) = \frac{1}{N}\sum_{i=1}^{N}C(\gamma_i); \quad C(\gamma_i) = \log_2(1 + Q\gamma_i) \tag{2}$$

The parameter Q in Eq. 2 is a channel-specific correction factor that accounts for the variations in SINR between transmissions. The authors of [6] refine this method by removing the correction factor and replacing the Gaussian-signaling capacity formula with the maximum mutual information achieved under BPSK signaling (as used on the R-SCH):

$$I_{MI}(\gamma_{eff}) = \frac{1}{N}\sum_{i=1}^{N}I_{MI}(\gamma_i) \tag{3}$$

In [7], an exponential mapping method is considered for OFDM systems to convert the various SINR levels on the subcarriers to an effective SINR value, for comparison with AWGN error curves, as follows:

$$\gamma_{\mathit{eff}} = -\beta \ln\left(\frac{1}{N}\sum_{i=1}^{N}\exp\left(\frac{-\gamma_i}{\beta}\right)\right) \tag{4}$$

Here, N refers to the number of useable subcarriers instead of the number of transmissions. As with the Q parameter seen earlier, the value of β depends on the channel, modulation, and code rate being considered. This formula is likely a derivative of the exponential mapping shown in [6], which is based on the pairwise error probability of a convolutional code:

$$\gamma_{\mathit{eff}} = -\ln\left(\sum_{i=1}^{N}p_i \exp(-\gamma_i)\right) \tag{5}$$

Here, p_i is the probability of the SINR γ_i occurring in the channel being considered.

When considering how to map multiple symbols with a single transmission, it is reasonable to say that each symbol in fact represents a separate transmission. Those transmissions just happen to occur immediately after each other (or possibly in parallel, in a multi-carrier system). Hence, one should just be able to use the existing methods treating each symbol, each with its own SINR, as a transmission.

Dealing with adaptive modulation is somewhat less simple. Of the above methods, the mutual information (MI) method seems the easiest to modify to account for adaptive modulation. For each symbol, one can find the corresponding mutual information using the MI vs. SINR curve for the modulation scheme of that symbol. Then, when de-mapping the average MI value, one can use a curve that is the average of the curves used for the symbols. For example, if 1/3 of the transmitted symbols use BPSK and 2/3 use QPSK, one would use a curve that is $2/3 \times MI_{QPSK}(\gamma) + 1/3 \times MI_{BPSK}(\gamma)$ to obtain γ_{eff}. Adapting the linear mapping could also be done by averaging the SINR values of the demodulated bits instead of the modulated symbols. However, given the known poor performance of the linear mapping method (see for example [6]), investigating this option is likely not worthwhile.

3. SYSTEM AND SIMULATION PARAMETERS

To investigate the various mapping methods, we consider a simple system using a two-frame transmission per packet. Incremental redundancy was employed on the downlink of the MC-CDMA system used in the 4MORE project [8],[9]. Relevant details of this system are shown in Table 1. A rate 1/5 turbo encoder with a random interleaver was used as the base encoder. 732 bits (plus 3 tail bits) were encoded for each user in each packet, with 3675 bits output in total. The output of the turbo encoder was then interleaved based on the forward packet data channel (F-PDCH) interleaver specified in the IS-2000 standard [4]. The systematic-then-parity structure output of the interleaver can be punctured by simply transmitting only the first portion of the output, with later transmissions (if required) simply picking up the sequence where the previous transmission left off.

Table 1. MC-CDMA System Parameters

Parameter	Value
Channelization bandwidth	50 MHz
Occupied Bandwidth	41.46 MHz
FFT size	1024
Number of available carriers	736
Subcarrier spacing	40.49 kHz
Sampling frequency	57.6 MHz = 15*3.84 MHz
Frame duration	0.666 ms
OFDM symbols per frame	30
User data symbols per OFDM symbol	16
Samples per frame	38400
Symbol duration	21.5 µs
Guard interval	3.75 µs (216 samples)
Frequency interleaving	Random
Spreading code length	16
Receiver detection method	MMSE
Number of active users	16 (1 primary, 15 interfering)
Number of base station antennas	1

Each frame of the packet consisted of 480 modulated symbols. The first frame consisted of 480 8PSK symbols, or 1440 coded bits, for a code rate of about 0.508. The second frame consisted of 480 QPSK symbols, for an additional 960 parity bits, dropping the effective code rate to 0.305. An independent Rayleigh channel was used, whereby the fading was uncorrelated between subcarriers and between adjacent OFDM symbols, as such a channel should represent the worst case for mapping. Perfect channel estimation was assumed at the receiver, both for the channel fading coefficients and the AWGN noise variance. The decoder used Max-Log-MAP decoding with a maximum of 8 iterations.

A total of 50000 packets were simulated for each data point. The resulting packet error rate curves are shown in Figure 1. We define the SNR E_S/N_0 as $1/\sigma^2$, where the modulated symbol energy is normalized to 1, and σ^2 is the total variance of the complex AWGN noise (i.e. a variance of $\sigma^2/2$ per dimension). For each despread and recombined symbol in the MMSE receiver, the SINR is calculated as $|\vartheta'|^2/2(\sigma^2_{MAI}+\sigma^2_n)$, where $|\vartheta'|$ is the attenuation, σ^2_n is the real-valued effective noise variance, and σ^2_{MAI} is the real-valued multiple-access interference variance (which is modeled as additive Gaussian noise), each as defined in [1].

4. SIMULATION RESULTS

The ultimate goal of the SINR mapping is to convert the 2- and 3-dimensional Rayleigh PER curves seen in Figure 1 into the 2-dimensional AWGN PER curves for the same system. To begin, we looked at the 1-transmission case. Figure 2a shows the curves that result from mapping the SINR values per symbol using the linear and the MI mapping methods.

As can be seen, both methods yield curves that are significantly off from the AWGN curve. Hence, we decided to try reintroducing a correction multiplication factor into the methods, similar to that in Eq. (2). Note that for the linear mapping, this is equivalent to simply shifting the curve by Δ dB, where $\Delta = 10\log(Q)$. The proper value for Q was determined by the value that minimized the mean squared error between SNR values on the AWGN PER curve and the SINR values on the mapped Rayleigh curve at the same PER levels (i.e. the MMSE Q value), similar to the method in [7]. We consider PER points in the range 0.8 to 0.003, as indicated by the X's in Figure 2b. This same method was used to find the appropriate value of β for the exponential mapping.

Figure 2b shows the results of the mapping with the correction factor, as well as the mapping for the exponential method. The parameters for the methods were $\Delta = 0.494$ dB, $Q = 1.175$, and $\beta = -1.95$. These values result in a mean squared mapping error for the linear, mutual information, and exponential methods of 3.267×10^{-4}, 3.668×10^{-5}, and 5.623×10^{-3}, respectively. As can be seen from both the figure and the numbers, the modified MI method provides the best mapping, while the exponential method is the worst. The relatively poor performance of the latter method may be due to its origins in use with convolutional codes, whereas a turbo code is used here. Unexpectedly, the required value for β was negative, whereas in [7], the β values were positive. This may have to do with the fact that the SINR levels in our system were

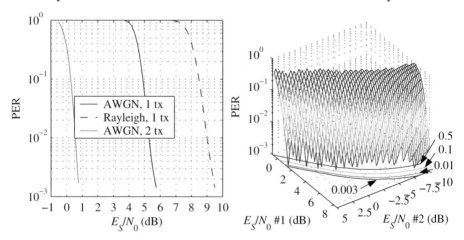

Figure 1. (a) PER vs. SNR for 1 and 2 transmissions in an AWGN channel and 1 transmission in a Rayleigh channel; (b) PER vs. SNR for two transmissions in a Rayleigh channel.

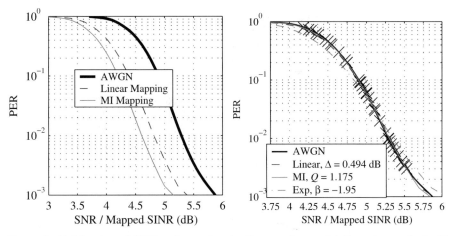

Figure 2. (a) Mapping of SINR values using the linear and mutual information methods. (b) Mapping of SINR values using the linear and mutual information methods with correction factors, and the exponential method (with no correction factor). An X on the AWGN line indicates a PER level where the AWGN SNR and the mapped SINR are compared.

measured after MMSE detection and signal despreading. In addition, the system analyzed in [7] did not spread modulated symbols across multiple subcarriers; only one symbol per subcarrier was used.

Since the MI method provided the best mapping, and because it can readily encompass adaptive modulation, we focus on this method for the two-transmission case. There are two possibilities we consider to obtain a correction factor for this case:

1. Use the same Q value for both the 8PSK and the QPSK symbols.
2. Use the previously obtained Q value of 1.175 with the 8PSK symbols, and find a second Q value to use with the QPSK symbols.

The second method has the advantage that, once the effective SINR value is found for the first transmission, the actual SINR values for the first transmission are no longer required to perform the mapping. It can be assumed that the first transmission took place in an AWGN channel with all symbols having the previously determined effective SINR.

Using these two methods, we attempted to map the points on the 3D PER curve in Figure 1b to the AWGN 2-transmission curve in Figure 1a. The MMSE Q value for the two methods was found to be 1.142 for the first method and 1.112 for the second. The results of the mapping are shown in Figure 3.

Clearly, having a constant correction factor does not yield good results. Although the Q values used minimized the overall mean square error in the SINR mapping in the two methods, there is still as much as a 0.36 dB difference in the mapped SINR value from the AWGN value and, more importantly, up to around an order of magnitude error in the PER. For example, if the mapped SINR indicates a PER of 0.02 based on the AWGN curve, the actual PER could

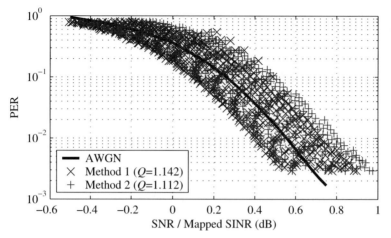

Figure 3. Mapping of two-transmission SINR values with a constant correction factor.

be anywhere from 0.005 to 0.12. Thus, a variable Q value seems necessary. The obvious choice would be to make the value of Q a function of the SNR for the two transmissions. Figure 4 shows the required Q values for the two methods to make the mapped SINR values virtually identical to the AWGN SNR values for a given PER.

Unfortunately, requiring knowledge of a Q value for all combinations of SNR#1 and SNR#2 really is not less complex than having to know the value of the PER itself for all those combinations. However, the Q value in this case can be reasonably approximated by a relatively less complex polynomial equation $Q(\gamma_1,\gamma_2)$ of the following form, where γ_1 and γ_2 are in units of dB:

$$Q \approx a_1\gamma_1^2\gamma_2^2 + a_2\gamma_1^2\gamma_2 + a_3\gamma_1^2 + a_4\gamma_1\gamma_2^2 + a_5\gamma_1\gamma_2 + a_6\gamma_1 + a_7\gamma_2^2 + a_8\gamma_2 + a_9 \qquad (6)$$

The MMSE values for the coefficients of the above equation are given in Table 2 for the two methods, while SINR values that result from this mapping are seen in Figure 5.

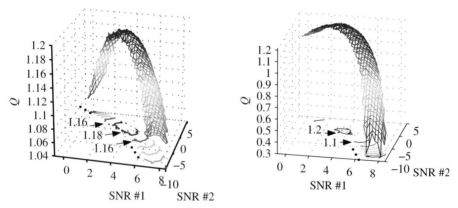

Figure 4. Required correction factor values as a function of SNR #1 and SNR #2 for (a) Method 1 and (b) Method 2.

Table 2. Q value equation coefficients ($\times 10^{-5}$)

Coefficient ($\times 10^{-5}$)	a_1 a_6	a_2 A_7	a_3 a_8	a_4	a_5 a_9
Method 1	-4.04218 5268.39	52.7559 -651.414	-676.589 3014.67	107.233	-885.334 108309
Method 2	-19.3166 31803.5	298.093 -2691.37	-4004.64 22118.4	295.201	-6607.84 58090.8

5. DISCUSSION/CONCLUSIONS

The results obtained for this 2-transmission system are quite encouraging. The methods introduced in this paper can be readily extended to a system with more than two transmissions. The second mapping method described in the previous section seems particularly favorable to extension. For example, when mapping a third transmission, the mapping could be done treating the previous two transmissions as a single combined AWGN transmission at the appropriate effective SINR value, as found from the two-transmission case. The mapping would essentially then just depend on that combined equivalent transmission and the SINR values for the third transmission. It should also be possible to express the approximation for the Q correction factor associated with the third transmission still as a function of just two variables: the equivalent SINR for the previous transmissions and the SNR for the current transmission. Hence, adding additional transmissions to the system has something of a recursive effect on the calculations for the mapping.

Considering a system that has correlation in time or between subcarriers should make the mapping process somewhat simpler. Correlated adjacent subcarriers or symbols would have very similar SINR values. Because of this, the number of SINR values that would need to be considered could be reduced.

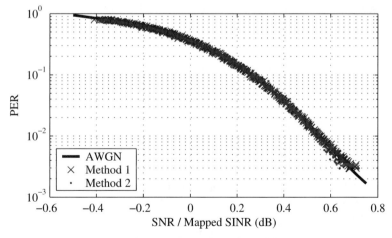

Figure 5. Mapped SINR values using polynomial Q approximation.

Instead of mapping with the SINR value for each symbol, a subset of the values that is representative of the SINRs in each of the correlated groups can be used instead, thereby speeding up the computation process. In the extreme case when all symbols in a transmission experience about the same fading conditions, the methods presented in this paper in fact reduce to the methods for block fading covered in previous work.

One drawback to the proposed scheme is that the Q values calculated are likely channel dependent. Thus, if a different channel were to be considered, the Q values would have to be recalculated in order to perform the mapping. Nevertheless, the overall performance of the methods discussed herein for mapping SINR values to their equivalent AWGN SNR levels seem beneficial enough to outweigh that drawback.

ACKNOWLEDGEMENTS

The authors wish to acknowledge funding for this work provided by the Natural Sciences and Engineering Research Council of Canada (NSERC), the Alberta Informatics Circle of Research Excellence (iCORE), the Alberta Ingenuity Fund, TRLabs, the German Aerospace Center (DLR), and the European IST project 4MORE.

REFERENCES

[1] K. Fazel and S. Kaiser, *Multi-Carrier and Spread Spectrum Systems*, Chichester, West Sussex, England: John Wiley & Sons Ltd., 2003.

[2] S.B. Wicker, *Error Control Systems for Digital Communication and Storage*, Englewood Cliffs, N.J.: Prentice Hall, 1995.

[3] TIA/EIA/IS-856, *cdma2000 High Rate Packet Data Air Interface Specification*, Telecommunications Industry Association, Arlington, Va., Nov. 2000.

[4] 3GPP2 C.S0002-D, *Physical Layer Standard for cdma2000 Spread Spectrum Systems – Release D*, 3rd Generation Partnership Project 2 (3GPP2), Mar. 2004.

[5] J. Kim, A. Ashikhmin, A. van Wijngaarden, E. Soljanin, and N. Gopalakrishnan, "Reverse link hybrid ARQ: link error prediction methodology based on convex metric," 3GPP2 submission C30-20030401-020, Apr. 1, 2003.

[6] S.S. Tsai and A.C.K. Soong, "Effective-SNR mapping for modeling frame error rates in multiple-state channels," 3GPP2 submission C30-20030429-010, Apr. 29, 2003.

[7] Nortel Networks, "OFDM exponential effective SIR mapping validation, EESM simulation results for system-level performance evaluations, and text proposal for section A.4.5 of TR 25.892," 3GPP submission R1-04-0089, Jan. 2004.

[8] S. Kaiser, Y. Durand, L. Hérault, J.-F. Hélard, D. Mottier, A. Gameiro, J. Rodriguez, C. Barquinero, F. Berens, F. Bauer, R. Rabineau, and F.J. Casajus, "4G MC-CDMA multi antenna system on chip for radio enhancements (4MORE)," in *Proc. IST Mobile & Wireless Commun. Summit 2004*, Lyon, France, June 2004, 5 pages.

[9] 4MORE Project website, http://www.ist-4more.org.

COMBINATION OF H-ARQ AND ITERATIVE MULTI-USER DETECTION FOR OFDMA-CDM

Alexander Arkhipov, Ronald Raulefs, Michael Schnell
German Aerospace Center (DLR)
Institute of Communications and Navigation
Oberpfaffenhofen, 82234 Wessling, Germany
{alexander.arkhipov,ronald.raulefs,michael.schnell}@dlr.de

Abstract   In this paper, superimposed packet allocation for orthogonal frequency-division multiple-access code-division multiplexing (OFDMA-CDM) is presented, where each transmitted packet is associated with one spreading code. An iterative algorithm which is a combination of parallel interference cancellation (PIC) and hybrid automatic repeat request (H-ARQ) based on soft value combining (SVC) is proposed, and its performance is studied and compared with other existing H-ARQ schemes. The interference of correctly received packets is ideally reconstructed and subtracted; thus, the overall system performance improves iteratively. As a result, the proposed algorithm outperforms conventional H-ARQ based on SVC as well as H-ARQ based on maximum ratio combining (MRC).

1. Introduction

Wireless systems of the next generation must provide high spectral efficiency, offer high data rates and high user capacity. The multicarrier modulation scheme realized by a combination of orthogonal frequency-division multiple-access (OFDMA) [1] with code-division multiplexing (CDM) attracts significant interest because of its robustness to multipath propagation and its high spectral efficiency [2]. In OFDMA-CDM, each transmitted symbol is spread over several subcarriers and the CDM component is used to transmit several symbols in parallel on the same subcarriers. The OFDMA component assures orthogonal user discrimination by assigning different users separate sets of subcarriers. Thus, multiple-access interference is avoided. Nevertheless, self-interference (SI) occurs in frequency-selective fading channels due to the loss of orthogonality of spreading codes. In order to cope with SI, a soft parallel interference cancellation (PIC) scheme [2] [3] is used as an efficient data detection and decoding technique.

In this contribution, we propose to associate each transmitted packet with one spreading code and develop an efficient algorithm that is a combination of SVC based H-ARQ and PIC. In the proposed algorithm, reliability information of erroneously received copies of the same data packet is used in order to improve the quality of interference cancellation.

The remainder of this paper is organized as follows: Section II contains the general description of the transmission structure, while Section III describes the advanced parallel interference cancellation (APIC) scheme, which is a combination of H-ARQ based on SVC and a soft PIC scheme. The simulation results are given in Section IV. Finally, Section V concludes the work.

2. OFDMA-CDM Transmission Scheme

In an OFDMA-CDM system, different users utilize separate sets of subcarriers for data transmission. Thus, all users are orthogonal to each other and this allows us to consider the data transmission of a single user in the following. For convenience, indices which distinguish different users are omitted.

The proposed OFDMA-CDM transmitter is capable to transmit L packets of equal size simultaneously by applying spreading codes of length L within the CDM component. At the end of each packet, cyclic redundancy check (CRC) bits are appended. The packet is decoded at the receiver and is considered error-free if the CRC passes. In this case, a positive acknowledgement (ACK) is generated and sent back to the transmitter. Otherwise, a negative acknowledgement (NACK) is sent back and the transmission of the erroneously decoded packet is repeated. It is assumed that acknowledgements from the receiver to the transmitter are error-free and buffer overflows occur neither at the transmitter nor at the receiver. The maximum allowed number of transmissions is limited to N_{tr}. If the packet can not be decoded in N_{tr} transmissions, the transmitter discards the packet. In the following, for each packet $l, l = 1, \ldots, L$, the total number of occurred transmissions is defined as $M_l \leq N_{tr}$. At the physical layer, the L packets are simultaneously transmitted within one frame, which consists of N_{fr} OFDMA-CDM symbols.

After channel encoding and outer interleaving Π_{out} the coded bits $b^{(l)}, l = 1, \ldots, L$, are symbol mapped, yielding complex-valued data symbols. The outer interleaving is constructed in such a way that it performs independent random interleaving between the retransmissions, which allows the system to exploit time and frequency diversity between retransmissions.

Data symbols from different packets are transmitted in groups of L symbols. In each group only one data symbol from each packet is transmitted. The number of groups transmitted within one OFDMA-CDM symbol is denoted as Q, in the sequel. Thus, Q symbols of each packet are transmitted within one OFDMA-CDM symbol, using LQ subcarriers. Assume that vector $\mathbf{d} = (d^{(1)}, d^{(2)}, \ldots, d^{(L)})^T$ represents the data symbols of one group. The sequence $\mathbf{s}$ represents spread data symbols of one group and is written as

$$\mathbf{s} = \mathbf{C}_L \mathbf{d}, \tag{1}$$

where a Walsh-Hadamard transformation

$$\mathbf{C}_L = \begin{pmatrix} \mathbf{C}_{L/2} & \mathbf{C}_{L/2} \\ \mathbf{C}_{L/2} & -\mathbf{C}_{L/2}, \end{pmatrix} \forall L = 2^\kappa, \kappa \geqslant 1, \mathbf{C}_1 = 1, \tag{2}$$

is applied to perform spreading. The resulting L columns $\mathbf{c}^{(l)}, l = 1, \ldots, L$, of matrix $\mathbf{C}_L$ represent the orthogonal spreading codes. This transformation allows one orthogonal Walsh-Hadamard (WH) spreading code $\mathbf{c}^{(l)}$ to be associated with each transmitted packet $l, l = 1, \ldots, L$.

The resulting sequence $\mathbf{s}$ is transmitted within one OFDMA-CDM symbol. After inner frequency interleaving Π_{in}, the elements of $\mathbf{s}$ are transmitted on separate subcarriers by performing an OFDM modulation. A guard interval Δ, which is larger than the maximum delay of the transmission channel, is added. The guard interval Δ avoids inter-symbol interference with the preceding OFDMA-CDM symbol.

In OFDMA-CDM systems, each subcarrier is exclusively used by a single user, and no multiple-access interference occurs. The total number of subcarriers is $N_c = K_{\max} LQ$, where $K_{\max}$ is the maximal number of simultaneously active users. The OFDMA-CDM symbols are transmitted over a frequency-selective mobile radio channel, where the orthogonality of the spreading codes is lost. At the receiver an inverse OFDM operation is performed. After frequency deinterleaving, the received sequence $\mathbf{r}$ in the frequency domain is given by

$$\mathbf{r} = \mathbf{H}\mathbf{s} + \mathbf{n}, \tag{3}$$

where $\mathbf{H} = diag(H_1, \ldots, H_L,)$ is a diagonal matrix, whose complex-valued diagonal elements represent the fading on the subcarriers on which $\mathbf{s}$ has been transmitted. Additive white Gaussian noise with variance $\sigma^2/2$ per dimension is denoted by the L-dimensional vector $\mathbf{n}$.

The received vector $\mathbf{r}$ is passed to the APIC block, where the algorithm is applied as described in the next section. At the output of the APIC, 'ACK' or 'NACK' is generated for each received packet.

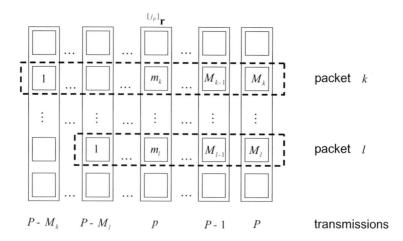

Figure 1. The content of the buffer of received values.

3. Advanced Parallel Interference Cancellation Scheme

3.1 Re-encoded Values Combining

The received vector $\mathbf{r}$ and matrix $\mathbf{H}$ are saved into a buffer of received values and fading values, respectively. Without loss of generality, we assume that $P-1$ vectors are already stored in each buffer. Thus, the received vector $\mathbf{r}$ and $\mathbf{H}$ represent the P-th elements in the buffers. In the following, we introduce the subscript $p = 1, \ldots, P$, which denotes the index number of the elements in the buffers.

In Fig. 1 the matrix-like structure demonstrates the content of the buffer of received values, where the columns denote the received vectors and the rows define the transmitted packets.

The proposed algorithm is performed in multiple iterations. In each iteration up to L packets can be decoded. In the following, index $j_p, j_p = 1, \ldots, L$, denotes the number of packets, whose contributions are ideally subtracted from $^{[j_p]}\mathbf{r}$. The notation $^{[j_p]}\mathbf{r}$ refers to the vector received after the p-th transmission from which the interference of j_p other packets have been removed.

For convenience, we introduce the decision status values $^{(p)}DS^{(l)} \in \{true, false\}$, $l = 1, \ldots, L$, and $p = 1, \ldots, P$. The expression $^{(p)}DS^{(l)} = true$ means that packet l has been successfully decoded and its contribution is ideally removed from $^{[j_p]}\mathbf{r}$.

We illustrate the improved interference cancellation scheme by considering the transmission p in Fig. 1. The packet l will be decoded by removing the interference of packets $k = 1, \ldots, L, k \neq l$. The scheme for the interference substraction is depicted in Fig. 2. In this algorithm the data symbols of the packets $k = 1, \ldots, L, k \neq l, {}^{(p)}DS^{(k)} = false$, are detected and decoded. Finally, the interference contribution of the m_k-th transmission of packet k, is reconstructed and subtracted from the received signal $^{[j_p]}\mathbf{r}$. The data symbols of the m_l-th transmission of packet l are detected in the lowest path of Fig. 2.

A minimum-mean square error (MMSE) detector is used to combat the phase and amplitude distortions caused by multipath propagation on the subcarriers. The matrix of equalization coefficients of the transmission p is denoted as ${}^{(p)}\mathbf{G}$, whose diagonal elements ${}^{(p)}G_i, i = 1, \ldots, L$, are given by

$${}^{(p)}G_i = \frac{{}^{(p)}H_i^*}{|{}^{(p)}H_i|^2 + \sigma^2}. \tag{4}$$

The output of the outer deinterleaver Π_{out}^{-1} delivers a soft estimate ${}^{(m_k)}w^{(k)}$ for the transmitted code bit $b^{(k)}$ within the m_k-th transmission of packet k. The log-likelihood ratio (LLR) of the transmitted bit $b^{(k)}$ given ${}^{(m_k)}w^{(k)}$ is [4]

$${}^{(m_k)}\theta^{(k)} = \ln\left(\frac{P(b^{(k)} = +1|{}^{(m_k)}w^{(k)})}{P(b^{(k)} = -1|{}^{(m_k)}w^{(k)})}\right) \tag{5}$$

and can take on values in the interval $(-\infty, +\infty)$. For OFDMA-CDM with MMSE detector the LLR in (5) can be represented as [2]

$${}^{(m_k)}\theta^{(k)} \approx \frac{4}{\sigma^2 L} {}^{(m_k)}w^{(k)} \sum_{i=1}^{L} |{}^{(p)}H_i|, \tag{6}$$

where ${}^{(p)}H_i, i = 1, \ldots, L$, describes the fading on the L subcarriers on which bit $b^{(k)}$ has been transmitted.

The LLR ${}^{(m_k)}\theta_{re}^{(k)}$ of a re-encoded bit is defined as [2] [3]

$${}^{(m_k)}\theta_{re}^{(k)} = \ln\left(\frac{P\{b^{(k)} = +1|{}^{(m_k)}\mathbf{w}^{(k)}\}}{P\{b^{(k)} = -1|{}^{(m_k)}\mathbf{w}^{(k)}\}}\right), \tag{7}$$

where the vector ${}^{(m_k)}\mathbf{w}^{(k)}$ represents the sequence of all estimates within the m_k-th transmission of packet k. For simplicity, we have omitted an index in the notation ${}^{(m_k)}w^{(k)}$ which indicates a certain soft estimate within the vector ${}^{(m_k)}\mathbf{w}^{(k)}$.

According to (7), the LLR ${}^{(m_k)}\theta_{re}^{(k)}$ which reflects the reliability of the re-encoded bit is calculated taking into account all soft estimates ${}^{(m_k)}\mathbf{w}^{(k)}$ of packet k.

In the following, soft decided values $^{(m_k)}\mathbf{w}^{(k)}$, $1, \ldots, M_k$, of the M_k available transmissions of packet k are exploited in order to improve the estimate of the re-encoded values. In comparison to (7), we define the overall LLR value as

$$\theta_{tot}^{(k)} = \ln\left(\frac{P\{b^{(k)} = +1 |^{(1)}\mathbf{w}^{(k)}, \ldots, ^{(M_k)}\mathbf{w}^{(k)}\}}{P\{b^{(k)} = -1 |^{(1)}\mathbf{w}^{(k)}, \ldots, ^{(M_k)}\mathbf{w}^{(k)}\}}\right). \quad (8)$$

The soft values $^{(m_k)}\mathbf{w}^{(k)}$ can be assumed conditionally independent given $b^{(k)}$, since the measurement errors introduced by the multipath channel and noise are assumed independent. This assumption can be made, if random outer interleaving between different transmissions is applied. In this case, (8) can be further developed as

$$\begin{aligned}
\theta_{tot}^{(k)} &= \ln\left(\frac{P\{b^{(k)} = +1 |^{(1)}\mathbf{w}^{(k)}, \ldots, ^{(M_k)}\mathbf{w}^{(k)}\}}{P\{b^{(k)} = -1 |^{(1)}\mathbf{w}^{(k)}, \ldots, ^{(M_k)}\mathbf{w}^{(k)}\}}\right) \quad (9)\\
&= \ln\left(\frac{P\{b^{(k)} = +1\}}{P\{b^{(k)} = -1\}}\right)\\
&\quad + \sum_{u=1}^{M_k} \ln\left(\frac{P\{^{(u)}\mathbf{w}^{(k)} | b^{(k)} = +1\}}{P\{^{(u)}\mathbf{w}^{(k)} | b^{(k)} = -1\}}\right)\\
&= \sum_{u=1}^{M_k} {}^{(u)}\theta_{re}^{(k)}.
\end{aligned}$$

We assume that both realizations of $b^{(k)}$ are equally probable and, thus

$$\ln\left(\frac{P\{b^{(k)} = +1\}}{P\{b^{(k)} = -1\}}\right) = 0. \quad (10)$$

To transfer the LLR value $\theta_{tot}^{(k)}$ into the bit domain, the average value of $b^{(k)}$, or the so-called "soft bit" [5], is used

$$\begin{aligned}
w_{re}^{(k)} &= E\{b^{(k)} |^{(1)}\mathbf{w}^{(k)}, \ldots, ^{(M_k)}\mathbf{w}^{(k)}\} \quad (11)\\
&= (+1) P\{b^{(k)} = +1 |^{(1)}\mathbf{w}^{(k)}, \ldots, ^{(M_k)}\mathbf{w}^{(k)}\}\\
&\quad + (-1) P\{b^{(k)} = -1 |^{(1)}\mathbf{w}^{(k)}, \ldots, ^{(M_k)}\mathbf{w}^{(k)}\}.
\end{aligned}$$

With (8), the conditional probabilities required in (11) can be written as

$$P\{b^{(k)} = +1 |^{(1)}\mathbf{w}^{(k)}, \ldots, ^{(M_k)}\mathbf{w}^{(k)}\} = \frac{e^{\theta_{tot}^{(k)}}}{1 + e^{\theta_{tot}^{(k)}}} \quad (12)$$

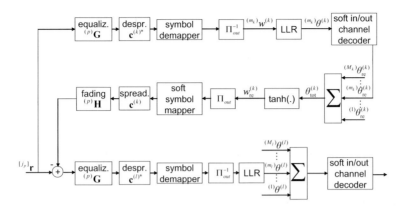

Figure 2. Proposed interference cancellation scheme with re-encoded values combining.

and

$$P\{b^{(k)} = -1|^{(1)}\mathbf{w}^{(k)}, \ldots, ^{(M_k)}\mathbf{w}^{(k)}\} = \frac{1}{1 + e^{\theta_{\text{tot}}^{(k)}}}. \quad (13)$$

Thus, with (12) and (13) the soft bit $w_{\text{re}}^{(k)}$ can be defined as

$$\begin{aligned} w_{\text{re}}^{(k)} &= \frac{e^{\theta_{\text{tot}}^{(k)}/2} - e^{-\theta_{\text{tot}}^{(k)}/2}}{e^{\theta_{\text{tot}}^{(k)}/2} + e^{-\theta_{\text{tot}}^{(k)}/2}} \\ &= \tanh(\theta_{\text{tot}}^{(k)}/2). \end{aligned} \quad (14)$$

The soft bit can take on values in the interval $w_{\text{re}}^{(k)} \in [-1, 1]$. After re-interleaving, the soft bits are modulated such that the reliability information is maintained in the obtained complex-valued data symbols. The obtained symbols are spread, weighted with the channel coefficients and subtracted from the received vector $^{[j_p]}\mathbf{r}$.

The summation of re-encoded values obtained in each retransmission according to (9) increases the absolute value of $\theta_{\text{tot}}^{(k)}$, and thus, the feedback loop uses the more reliable estimates of transmitted bits and avoids error propagation.

After interference cancellation of all interfering packets, the data symbols of packet l are detected by applying a single-user detection technique. In contrast to (4), the equalizer coefficients are adopted to the quasi-SI free case and maximum ratio combining with equalization coefficients given by $^{(p)}G_i = ^{(p)}H_i^*$ is used.

The output of the OFDMA-CDM demodulator delivers a complex-valued vector of the symbols of packet l. The received symbols are demapped, and the obtained bits are deinterleaved. Similar to (6), the reliability estimator delivers the LLR values $^{(m_l)}\theta^{(l)}$. Since in all transmissions of packet l, copies of the same data are sent, the obtained LLR values are combined to improve the performance of H-ARQ scheme.

3.2 H-ARQ Based on SVC

SVC combines several repeated packets encoded with a code rate R [6]. The output of a soft value combiner delivers a value $\sum_{u=1}^{M_l} {}^{(u)}\theta^{(l)}$ for each code bit. The values at the output of a soft value combiner represent the data of packet l encoded with a more powerful error-correcting code of rate R/M_l.

3.3 Ideal Interference Cancellation

Assume that the CRC verifies an error-free transmission of packet l. In this case, the soft bit is equal to $w_{\text{re}}^{(l)} = 1$ if $b^{(l)} = 1$ is transmitted or $w_{\text{re}}^{(l)} = -1$ otherwise. Fig. 3 illustrates the obtained simplified interference cancellation scheme. The soft bits $w_{\text{re}}^{(k)}$ or equivalently the correctly decided code bits $b^{(l)}$ within packet l are interleaved and at the output of the symbol mapper the complex-valued data symbols are obtained. The data symbols are spread and the obtained chips are weighted with the appropriate fading coefficients. Then, the reconstructed interference is subtracted from the received vector $^{[j_p]}\mathbf{r}$. Finally, the obtained vector $^{[j_p+1]}\mathbf{r}$ updates the value $^{[j_p]}\mathbf{r}$ in the buffer of the received values.

3.4 Proposed Algorithm

The proposed algorithm is carried out in multiple iterations. Up to L packets can be decoded in one iteration. If any packet has been decoded in the iteration, the next iteration starts, since ideal interference removal can improve the decoding probability of undecoded packets.

If no additional packets can be decoded in the iteration, the algorithm stops and notifies the transmitter about the current decoding status of each transmitted packet. If the packet can not be decoded, a 'NACK' is generated, otherwise an 'ACK' is generated.

If index $j_p, p = 1, \ldots, P$, reaches the maximum value L, all packets are decoded in $^{[L]}\mathbf{r}$. Thus, $^{[L]}\mathbf{r}$ contains no more useful information. The vector $^{[L]}\mathbf{r}$ and the corresponding matrix $^{(p)}\mathbf{H}$ are removed from the buffer of the received values and from the buffer of the fading values, respectively.

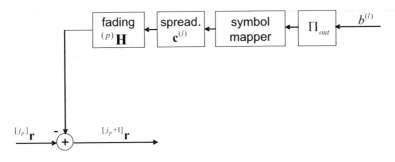

Figure 3. Ideal interference reconstruction and substraction.

4. Simulation Results

The transmission system under investigation has a bandwidth of $BW=$ 50 MHz and the carrier frequency is located at 5 GHz. The total number of subcarriers is $N_c = 1024$. The resulting subcarrier spacing is 48.82 kHz and the OFDMA-CDM symbol duration is $20.4\mu s$. The modulation is either QPSK, 16-QAM or 64-QAM. The number of simultaneously active users is 8 and the number of groups is $Q = 16$. Walsh-Hadamard codes of length $L = 8$ are applied, which is a good compromise between complexity and performance. A convolutional channel encoder with code rate $R = 1/2$ and memory $m_e = 6$ is used for the simulations. In the following, Rayleigh fading is assumed to be independent between adjacent subcarriers and OFDMA-CDM symbols [4].

The performance of a conventional OFDMA system with comparable parameter setting as for OFDMA-CDM is used as a reference. For OFDMA, maximum ratio combining equalization [2] is employed together with H-ARQ based on SVC.

As another performance reference, an OFDMA-CDM system is used where several subsequent OFDMA-CDM symbols are employed for the transmission of data symbols of one packet. In this case, all spreading codes are assigned for the transmission of one packet. This allocation allows to exploit H-ARQ based on MRC, since in each retransmission the same data symbols are transmitted. As data detection and decoding technique, soft PIC is applied.

For all considered systems, transmitted packets have the same size of 2042, 4090 and 3066 bits for QPSK, 16-QAM and 64-QAM, respectively. The number of OFDMA-CDM symbols in the frame N_{fr} is equal to 128 for QPSK and 16-QAM, and 64 for 64QAM.

We evaluate the system performance as a function of normalized throughput T_{thr} versus SNR. The normalized throughput is defined as the reciprocal value of the average number of transmissions needed to successfully decode the transmitted packet.

The SNR corresponding to $T_{thr} = 0.95$ is referred as the working point in the following. In the simulations, we assume $N_{tr} = 10$.

In Fig. 4 the normalized throughput as a function of the SNR is depicted. The modulation alphabet is QPSK. One can see that OFDMA-CDM system with APIC performs better than any other technique. At the working point of OFDMA-CDM with APIC, the gain in throughput and bandwidth efficiency is 17.2% compared to OFDMA-CDM with PIC and H-ARQ based on SVC. The bandwidth efficiency gap between OFDMA-CDM with APIC and OFDMA is around 45.2%. The OFDMA-CDM with H-ARQ based on MRC performs 27.1% worse than OFDMA-CDM with APIC. At lower SNR, the performance of APIC is comparable with MRC and SVC. Because of the spreading, and thus, frequency diversity, the OFDMA-CDM system outperforms OFDMA at high SNR. At SNR values lower than 3dB, the soft PIC can not cope with SI and OFDMA-CDM performs the same as conventional OFDMA.

In Fig. 5 the modulation is increased to 16-QAM. Again, it can be observed that the OFDMA-CDM scheme with APIC outperforms all other techniques. At the working point, the gain in bandwidth efficiency is 26% in comparison with OFDMA-CDM and H-ARQ based on SVC.

In Fig. 6 the modulation cardinality is further increased to 64-QAM, which leads to increased SI. Therefore, the OFDMA system outperforms OFDMA-CDM with H-ARQ based on SVC and OFDMA-CDM with H-ARQ based on MRC. The explanation of this effect has been given in [7]. In an OFDMA-CDM system, high level of interference dominates over the effect of achieved frequency diversity. Due to the high level of SI, the performance of the OFDMA-CDM system can not be further improved even with soft PIC.

The analysis of the obtained results demonstrate, that neither OFDMA-CDM with H-ARQ based on SVC nor H-ARQ based on MRC nor OFDMA can be chosen as a dominant scheme, since their performances depend on the modulation alphabet. With APIC, an OFDMA-CDM system with adaptive modulation can handle all symbol mapping schemes from QSPK to 64-QAM to increase the data rate so that the multiplexing scheme need not be changed to conventional OFDMA to guarantee optimum performance.

5. Conclusions

Superimposed packet allocation for OFDMA-CDM transmission has been proposed. An iterative APIC algorithm, which is a combination

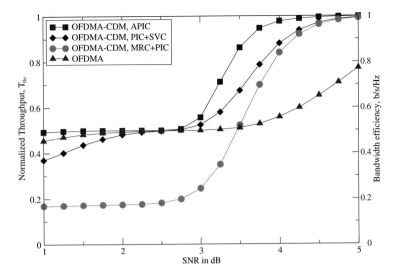

Figure 4. Normalized throughput for different systems and different H-ARQ schemes versus SNR; QPSK modulation used.

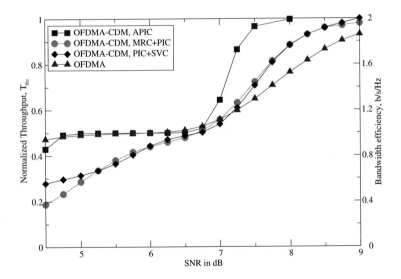

Figure 5. Normalized throughput for different systems and different H-ARQ schemes versus SNR; 16-QAM modulation used.

between H-ARQ based on SVC and PIC has been presented. The simulation results show that the proposed algorithm outperforms an OFDMA system with H-ARQ based on SVC, an OFDMA-CDM system with PIC and H-ARQ based on SVC, and an OFDMA-CDM system with H-ARQ based on MRC and PIC. The ideal interference cancellation scheme

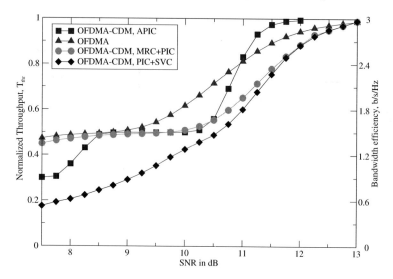

Figure 6. Normalized throughput for different systems and different H-ARQ schemes versus SNR; 64-QAM modulation used.

allows the receiver to remove the interference at high SNR, where the packet error rate does not exceed 5%. The advantages of re-encoded values combining appear especially at lower SNR, where the number of retransmissions is large.

References

[1] K. Fazel and S. Kaiser. *Multi-Carrier and Spread Spectrum Systems*. John Wiley and Sons Ltd., West Sussex, England, 2003.

[2] S. Kaiser. OFDM code-division multiplexing in fading channels. *IEEE Trans. on Commun.*, 50(8):1266–1273, Aug. 2002.

[3] S. Kaiser and J. Hagenauer. Multi-carrier CDMA with iterative decoding and soft-interference cancellation. *in. Proc. IEEE Global Telecommun. Conf. (GLOBECOM'97)*, pages 6–10, Nov. 1997.

[4] J.G. Proakis. *Digital Communications*. McGraw-Hill, New York, 4th. edition, 2001.

[5] J. Hagenauer. Forward error correcting for CDMA systems. *in Proc. IEEE Fourth Int. Symp. Spread Spectrum Tech. Applicat. (ISSSTA '96)*, pages 566–569, Sep. 1996.

[6] D. Chase. Code combining - a maximum-likelihood decoding approach for combining an arbitary number of noisy packets. *IEEE Trans. on Commun.*, COM-33(5):385–393, May 1985.

[7] R. Raulefs, A. Dammann, S. Kaiser, and G. Auer. Comparing multicarrier based broadband systems for higher modulation cardinalities. *in Proc. of Vehicular Technology Conference (VTC 2004 - Fall)*, Los Angeles, CA, USA, Sep. 2004.

Section VI

SYSTEM PERFORMANCE AND IMPLEMENTATION ASPECTS

A COMPARATIVE ANALYSIS OF CDM-OFDMA AND MC-CDMA SYSTEMS

Wei Zhang and Jürgen Lindner
Department of Information Technology, University of Ulm
Albert-Einstein-Allee 43, 89081 Ulm, Germany
{wei.zhang,juergen.lindner}@uni-ulm.de

Abstract   In this paper we investigate CDM-OFDMA and MC-CDMA systems comparatively. The impacts of frequency offsets on both systems are studied. Conventional CDM-OFDMA, interleaved CDM-OFDMA, conventional MC-CDMA, interleaved CDM-OFDMA and MC-CDMA with maximum frequency diversity are considered. Three aspects are taken into account: system structure, parameter estimation and system performance. It is shown that interleaved CDM-OFDMA is possibly more suitable for uplink transmission than MC-CDMA because of its less parameter estimation complexity.

1.   Introduction

CDM-OFDMA (Code Division Multiplexing - Orthogonal Frequency Division Multiple Access, also called SS-MC-MA by S. Kaiser) and MC-CDMA are the combination of multi-carrier (OFDM) and spread spectrum techniques in different ways [1]. In CDM-OFDMA users are separated in frequency. The simultaneously transmitted symbols of an individual user are spread over his own subcarriers to attain frequency diversity. In MC-CDMA systems, however, all subcarriers are available for any individual user such that the maximum frequency diversity can be obtained. User separation is realized by distinct spreading codes.

In this paper the systems are compared in the presence of frequency offsets. For simplicity time synchronization is assumed. A typical channel model in case of wireless communication, frequency selective fading channel, is considered in the simulation. By comparing CDM-OFDMA with MC-CDMA in terms of system structure, parameter estimation and system performance, we try to find out which system should be preferred with respect to implementation.

The reminder of the paper is organized as follows. In section 2 the system models of CDM-OFDMA and MC-CDMA are described, and influences of frequency offsets and frequency selective fading channels are studied briefly. In section 3 we compare both systems and section 4 concludes the paper.

2. System Model

2.1 Downlink Transmission

A vector-valued downlink transmission model of CDM-OFDMA in the presence of frequency offset can be expressed as [1]:

$$\tilde{\mathbf{x}}_{*,i} = \exp(j2\pi i\epsilon(1 + \frac{N_g}{N_f}))\hat{\mathbf{E}}\mathbf{H}\mathbf{T}\mathbf{U}\mathbf{x}_{*,i} + \mathbf{n}_i, \quad (1)$$

where $\mathbf{x}_{*,i}$ contains the transmit symbols from all active users, which are convolutionally encoded and then PSK-mapped. $\mathbf{H}$ is the channel matrix in frequency domain with channel transfer functions on its main diagonal. ϵ represents the frequency offset (FO) normalized to subcarrier spacing. $\hat{\mathbf{E}}$ stands for the influence of a frequency offset on OFDM in frequency domain, and $\exp(j2\pi i\epsilon(1+\frac{N_g}{N_f}))$ is the time-variant phase error due to frequency offset. N_f and N_g are the length of $\mathbf{x}_{*,i}$ and the length of guard interval, respectively. $\mathbf{U}$ represents a spreading matrix. Let $\mathbf{U}_k$ denote the user-specific spreading matrix of user k with the structure

$$\mathbf{U}_k = \begin{pmatrix} 0 & & 0 \\ & \ddots & \\ & \mathbf{W}_k & \\ & & \ddots \\ 0 & & 0 \end{pmatrix}_{N_f \times N_f} \quad (2)$$

where $\mathbf{W}_k$ is called the user-specific spreading submatrix of size $P \times P$. Each column of $\mathbf{W}_k$ is a spreading codeword, e.g., from an Walsh-Hadamard code. For any l and k, $1 \le l, k \le K$, $\mathbf{W}_l$ and $\mathbf{W}_k$ can be either the same or distinct. So $\mathbf{U} = \sum_{k=1}^{K} \mathbf{U}_k$ is a block diagonal matrix with $\{\mathbf{W}_k\}_{k=1}^{K}$ on the main diagonal. In (1) the square matrix $\mathbf{T}$ specifies the subchannel assignment scheme. It works as a subcarrier interleaver. In conventional CDM-OFDMA $\mathbf{T} = \mathbf{I}$ is an identity matrix. Interleaved CDM-OFDMA is defined, if subcarriers of a certain user are selected with an equidistance greater than one.

The expression in (1) is also suitable for MC-CDMA, except that some notations should be redefined. First of all, $\mathbf{x}_{*,i}$ symbols from distinct users are assigned in a interleaved order in comparison to that in CDM-OFDMA. Secondly, although $\mathbf{U} = \sum_{k=1}^{K} \mathbf{U}_k$ holds, the structure of $\mathbf{U}_k$

is redefined as

$$\mathbf{U}_k = \begin{pmatrix} \mathbf{W}_{k,1} & 0 & \cdots & 0 \\ 0 & \mathbf{W}_{k,2} & \cdots & 0 \\ 0 & \vdots & \ddots & \vdots \\ 0 & 0 & \cdots & \mathbf{W}_{k,P} \end{pmatrix}_{N_f \times N_f}, \quad (3)$$

where the user-specific spreading submatrix $\mathbf{W}_{k,p}$ is a $K \times K$ matrix. Its kth column is a spreading sequence of user k on the pth subchannel. All elements in other columns are zero. For any p and q, $\mathbf{W}_{k,p}$ and $\mathbf{W}_{k,q}$ can be the same or different, but $\mathbf{W}_{l,p}$ and $\mathbf{W}_{k,q}$ must be different for any $l \neq k$. An interleaved scheme is defined, if subcarriers assigned to a symbol of a certain user are distributed in a interleaved way. An alternative is that each symbol is spread over all available subcarriers, and thus the maximum frequency diversity can be obtained. In such case each symbol has an individual spreading code to distinguish the symbols of a certain user.

2.2 Uplink Transmission

The general description for the uplink transmission of CDM-OFDMA is given by [1]

$$\tilde{\mathbf{x}}_{*,i} = \sum_{k=1}^{K} \exp(2\pi i \epsilon_k (1 + \frac{N_g}{N_f})) \hat{\mathbf{E}}_k \mathbf{H}_k \mathbf{T} \mathbf{U}_k \mathbf{x}_{*,i} + \mathbf{n}_i. \quad (4)$$

The joint operation of $\mathbf{TU}_k$ extracts the $\mathbf{x}_{k,i}$ from $\mathbf{x}_{*,i}$, spreads them and assigns them to the subcarriers belonging to user k. It ensures that the vector after spreading ($\mathbf{U}_k$) has the same form as the unspread vector in OFDMA. A simplified expression can therefore be given (see [1]). If (4) is used for MC-CDMA, according to (3) each individual user will occupy all available subcarriers. Eq. (4) can not be simplified then.

2.3 Influence of Frequency Offset and Frequency Selective Fading Channel

Consider the downlink transmission of MC-CDMA and CDM-OFDMA in the presence of a frequency offset. In case of an ideal channel, according to $\hat{\mathbf{E}}$, the orthogonality between OFDM-subcarriers is destroyed by the frequency offset, which gives rise to amplitude reduction and phase rotation of desired OFDM-symbols as well as intercarrier interference (ICI). Furthermore, the ICI destroys the orthogonality between code-division subchannels [1, 7]. The resulting interference can be classified to self-interference (SI) from the same user and multiuser interference (MUI) from other users, which degrade the system performance. In addition, if the frequency offset is not compensated, a time-variant phase error $\exp(j2\pi i\epsilon(1 + \frac{N_g}{N_f}))$ should be taken into account.

In a multipath propagation scenario, the frequency selective behavior of the channel also destroys the orthogonality of code division subchannels. In the presence of frequency offset, channel transfer functions scale the desired OFDM-symbols and the amount of the crosstalk between OFDM-subcarriers. After despreading the interference is caused by the joint effects of frequency offset and frequency selective fading channel.

3. Comparison of CDM-OFDMA and MC-CDMA

3.1 Comparison of System Structure

Both MC-CDMA and CDM-OFDMA are the combination of multi-carrier and spread spectrum techniques. From the system description above we can see that the structure of MC-CDMA is similar to that of CDM-OFDMA. The essential difference between CDM-OFDMA and MC-CDMA is their multiple access mode: in the former case users are separated in frequency (OFDMA) and in the latter case by distinct spreading codes (CDMA). Compared with single user OFDM transmission, the system complexity of CDM-OFDMA and MC-CDMA will increase.

CDM-OFDMA is an extension of the OFDMA. It takes advantage of spread spectrum to achieve frequency diversity in a given subcarrier group. The simultaneously transmitted symbols of a certain user are separated by different spreading codes. Therefore, the power efficiency is limited by the number of subcarriers belonging to each individual user.

MC-CDMA, however, is an extension of CDMA with spreading in frequency domain. the maximum frequency diversity can be obtained if a symbol is spread over all available subcarriers.

3.2 Comparison of Parameter Estimation

Parameter estimation in communication systems is of importance, because estimation errors can degrade the overall system performance. In the following the estimation complexity in CDM-OFDMA and MC-CDMA will be analyzed. We begin with the downlink transmission.

In the downlink, at the mobile terminal receiver received symbols experience the same physical channel and have the same frequency offset. Therefore, the synchronization and channel estimation approaches

Figure 1. Frequency offsets distribution in the uplink of CDM-OFDMA (left) and MC-CDMA (right) systems.

for single user OFDM can be implemented in CDM-OFDMA and MC-CDMA downlink without or with slight change according to the special transmission schemes.

In the uplink, received signals at the base station receiver are from all active users which undergo different different channels and possibly have different frequency offsets. The distinction of multiple access mode determines that different parameter estimation complexity in CDM-OFDMA and MC-CDMA uplink. Figure 1 illustrates the distribution of frequency offsets in the uplink of both systems. On any individual subcarrier, in CDM-OFDMA only one frequency offset plays the main role, since users are roughly separated in frequency and each subcarrier carries the signals from an individual user; whereas in MC-CDMA multiple FOs belonging to different users are of the same importance. Therefore, in the former case the estimation of frequency offsets will possibly be simpler than in the latter case.

For the same reason, in the CDM-OFDMA uplink channel estimation for a certain user can be done simply in frequency domain under assumption of perfect synchronization. This implies that synchronization and parameter estimation approaches developed for OFDMA uplink can be implemented without change or with slight change. Furthermore, phase rotation due to frequency offsets can still be detected even if the interference from other users exists.

In MC-CDMA uplink after FFT on each subcarrier the received signal is the superposition of symbols weighted by different channel transfer functions and possibly with different frequency offsets. This coexistence leads to the increase of parameter estimation complexity. After despreading users are roughly separated from one another, but the information of channel characteristics is also destroyed, especially in the presence of synchronization error. The parameter estimation hereby becomes a crucial task in the uplink of MC-CDMA.

3.3 Comparison of System Performance

To compare the system performance, a linear minimum mean square error multiuser detector (LMMSE-MUD) is implemented in the mobile

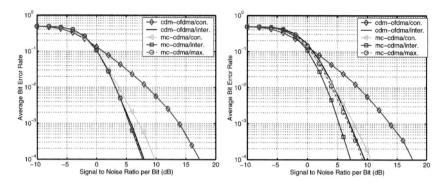

Figure 2. The average BER performance in the downlink transmission over frequency selective fading channel with (right) and without (left) frequency offset.

terminal and base station receiver. The symbol vector before the decision becomes then

$$\tilde{\mathbf{r}}_{*,i} = (\mathbf{R}^H \mathbf{R} + \frac{\sigma_n^2}{\sigma_s^2} \mathbf{I})^{-1} \mathbf{R}^H \tilde{\mathbf{x}}_{*,i}. \tag{5}$$

$\mathbf{R} = \sum_{k=1}^{K} \hat{\mathbf{E}}_k \mathbf{H}_k \mathbf{T} \mathbf{U}_k$ is the equivalent channel matrix. σ_s^2/σ_n^2 represents signal to noise ratio (SNR).

In the simulations perfect timing is assumed. Furthermore, we assume that perfect knowledge of channel state information and frequency offsets are available. The channels are supposed to be frequency selective Rayleigh fading with impulse responses generated according to a 8-tap Rayleigh fading power delay profile [7]. The channel impulse responses (CIR) keep constant during one vector duration, i.e., block fading is assumed. The normalized frequency offsets are uniformly distributed in the range of [-0.33, 0.33]. The transmit symbols are first encoded by convolutional encoder C(133,171) of rate $R = 1/2$, and then QPSK-mapped. Walsh-Hadamard codes are used as spreading codes. In the simulation $N_f = 64$, $N_g = 16$ and $K = 4$. The time-variant phase error due to frequency offset is not taken into account in the simulation.

Figure 2 illustrates the average bit error rate (BER) performance of downlink transmission over frequency selective fading channel. It can be seen that without frequency offset the similar performance is obtained by interleaved CDM-OFDMA, interleaved MC-CDMA and MC-CDMA with the maximum frequency diversity (marked 'mc-cdma/max.' in the figure). Compared with transmission over AWGN channel (left figure in Fig. 3), about $5 \sim 6$ dB performance loss results from frequency selective behavior of the channel. In the presence of frequency offsets, the performance of different transmission systems degrade differently. Conventional MC-CDMA, MC-CDMA with the maximum frequency

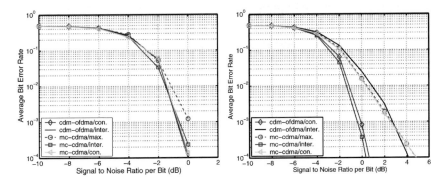

Figure 3. The average BER performance in the uplink transmission over AWGN channel with (right) and without (left) frequency offsets.

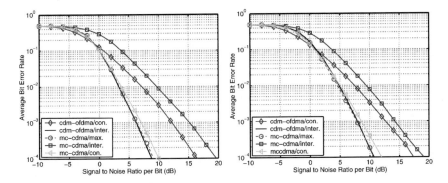

Figure 4. The average BER performance in the uplink transmission over frequency selective fading channels with (right) and without (left) frequency offsets.

diversity and interleaved CDM-OFDMA have about 1.3 dB performance loss due to frequency offsets at a given BER of 10^{-3}. Conventional CDM-OFDMA has the worst performance in downlink transmission.

Figure 3 illustrates the average BER performance in AWGN channel with and without frequency offsets. It is shown that the similar performance is obtained for all proposed systems if only white noise degrades the system performance. In the presence of frequency offsets, the performance of different transmission systems degrade differently. Conventional MC-CDMA, MC-CDMA with the maximum frequency diversity and interleaved CDM-OFDMA have about 2.5 ~ 3 dB performance loss due to frequency offsets in comparison to interleaved MC-CDMA, although FOs are compensated by LMMSE-MUD.

Figure 4 depicts the uplink transmission in an multipath scenario. A severe performance degradation occurs for interleaved MC-CDMA, whereas the best average BER is obtained by interleaved CDM-OFDMA and MC-CDMA with the maximum frequency diversity. With less frequency diversity, conventional MC-CDMA has a slight performance loss. It should be noted that performance loss due to frequency offsets is less than for AWGN channel. Comparing Fig. 3 with 4, it is easy to see that the channel characteristics are the dominating factor that degrades the system performance. In general, interleaved CDM-OFDMA is one of the most promising candidates for the wireless uplink transmission.

4. Conclusions

CDM-OFDMA and MC-CDMA have been compared in terms of system structure, parameter estimation and BER performance. It has been shown that MC-CDMA with the maximum diversity and interleaved CDM-OFDMA are good candidates for future wireless communication systems. Furthermore, interleaved CDM-OFDMA is possibly more suitable for uplink transmission than MC-CDMA because of its less parameter estimation complexity.

References

[1] W. Zhang and J. Lindner, "Modeling of OFDM-based systems with frequency offsets and frequency selective fading channels", *Proceeding of VTC-Spring'05*, May/Juni 2005, Stockholm.

[2] H. Steendam and M. Moeneclaey, "The effect of carrier frequency offsets on downlink and uplink MC-DS-CDMA", *IEEE Trans. Journal. on select. Areas*, Vol. 19, No. 12, pp. 2528-2536, Dec. 2001.

[3] J. J. van de Beek, P.O. Börjessen, and M. L. Boucheret, "A time and frequency synchronization scheme for multiuer OFDM", *IEEE Trans. Signal Processing*, vol. 45, No. 7, pp. 1800-1805, July 1997.

[4] J. Lindner, "MC-CDMA in the context of general multiuser/multisubchannel transmission methods", *Europen Trans. Telecomm.*, vol. 10, No. 5, pp. 351-367, July/Aug. 1999.

[5] J. Lindner, *Informationsübertragung*, Springer-Verlag, 2005.

[6] W. Zhang and J. Lindner, "A Frequency Offset Tracking Scheme for the Uplink of OFDMA Systems", *Proceeding of InOWo2005*, 10th International OFDM-Workshop, August/September 2005, Hamburg.

[7] W. Zhang and J. Lindner, "MC-CDMA in the uplink: a transmission scheme which tolerates frequency offsets and freuquency selective channels", *PIMRC2004*, Barcelona, 2004.

MULTI-USER TRANSMIT POWER CONTROL FOR MULTI-CARRIER MODULATION SYSTEMS IN QUASI-SYNCHRONOUS UPLINK CHANNEL

Masahiro Fujii, Makoto Itami and Kohji Itoh
Department of Applied Electronics, Tokyo University of Science,
2641 Yamazaki, Noda-Shi, Chiba, Japan
{fujii,itami,itoh}@itlb.te.noda.sut.ac.jp

Abstract Multi-carrier modulation systems with Transmit Power Control (TPC) in a quasi-synchronous up-link channel is investigated in this paper. When the channel state information for all the mobile stations are available at a base station, the base station instructs each of the mobile stations to allocate its fixed transmit power to a carrier channel to minimize the average pairwise error probability in the up-link. We consider the TPC for both Multi-Carrier/Code Division Multiple Access (MC/CDMA) and Orthogonal Frequency Division Multiple Access (OFDMA) systems. We show a theoretical evidence for the fact that the adaptive TPC OFDMA outperforms MC/CDMA. Moreover, we propose a simple algorithm to obtain sub-optimum solution to the assignment problem in OFDMA with TPC to reduce the computational complexity of the conventional optimization algorithm to solve the problem. By computer simulation, we found the proposed algorithm was able to achieve a near-optimum performance for the number of simultaneous users less than about 80% of the fully loaded system and a significant improvement of the performance in comparison with that of the single user bound to the maximum likelihood reception in MC/CDMA without TPC.

1. Introduction

An impressive interest has been devoted to the studies of Multi-Carrier (MC) modulation systems for the next generation wireless communication systems. The MC modulation systems realized by Orthogonal Frequency Division Multiplexing (OFDM) techniques have many advantages compared with systems designed in time domain [1]. On the other hand, the MC/Code Division Multiple Access (CDMA) system is a spread spectrum technique combined with OFDM and is extensively investigated [2] [3]. The MC/CDMA system is able to enjoy the frequency diversity effect owing to the scheme in which each data symbol is spread over the carrier channels by modulating the OFDM signals according to a respective signature sequence, but suffers from Multiple Access Interference (MAI) caused by the simultaneous users sharing the same carrier channels. An alternative to MC/CDMA system is Orthogonal Frequency Division Multiple Access (OFDMA) system. In contrast to MC/CDMA whose carriers are shared by the users, OFDMA system is realized by assigning at least one distinct carrier channel to one user [4] [1].

On the other hand, several studies have been made on the adaptive transmission whose transmit power and number of bit per carrier channel in MC modulation system is adapted to the Channel State Information (CSI), assuming the knowledge at the transmitter [5]. In [6], the transmitter and receiver optimization in the up-link MC/CDMA system has been proposed to minimize the total transmit power for all the users under the condition that Mean Squared Error (MSE) is greater than a required value.

In this paper, we focus on the optimization of TPC for MC/CDMA systems in a quasi-synchronous up-link channel. We consider how to allocate the transmit power to the carrier channels at each user to minimize the average Pairwise Error Probability (PEP) as the union bounded to error rate assuming the Maximum Likelihood (ML) detector which yields the minimum achievable probability of error in CDMA channel under a fixed total transmit power for each user in contrast to the optimization scheme in [6]. We investigate the optimum TPC schemes in both MC/CDMA and OFDMA aspects.

2. System Model

In this paper, we assume that the orthogonal MC modulation system is able to make use of N carrier channels. In quasi-synchronous up-link channel, figure 1 shows the block diagram of the multi-user transmission system applying the orthogonal MC signal design.

The received signal vector $\underline{Y}$ is expressed by $\underline{Y} = \boldsymbol{G}\underline{d} + \underline{\eta}$ where $\underline{d} = [d_1, \ldots, d_M]^T$ whose mth element d_m is the data symbol for the mth user and $\underline{\eta} = [\eta_1, \ldots, \eta_N]^T$ whose nth element η_n is the additive white Gaussian noise at the nth carrier channel. The mth column $\underline{G}_m$ of $N \times M$ matrix $\boldsymbol{G}$ is expressed by $\underline{G}_m = \boldsymbol{H}_m \underline{W}_m$ where $N \times N$ diagonal matrix $\boldsymbol{H}_m = \text{diag}[H_{m,1}, \ldots, H_{m,N}]$ and $N \times 1$ vector $\underline{W}_m$ are, respectively, a channel frequency response matrix and a transmit weight vector for the mth user. The channel frequency response $H_{m,n}$

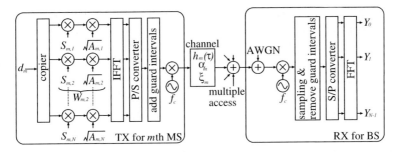

Figure 1. Multi-user transmission system applying orthogonal multi-carrier signal design

at the nth carrier frequency f_n is defined by

$$H_{m,n} = \sqrt{\alpha_m} e^{-j2\pi f_n \xi_m} \int h_m(\tau) e^{-j2\pi f_n \tau} d\tau \qquad (1)$$

where $\sqrt{\alpha_m}$, ξ_m and $h_m(\tau)$ are, respectively, the amplitude loss associated with the distance between the mobile and base stations, the slight time lag error beyond the transmit timing control and the channel impulse response for the mth user. And $N \times 1$ transmit weight vector $\underline{W}_m$ is defined by $\boldsymbol{A}_m \underline{S}_m$ where $N \times N$ diagonal matrix $\boldsymbol{A}_m = \mathtt{diag}\,[\sqrt{A_{m,1}}, \cdots, \sqrt{A_{m,N}}]$ and $N \times 1$ vector $\underline{S}_m$ are, respectively, a transmit amplitude matrix and a spreading sequence vector for the mth user.

3. Pairwise Error Probability

In this section, let us consider the optimum method to assign the transmit power to the carrier channels for each user under the condition that the maximum available transmit power of each user is constant value P_{TX}. This method is intended to minimize the average PEP as the union bound to symbol error rate assuming use of the ML detector.

When ideal CSI (i.e. $\boldsymbol{G}$) is available at the receiver, the PEP is defined as the probability $P(\underline{d} \to \underline{\hat{d}})$ that the ML detector prefers the symbol vector $\underline{\hat{d}}$ to the transmitted symbol vector $\underline{d}$, and is given by $P(\underline{d} \to \underline{\hat{d}}) = \mathtt{erfc}\,(\sqrt{\|\boldsymbol{G}(\underline{d}-\underline{\hat{d}})\|^2/(4\sigma_\eta^2)})/2$ where $\mathtt{erfc}\,(x)$ is the complementary error function and $\|\underline{x}\|^2 = \underline{x}^H \underline{x}$ where $\cdot^H$ denotes the Hermitian transposition. The average PEP is defined by averaging over L possible transmitted symbol vectors and can be upper-bounded by a union bound as

$$\bar{P}_{PEP} \leq \sum_{i=1}^{L} P(\underline{d}) \sum_{j=1, j \neq i}^{L} \frac{1}{2} \mathtt{erfc}\left(\sqrt{\|\boldsymbol{G}(\underline{d} - \hat{\underline{d}})\|^2 / (4\sigma_\eta^2)}\right) \quad (2)$$

where $P(\underline{d})$ is the a priori probability of $\underline{d}$ being transmitted and is usually equal to $1/L$. The difference vector $\underline{d} - \hat{\underline{d}}$ between the transmitted vector $\underline{d}$ and the detected vector $\hat{\underline{d}}$ by the ML detector can be classified into K error vectors. Defining error vectors $\underline{e}_k = [e_{k,1}, \cdots, e_{k,M}]^T$ for $1 \leq k \leq K$, we can rewrite the average PEP in the form $\bar{P}_{PEP} \leq \frac{1}{2L} \sum_{k=1}^{K} C(\underline{e}_k) \mathtt{erfc}\left(\sqrt{\|\boldsymbol{G}\underline{e}_k\|^2 / (4\sigma_\eta^2)}\right)$ where $C(\underline{e}_k)$ denotes the tautological number of $\underline{e}_k$.

Let $\mathcal{E}$ denote the set which consists of all the possible error vectors $\underline{e}_k$ and $\mathcal{E}^{(i)}$ denote the set which consists of the set of i-fold error vectors $\underline{e}_k^{(i)}$ whose i elements of $\underline{e}_k^{(i)}$ are non-zero and the other elements are zero. Since $\mathcal{E}^{(i)} \cap \mathcal{E}^{(j)} = \emptyset$ $(i \neq j)$, we can rewrite the average PEP by

$$\bar{P}_{PEP} \leq \frac{1}{2L} \sum_{i=1}^{M} \sum_{k=1}^{K^{(i)}} C(\underline{e}_k^{(i)}) \mathtt{erfc}\left(\sqrt{\|\boldsymbol{G}\underline{e}_k^{(i)}\|^2 / (4\sigma_\eta^2)}\right) \quad (3)$$

where $K^{(i)}$ is the number of sample error vectors $\underline{e}_k^{(i)}$ which constitute i-fold error sets $\mathcal{E}^{(i)}$ and $\sum_{i=1}^{M} K^{(i)} = K$ holds.

4. Transmit Power Allocation

In this section, we propose a new algorithm to decide the transmit weight $\underline{W}_m$ for each user to minimize the upper-bound of the average PEP $\bar{P}_{PEP}$ in equation (3) under the condition that the total transmit power for each user is constant value and is defined as $\|\underline{W}_m\|^2 = P_{TX}$.

4.1 Single Error Set $\mathcal{E}^{(1)}$

Let us consider the error vectors $\{\underline{e}_k^{(1)} : k = 1, \cdots, K^{(1)}\}$ which constitute the single error set collection $\mathcal{E}^{(1)}$. Consider a sample error vector $\underline{e}_k^{(1)} = [0, \cdots, 0, e_{k,i}, 0, \cdots, 0]^T$ whose ith element is non-zero and the other elements are zero. For this set, the squared Euclidean distance $\|\boldsymbol{G}\underline{e}_k^{(1)}\|^2$ in the complementary error function $\mathtt{erfc}(\cdot)$ in equation (3) can be rewritten as $\|\boldsymbol{G}\underline{e}_k^{(1)}\|^2 = |e_{k,i}|^2 \|\boldsymbol{G}_i\|^2$. Since the complementary function is a monotonic decreasing function, in order to minimize the average PEP union bound on the single error sets $\mathcal{E}^{(1)}$, it is seen that the optimum transmit power allocation is achieved by maximizing for all the users, $1 \le i \le M$, $\|\underline{G}_i\|^2 = \|\boldsymbol{H}_i\underline{W}_i\|^2$ under $\|\underline{W}_i\|^2 = P_{TX}$. It is easy to solve this problem by introducing the Lagrangian multiplier method. The optimum transmit power allocation is obtained by

$$W_{i,n} = S_{i,n}\sqrt{A_{i,n}} = \begin{cases} \sqrt{P_{TX}} & , \quad (n = n_i) \\ 0 & , \quad (n \ne n_i) \end{cases}, (1 \le i \le M) \quad (4)$$

where $n_i = \arg\max_n\{|H_{i,n}|^2\}$.

From what has been discussed above, we can conclude that the MC modulation system with the optimum power allocation is OFDMA system in which each user transmits the data symbol using a carrier channel with the best channel gain when only the single error vectors are taken into account.

4.2 Double Error Set $\mathcal{E}^{(2)}$

Next, we focus on the error vectors $\{\underline{e}_k^{(2)} : k = 1, \cdots, K^{(2)}\}$ which constitute the double error set collection $\mathcal{E}^{(2)}$. Let us consider a sample error vector $\underline{e}_k^{(2)} = [0, \cdots, 0, e_{k,i}, 0, \cdots, 0, e_{k,j}, 0, \cdots, 0]^T$ whose ith and jth elements are non-zero and other elements are zero. Then, in order to maximize the squared Euclidean distance $\|\boldsymbol{G}\underline{e}_k^{(2)}\|^2$ given by $\|\boldsymbol{G}\underline{e}_k^{(2)}\|^2 = \|e_{k,i}\underline{G}_i + e_{k,j}\underline{G}_j\|^2$ under the condition $\|\underline{W}_i\|^2 = \|\underline{W}_j\|^2 = P_{TX}$, we find the transmit weight vectors $\underline{W}_i$ and $\underline{W}_j$. It may be difficult to obtain a closed-form solution if we apply the Lagrangian multiplier method and its iterative approach. Let us consider this problem form a different point of view. We rewrite the squared Euclidean distance $\|\boldsymbol{G}\underline{e}_k^{(2)}\|^2$ as $a + b$ where

$$a = |e_{k,i}|^2\|\underline{G}_i\|^2 + |e_{k,j}|^2\|\underline{G}_j\|^2 \ge 0 \text{ and} \quad (5)$$

$$b = 2\Re\{e_{k,i}^*e_{k,j}\underline{G}_i^H\underline{G}_j\}. \quad (6)$$

On the other hand, there is certainly another sample error vector $\underline{e}_l^{(2)} = [0, \cdots, 0, e_{l,i}, 0, \cdots, 0, e_{l,j}, 0, \cdots, 0]^T$ where $e_{l,i} = e_{k,i}$ and $e_{l,j} = -e_{k,j}$ (i.e. the only difference between $\underline{e}_k^{(2)}$ and $\underline{e}_l^{(2)}$ is the inversion of polarity of the jth element of the error vector). The squared Euclidean distance with respect to $\underline{e}_l^{(2)}$ is expressed in the form $\|\boldsymbol{G}\underline{e}_l^{(2)}\|^2 = a - b$. Since $C(\underline{e}_k^{(2)}) = C(\underline{e}_l^{(2)})$, the average PEP union bound over $\underline{e}_k^{(2)}$ and $\underline{e}_l^{(2)}$ satisfies Jensen's inequality $\texttt{erfc}(\sqrt{(a+b)/(4\sigma_\eta^2)}) + \texttt{erfc}(\sqrt{(a-b)/(4\sigma_\eta^2)}) \ge 2\,\texttt{erfc}(\sqrt{a/(4\sigma_\eta^2)})$ with equality $\underline{G}_i^H\underline{G}_j = (\boldsymbol{H}_i\underline{W}_i)^H\boldsymbol{H}_j\underline{W}_j = 0$ (i.e. $b = 0$).

This result leads us to the conclusion that the MC modulation system without the cross-correlation among the simultaneous users outperforms the other systems. Therefore, it is preferable to design the MC modulation system which satisfies $\underline{G}_i^H \underline{G}_j = \underline{W}_i^H \boldsymbol{H}_i^H \boldsymbol{H}_j \underline{W}_j = 0$. If we abandon the transmit phase control scheme, a possible scheme to realize it is to assign the transmit power of the ith and jth user to mutually exclusive carrier channels. When we consider all of the users, a semi-optimum power control, therefore, will be to assign one carrier channel with the larger $|H_{i,n}|$ to each user, i.e. OFDMA scheme, avoiding collision of the carrier channel assignment among the simultaneous users. This conclusion on the double error set $\mathcal{E}^{(2)}$ thus corresponds to the one on the single error set $\mathcal{E}^{(1)}$ in section 4.1. Moreover, this result holds on the i-fold error sets $\mathcal{E}^{(i)}$ for $3 \leq i \leq M$ in a similar fashion. This theoretical explanation of the possibility of OFDMA outperforming MC/CDMA in the designs of the up-link MC system corresponds to the result observed from the simulations in [6]. In the next section, we focus on optimum carrier channel assignments for OFDMA.

4.3   Carrier Channel Assignment for OFDMA

The carrier channel assignment problem in OFDMA can be formulated as follows. In OFDMA, minimizing the average PEP $\bar{P}_{PEP}$ is equivalent to minimizing the average Symbol Error Probability (SEP) $\bar{P}_{SEP}$ given by

$$\bar{P}_{SEP} = \frac{1}{M} \sum_{m=1}^{M} \sum_{n=1}^{N} v_{m,n} P_{SEP}(|H_{m,n}|^2) \qquad (7)$$

under the conditions that

$$\begin{cases} \sum_{m=1}^{M} v_{m,n} = 1, & (1 \leq n \leq N) \\ \sum_{n=1}^{N} v_{m,n} = 1, & (1 \leq m \leq M) \end{cases}, (v_{m,n} \in \{0,1\}) \qquad (8)$$

where $P_{SEP}(|H_{m,n}|^2)$ denotes the instantaneous SEP as a function of the channel gain $|H_{m,n}|^2$. Although it is necessary to compare the average SEP's of $_N P_M$ possible assignments to solve the problem, we can fortunately make use of the Hungarian method with a computational complexity in polynomial time [7]. We are forced in the Hungarian method, however, to resort to an iterative process, and we propose a simple solution for the assignment problem which turns out to be effectively optimal.

In general, it seems that the higher instantaneous SEP of the user with smaller overall channel gain dominates the average SEP. We propose a new solution which assigns, by priority, carrier channels with the

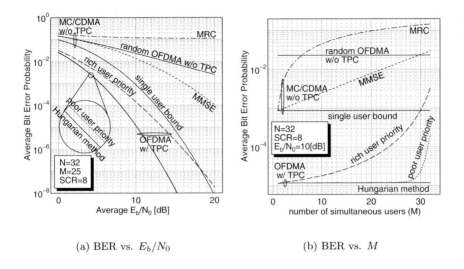

Figure 2. Average BER performances

better gain to the user with the lower average channel gain expressed by $\text{tr}\{\boldsymbol{H}_m^H \boldsymbol{H}_m\}/N$. In a similar fashion, we preferentially allot the carrier channels to the simultaneous users in the ascending order of the average channel gains.

5. Numerical Examples

In this section, we show numerical examples to evaluate performances of the MC modulation systems with the proposed TPC by computer simulations. We assume that the delay-power profile of different users is all equal to $\rho(\tau)$ which decays exponentially. In our computer simulations, we assume that QPSK modulation is adopted for the data symbols d_m of each user and the average signal energy per bit at the receiver for the mth user is defined by $E_m = P_{TX} \, \text{E}\{|d_m|^2 \, \text{tr}\{\boldsymbol{H}_m^H \boldsymbol{H}_m\}\} T/2 = P_{TX} \alpha_m T/2$. The number of carrier channels N is 32. We assume that the CSI for all mobile stations is ideally available at the base station and the fading of channel is invariant during the frame period for TPC.

In Figure 2(a), we show the average Bit Error Probability (BEP) performance versus the average E_b/N_0, which is equal to $E_m/(\sigma_\eta^2 T)$ when E_m is equal for all the users, for $N = 32$, $M = 25$ and SCR= 8 where SCR is System bandwidth to Coherent bandwidth Ratio. In MC/CDMA without TPC, each user uses a signature sequence chosen randomly from Walsh-Hadamard sequences with length N, and MRC or MMSE [2] receivers are adopted. OFDMA with TPC provides best BEP performance and the proposed solution (labeled 'poor user priority') in section 4.3 is able to achieve very close performance to the optimum Hungarian method. An important point to be emphasized is the fact that the BEP curves of OFDMA using the poor priority or the Hungarian method have a similar slope of the single user bound and results in the improvement about 4[dB].

The average BEP performance is illustrated in Figure 2(b) as a function of the number of users M for $N = 32$, SCR= 8 and $E_b/N_0 = 10$[dB]. In CDMA, the performance is deteriorated with the increasing M due to the cross-correlation among the simultaneous users regardless of the multi-user detectors. OFDMA with TPC in general depends, however, on varying M because the probability of occurrence of the cases in which different users require the same carrier channel grows with the increasing M. On the contrary, even the solution with the poor user priority is able to achieve a performance very close to that of the optimum algorithm for less than about 80% users in the fully loaded system which allows multiple access for N users on N carrier channels.

6. Conclusions

Assuming that the CSI for all the mobile stations are available at the base station, the base station instructs the mobile stations to allocate the fixed transmit power to each carrier channel to minimize the average PEP, a bound to the symbol error probability assuming the maximum likelihood detector in the up-link. It was concluded that the OFDMA system with TPC might outperform MC/CDMA system with TPC. By the computer simulation, we showed the proposed algorithm could achieve a performance very close to that of the optimum algorithm for less than 80% users of the number of carrier channels.

References

[1] R. van. Nee and R. Parasad, *OFDM for Wireless Multimedia Communications*, Artech House Publishers, 2000.
[2] S. Hara and R. Parasad, "Overview of Multicarrier CDMA", *IEEE Commun. Mag.*, vol. 35, pp. 126-133, Dec.,1997.
[3] K. Fazel and S. Kaiser, *Multi-Carrier and Spread Spectrum Systems*, John Wiley & Sons, 2003.
[4] H. Sari and G. Karam, "Orthogonal Frequency-Division Multiple Access and its Application to CATV Networks", *European Trans. Commun.*, vol. 9, no. 6, pp. 507-516, Nov. /Dec., 1998.
[5] T. Keller and L. Hanzo, "Adaptive Modulation Techniques for Duplex OFDM Transmission", *IEEE Trans. Veh. Technol.* vol. 49, no. 5, pp. 1893-1906, Setp., 2000.
[6] T. M. Lok and T. F. Wong, "Transmitter and Receiver Optimization in Multicarrier CDMA System", *IEEE Trans. Commun.*, vol. 48, no. 7, pp. 1197-1207, July, 2000.
[7] C. H. Kuhn, "The Hungarian method for the assignment problem", *Naval Research Logistics Quarterly 2*, pp. 83-97, 1955.

ON THE COMPENSATION OF IQ IMBALANCE IN AN SC/FDE SYSTEM

C. Wicpalek[1], H. Witschnig[2] and A. Springer[1]
[1]*Institute for Communications and Information Engineering, University of Linz, Austria,*
{a.springer, c. wicpalek}@icie.jku.at
[2]*Philips Austria, harald.witschnig@philips.com*

Abstract: Robust communication concepts and/or synchronization algorithms represent one of the most important aspects for high rate wireless communication systems. In this work efficient concepts and algorithms for the estimation and compensation of IQ imbalance are developed. The investigations are based on the concept of a Single Carrier System with Frequency Domain Equalization (SC/FDE), as the underlying data structure is optimally suited for synchronization purposes. Novel algorithms are developed in the time and frequency domain, which allow to estimate and compensate the IQ imbalance almost completely – for the case of AWGN as well as multi-path propagation.

Key words: SC/FDE, IQ imbalance, synchronisation.

1. INTRODUCTION

High rate wireless communication systems have to face severe challenges in the area of equalization and synchronization. In terms of equalization it is the quadratically growth of the processing effort in the time domain, which represents one of the most decisive factors. Concepts that implement this equalization in the frequency domain such as OFDM (Orthogonal Frequency Division Multiplexing) and SC/FDE (Single Carrier System with Frequency Domain Equalization) are able to fulfill both criteria, good performance and reduced implementation effort as the implementation effort grows only slightly more than linear [1],[2],[3],[4]. Besides this synchronization and channel estimation schemes are indispensable criteria for high rate radio transmission. Therefore it is essential to develop efficient strategies to compensate for different kinds of imperfections – as it is done in this work for the IQ imbalance.

But before going into details of IQ imbalance (section 2), its estimation and compensation for the case of AWGN (section 3) and multipath-propagation (section 4), the underlying SC/FDE concept is briefly introduced.

Frequency domain equalization for single carrier systems is based on the equivalence between the convolution of two sequences in the time domain and the product of their Fourier transforms. In [1] it has been proposed to perform a blockwise transmission similar to OFDM and also to insert a Cyclic Prefix (CP) between successive blocks. Due to the cyclic extension of the transmitted blocks, the convolution of one cyclically extended transmitted block and the channel impulse response can be calculated by a circular convolution or equivalently by multiplication of the transfer functions in the frequency domain. It has to be mentioned that the advantageous implementation of the SC/FDE system is based on fulfilling the theorem of cyclic convolution, at the expense of a cyclic prefix. Figure 1 shows another possible data structure based on the so called unique word (UW), which also fulfills the theorem of cyclic convolution, but shows significant advantages in terms of synchronization and equalization [3],[4]. The advantages of this UW may be summarized as follows: the UW is known in advance and it can be chosen in a preferred way.

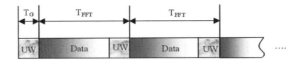

Figure 1. SC/FDE structure based on UW.

2. IQ IMBALANCE IN A ZERO-IF RECEIVER

The Zero-IF receiver architecture (also called homodyne architecture) shows by most the highest integration potential and is therefore the dominant architecture in modern wireless systems. Nevertheless, the homodyne topology entails a number of issues that do not exist or are not as serious as in a heterodyne receiver. One of these is the IQ imbalance. The IQ imbalance may be described as follows: For the IQ demodulation in a QPSK system two mixers and one phase shifter are needed. Figure 2 depicts this structure. In a real system it is not possible to generate an exact 90° phase shift because there are tolerances and/or thermal drifts of the electronic com-

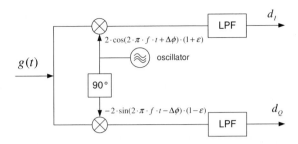

Figure 2. Zero IF-receiver.

-ponents. On the one hand there is a phase mismatch $\Delta\Phi$ and on the other hand an amplitude imbalance ε exists. With the received signal

$$g = r_I \cdot \cos(2\cdot\pi\cdot f \cdot t) - r_Q \cdot \sin(2\cdot\pi\cdot f \cdot t), \qquad (1)$$

the demodulated and distorted I- and Q-path signals (as shown in figure 2) are given by

$$\begin{aligned} d_I &= r_I \cdot (1+\varepsilon) \cdot [\cos(\Delta\Phi)] + r_Q (1+\varepsilon) \cdot [\sin(\Delta\Phi)] \\ d_Q &= r_I \cdot (1-\varepsilon) \cdot [\sin(\Delta\Phi)] + r_Q (1-\varepsilon) \cdot [\cos(\Delta\Phi)] \end{aligned}. \qquad (2)$$

A possible way to compensate the IQ imbalance is consequently to estimate ε and $\Delta\Phi$, which may be based on two consecutive symbols. Nevertheless, the strong non-linearity of the underlying equations prevents an easy, efficient and fast implementation. One strategy has been introduced for OFDM in [5],[6],[7]. This method is based on solving linear equations based on two OFDM symbols in the frequency domain. This is not directly applicable for the SC/FDE structure since the SC/FDE symbols are received in the time domain, and each symbol is spread over the entire system bandwidth. In the following a novel method will be derived which is also based on solving a set of linear equations. The combination of a complex baseband signal $r = r_I + j\, r_Q$ and Eq. (2) results in

$$\begin{aligned} d &= (\cos(\Delta\Phi) - j\cdot\varepsilon\cdot\sin(\Delta\Phi))\cdot r + (\varepsilon\cdot\cos(\Delta\Phi) + j\cdot\sin(\Delta\Phi))\cdot r^* \\ &= \alpha\cdot r + \beta\cdot r^* \end{aligned}. \qquad (3)$$

Here we introduce two new parameters α and β which describe the IQ-imbalance in mathematically more convenient way. As it can be seen from Eq. (3), this results in a complex linear equation. Note that it is not necessary to derive $\Delta\Phi$ and ε subsequently because the compensation can be done directly with α and β.

3.   ESTIMATION OF α AND β FOR THE CASE OF AWGN

The Unique Word structure fits very well for estimation of α and β as the symbols of the UW are known in advance and therefore allow a continuous tracking. For the following derivation d_1 and d_2 describe two received consecutive symbols and r_1 and r_2 stand for the known transmitted symbols. The received symbols are distorted by IQ imbalance and noise. Based on Eq. (3) we obtain a system of equations for two consecutive symbols of the form:

$$d_1 = \alpha \cdot r_1 + \beta \cdot r_1^*$$
$$d_2 = \alpha \cdot r_2 + \beta \cdot r_2^*$$
(4)

Solving these set of complex equations for α and β yields

$$\alpha = j \cdot \frac{d_1 \cdot r_2^* - d_2 \cdot r_1^*}{2 \cdot (\Re\{r_1\}\Im\{r_2\} - \Re\{r_2\}\Im\{r_1\})}$$
$$\beta = j \cdot \frac{d_2 \cdot r_1 - d_1 \cdot r_2}{2 \cdot (\Re\{r_1\}\Im\{r_2\} - \Re\{r_2\}\Im\{r_1\})}$$
(5)

Based on the received symbol (d) as well as the knowledge of α and β a procedure of how to evaluate a corrected symbol (r_c) is developed. Due to Eq. (3) it can be formulated

$$d = \alpha \cdot r_c + \beta \cdot r_c^* .$$
(6)

Solving the system for r_c leads to Eq. (7) which represents the formulation for the corrected symbol in time domain:

$$r_c = \frac{\alpha^* \cdot d - \beta \cdot d^*}{|\alpha|^2 - |\beta|^2} .$$
(7)

Based on the model described by Eq. (3), IQ imbalance is applied to the transmission sequence. The value of the parameter ε is chosen in a range of

0% to 20% and the parameter $\Delta\Phi$ is chosen up to 15°. By evaluating Eq. (5), an accurate estimation of the parameters α and β can be carried out over the entire UW interval. The simulation results are summarized in figure 3, showing the BER behavior for $\varepsilon = 20\%$ and $\Delta\Phi = 15°$. It gets obvious that the performance loss due to IQ imbalance is about 3 dB at a BER of 10^{-5} – this significant loss is reduced to about 0.5 dB based on the introduced algorithm, proving that the IQ imbalance is compensated almost completely. Nevertheless, figure 3 also demonstrates that this method fails for multi-path propagation. As shown, it is not feasible to estimate the parameters α and β in a multi-path environment with the algorithm described above.

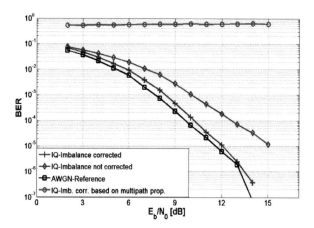

Figure 3. BER, IQ-Imbalance for an AWGN- channel.

4. ESTIMATION OF α AND β IN A MULTI- PATH ENVIRONMENT

For the following investigations we used Zadoff-Chu training sequences of a length comparable to the block-size. The sequences are composed to a preamble similar to the IEEE 802.11a standard. We adapt Eq. (3)

$$d = \alpha \cdot (r * c) + \beta \cdot (r * c)^*, \qquad (8)$$

where r now describes a long training sequence, c stands for the impulse response of the radio channel and d describes the received distorted signal sequence. Solving Eq. (8) for the parameters α and β is obviously more difficult than in the investigated AWGN case – nevertheless a significant simplification is achieved by carrying out this description in the frequency domain as follows:

$$D = \alpha \cdot R \cdot C + \beta \cdot R_m^* \cdot C_m^*. \qquad (9)$$

Note that due to the transformation of a complex conjugate the index m stands for $m = mod(N-i+2,N)$, where i describes the positive index in the FFT vector and N describes the length of the transformed sequence. If we neglect noise and the influence of IQ imbalance then the interrelationship between the frequency domain representation of the transmitted and the received long training symbols, R and D, respectively, is given by $C = (RC)/R = D/R$. Therefore $H = D/R$ is often used as a first channel estimation. H stands for the channel estimation, while C describes the real channel in the frequency domain. Substituting $H = D/R$ into Eq. (9) results in:

$$H = \alpha \cdot C + \beta \cdot \frac{R_m^*}{R} \cdot C_m^*. \qquad (10)$$

The right hand side of Eq. (10) stresses the (negative) influence of α and β. Investigations show that the estimated channel transfer function suffers from leaps due to the IQ imbalance, while the real channel transfer function is comparably smooth – this effect may be taken into account for an estimation procedure for α and β. While two adjacent taps may show significant difference for the estimated transfer function, these adjacent taps do not vary significantly for the real channel. Based on that typical behavior, the following method was developed to estimate the parameters α and β: Due to Eq. (10) a system of equations for two adjacent taps is given by

$$H_i = \alpha \cdot C_i + \beta \cdot \frac{R_{m\_i}^*}{R_i} \cdot C_{m\_i}^*$$

$$H_j = \alpha \cdot C_j + \beta \cdot \frac{R_{m\_j}^*}{R_j} \cdot C_{m\_j}^* \qquad (11)$$

The difference of H_i and H_j results in:

$$H_i - H_j = \alpha \cdot (C_i - C_j) + \beta \cdot \left(\frac{R_{m\_i}^*}{R_i} \cdot C_i^* - \frac{R_{m\_j}^*}{R_j} \cdot C_j^* \right). \qquad (12)$$

Due to the previous description, the difference between H_i and H_j will be significantly larger than the difference between C_i and C_j. Therefore the term $\alpha (C_i-C_j)$ is neglected. But Eq. (12) still comprises the unknown, real channel C. An additional simplification for the case of small values of $\Delta\Phi$ (resulting in $\sin(\Delta\Phi) \approx 0$ and $\cos(\Delta\Phi) \approx 1$) leads to the assumption that $H_{m\_i}^*$ represents an approximation for $C_{m\_i}^*$. Therefore we substitute $C_{m\_i}^*$ by $H_{m\_i}^*$ in Eq. (12). Now the equation can be solved for β:

$$\beta = \frac{H_i - H_j}{\frac{R_{m\_i}^*}{R_i} \cdot H_{m\_i}^* - \frac{R_{m\_j}^*}{R_j} \cdot H_{m\_j}^*}. \tag{13}$$

Based on the estimation for β, Eq. (3) allows to solve for α.

$$\alpha = \sqrt{1 - \Im^2\{\beta\}} - j \cdot \frac{\Re\{\beta\} \cdot \Im\{\beta\}}{\sqrt{1 - \Im^2\{\beta\}}}. \tag{14}$$

As proposed, only the transmitted (undistorted) and the received (distorted) signals are required for the estimation process according to Eqs. (13) and (14). The IQ imbalance compensation itself may be implemented in the time domain again.

Based on the derivation above a performance characterization is carried out. Figure 4 shows the BER for an exemplary radio channel (the radio channel has been modeled as a tapped delay line, each tap with random uniformly distributed phase and Rayleigh distributed magnitude and with the power delay profile decaying exponentially) with deep spectral fades (channel B) as well as for a radio channel with low spectral fading (channel A). Based on a phase offset of $\Delta\Phi = 15°$ and an amplitude offset of $\varepsilon = 20\%$, the simulation results demonstrate the efficiency of the introduced algorithm. Notice that especially at low E_b/N_0 ratios estimation errors may occur, indicated by deviations of the ideal behavior – aspects, which are topic of further research. Nevertheless, these simulations results demonstrate impressively the quality of the introduced algorithms to estimate and compensate IQ imbalance in the multi-path case.

5. SUMMARY AND CONCLUSION

Besides an efficient and powerful equalization strategy the concept of SC/FDE shows robustness against imperfect synchronization. Additionally it is the underlying blockwise structuring based on a UW which allows to implement efficient algorithms for synchronization purposes – as demonstrated in this paper. Efficient algorithms for the estimation of I/Q imbalance were developed – in time and in the frequency domain – based on a preamble as well as on the specific (UW) structure of the investigated SC/FDE system. Based on these strategies it has been demonstrated that the effect of the IQ imbalance is compensated almost completely.

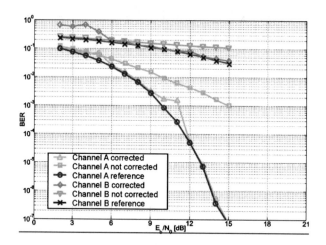

Figure 4. Parameter estimation in the frequency domain.

References

[1] H. Sari, G. Karam, I. Jeanclaude. "An Analysis of Orthogonal Frequency-Division Multiplexing for Mobile Radio Applications". *Proc. of the IEEE Vehicular Technology Conference (VTC '94)*, Stockholm, Sweden, pp. 1635-1639, June 1994

[2] D. Falconer, S.L. Ariyavisitakul, A. Benjamin-Seeyar, B. Eidson, "Frequency Domain Equalization for Single-Carrier Broadband Wireless Systems", *IEEE Communications Magazin*, Vol. 40, no. 4, pp. 58-66, April 2002

[3] N. Benvenuto, S. Tomasin, "On the Comparison between OFDM and Single Carrier Modulation with a DFE using Frequency Domain Feedforward Filter", *IEEE Transactions on Communications*, Vol. 50, no 6, pp. 947-955, June 2002

[4] H. Witschnig, T. Mayer, A. Springer, M. Huemer, "A Different Look on Cyclic Prefix for a SC/FDE System", 13th *IEEE International Symposium on Personal, Indoor and Mobile Communications*, 5 pages on CD, Lisbon, Portugal, September 2002

[5] J. Tubbax, B. Come, L. van der Perre, S. Donnay, M. Engels, "IQ Compensation for OFDM Systems", *IEEE International Conference on Communications (ICC)*, May 2003, pp. 3403-3407

[6] J. Tubbax, B. Come, L. van der Perre, S. Donnay, M. Engels, "Joint Compensation of IQ Imbalance, Frequency Offset and Phase Noise", *International Workshop on Multi Carrier Spread Spectrum and Related Topics*, Oberpfaffenhofen, Germany, Sept. 2003, pp. 243-250

[7] J. Tubbax, A. Fort, L. van der Perre, S. Donnay, M. Engels, M. Moonen, H. de Man, "Joint Compensation of IQ Imbalance and Frequency Offset in OFDM Systems", *GLOBECOM 2003*, Vol. 22, No. 1, Dec. 2003, pp. 2365-2369

ON IMPLEMENTATION ASPECTS OF UPLINK MC-CDMA MULTI-USER DETECTION

Aki Happonen[1], Franziskus Bauer[2], and Adrian Burian[3]
[1]*Nokia Technology Platforms, Elektroniikkatie 3, FIN-90570 Oulu, Finland, aki.p.happonen@nokia.com;* [2]*Nokia Research Center, Meesmannstrasse 103, D-44807 Bochum, Germany, franziskus.bauer@nokia.com;* [3]*Nokia Research Center, Visiokatu 1, FIN-33721Tampere, Finland, adrian.burian@nokia.com*

Abstract: This paper considers the implementation of multi-user detector in MC-CDMA receivers using fixed-point matrix inversion algorithms. The fixed-point word length analysis is based on the matrix condition number analysis and residual errors. The obtained bit error results have been compared to floating point matrix inversion results.

Key words: Condition Number, Implementation, Matrix Inversion, MC-CDMA, Multi-User Detection

1. INTRODUCTION

A modulation technique called multi-carrier code division multiple access (MC-CDMA) has been proposed by Yee et al. for multimedia services in high data rate wireless networks[1]. This technique has high bandwidth efficiency and combines CDMA as a multiple access technique with the orthogonal frequency division multiplexing (OFDM) as a multi-carrier transmission system. Its performances in the uplink are limited because of the multiple access interference (MAI).

In this paper we consider implementation requirements for zero-forcing (ZF) multi-user detector (MUD), emphasizing the fixed-point matrix inversion implementation needed by the detector, and its effects on the bit error rate (BER) performance.

We assume an MC-CDMA system with Walsh-Hadamard spreading in frequency domain on top OFDM. Its modulation and spreading block components are illustrated in Figure 1.

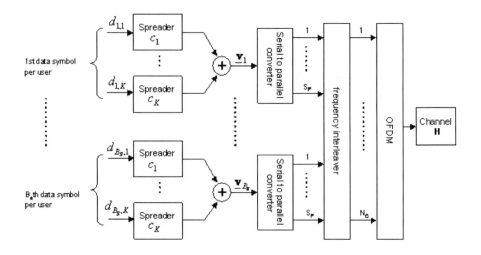

Figure 1. Modulation and Spreading blocks.

We can express the received signal **y** at the base station with

$$\underline{y} = \sum_{k=1}^{K} \underline{H}_k \underline{M}_k \underline{c}_k x_k + \underline{n},$$

where **y** is the base station received signal vector of dimension (N_cx1) with N_c being the available set of carriers for the link, $\underline{H}_k$(N_cxN_c) is a diagonal matrix collecting the user specific channel response, $\underline{M}_k$(N_cxN_a) is the frequency multiplexing matrix remarking the specific set of carriers assigned to the user with N_a the number of available carriers per user, $\underline{c}_k$(N_{SF}x1) is a vector with N_{SF} the spreading factor (SF) and x_k represents the transmitted symbol[2]. For initial equalization we have used a maximum ratio combining (MRC) for each user and ZF estimator in multi-user detection. The MRC equalization coefficients $g_{n,k}$ are given by

$$g_{n,k} = \frac{h_{n,k}^*}{\sum_{m=0}^{SF-1} h_{m,k} h_{m,k}^*},$$

where $h_{n,k}$ is the *nth* channel coefficient, and appears on the diagonal of the channel matrix $\underline{H}_k$, and $h_{n,k}^*$ is its complex conjugate. The interference $R_{k,l}$ generated by one user *l* on the equalized symbol r_k of user *k* is computed using

$$R_{k,l} = x_l \frac{\sum_{m=0}^{SF-1} h_{m,k}^* h_{m,l} c_{m,k} c_{m,l}}{\sum_{m=0}^{SF-1} h_{m,k} h_{m,k}^*}.$$

These interference terms are collected in an interference matrix $\underline{R}$ with the structure

$$\underline{R} = \begin{pmatrix} 0 & R_{1,2} & \cdots & R_{1,SF-1} \\ R_{2,1} & 0 & & \\ & & \cdots & \\ R_{SF-1,1} & & & 0 \end{pmatrix}.$$

The received estimated data vector $\hat{\underline{x}}$ for K users equals

$$\hat{\underline{x}} = (\underline{I} + \underline{R})\underline{x} + \underline{\tilde{n}} = \underline{A}\underline{x} + \underline{\tilde{n}} \qquad (1)$$

The ZF MUD multiplies the received and equalized signal vector $\hat{\underline{x}}$ with the inverse matrix $\underline{A}^{-1}$ to yield a new estimate $\hat{\underline{x}}_2$ with

$$\hat{\underline{x}}_2 = \underline{A}^{-1} \hat{\underline{x}} = \underline{x} + \underline{A}^{-1} \underline{\tilde{n}} \qquad (2)$$

The key role in the implementation is played by the required matrix inversion (eq. (2)), or by the need to solve the system of equations (1). In this paper we will consider the results we have obtained when implementing the matrix inversion in fixed-point using Cholesky decomposition.

2. MATRIX INVERSION AND CONDITION NUMBER

By using Cholesky decomposition, any nonsingular matrix $\underline{A}(n \times n)$ can be factored in the form of a product of two triangular matrices, if it is positive definite. We note that in order to be able to apply Cholesky decomposition, the positive definitiveness of input matrices is a necessary condition. Otherwise, two supplementary matrix multiplications are needed: if $\underline{B}$ is a general dense nonsingular matrix then $\underline{A} = \underline{B}^T\underline{B}$ is positive definite, and $\underline{B}^{-1} = (\underline{B}^T\underline{B})^{-1} \underline{B}^T = \underline{A}^{-1}\underline{B}^T$. So, if $\underline{A}$ is positive definite, then it can be factored in the form $\underline{A} = \underline{L}^T\underline{L}$, where $\underline{L}$ is lower triangular. If, in addition, we require the diagonal elements of $\underline{L}$ to be positive, the decomposition is unique and is called the Cholesky decomposition or the Cholesky factorization of matrix $\underline{A}^3$.

The complexity and stability of numerical algorithms are usually estimated in terms of the problem instance dimension and of a 'condition number'. So, the complexity of solving an $n \times n$ linear system $\underline{\mathbf{A}}\underline{\mathbf{x}} = \underline{\mathbf{b}}$ is usually estimated in terms of the dimension n and of the condition number of matrix $\underline{\mathbf{A}}$. The condition number (with respect to the 2-norm) of a matrix $\underline{\mathbf{A}}$ with respect to inversion is defined as

$$\kappa(\underline{\mathbf{A}}) = \|\underline{\mathbf{A}}\| \cdot \|\underline{\mathbf{A}}^{-1}\|.$$

3. ARCHITECTURE OVERVIEW

In this section, we briefly introduce implementation of the used matrix inversion engine. This engine is implemented using the reconfigurable processing element illustrated in Figure 2. The used architecture has fixed latency for following mathematical operations: addition, subtraction, multiplication, division root and square root. Additional to these fundamental operations, conjugated equations, e.g. $\mathbf{C} = \mathbf{A}/\mathbf{B}^{1/2}$, can be executed every clock cycle with at most three clock cycles latency.

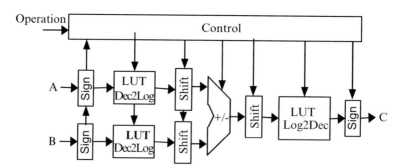

Figure 2. Architecture block diagram.

The processing element architecture mixes logarithm and decimal arithmetic. Look-up-tables (LUT) are used to perform conversion between decimal and logarithm scale. By using the logarithmic arithmetic for multiplication and division, the implementation uses only adders. The implementation of square and square root uses only shifting operations. Adding and subtracting operations are done in decimal arithmetic and for these cases the LUTs are bypassed. The used architecture and its implementation details are presented in more detailed in[4].

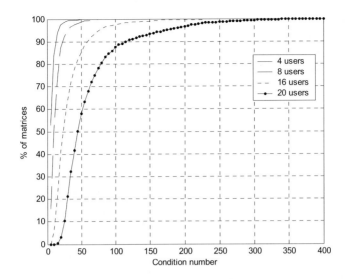

Figure 3. Condition number distribution for different number of users.

Table 1. Minimum, maximum, and mean for condition numbers of different number of users.

Number of users	4 users	8 users	16 users	20 users
Min	1.15	2.16	5.01	12.13
Max	122.88	90.71	270.91	398.33
Mean	6.52	12.41	31.11	61.18

Table 2. Residual errors in matrix inversion for different number of users.

Error / Number of users	4 users	8 users	16 users	20 users
ε_0	100	100	88.0	19.4
ε_1	99.6	99.0	55.7	5.1
ε_2	98.7	94.2	12.9	0
ε_3	96.4	79.5	0.4	0
ε_4	84.4	47.8	0	0
ε_5	43.0	5.3	0	0
ε_6	1.25	0		

4. SIMULATION RESULTS

As a first stage of our work, we have analyzed the practical condition numbers of the matrices to be inverted. We are targeting the most economical and fast fixed-point implementation, but with acceptable performances for the overall system. The influences of using a fixed-point implementation with different word-lengths are compared by computing the obtained bite error rates (BERs) in different practical situations. We have used these obtained condition numbers to establish initial values for the word length. The obtained practical condition number distribution of matrix **A** when having 4, 8, 16, and 20 users are shown in Figure 3. The statistical properties of the obtained condition numbers are given in Table 1. In this table, the minimum, maximum and average values of the condition numbers for different number of users are given.

The quality of an approximation **Y** ≈ **A**$^{-1}$ can be assessed by looking at the right and left residuals, *AY* — *I* and *YA* — *I*. For reporting the results, only the right residual *Z*=|*A*Y-I*| has been used, because since our input matrices are positive definite, the left residual is not needed. The computation has been considered successful if the 2-norm of the residual is less than a predefined error level. The used error levels have been $\varepsilon_k = 2^{-k}$, k = 0,1,…,5. Table 2 shows residual errors as percentages of successful matrix inversions. Based on our previous results[5,6] and the obtained condition number distribution, we have decided to start by using a 16 bits fixed-point implementation. Also, for this word-length other algorithms besides Cholesky decomposition have been implemented and tested. We have also tested the more stable but also more complex and slow Newton's iteration algorithm. We showed[6] that this iterative algorithm offers more stability to matrix inversion, when the used word-lengths have small values.

For 4, 8, 16 and 20 users we compare the bit error rates with fixed-point matrix inversion implementation to the bit error rates of floating point multi-user detection with ZF and with single user detection. The obtained BER results for different number of users are presented in Figures 4-7. The parameters for the simulation are shown in Table 3. Our results show that for low number of users MUD does not improve BER performance compared to SUD. For higher number of users MUD starts to outperform SUD. Our 16-bits implementation for matrix inversion starts to introduce noise floor for in higher SNR conditions. These residual errors become visible when lower number of matrix inversions may be considered to be successful. For higher number of users 16-bits implementation does not fulfill overall system requirements. In these cases overflow errors occurs due the higher number of elements in matrices.

Table 3. Parameters for BER simulation.

FFT size	1024 points	Mobile speed	60km/h
Sampling frequency	57.6 MHz	Number of users K	1…32
Slot size	32 OFDM symbols	Modulated carriers	736
Guard Interval	216 chips	Combining type	MRC
Spreading factor SF	32	Channel coding	turbo code (R=1/2)
Multipath Channel	BRAN E	Multi User Detection	Zero Forcing

Implementation Aspects of Uplink MC-CDMA Multi-User Detection   415

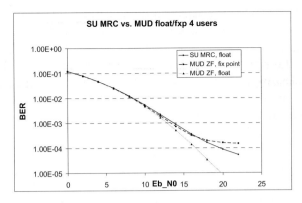

Figure 4. BER of fixed- and floating-point implementations (4 users).

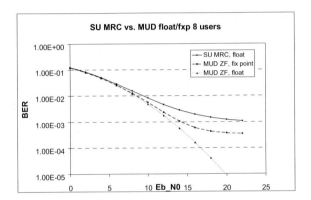

Figure 5. BER of fixed- and floating-point implementations (8 users).

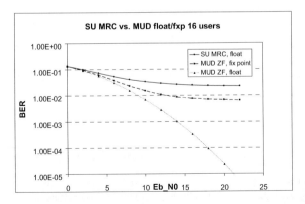

Figure 6. BER of fixed- and floating-point implementations (16 users).

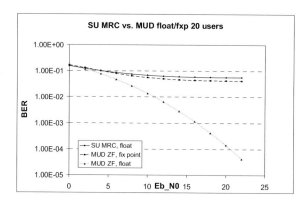

Figure 7. BER of fixed- and floating-point implementations (20 users).

5. CONCLUSIONS

Implementation issues of Uplink MC-CDMA Multi-User Detector have been considered. A fixed-point implementation of required matrix operations is considered. Our preliminary analysis has shown that word-lengths greater than 16 bits are needed if acceptable values for BER performance are to be obtained. The dynamic range of the involved signals is so great, that only the usage of floating point implementations (32 bits) or fixed point with quite large word-lengths (at least double when compared to floating point) do not decrease the overall performance of the system.

REFERENCES

1. Yee N., Linnartz J.P., Fettweis G., "Multicarrier CDMA in indoor wireless radio networks", Proceedings of IEEE PIMRC'93, pp 109-113, Sep. 1993, Yokohama, Japan.
2. S. Kaiser, Multi Carrier CDMA Mobile Radio Systems – Analysis and Optimisation of Detection, Decoding and Channel Estimation. PhD Thesis published with VDI Verlag, Düsseldorf, Germany, 1998.
3. N.J. Higham, "Accuracy and Stability of Numerical Algorithms", SIAM, Philadelphia, 1996.
4. A. Happonen, A. Burian and E. Hemming, "A Reconfigurable Processing Element Implementation for Matrix Inversion Using Cholesky Decomposition", Proceedings of the International Conference on Information Technology, ICIT2004, pp. 233-236. December 17-19, 2004, Istanbul, Turkey
5. A. Happonen, E. Hemming, and M.J. Juntti, "A novel coarse grain reconfigurable processing element architecture", Proceedings of the 46th International Midwest Symposium on Circuits and Systems MWSCAS, December 27-30, Cairo, Egypt.
6. M. Ylinen, A. Burian, and J. Takala, "Direct versus iterative methods for fixed-point implementation of matrix inversion", Proceedings of the 2004 International Symposium on Circuits and Systems, vol. 3, 23-26 May 2004, pp. III - 225-8.

ANALYTICAL PERFORMANCE OF A FREQUENCY OFFSET MULTI-USER MULTI- ARRAY SYSTEM

Adrien Renoult[1,2], Inbar Fijalkow[1] and Marc Chenu-Tournier[2]

[1]*ETIS/ENSEA - Univ de Cergy-Pontoise - CNRS*
6 av du Ponceau, 95014 Cergy-Pontoise France
[2]*Thales Communications*
160 bld de Valmy BP 92, 92704 Colombes Cedex, France
adrien.renoult@fr.thalesgroup.com, fijalkow@ensea.fr, marc.chenu@fr.thalesgroup.com

1. Introduction

Orthogonal Frequency Division Multiplexing (OFDM) modulation is a technique commonly used to deal with scattering environment. This is mainly due to the fact that it does not involve any complicated channel equalization, since each sub-carrier undergoes a flat fading channel. However this modulation is very sensitivity to frequency offset in the channel. The frequency offset causes two bad effects: the first is the reduction of the signal amplitude of each sub-carrier and the second is introduction of Inter Carrier Interferences (ICI) from the others carriers which are now no longer orthogonal. In a multi-users system the different transmitters got a independent local oscillator, then each transmitter is affected by a different frequency offset (in comparison to the receiver). The usual solution to deal with this problem in the single user case is to lock the frequency of the transmitter and the receiver using a phase lock loop. But in the case of multi-users transmission we cannot do it this way.

The aim of this paper are twofold. First we first give an improved multi-users model including the frequency offset effects. Then using this model we derive the analytical performance at the channel decoder output.

This paper is subdivided in three parts. First we describe our system. This model takes into account the effect of the frequency offset. Then by derivation of this model we describe the effect of the frequency offset as a noise floor, and we explain the analytical performance of our system according to the frequency offset of the different transmitters. In a

2. Model of multi-users system with frequency offset

In this section we propose a model for a multi-users system with frequency offset. This model is a non-equipower multi-users system over Rayleigh fading channel with additive white Gaussian. In this model the power of a given user depends of the channel realization of this user and the amount of frequency offset of this user. The power of the noise depend of the power of the thermal noise the channel realization of every users and the frequency offset of the users.

In the case of frequency offset between the different users the model is then:

$$\mathbf{y}(k,n) = \sum_{s=1}^{N_T} x_s(k,n) \mathbf{h}_s(k,n) \frac{\sin(\pi(\Delta f_s))}{\sin(\frac{\pi}{N}(\Delta f_s))}$$

$$+ \sum_{s=1}^{N_T} \sum_{\substack{i=0 \\ i \neq k}}^{K-1} x_s(i,n) \mathbf{h}_s(i,n) \frac{\sin(\pi(k + \Delta f_s - i))}{\sin(\frac{\pi}{N}(k + \Delta f_s - i))} + \mathbf{b}(k,n) \quad (1)$$

and we want to arrive to the model

$$\mathbf{y}(k,n) = \sum_{s=1}^{N_T} x_s(k,n) \mathbf{h}_s(k,n) \frac{\sin(\pi(\Delta f_s))}{\sin(\frac{\pi}{N}(\Delta f_s))} + \mathbf{n}(k,n)$$

To do this we will prove that $\mathbf{n}(k,n)$ is a sum of complexe Gaussian process. By definition $\mathbf{b}(k,n)$ and $\mathbf{h}_s(i,n)$ $\forall (i,n)$ are complexe Gaussian process. Then as $\frac{\sin(\pi(k+\Delta f_s - i))}{\sin(\frac{\pi}{N}(k+\Delta f_s - i))}$ is a deterministic process then $\mathbf{h}_s(i,n) \frac{\sin(\pi(k+\Delta f_s - i))}{\sin(\frac{\pi}{N}(k+\Delta f_s - i))}$ $\forall (i,n)$ are Gaussian processes too. Then we prove than $\mathbf{n}(k,n)$ as being a sum of complexe Gaussian process is a Gaussian process. It is straightforward to calculated the power of the noise. Using the fact that: $\sum_{i=0}^{N-1} \left(\frac{\sin(\pi(k+\Delta f_s - i))}{\sin(\frac{\pi}{N}(k+\Delta f_s - i))} \right)^2 = 1$ we obtain that the power of the noise is:

$$\sigma_i^2 = \frac{1}{N_T} \sum_{s=1}^{N_T} \left(1 - \left(\frac{\sin(\pi(k + \Delta f_s - i))}{\sin(\frac{\pi}{N}(k + \Delta f_s - i))} \right)^2 \right) + \sigma^2$$

We then obtain the model:

$$\mathbf{y}(k,n) = \sum_{s=1}^{N_T} x_s(k,n) \mathbf{h}_s(k,n) \frac{\sin(\pi(\Delta f_s))}{\sin(\frac{\pi}{N}(\Delta f_s))} + \mathbf{n}(k,n) \quad (2)$$

In this model we consider the noise $\mathbf{n}(k)$ as constituted of the thermal noise and of the intercarriers interfrence effect of the frequency offset. Under some asymptions, $\mathbf{n}(k)$ is a white Gaussian noise of power σ_i^2.

When considering the model (2) without frequency offset we obtain the usual flat fading model:

$$\mathbf{y}(k,n) = \sum_{s=1}^{N_T} \mathbf{h}_s(k,n) x_s(k,n) + \mathbf{b}(k,n) \quad (3)$$

We notice that in the contrary of model (3) in model (2) the noise is correlated with the signal of interest.

3. Performance of multi-users system with frequency offset

In this part we want to derive the analytical performance for this system where there is a frequency offset. We derive the probability of error at the Viterbi decoder output. To this aim we use the Viterbi bound $P_b \leq \sum_{d=d_{\text{free}}}^{\infty} c_d P_d$, where c_d equals the mean of the bit error for all the paths at Hamming distance d from the correct path. It depends only on the code and can be find using the generating function of the code and can be found in [5]. We have to calculated the probability P_d of choosing a wrong path at Hamming distance d. In the model (2) thanks to the interleaver we can consider that the channel realization is independent from two following symbol in the path. But the noise is now not independent of the channel realization. However to allowed the derivation of P_d we will consider the noise as independent. Then by extended the result of [3] we first derived the the expression of P_d. P_d is the probability of choosing a wrong sequence at Hamming distance d from the transmited sequence. The event is equivalent to the statement: the incorrect sequence is preferably choosen by the channel decoder instead of the transmited sequence. The event "a sequence i is choosen instead of the transmited sequence" is equivalent to the fact that the metric for the sequence i is smaller than the metric for the transmited sequence. We will note this event $D_i \leq 0$ where D_i is the metric difference between the sequence i and the correct sequence (the all one sequence). Then we obtain that the event "choosing one of the wrong sequences at Hamming distance d" is equivalent to:

$$\bigcup_{i \in \mathcal{I}_d} (D_i \leq 0) \qquad (4)$$

where $\mathcal{I}_d$ represents all sequences of Hamming distance d for the user one. Note that the Hamming distance can be different for the interfering users. To obtain the probability P_d, we have to find the probability of the event (4) averaging over the channel statistics. We then get:

$$P_d = \int_{\mathbf{H}} P\left(\bigcup_{i \in \mathcal{I}_d} (D_i \leq 0) \,|\, \mathbf{H}\right) p(\mathbf{H}) \mathrm{d}\mathbf{H} \qquad (5)$$

Using the union bound $(P(\bigcup_i u_i) \leq \sum_i P(u_i))$, we obtain:

$$P_d \leq \sum_{i \in \mathcal{I}_d} \int_{\mathbf{H}} P(D_i \leq 0 \,|\, \mathbf{H}) p(\mathbf{H}) \mathrm{d}\mathbf{H} \qquad (6)$$

To continue the derivation of P_d, we must consider two steps. The first step consists in finding the expression of D_i and the second step consists in integrating the probability over the different channel realisations. The expression of D_i does not depend on the number of users and is provided at the end of this section. The derivation of the expression of equation (6) depends on the number of users and is derived in section 3 for the single-user case and in section 4 for the multi-user case.

$D_i \leq 0$ is defined as choosing the wrong sequence i according to a maximum likelihood detector (Viterbi algorithm). This algorithm introduced by [4] finds a sequence through a code treillis. The chosen sequence through the trellis is the one with the maximum probability. Given $\mathcal{C}_i$ the couples of indexes (k, n) of the sequence i through the treillis, $\{\mathbf{y}(k, n)\}_{(k,n) \in \mathcal{C}_i}$ is the received signal, $\{\mathbf{H}(k, n)\}_{(k,n) \in \mathcal{C}_i}$ is the channel and $\{\mathbf{x}^i(k, n)\}_{(k,n) \in \mathcal{C}_i}$ is a sequence through the treillis (the coded bits corresponding to this sequence). To have the best performance each channel coefficient associated with each couple has to be as uncorrelated as possible from its neighbours. We assume this is true when using a interleaver. Then the Viterbi algorithm choses the path $\{\mathbf{x}^i(k, n)\}_{(k,n) \in \mathcal{C}_i}$ that maximizes:

$$P_r\left(\{\mathbf{y}(k, n)\}_{(k,n) \in \mathcal{C}_m}, \{\mathbf{H}(k, n)\}_{(k,n) \in \mathcal{C}_m} \,\Big|\, \{\mathbf{x}^m(k, n)\}_{(k,n) \in \mathcal{C}_m}\right)$$

Because of the hypothesis on the stastical independence between the different couples (k, n), this is equivalent to find the maximum over all m of:

$$\sum_{(k,n)\in \mathcal{C}_m} \log\left(P_r\left(\mathbf{y}\left(k,n\right), \mathbf{H}\left(k,n\right) | \mathbf{x}^m\left(k,n\right)\right)\right) \qquad (7)$$

where $\mathcal{C}_m$ is the ensemble of the couples (k, n) corresponding to the sequence m. Under a maximum likelihood criteria and by defining D_i as the difference of metric between the sequence i and the all one sequence:

$$D_i = \sum_{(k,n)\in \mathcal{C}_i} \left\| \mathbf{y}\left(k,n\right) - \mathbf{H}\left(k,n\right) \mathbf{x}^i\left(k,n\right) \right\|^2 - \sum_{(k,n)\in \mathcal{C}_i} \left\| \mathbf{y}\left(k,n\right) - \mathbf{H}\left(k,n\right) \mathbf{1} \right\|^2 \qquad (8)$$

or:

$$D_i = \sum_{(k,n)\in \mathcal{C}_i} \left\| \mathbf{y}(k,n) - \sum_{t=1}^{N_T} \mathbf{h}_t(k,n) x_t^i(k,n) \right\|^2 - \sum_{(k,n)\in \mathcal{C}_i} \left\| \mathbf{y}(k,n) - \sum_{t=1}^{N_T} \mathbf{h}_t(k,n) 1 \right\|^2 \qquad (9)$$

D_i can be expressed as a Hermitian quadratic form as:

$$D_i = \sum_{(k,n)\in \mathcal{C}_i} \sum_{r=1}^{N_R} \mathbf{z}_r^\dagger(k,n) \mathbf{F}_i(n,k) \mathbf{z}_r(k,n) \qquad (10)$$

with $\mathbf{z}_r(k,n) = (y_r(k,n), h_{1,r}(n,k), \cdots, h_{N_T,r}(n,k))^t$ and $\mathbf{F}_i(n,k)$ the rank two matrix defined by $\mathbf{u}_i(n,k)\mathbf{u}_i(n,k)^\dagger - \mathbf{1}\mathbf{1}^\dagger$ where $\mathbf{u}_i(n,k) = (1, -x_1^i(n,k), \cdots, -x_{N_T}^i(n,k))^t$ and $\mathbf{1} = (1, -1, \cdots, -1)^t$. As the channel stastics in $\mathbf{z}_r(k,n)$ and the data statistics in $\mathbf{F}_i(n,k)$ are independant. Each term of D_i is a Hermitian quadratic form of the N_T+1 zero-mean complex Gaussian variables in $\mathbf{z}_r(k,n)$.

To calculate the probability P_i we have now to derive the probability for D_i to be a negative valued. The statistics of quadratic forms as in (10) knowing the ratio of the two eigen values of $\mathbf{R}\mathbf{F}_i$ ($\mathbf{R} = \frac{1}{2}E\left(\mathbf{z}_r(k,n)\mathbf{z}_r^\dagger(k,n)\right)$), this ratio is obtained using the Bôcher formula [1]. $\mathbf{F}_i(n,k)$ has to be independent of the couple (n,k) and then we will have to limit the path i to the one where $\mathbf{F}_i(n,k)$ is constant for $(n,k) \in \mathcal{B}_i$.

Assuming the interfering users are erroneously decoded only if the user of interest is erroneously decoded and $\mathbf{F}_i(n,k)$ constant regarding

$(k,n) \in \mathcal{B}_i$,

$$P_i = \frac{1}{(1-\mu_i)^{2dN_R-1}} \sum_{r=0}^{dN_R-1} \binom{2dN_R-1}{r}(-\mu_i)^r \qquad (11)$$

with $\mu_i = \frac{a_i\gamma + \sqrt{b_i\gamma^2 + 2a_i\gamma}}{a_i\gamma - \sqrt{b_i\gamma^2 + 2a_i\gamma}}$,

$a_i = \sum_{s=1}^{N_T} \alpha_s^2 (1 - x_s^i)$,

$b_i = 2\sum_{s=1}^{N_T}\sum_{t=1}^{N_T} \alpha_s^2 \alpha_t^2 (1-x_s^i) + \left(\sum_s^{N_T} \alpha_s^2 x_s^i\right)^2 - \left(\sum_s^{N_T}\alpha_s^2\right)^2$,

$\gamma = \frac{E_b}{\sum_{s=1}^{N_T}(1-\alpha_s^2) + N_0}$ and $i \in \mathcal{J}_d$. x_t^i is the symbol of user t for the path i.

The derivation of the expression of P_d is then different form what is presented in [3] because of the non equipower of the different users. In the final paper we will explain how by using the result of [2] of performance of non equipower multi-user system we will obtain that:

$$P_b \leq \sum_{d=d_{free}}^{\infty} c_d \sum_{i \in \mathcal{J}_d} \frac{1}{(1-\mu_i)^{2dN_R-1}} \sum_{r=0}^{dN_R-1} \binom{2dN_R-1}{r}(-\mu_i)^r \qquad (12)$$

with $\mu_i = \frac{a_i\gamma + \sqrt{b_i\gamma^2 + 2a_i\gamma}}{a_i\gamma - \sqrt{b_i\gamma^2 + 2a_i\gamma}}$,

$a_i = \sum_{s=1}^{N_T} \alpha_s^2 (1 - x_s^i)$,

$b_i = 2\sum_{s=1}^{N_T}\sum_{t=1}^{N_T} \alpha_s^2 \alpha_t^2 (1-x_s^i) + \left(\sum_s^{N_T} \alpha_s^2 x_s^i\right)^2 - \left(\sum_s^{N_T}\alpha_s^2\right)^2$,

$\gamma = \frac{E_b}{\sum_{s=1}^{N_T}(1-\alpha_s^2) + N_0}$ and $i \in \mathcal{J}_d$. x_t^i is the symbol of user t for the path i.

4. Simulation

The figure 4 give the comparison between the analytical performance and the simulations for the single-user single-receiver case. This performance are for a delta of frequency of 20% and using convolutional code of generator polynomial (5,7). This figure highlight the high accuracy between the analytical performance and the simulations.

The figure 1 give the comparaison between the analytical performance and the simulations for the two-users two-receivers case. This performance are obtained for a delta of frequency of 35% for the first user and 30% for the second user. Both users use a independant convoulutional code of generator polynomial (5,7). In this figure we higlight the gap between the analytical performance and the simulations. This gap are

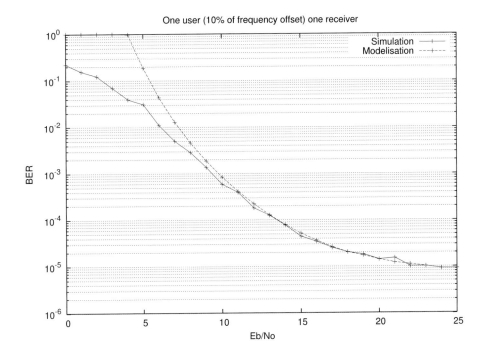

Figure 1. Effects of frequency offset on a single user system

caused by the differents assuption we made in the aim of deriving the performance.

5. Conclusion

In this paper we obtain the analytical performance of a multi-user system affected by frequency offset. The aim of deriving this performance are to shortenning the simulation time. The analytical performance during this paper are presented for the single-user and the multi-user case. We then qualify the accuracy of our analytical performances using monte-carlo simulations.

In the single-user we higlight the high accurancy between the analytical performance and the simulation.

In the multi-user case, we obtain an expression of the BER at the decoder output but in this case the analytical expression does not match perfectly with the simulations results. This could be caused by the two union bounds necessary to derive the analytical performance or to the simplyfing assumption we made in the derivation.

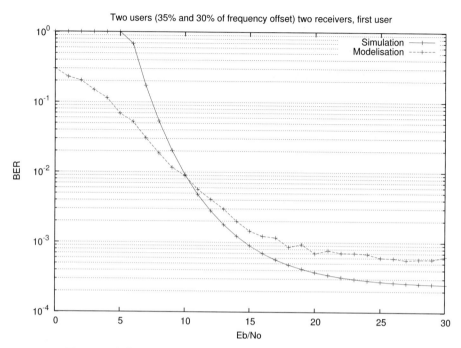

Figure 2. Effects of frequency offset on a multi users system

References

[1] P. DeRusso, R. Roy and C. Close *State variables for engineers* Wiley, 1965
[2] Stephen J. Grant and James K. Cavers. Performance enhancement through joint detection of cochannel signals using diversity arrays. *IEEE Transactions on Communications* August 1998
[3] Adrien Renoult, Inbar Fijalkow and Marc Chenu-Tournier On the performances of coded OFDM multi-users system without orthogonal access. *Submitted to IEEE Transactions on Signal Processing* 2005
[4] Andrew J. Viterbi. Error bounds for convolutional codes and a asymptotically optimum decoding algorithm *IEEE Transcation on Information Theory* April 1967
[5] Andrew J. Viterbi and J. K. Omura *Principles of digital communications* Addison-Wesley 1979

EFFECTS OF SUBCARRIER INTERLEAVING ON LDPC CODED MC-CDMA SYSTEMS

Yusung Lee, Kwanghoon Kim, and Hyuncheol Park
Information and Communications University (ICU)
119 Munjiro, Yuseong-gu, Daejeon, 305-732, Korea
{diotima, hoon0217, hpark}@icu.ac.kr

Dongseung Kwon
Electronics and Telecommunications Research Institute (ETRI)
161, Gajeong-dong, Yuseong-gu, Daejeon, 305-350, Korea
dskwon@etri.re.kr

Abstract This paper describes the effect of subcarrier interleaving on low density parity check (LDPC) coded multi-carrier code division multiple access (MC-CDMA) system. Subcarrier interleaving in MC-CDMA system can achieve the randomization effect of burst error in frequency domain, but it generates the distortion among the spreading codes due to the destruction of orthogonality. In channel-coded system, however, the achievable frequency diversity effect associated with the subcarrier interleaving in frequency domain also depends on the channel coding gain. So, there is a trade off between frequency diversity gain via subcarrier interleaving or channel coding, and destruction of orthogonality. From the simulation results, frequency diversity gain by channel coding is more effective than subcarrier interleaving, especially for system with high order modulations and with low code rates.

1. Introduction

Multicarrier-code division multiple access (MC-CDMA) is a multiplexing technique that combines orthogonal frequency division multiplexing (OFDM) with CDMA. Due to its advantages such as high spectral efficiency and robustness to frequency selectivity, MC-CDMA has been considered as one the most promising candidates for mobile radio communications [1]. In MC-CDMA system, modulated signals of each user are spread by user-specific spreading code. Spreading is done in the frequency domain and the spread data symbols are assigned to allocated subcarriers and transmitted.

For obtaining the diversity gain, MC-CDMA may use subcarrier interleaving technique. Subcarrier interleaving in MC-CDMA system can achieve the frequency diversity gain by making the randomization effect of burst error in frequency domain. If transmission channel offers the frequency selectivity fully, MC-CDMA can nearly achieve the diversity order of the number of chips in the single user case [2] [3]. In channel coded MC-CDMA system, it can also get the frequency diversity gain through the channel coding. Therefore, the achievable frequency diversity effect associated with the subcarrier interleaving depends on the channel coding gain.

To reduce the orthogonality degradation, effective data detection techniques are needed. Minimum mean square error (MMSE) equalization, which reduces the intercode interference (or multiple access interference, MAI) and utilizes the diversity of the frequency selective channel, is a good solution for single-user detection [4]. MMSE equalization can not remove the intercode interference completely. Therefore, during the MMSE equalization, orthogonality loss caused by the randomization of transmission channel due to subcarrier interleaving operates as interferences in the channel decoding process. Therefore, a thorough understanding of the relation between frequency diversity gain by subcarrier interleaving or channel coding and destruction of orthogonality is required.

In [5], three interleaving methods for variable spreading factor-orthogonal frequency and code division multiplexing (VSF-OFCDM) system employing turbo coding have been fully investigated. In [6], the relation between channel coding and spreading as comparing MC-CDMA system with multicarrier-frequency division multiple access (MC-FDMA) system is shown and in [7], the trade-off between spreading and channel coding in OFDM-code division multiplexing (OFDM-CDM) system is shown.

In this paper, we describe the diversity effects of subcarrier interleaving on MC-CDMA system employing low density parity check (LDPC) codes with different code rates.

This paper is organized as follows. The system model and channel model are described in Section 2. In Section 3, we introduce the MMSE detection technique and the derivation of LLRs for LDPC coded MC-CDMA system with MMSE detection. Section 4 gives simulation environments for evaluation and the simulation results are presented. Finally, we conclude the paper in Section 5.

2. System Model

We consider a downlink MC-CDMA system in which the total number of subcarriers, N_c, is divided into the spreading code length, K [3]. As shown in Figure 1, each user transmits simultaneously M data symbols per OFDM symbol and each individual data symbol is spread by user-specific spreading code, $\mathbf{c}^{(q)}$, for which Walsh-Hadamard code is used. The total number of subcarrier is $N_c = M \times K$. After the frequency domain spreading, K chips are interleaved chip-by-chip to assign the successive chips in the frequency domain to separated subcarriers. The subcarrier interleaving pattern is described in Figure 1.

At the receiver as shown in Figure 2, the incoming signal is first sampled with period T_s and the cyclic prefix is removed. For notational simplicity we ignore subcarrier interleaving process. Assuming that the fading on each subchannel is flat and the guard interval duration exceeds the delay spread of the multipath channel, the discrete Fourier transform (DFT) output of the m-th block, $\mathbf{y}_m = [y_{m,1}, y_{m,2}, \ldots, y_{m,K}]^T$, can be represented as:

$$\mathbf{y}_m = \mathbf{H}_m \mathbf{C} \mathbf{d}_m + \mathbf{n}_m \tag{1}$$

where $K \times K$ diagonal matrix $\mathbf{H}_m$ represents frequency response over each subcarrier of the m-th block in which channel response, $H_{m,k}$, over the k-th subcarrier of the m-th block is:

$$H_{m,k} = \sum_{l=1}^{L} h_m(l) e^{-j2\pi(l-1)i_k/N}, \tag{2}$$

where $h_m(l)$ is the complex fading envelope of the l-th path, L represents the maximum channel length sampled by T_s, and i_k, $\{i_k : 1 \leq k \leq K\}$, is the subchannel index in which k-th spread data symbol of each block is transmitted. $\mathbf{C} = [\mathbf{c}^{(1)}, \mathbf{c}^{(2)}, \ldots, \mathbf{c}^{(Q)}]$ is $K \times Q$ matrix that consists

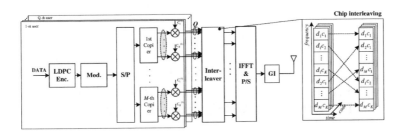

Figure 1. Block diagram of LDPC coded MC-CDMA transmitter with subcarrier interleaving.

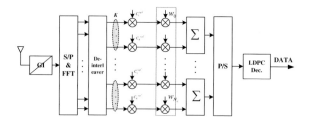

Figure 2. Block diagram of LDPC coded MC-CDMA receiver for user q.

of spreading code vector of each user with elements, $c_i^{(q)} \in \{\frac{1}{\sqrt{K}}, -\frac{1}{\sqrt{K}}\}$, and $\mathbf{d}_m$ is a Q-dimensional vector consisting of the channel coded and modulated data symbol of each user. Finally, the vector $\mathbf{n}_m$ is a white Gaussian noise with zero mean and variance σ_n^2 on K subcarriers.

3. MC-CDMA with LDPC channel coding

In MC-CDMA systems, due to the frequency and time selective fading caused by time-variant multipath channel, all the received subcarriers have different amplitude levels and different phase shifts, which result in the distortion of the orthogonality among code-multiplexed user data. Therefore, efficient data detection techniques have to be applied to mitigate the distortion caused by the fading channel.

In this paper, we use MMSE equalization for data detection because it can reduce the multiple access interference and can utilize the diversity of the frequency selective channel [4]. For notational simplicity we ignore the block index m. From (1), the detection process can be expressed as:

$$\mathbf{u} = \mathbf{Wy} = \mathbf{WHCd} + \mathbf{Wn} \qquad (3)$$

where $\mathbf{W}$ is $K \times K$ diagonal matrix which gives equalization coefficients. An MMSE equalization coefficient W_k, the diagonal elements of $\mathbf{W}$, is calculated as [3]:

$$W_k = \frac{H_k^*}{|H_k|^2 + 1/\gamma_c} \qquad (4)$$

where γ_c is actual signal to noise ration (SNR) per subcarrier and $(\cdot)^*$ denotes complex conjugate. Finally, at combining output, the data symbol of user q can be written as:

$$z^{(q)} = \mathbf{c}^{(q)^H}\mathbf{u} = \mathbf{c}^{(q)^H}\mathbf{WHCd} + \mathbf{c}^{(q)^H}\mathbf{Wn} \qquad (5)$$

where $(\cdot)^H$ denotes the conjugate transpose.

From (4) and (5), we can obtain the soft values for log-likelihood ratio (LLR) for LDPC coded MC-CDMA system with MMSE detection. The LLR of the ν-th coded bit of the q-th user data symbol, $b_\nu^{(q)}$, can be calculated as:

$$\Lambda(b_\nu^{(q)}) = log\left(\frac{Pr(z^{(q)}|b_\nu^{(q)}=1)}{Pr(z^{(q)}|b_\nu^{(q)}=0)}\right)$$
$$= log\left(\frac{\sum_{d\in\varphi^{-1}(b_\nu^{(q)}=1)} exp\left[-\frac{1}{2\sigma^2}\left(z^{(q)}-\frac{1}{K}\sum_{k=1}^{k=K} H_k W_k d\right)^2\right]}{\sum_{d\in\varphi^{-1}(b_\nu^{(q)}=0)} exp\left[-\frac{1}{2\sigma^2}\left(z^{(q)}-\frac{1}{K}\sum_{k=1}^{k=K} H_k W_k d\right)^2\right]}\right) \quad (6)$$

where φ is modulation function, W_k is MMSE detection coefficient defined in (4), and the variance of MAI plus noise, σ^2 can be obtained according to (5) [3]:

$$\sigma^2 = \sigma_{MAI}^2 + \sigma_{noise}^2$$
$$= (Q-1)\left(\frac{1}{K^2}\sum_{k=1}^{K}|W_k H_k|^2\right) + \frac{\sigma_n^2}{2K}\sum_{k=1}^{K}|W_k|^2. \quad (7)$$

4. Simulation Results

Computer simulations have been carried out to analyze the effects of subcarrier interleaving on LDPC coded system with MMSE detection in terms of bit error rate (BER). In the simulations, as a channel model, we consider the Rayleigh faded exponential decay model. We assume that channel is perfectly estimated at the receiver. The system parameters are given in Table 1.

Figure 3 shows average BER versus SNR per bit for QPSK and 64-QAM with different code rates. To observe the interleaving effect, we compare two cases: with subcarrier interleaving and without interleaving. First, in QPSK modulation, the subcarrier interleaving method has

Table 1. Simulation parameters

Parameters	Values
Bandwidth @ carrier frequency	16.894MHz @ 2.4GHz
FFT/IFFT size, N	2048
Number of subcarriers, N_c	1728 (9.766 KHz subcarrier spacing)
OFDM symbol duration	$102.4 + 12.8$ μs
Data modulation	QPSK/64QAM
Spreading code	Walsh-Hadamard sequence
Channel coding	Irregular LDPC code ($R = 1/2$, 1056 coded bits and $R = 3/4$, 1056 coded bits)
Decoding algorithm	Belief propagation (BP) with max. iteration 50

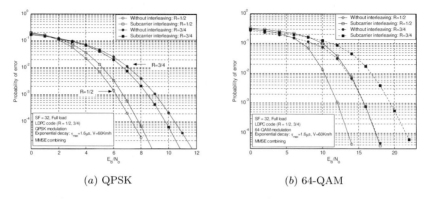

(a) QPSK (b) 64-QAM

Figure 3. Average BER versus SNR per bit for LDPC coded MC-CDMA using MMSE combining; code rate=1/2, 3/4; exponential decay channel ($\tau_{max} = 1.6\mu s$, 32 path).

slight better performance than without interleaving with code rate of 3/4. With code rate of 1/2, however, the system without interleaving has better performance. From this result, we can see that the channel coding of lower code rate such as $R = 1/2$ does not improve the performance of subcarrier interleaved system compared with the case without interleaving. The compensation of the orthogonality destruction is more important in subcarrier interleaved system in this case. In 64-QAM, the performance of subcarrier interleaving is degraded as compared with the case without interleaving. With code rate of 3/4, the system without interleaving has about 4.1 dB better performance than the subcarrier interleaving, contrary to the result of QPSK. The reason is that higher order modulations such as 64-QAM are very sensitive to inter-code interference. In spite of higher frequency diversity gain, the performance mostly depends on the orthogonality among spreading codes. This fact is also described in [5].

Figure 4 shows the required E_b/N_0 at BER of 10^{-4} as a function of the number of multiplexed users (active users) for QPSK and 64-QAM. If the number of multiplexed users is small, which implies that the MAI is so small, the effects of subcarrier interleaving is so effective. Due to the subcarrier interleaving and channel coding, the systems with subcarrier interleaving have higher diversity gain in region of the smaller number of active user, but on the other hand, as the number of active users increases, the performance of the subcarrier interleaving method becomes severely degraded. This is because the subcarrier interleaving generates the distortion among the spreading codes due to the destruction of orthogonality when MMSE detection technique is used. In the

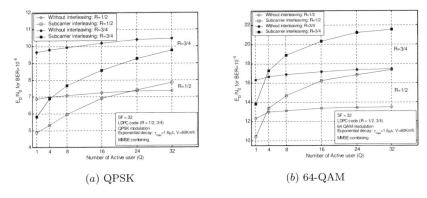

(a) QPSK (b) 64-QAM

Figure 4. Required SNR per bit at BER of 10^{-4} as a function of the number of multiplexed users.

case of the system without interleaving, it has lower diversity gain than the subcarrier interleaving method. This is due to the fact that almost all the spread chips transmit through the nearly correlated channel. In the case of 64-QAM as shown in Figure 4-(b), we can obtain the similar results with QPSK case. As we considered before, we can see that higher order modulations such as 64-QAM are very sensitive to inter-code interference.

The effects of the employed spreading factor on the interleaving methods are shown in Figure 5. For this results, we assume fully loaded system ($Q = K$) and we consider the systems for QPSK with $R = 3/4$ and 64-QAM with $R = 1/2$. Simulation results show that all the systems have similar pattern. When the K is smaller than 16, the performance of both systems is improved as K increases, but the performance is little changed when $K \geq 16$. From these results, we can see that there is a performance limit of achievable diversity gain. As the spreading factor increases, the performance is not improved. For larger K, due to the increase of MAI, the performance may be degraded as the spreading factor increases.

5. Conclusions

In this paper, we investigated the effects of subcarrier interleaving method on LDPC coded MC-CDMA systems. In the full loaded systems, the results show that the system without interleaving has better performance than the subcarrier interleaving method in the case of lower coding rate such as $R = 1/2$ and of higher order modulation such as 64-QAM. Also, the simulation results show that the subcarrier interleaving method can achieve higher diversity gain but it generates the distortion

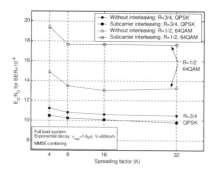

Figure 5. Required E_b/N_0 at BER of 10^{-4} as a function of the spreading factor.

among the spreading codes due to MMSE equalization. As the number of active users increases, the performance of the subcarrier interleaving method becomes severely degraded.

In channel coded MC-CDMA systems, we can see that frequency diversity gain via channel coding is more effective than subcarrier interleaving, especially for the system which higher order modulation and with lower code rate. In subcarrier interleaving method, the orthogonality loss due to the interference among the multiplexed users is larger than the diversity gain caused by the randomization effect.

References

[1] S. Hara and R. Prasad, "Overview of multicarrier CDMA," *IEEE Communications Magazine*, vol. 12, pp. 126-133, 1997.

[2] M. Schnell and S. Kaiser, "Diversity considerations for MC-CDMA systems in mobile communications," *IEEE Fourth International Symposium on Spread Spectrum Techniques and Applications (ISSSTA'96)*, pp. 131-135, Sept. 1996.

[3] S. Kaiser, *Multicarrier CDMA mobile radio system-analysis and optimization of detection, decoding, and channel estimation*, Ph.D. thesis, VDI-Verlag, 1998.

[4] G. Auer, S. Sand, A. Dammann, and S. Kaiser, "Analysis of cellular interference for MC-CDMA and its impact on channel estimation," *European Transactions on Telecommunications*, vol. 15, issue 3 pp. 173-184, May/June 2004.

[5] N. Maeda, H. Atarashi, and M. Sawahashi, "Performance comparison of channel interleaving methods in frequency domain for VSF-OFCDM broadband wireless access in forward link," *IEICE Transaction on Communications*, vol. E86-B, no. 1, Jan. 2003.

[6] S. Kaiser, "Trade-off between channel coding and spreading in multi-carrier CDMA systems," *IEEE Fourth International Symposium on Spread Spectrum Techniques and Applications (ISSSTA'96)*, pp. 1366-1370, Sept. 1996.

[7] S. Kaiser, "OFDM code-division multiplexing in fading channels," *IEEE Transactions on Communications*, vol. 50, no. 8, pp. 1266-1273, August 2002.

ONE-SHOT AND ITERATIVE SYMBOL PREDISTORTION TECHNIQUES FOR PAPR REDUCTION IN OFDM SYSTEMS

Serdar Sezginer and Hikmet Sari
Telecommunications Department, Supélec, F-91192, Gif-sur-Yvette, France

Abstract: In this paper, we investigate different variants of metric-based symbol predistortion for peak-to-average power ratio (PAPR) reduction for OFDM systems. They mainly consist of predistorting a set of input symbols per block using simple metrics, and they do not require transmitting any side information to the receiver. The metrics measure how much each symbol contributes to the output signal samples of large magnitude, and the predistorted symbols are selected as those with the largest metrics. The proposed techniques can be implemented as a one-shot or an iterative procedure.

Key words: OFDM; PAPR reduction; symbol predistortion.

1. INTRODUCTION

After its adoption for digital audio broadcasting (DAB) and terrestrial digital video broadcasting (DVB-T) over a decade ago, orthogonal frequency-division multiplexing (OFDM) has recently become very popular in wireless communications. Indeed, the IEEE 802.11a specifications for wireless local area networks (LANs) and the principal mode of the IEEE 802.16 specifications for broadband wireless access at frequencies below 11 GHz are based on this technique, and OFDM also appears today as a very strong candidate for the fourth generation of mobile cellular systems.

But the flexibility and other attractive features of OFDM may be counter balanced by its high peak-to-average power ratio (PAPR), which is substantially higher than that of single-carrier transmission. Its increased

PAPR requires a large back off of the transmit power amplifier from its output saturation point, and this back off leads to a very inefficient use of the available power. This problem is particularly important on the uplink, because of the stringent low-cost and power consumption requirements on user terminals.

A number of techniques are available in the literature for peak power reduction in OFDM systems. This includes coding [1], phase optimization [2], selective mapping [3], partial transmit sequences [4], and others. Coding reduces the useful data rate and other techniques require the transmission of side information to the receiver, both of which are undesirable. More recently, peak power reduction techniques were developed which are based on modifying the signal constellation, introducing new constellations, or inserting pilot signals [5]–[7].

A simple metric-based symbol predistortion technique to reduce OFDM peak power was recently introduced in [8] and [9]. It consists of predistorting the amplitude of a subset of each input symbol block and involves the computation of the symbol metrics which give a measure of the contribution of each symbol to the output signal samples of large magnitude. This metric-based structure makes the algorithm completely different from the previous attempts [6][7], where the symbol predistortion concept first appears. The parameters inherent in its structure give high flexibility which enables various tradeoffs between performance and complexity. This technique can be used as a one-shot or an iterative procedure.

This paper presents the general principle and describes three different variants of this symbol predistortion technique. Section 2 gives the definitions relating to the PAPR problem in OFDM transmission. In Section 3, we describe the proposed PAPR reduction techniques. Section 4 is devoted to performance results of these techniques for different signaling formats. Finally, conclusions are drawn in Section 5.

2. THE PAPR PROBLEM IN OFDM

In OFDM transmission, the complex data symbol block $\mathbf{a} = (a_0, a_1, ..., a_{N-1})$ is passed through an N-point inverse fast Fourier transform (IFFT) to obtain the discrete time-domain samples to be transmitted. The transmitted signal samples can be written as

$$b_n = \frac{1}{\sqrt{N}} \sum_{m=0}^{N-1} a_m e^{j2\pi nm/N} , \qquad (1)$$

where a_m is the data symbol transmitted over the mth subcarrier. The PAPR of the time-domain sample sequence $\mathbf{b} = (b_0, b_1, \ldots, b_{N-1})$ is defined as

$$PAPR(\mathbf{b}) = \frac{\|\mathbf{b}\|_\infty^2}{E\left\{\|\mathbf{b}\|_2^2\right\}/N}, \qquad (2)$$

where $\|\cdot\|_p$ denotes the p-norm of the enclosed vector.

In order to investigate the performance of a PAPR reduction algorithm, it is more relevant to examine the PAPR of an oversampled signal closely approximating the transmitted continuous-time signal. In this paper, this is performed by employing the IFFT of the zero-padded input data sequence of length QN, where Q denotes the oversampling rate.

In the literature, the most common way to evaluate the performance is to determine the probability that the PAPR of a block is larger than a certain level γ^2. This is represented by the Complementary Cumulative Distribution Function (CCDF) of the $PAPR(\mathbf{b})$, which is a random variable, as

$$CCDF(PAPR(\mathbf{b})) = \Pr(PAPR(\mathbf{b}) > \gamma^2). \qquad (3)$$

Next section gives the description of the symbol predistortion principle and presents several different variants.

3. SYMBOL PREDISTORTION

Because of their interesting features in PAPR reduction, symbol predistortion techniques have recently drawn considerable attention. They mainly tend to play with the constellation intelligently and predistort the transmitted data symbol values without affecting the minimum distance (see, e.g. [6] and [7]). These methods actually increase the transmitted average signal power. However, since the high peaks occur with small probability, this increase will not be a problem and can be easily controlled.

Algorithms based on symbol predistortion need complex optimization to obtain the right parameters minimizing the PAPR [5][7]. Practical solutions were presented in [6] and [7], namely, the projection onto convex sets (POCS) and the smart gradient project (SGP) methods. In addition, simple metric-based amplitude predistortion techniques were introduced in [8] and [9]. Using QPSK, it was shown that it is also possible to obtain satisfactory reduction in peak power by simple scaling of some symbols per block. This type of predistortion only scales upward the magnitude of the transmitted symbols leaving their phase unchanged. It can be easily extended to higher-level *M*-state quadrature amplitude modulation (*M*-QAM) signal constellations by limiting the predistortion to the outer points, in the same way as discussed in [7]. Specifically, the corner points of the constellation can be expanded as in QPSK, while only the real or the imaginary parts of the side symbols can be expanded in order not to reduce the minimum distance in the signal space.

We will now describe the methods based on amplitude predistortion and propose a new metric-based symbol predistortion technique.

3.1 Amplitude Predistortion

In the schemes proposed in [8] and [9], a metric is computed for each input data symbol which measures how this symbol contributes to the IFFT output samples with large values. This metric gives an idea for the scaling of the input data symbols. In its general form, such a metric is defined as

$$\mu_m = \sum_n w(n) f(n,m), \qquad (4)$$

where $f(n,m)$ is a function which gives a convenient measure of the contribution of symbol a_m to the IFFT output sample b_n and $w(n)$ is a weighting function of b_n. An appropriate choice for $f(n,m)$ is

$$f(n,m) = -\cos(\varphi_{nm}), \qquad (5)$$

where φ_{nm} is the angle between $a_m e^{j2\pi nm/N}$ and b_n. This function takes its maximum value when $a_m e^{j2\pi nm/N}$ and b_n are almost in opposite phase and this indicates that the symbol a_m can be predistorted to reduce the magnitude of b_n without reducing minimum Euclidean distance.

The weighting function $w(n)$ is defined to reflect the importance of the output samples on the metric μ_m. A reasonable and simple weighting function is $w(n) = |b_n|^p$, where p is an appropriately selected parameter. Using the definition $\cos(\varphi_{nm}) = \text{Re}\{b_n a_m^* e^{-j2\pi nm/N}\}/|b_n||a_m|$, the final metric becomes

$$\mu_m = \frac{-1}{K|a_m|} \sum_{n \in S_K} |b_n|^{p-1} \text{Re}\{b_n a_m^* e^{-j2\pi nm/N}\}, \qquad (6)$$

where K is a normalization factor and denotes the size of the set S_K whose elements are the indices of the output samples larger than a predetermined threshold value A.

Next, the metrics are computed for all input symbols of the block. Then, the L symbols with greatest positive metrics are determined to be predistorted with their corresponding scaling factors $d_k > 1$. Finally, the output samples are updated as

$$\overline{b}_n = b_n + \frac{1}{\sqrt{N}} \sum_{m \in S_L} (d_m - 1) a_m e^{j2\pi nm/N}, \qquad (7)$$

where S_L is a set of size L whose elements are the indices of the expanded symbols in the input sequence. Given the scaling factor and A, the parameter L is determined by observing the output peak power reduction. It is ideally taken as the value after which peak output power stops decreasing.

The algorithm involves a parameter set $\{A, p, d_m, L\}$, which makes it highly flexible. All of these parameters can be determined beforehand to achieve a target PAPR level and the desired tradeoff between complexity and performance.

Two variants of this symbol predistortion are simple amplitude predistortion [8] and multilevel amplitude predistortion [9].

3.1.1 Simple Amplitude Predistortion

Simple amplitude predistortion, as described in [8], is performed by using a constant scaling factor α with a value greater than 1. Thus, the real and imaginary parts of the input symbols to be predistorted are expanded with the same constant which makes the implementation considerably simple. As described above, for a given value of A, we pick the parameters α and L for which we observe sufficiently large PAPR decrease on average. The metric is evaluated as in (5) and the output samples are obtained as

$$\overline{b}_n = b_n + \frac{\alpha - 1}{\sqrt{N}} \sum_{m \in S_L} a_m e^{j2\pi nm/N}. \qquad (8)$$

The procedure can be repeated for further iterations, by using either the same or a separate $\{\alpha, L\}$ pair at each iteration.

3.1.2 Multilevel Amplitude Predistortion

The multilevel predistortion presented in [9] involves the expansion with a scaling factor d_m which differs from symbol to symbol. The expansion factor in this technique is defined as:

$$d_m = (1 + \alpha\sqrt{\mu_m^+}), \qquad (9)$$

where α is a positive real number. The notation $(\cdot)^+$ is used to indicate the positive values of the metric. The choice of the parameters α and L has a strong impact on the PAPR reduction performance of this technique and these parameters are determined as in the previous case. Finally, the output samples are updated as

$$\overline{b}_n = b_n + \frac{\alpha}{\sqrt{N}} \sum_{m \in S_L} \sqrt{\mu_m^+} a_m e^{j2\pi nm/N}. \qquad (10)$$

Note that the metric information is directly used in the predistortion process. Hence, in the metric calculation, the choice of the parameters and the functions becomes more crucial. Although optimization of the metric may bring some improvement, it seems almost impossible and it is beyond the scope of this paper.

3.2 Complex Symbol Predistortion

Now, we extend symbol predistortion to the phase dimension. In the complex predistortion case, the contribution of the real and imaginary parts are evaluated using separate metrics and both parts are predistorted in a fashion similar to multilevel amplitude predistortion using the two metrics. We define the metrics of the real and imaginary parts respectively as

$$\mu_{m,R} = \frac{-1}{K_R |\text{Re}\{a_m\}|} \sum_{n \in S_{K_R}} |b_n|^{p-1} \text{Re}\{a_m\} \text{Re}\{b_n e^{-j2\pi nm/N}\} \qquad (11)$$

and

$$\mu_{m,I} = \frac{-1}{K_I |\text{Im}\{a_m\}|} \sum_{n \in S_{K_I}} |b_n|^{p-1} \text{Im}\{a_m\} \text{Im}\{b_n e^{-j2\pi nm/N}\}. \qquad (12)$$

Determination of proper values for the parameters which appear in these two expressions is quite involved. Although the parameters are set beforehand, it is much more convenient to utilize the same parameters in both metrics. This is not the optimum solution, but it is reasonable because of the symmetric distribution of symbols. Further simplifications result in the following metrics:

$$\mu_{m,R} = \frac{-\mathrm{sgn}(\mathrm{Re}\{a_m\})}{K} \sum_{n \in S_K} |b_n|^{p-1} \mathrm{Re}\{b_n e^{-j2\pi nm/N}\}, \qquad (13)$$

$$\mu_{m,I} = \frac{-\mathrm{sgn}(\mathrm{Im}\{a_m\})}{K} \sum_{n \in S_K} |b_n|^{p-1} \mathrm{Im}\{b_n e^{-j2\pi nm/N}\}. \qquad (14)$$

The parameters which appear in these expressions are the same as those of amplitude predistortion, so they are determined in a similar fashion. Finally, the scaling factors are defined as $d_{m,R} = (1 + \alpha \sqrt{\mu_{m,R}^+})$ and $d_{m,I} = (1 + \alpha \sqrt{\mu_{m,I}^+})$ and the output samples are obtained by

$$\overline{b}_n = b_n + \frac{\alpha}{\sqrt{N}} \sum_{m \in S_L} \left(\sqrt{\mu_{m,R}^+} \mathrm{Re}\{a_m\} + j \sqrt{\mu_{m,I}^+} \mathrm{Im}\{a_m\} \right) e^{j2\pi nm/N}. \qquad (15)$$

This type of symbol predistortion is similar to active constellation extension (ACE) presented in [7]. But the use of metrics makes the proposed method completely different from the previous ACE methods [6][7] which use clipping. As mentioned before, the parameters used in this technique give high flexibility and make it simple to control the performance and complexity tradeoff.

4. SIMULATION RESULTS

In the simulations, we used the complex representation of OFDM signals, with $N = 256$ subcarriers. We applied all methods when the PAPR is larger than 6 dB. The algorithms were applied to output samples oversampled by a factor of 2 and then the oversampling factor was increased to 8 by interpolation to approximate the analog signal. In the simple amplitude predistortion, a threshold level A of 3.9 dB above the average power was used and p was set to 6. The threshold level was increased to 4.7 and p was set to 5 for multilevel and complex predistortion techniques. For the ACE-SGP method [7], the clipping value was taken as 4.86 dB above the average power for all signaling formats.

In Fig. 1, we present the results in terms of the CCDF for QPSK signaling. The solid-line curve in this figure corresponds to OFDM with no PAPR reduction and the marked curves correspond to the first three iterations of the proposed PAPR reduction techniques. Also, we included the 3rd iteration of the ACE-SGP method [7] for comparison. All of the algorithms converge to their maximum reduction value almost in 3 iterations and no substantial improvement occurs after that. In the simple predistortion case, the improvement is on the order of 2 dB below the probability of 10^{-3} for the first iteration and increases to 3 dB at the probability of 10^{-3} and to 3.2 dB at the probability of 10^{-5} at the 3rd iteration.

Compared to simple amplitude predistortion, multilevel amplitude predistortion [9] achieves at the first iteration a gain of 0.4 dB at the probability of 10^{-3} and of 0.6 dB at the probability of 10^{-5}. At the second and third iterations, however, respective performances of the two techniques are similar. The interesting observation here is that at the second iteration, complex predistortion achieves the third-iteration performance of the ACE-SGP method and it outperforms it by 0.2 – 0.3 dB at the third iteration.

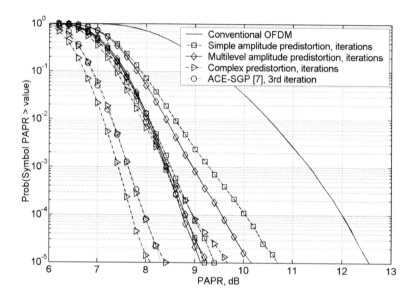

Figure 1. CCDF of PAPR for the symbol predistortion techniques using QPSK.

Next, Figs. 2 and 3 show similar performance results for the 16-QAM and 64-QAM signal constellations. The results are similar to those of Fig. 1. At the first iteration, multilevel amplitude predistortion outperforms simple amplitude predistortion by 0.4 dB below the probability of 10^{-3}, but the two techniques lead to similar performances at the second and third iterations. The next observation is that at the third iteration, the ACE-SGP method achieves a 0.3 dB improvement over these two techniques, and complex predistortion leads to better performance at the second and third iterations.

Comparing Figs. 2 and 3, we can see that the performance results corresponding to 64-QAM are very similar to those of 16-QAM. Again, the ACE-SGP method slightly outperforms amplitude predistortion techniques, and complex predistortion further improves performance.

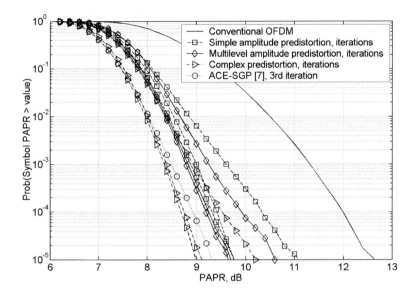

Figure 2. CCDF of PAPR for the symbol predistortion techniques using 16-QAM.

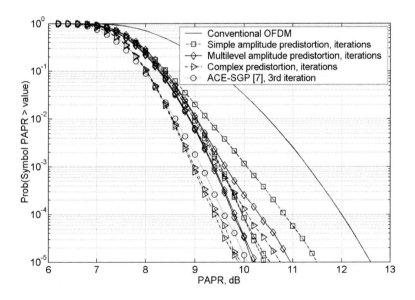

Figure 3. CCDF of PAPR for the symbol predistortion techniques using 64-QAM.

5. CONCLUDING REMARKS

Several variants of metric-based symbol predistortion have been presented. The algorithms commonly employ an appropriately defined metric for each input symbol that measures its contribution to the output samples of large magnitude. Once the metrics are computed for all symbols of the block, a set of symbols with the largest metrics are predistorted using either simple scaling, multilevel scaling, or complex predistortion. The computational complexity is low in all of these techniques and they are very flexible since the symbol metrics include several design parameters. The proposed techniques can be used as a single-shot procedure or the process can be iterated one or more times to further improve their performance.

References

[1] K. Patterson, "Generalized Reed-Muller Codes and Power Control in OFDM Modulation," *IEEE Trans. on Info. Theory*, vol. 46, pp. 104-120, Jan. 2000.

[2] C. Tellambura, "Phase Optimization Criterion for Reducing Peak-to-Average Power Ratio in OFDM," *Electronics Letters*, vol. 34, pp. 169-170, Jan. 1998.

[3] M. Breiling, S. H. Muller, and J. B. Huber, "SLM Peak-Power Reduction without Explicit Side Information," *IEEE Commun. Letters*, vol. 5, pp. 239-241, June 2001.

[4] S. H. Muller and J. B. Huber, "OFDM with Reduced Peak-to-Average Power Ratio by Optimum Combination of Partial Transmit Sequences," *Electronics Letters*, vol. 33, pp. 368-369, Feb. 1997.

[5] J. Tellado, *Multicarrier Modulation with Low Peak to Average Power Applications to xDSL and Broadband Wireless*, Boston/Dordrecht/London: Kluwer Academic Publishers, 2000.

[6] D. Jones, "Peak Power Reduction in OFDM and DMT via Active Channel Modifications," *in Proceedings of 33rd Asilomar Conf. on Signals, Systems and Computers*, pp. 1076-1079, 1999.

[7] B. S. Krongold and D. L. Jones, "PAR Reduction in OFDM via Active Constellation Extension," *IEEE Trans. on Broadcasting*, vol. 3, pp. 258-268, Sept. 2003.

[8] S. Sezginer and H. Sari, "Peak Reduction in OFDM Systems Using Dynamic Constellation Shaping," *in Proc. of EUSIPCO'05*, Antalya, Turkey, Sept. 2005.

[9] S. Sezginer and H. Sari, "OFDM Peak Power Reduction Using Metric-Based Amplitude Predistortion," *to appear in Proc. of GLOBECOM'05*, Saint Louis, Missouri, Nov. 2005.

ITERATIVE CORRECTION AND DECODING OF OFDM SIGNALS AFFECTED BY CLIPPING

Wolfgang Rave, Peter Zillmann, and Gerhard Fettweis
Dresden University of Technology, Vodafone Chair Mobile Communications Systems
Mommsenstrasse 18, 01062 Dresden, Germany

rave@ifn.et.tu-dresden.de

Abstract   We investigate iterative receive algorithms for orthogonal frequency division multiplexing (OFDM) transmission affected by clipping at the transmit power amplifier (PA). One algorithm aims at minimizing the Euclidean distance to the received sequence, while the other reconstructs the clipping noise. Both methods significantly reduce the BER compared to simple zero forcing, even for severe clipping. The iterative clipping noise correction is also investigated with a 'soft' sequence and a Soft-Input Soft-Output (SISO) decoder with coded OFDM. The gains with a soft sequence are moderate due to burst errors, but the number of iterations can be reduced compared to a correction based on hard decisions.

1. Introduction

The transmit signal in multicarrier data transmission systems is the superposition of many narrowband signals. This results in an approximately Gaussian distribution of the I- and Q-components of the complex baseband signal (central limit theorem). Consequently, multicarrier systems like OFDM are often impaired by clipping at the transmit PA. This paper presents several approaches to receiver-based iterative correction of clipped OFDM signals.

2. System Model

The system model used throughout the paper represents a simplified, discrete time, baseband equivalent OFDM system. A vector $\mathbf{X} = [X_0, X_1, \ldots, X_{N-1}]^T$ of length N is formed at the transmitter, and $X_k \in \mathcal{S} \; \forall \, k \in \mathbb{N}, 0 \leq k \leq N-1$, where the symbol set $\mathcal{S} = (S_0, S_1, \ldots, S_{M-1})$ contains all $M = 2^p, p \in \mathbb{N}$ possible complex transmit symbols.

X is then converted to the time domain vector $\mathbf{x} = [x_0, x_1, \ldots, x_{N-1}]^T$ by means of an *Inverse Discrete Fourier Transform* (IDFT) of length N. The elements of **x** are approximately complex Gaussian distributed, with variance P_x. The time domain vector is then distorted by a nonlinear function $g(\cdot)$ to deliver $\mathbf{z} = [z_0, z_1, \ldots, z_{N-1}]^T$. In this paper, the baseband equivalent memoryless nonlinearity $g(\cdot)$ is assumed to be a *soft limiter* (SL), which distorts the magnitude but not the phase of the elements of **x**. This models the baseband equivalent of an ideally predistorted nonlinear PA. The magnitude $r_{z,k} = |z_k|$ is then given by

$$r_{z,k}(r_{x,k}) = \begin{cases} r_{x,k} & r_{x,k} \leq A \\ A & r_{x,k} > A \end{cases}, \qquad (1)$$

where A is the clipping level of the nonlinear device. After the nonlinearity, a vector $\mathbf{n} = [n_0, n_1, \ldots, n_{N-1}]^T$ of complex AWGN further corrupts **z** to yield

$$\mathbf{y} = \mathbf{z} + \mathbf{n}. \qquad (2)$$

This system model is shown in Fig. 1. The complex Gaussian random

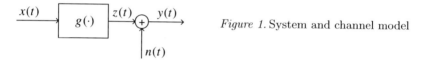

Figure 1. System and channel model

variables n_k are assumed zero-mean with variance $2\sigma_n^2$ (σ_n^2 per real dimension) and uncorrelated. The *Input Power Backoff* (IBO) for the soft limiter device is given by

$$\text{IBO} = A^2/P_x. \qquad (3)$$

It is well known that the output of a memoryless nonlinearity with Gaussian input can be described by means of the Bussgang decomposition [1]. The signal z_k can be written as

$$z_k = \alpha x_k + d_k, \qquad (4)$$

where α is a scaling factor depending on the nonlinearity and P_x, and d_k is a clipping noise term uncorrelated with x_k. For complex Gaussian input and real-valued $g(\cdot)$, α can be determined from

$$\alpha = \frac{1}{P_x} \int_0^\infty r_x\, p_{r_x}(r_x) g(r_x)\, dr_x, \qquad (5)$$

with r_x being the magnitude of x. The Rayleigh distribution for $p_{r_x}(r_x)$ leads, for the SL, to

$$\alpha = \left(1 - e^{-\text{IBO}}\right) + \frac{1}{2}\sqrt{\pi\,\text{IBO}}\,\text{erfc}(\sqrt{\text{IBO}})\,, \qquad (6)$$

where $\text{erfc}(\cdot)$ is the complementary error function. Furthermore, it is easily shown that, in our special case considered here $P_z = \left(1 - e^{-\text{IBO}}\right)P_x = \beta P_x$, from where the power of the clipping noise P_d can be computed as

$$P_d = P_z - \alpha^2 P_x = (\beta - \alpha^2)P_x\,. \qquad (7)$$

3. Clipping Correction Algorithms

3.1 Simplified ML Detection Algorithm

In the following, we assume that the nonlinear function $g(\cdot)$ is known at the receiver. The ideal ML detector for the communications system under consideration forms all possible transmit vectors $\hat{\mathbf{X}}_i$, where $0 \leq i \leq M^N - 1$. Then, these vectors are converted to time domain and distorted by $g(\cdot)$, resulting in the vectors $\hat{\mathbf{z}}_i$. Finally, the distance of each $\hat{\mathbf{z}}_i$ to the received vector $\mathbf{y}$ is computed, and the one with minimum distance is selected. Thus the ML detector solves the following problem:

$$\hat{\mathbf{X}} = \arg\min_{\hat{\mathbf{X}}_i} \left\| \mathbf{y} - g\!\left[\text{IDFT}\{\hat{\mathbf{X}}_i\}\right] \right\|^2. \qquad (8)$$

It is apparent that such a detector is prohibitively complex for practical values of N and M. The following approach is proposed: Instead of generating all possible hypotheses for the vector $\mathbf{X}$ in parallel, we look for the symbol which minimizes the above distance metric in the time domain for each element of $\mathbf{X}$ sequentially. The complete detection algorithm then consists of the following steps:

Sequential Mean Square Error Reduction

1. Convert the received time domain vector $\mathbf{y}/\alpha$ into the frequency domain vector $\mathbf{Y}$ by means of a *Discrete Fourier Transform* (DFT), with the scaling factor α depending on IBO. Then, quantize $\mathbf{Y}$ according to the decision boundaries corresponding to the symbol alphabet $\mathcal{S}$ (hard detection). This operation delivers $\hat{\mathbf{X}}_1$ as a first estimate of $\mathbf{X}$.

2. Generate M new frequency domain vectors $\hat{\mathbf{X}}_{1,1}^{(0)}$ to $\hat{\mathbf{X}}_{1,M}^{(0)}$ based on $\hat{\mathbf{X}}_1$ by inserting all possible symbols from $\mathcal{S}$ as the first vector element. The superscript denotes the index of the element which is varied.

3. For $1 \leq m \leq M$ compute the distances

$$d_{1,m}^{(0)} = \left\| \mathbf{y} - g\!\left[\text{IDFT}\{\hat{\mathbf{X}}_{1,m}^{(0)}\}\right] \right\|^2. \qquad (9)$$

and choose the symbol which results in the smallest distance.

4. Repeat steps 2 and 3 for all N elements of $\hat{\mathbf{X}}_1$. Thus, the symbol decisions are refined sequentially.

When no nonlinear distortion is present, this algorithm is equal to the complete ML solution, because the DFT and IDFT are orthogonal transforms, and reducing a possible error term in the frequency domain inevitably results in a reduced time domain error. The nonlinearity destroys orthogonality, which makes the above algorithm suboptimum. However, it reduces exponential to linear complexity. Only MN hypotheses are tested per OFDM symbol. The loss of orthogonality implies that the order, in which the subcarriers are processed, can affect the convergence behavior. Furthermore, several iterations of the whole algorithm can potentially improve performance. Note that, for each hypothesis, only a discrete increment in frequency domain has to be transformed into the time domain. A complete IFFT is not necessary.

Performance Results for AWGN Channel: The algorithm was tested for uncoded QAM transmission with $N = 64$. No guard or pilot subcarriers were simulated, and the transmission channel model was AWGN. The SL backoff was varied from 0-2 dB, modelling severe clipping (the clipping probability $p_{clip} = e^{-IBO}$ amounts to 0.368, 0.284 and 0.205 for IBO values of 0, 1 and 2 dB.). Fig. 1 (left) shows the performance gains after one iteration obtained for 16-QAM for these IBO values when the sequential MSE reduction algorithm is applied.

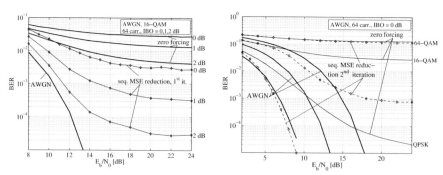

Figure 1. Performance of sequential MSE reduction for 16-QAM and IBO values of 0, 1 and 2 dB (left) and for QAM modulation with different constellation size (right).

On the right, the performance of a simplified version of the proposed algorithm for different constellation size after two iterations is compared. The hypotheses were now generated in a hierarchical manner by first selecting the quadrant which minimizes the MSE. Then, the actual QAM symbol in that quadrant was selected. Thus, only 8 instead of 16 (16-QAM) or 12 instead of 64 (64-QAM) hypotheses were tested per subcarrier. The figure shows how a 2nd iteration of the algorithm further improves performance. The hierarchical performance for 16-QAM and this IBO is identical to the full search.

It can be seen that the performance with clipping and sequential *Mean-Square Error* (MSE) reduction for QPSK actually surpasses linear AWGN performance in certain E_b/N_0 regions. Two effects are responsible for this behavior: First, clipping reduces the transmit power by a factor of $\beta = 1 - e^{-\text{IBO}}$, which results in a rescaling of the E_b/N_0 axis for a fair comparison. For an IBO of 0 dB, the resulting SNR gain is 2 dB. Secondly, the nonlinear distortion introduces dependencies between the frequency domain symbols, which help in recovering the correct symbols.

3.2 Iterative Correction with Hard Detection

Instead of approximating the complete search in some suboptimum way, an alternative is to estimate the clipping noise signal.

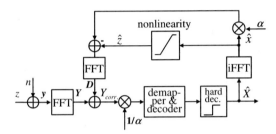

Tellado et al. [2], as well as *Chen* and *Haimovich* [3], proposed to compute the clipping noise term according to the Bussgang decomposition. This noise term is then subtracted in the frequency domain according to the following steps (see the sketch of the algorithm above for the uncoded case):

Iterative Clipping Correction

1. compute an estimate of the transmit sequence $\hat{\mathbf{x}}$ based on hard decisions $\hat{\mathbf{X}}$ derived from the demodulated equalized symbol vector $\mathbf{Y}$.

2. calculate the difference $\mathbf{d} = \alpha \hat{\mathbf{x}} - g(\hat{\mathbf{x}})$ between the clipped sequence $\hat{\mathbf{z}} = g(\hat{\mathbf{x}})$ and the attenuated unclipped sequence $\alpha \hat{\mathbf{x}}$.

3. add the Fourier transform $\mathbf{D}$ of this difference as a spectral correction to the demodulated symbols.

These steps can be applied iteratively to improve the reconstruction of the time domain sequence and the spectral correction respectively.

Performance Results for AWGN Channel: Again, severe clipping and a medium constellation (16-QAM with an IBO of 0 dB) is the point of reference. The correction capability of the algorithm for different IBOs is shown in Fig. 2 (left). Comparing the zero forcing solution to the first and third iteration of the algorithm, we note that for all IBO levels the performance is improved by two orders of magnitude.

The dependence on constellation size is demonstrated in Fig. 2 (right). The initial error rate for 64-QAM is high ($\simeq 12\%$), so that the algorithm does not improve the BER. On the other hand, for a robust constellation like QPSK, the first iteration step suffices to improve the zero forcing

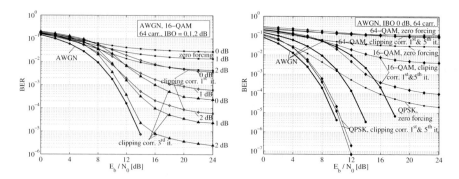

Figure 2. Performance of iterative clipping correction for varying IBO (16-QAM transmission, AWGN channel) (left) and different constellation size (right).

result greatly. A better than unclipped AWGN performance was not observed, though.

3.3 Modified Algorithm using a soft Sequence

As is well known from decision feedback algorithms and decoding, the performance of an algorithm can be improved if soft instead of hard decisions are used. In our case, other sequence hypotheses can be incorporated into the correction term used for feedback. In the uncoded case, we compute a weighted sum of sequences instead of a single hard decided sequence

$$\hat{\mathbf{x}}_{soft} = \frac{1}{\sum_n P_{dist,n}} \sum_{n=0}^{N_{cand}} P_{dist,n}(\hat{\mathbf{x}}_n)\, \hat{\mathbf{x}}_n\,. \quad (10)$$

The weights $P_{dist,n}(\hat{\mathbf{x}}_n)$ are calculated according to the euclidean distance between the sequence hypotheses and the observed sequence. The conditional probability for observing the estimated sequence of N symbols in complex gaussian noise including the clipping noise d_k with variance $N_0 = 2(\sigma_n^2 + \sigma_{nd}^2)$ is calculated according to

$$P_{dist} = \frac{1}{(\pi N_0)^N} \prod_{l=1}^{N} \exp\left\{-\frac{||g(\hat{x}_l) - y_l||^2}{N_0}\right\}. \quad (11)$$

$g(\hat{x}_l)$ and y_l denote samples of the clipped reconstructed sequence and the corresponding samples of the observed sequence. To select the candidate sequences we adopted the obvious idea to use metric values of the detected symbols to select those bits as candidates to be flipped, that have the smallest absolute likelihood values. Starting from the sequence based on hard decisions, again only a scaled DFT vector (according to the modified subcarrier symbol) has to be added to the current sequence to obtain the new hypothesis for which the euclidean distance is computed. In contrast to the recently proposed idea to check all possible combinations of a set of unreliable bits ('reduced state sequence estimation', see [4]), we found it more efficient to test the bits in the candidate set sequentially and to make immediately a step to the modified sequence, if the distance decreased. The list of bits is passed twice which led to improved performance. Steps in the wrong direction could be made less probable by testing the bits sorted according to their absolute L-values in ascending order.

Soft Iterative Clipping Correction (uncoded case)

Initialization: calculate metric values from the received symbols **Y** to obtain a list of unreliable bits together with their symbol and bit positions. Set the current and soft sequences to the reconstructed sequence based on hard decisions.

1. select the next bit in the list and compute a modified sequence with respect to the current best sequence and its euclidean distance.
2. add the modified sequence according to its weight to the soft sequence and update the total probability of all sequences accordingly.
3. replace the current sequence by the modified sequence, if a distance decrease was observed.
4. go to step 1, if end of list is not reached or the list is passed for the first time.
5. final step: normalize the soft sequence by the total probability (sum weight of all sequences).

Performance Results for AWGN Channel: Performance was tested for AWGN transmission of 16-QAM with IBO 0 dB as shown in Fig. 3. Sequence alternatives for the 64 smallest absolute metric values were tested (this number had been found to give a reasonable compromise between performance improvement and complexity). Two

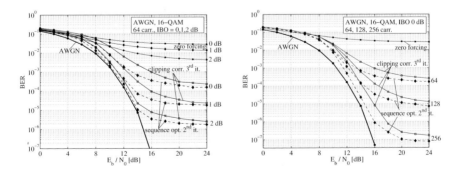

Figure 3. Comparison of hard and soft iterative clipping correction for different IBO (16-QAM, left) and different numbers of subcarriers (16-QAM, IBO 0 dB, right).

iterations with sequentially optimized sequences achieve a similar performance as the original algorithm with about three iterations. In addition a gain in the transition region above the error floor is visible. Similar results were obtained when the number of subcarriers was varied (see Fig. 3, right). Again two iterations with soft sequences perform slighly better than three iterations with the original algorithm.

4. Iterative Correction with Soft Decoding

Correction Algorithm: We extended the approach from the previous section to a coded system by using reliability information about the coded bits gained from a MAP decoder operating according to the BCJR algorithm [5]. A bit-interleaved coded OFDM system suffering from clipping distortion can be considered as a serial concatenation of a channel and an outer code. Therefore, a SISO [6] decoder can be employed to get reliability information not only about information bits but

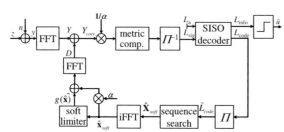

also about the *coded* bits that can be fed back to compute a soft estimate of the transmitted sequence. The soft sequence was again optimized sequen-

tially to avoid exponential complexity, generalizing slightly the approach of the uncoded case by incorporating a-priori information obtained from the decoder. The probability that the observed sequence $\hat{\mathbf{y}}$ is due to the estimated sequence $\hat{\mathbf{x}}$ after decoding is written as the product of the probability for the euclidean distance between hypothesis and observation and the likelihood derived from the decoder:

$$P_{seq}(\hat{\mathbf{x}}) = P_{dist}(\mathbf{y}|\hat{\mathbf{x}}) \cdot P_{flip}(\hat{\mathbf{x}}). \quad (12)$$

The conditional probability P_{dist} for observing the received sequence $\mathbf{y}$ after transmitting $\hat{\mathbf{x}}$ is computed as in eq. (11). The second factor in eq. (12) represents a-priori knowledge gained from decoding. It is calculated relative to the probability P_0 (set to $P_0 = 1$) of the most likely (hard decided) sequence $\hat{\mathbf{c}} = \text{sgn}(L_{code}(\mathbf{c}))$ after SISO decoding. Taking other candidate sequences into account reduces the sequence probability according to the probabilities of the coded bits c_l. The probability for a sequence in which one or more bits are flipped is

$$P_{flip}(\hat{\mathbf{x}}) = P_0 \cdot \prod_l \frac{1 - P(c_l = \hat{c}_l)}{P(c_l = \hat{c}_l)} = P_0 \cdot \prod_l e^{-|L(c_l)|}. \quad (13)$$

To limit the sequence search to tractable complexity, a set size N_{cand} of unreliable bits has to be fixed. Finally a 'soft sequence' is calculated as the weighted average of all candidate sequences:

$$\hat{\mathbf{x}}_{soft} = \frac{1}{\sum_l P_{seq,l}} \sum_{l=0}^{N_{cand}} P_{seq,l}(\hat{\mathbf{x}}_l)\, \hat{\mathbf{x}}_l. \tag{14}$$

Soft Iterative Clipping Correction (coded case)

1. compute metric values $\tilde{L}_{sig}$ from vector $\mathbf{Y}$; deinterleave to obtain vector L_{sig}.
2. soft decoding of L_{sig} to obtain reliability information L_{code} for coded bits and information bits L_{info}.
3. compute hard decided sequence $\hat{\mathbf{x}}_{hard}$ from interleaved vector $\tilde{L}_{code}$ and a weighted average of sequences $\hat{\mathbf{x}}_{soft}$, where the most unreliable bits are flipped as a soft estimate for the transmitted signal.
4. compute the spectrum $\mathbf{D}$ of the difference between clipped and non-clipped soft estimate of the time-domain sequence $\hat{\mathbf{x}}_{soft}$ and add it to $\mathbf{Y}$.
5. goto step 1, if the maximum number of iterations is not reached.
6. final step: compute hard decisions for information bits using L_{info}.

Performance Results for AWGN Channel: Results illustrating the performance of 16-QAM transmission for an AWGN channel and severe clipping are presented in Fig. 4. An IBO of 0 dB with uncoded transmission and ZF clipping correction leads to a BER floor at $\simeq 0.04$. A simple non-systematic terminated half rate code with constraint length 3 and a block size of one OFDM symbol was used with soft-in soft-out decoding.

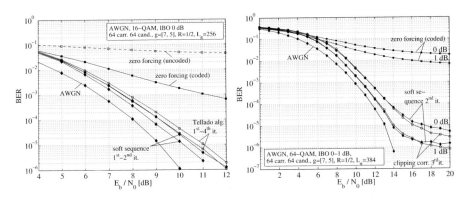

Figure 4. Clipping correction with soft or hard decided sequence for coded AWGN transmission of 16-QAM (left) and 64-QAM with different IBO 0 and 1 dB (right).

The performance after non-iterative decoding is still much worse than unclipped AWGN performance. The iterative correction algorithm from

[2, 3] for coded transmission with hard detection leads to a significant improvement. More than 2 iterations yield only marginal additional benefit due to error propagation. The sequential optimization of the soft sequence was carried out by testing the sequence alternatives for the 64 smallest L-values. Two loops over all subcarriers were made, reducing the gap to unclipped performance by about 0.5 dB at BER = 10^{-4}.

An even more severe clipping situation with 64-QAM and IBO values of 0 and 1 dB was studied for 64 subcarriers. Fig. 4 (right) shows the performance for both IBO values after the second iteration. It is comparable to three iterations with hard decisions; more iterations do not deliver additional benefits. One iteration in the decoder can be avoided at the expense of more effort in the detector.

5. Conclusions

Under the premise that deliberate clipping in baseband can faithfully be reproduced at the receiver, the bit error rate can be significantly reduced by clipping correction techniques. In particular, the clipping noise reconstruction algorithm appears to be a promising approach with reasonable complexity. Using a weighted sum of hypotheses for feedback, the error rate can be reduced, but an error floor remains, and the advantage with respect to hard decisions is small. Thus, the hard decision technique remains attractive due to its simplicity.

References

[1] A. Papoulis. *Probability, Random Variables and Stochastic Processes.* 3rd. Ed. McGraw-Hill Inc., 1991.

[2] J. Tellado, L. Hoo, and John M.Cioffi. Maximum-Likelihood Detection of Nonlinearly Distorted Multicarrier Symbols by Iterative Decoding. *IEEE Trans. Comm.*, 51(2):218–228, February 2003.

[3] Hangjun Chen and Albert M. Haimovich. Iterative Estimation and Cancellation of Clipping Noise for OFDM. *IEEE Comm. Lett.*, 7(7):305–307, July 2003.

[4] Huy D. Han and Peter Hoeher. Simultaneous Predistortion and Nonlinear Detection for Nonlinearly Distorted OFDM Signals. In *Proc. IST Summit 2005.*

[5] L. Bahl, J. Cocke, F. Jelinek, and J. Raviv. Optimal Decoding of Linear Codes for Minimizing Symbol Error Rate. *IEEE Trans. Inf. Theory*, IT-20:284–287, 1974.

[6] S. Benedetto, G. Montorsi, D. Divsalar, and F. Polara. Soft-Input Soft-Output Modules for the Construction and Distributive Iterative Decoding of Code Networks. *Europ. Trans. Telecomm.*, Vol. 9, pages 155–172, March 1998.

PAPR REDUCTION OF MC-CDMA SIGNALS BY SELECTED MAPPING WITH INTERLEAVERS

Masato Saito, Akihiro Okuda, Minoru Okada and Heiichi Yamamoto
Nara Institute of Science and Technology (NAIST)
Graduate, School of Information Science
Ikoma, Nara, 630-0192 Japan
saito@is.naist.jp

Abstract: In this study, we propose to apply interleavers as a mapping function of SLM (Selected Mapping) to reduce PAPR (Peak-to-Average Power Ratio) of MC-CDMA signals. We also propose a method to send information about selected interleaver without explicit side information. The SLM with pseudo-random interleavers obtains the same PAPR reduction ability as SLM with random phase rotation, i.e., conventional SLM. We confirm that by the proposed method receivers can sufficiently accurately detect interleavers without degradation of bit error rate.

1. Introduction

MC-CDMA (Multi-Carrier Code Division Multiple Access) systems have been a focus of constant attention for beyond 3G or next generation mobile communications systems [1]–[4]. Since MC-CDMA systems possess benefits of both OFDM (Orthogonal Frequency Division Muliplexing) and DS-CDMA (Direct-Sequence CDMA), the systems can provide high data rate even in multipath fading channel as well as cellular environment.

In this paper, we treat peak power problem, which is measured by PAPR (Peak-to-Average Power Ratio), of MC-CDMA systems. Effectively reducing PAPR of MC-CDMA signals has been a challenging problem and leads to cost reduction of base stations when MC-CDMA system is employed in the downlink.

We notice SLM (Selected Mapping) proposed by R.W. Bäuml et al. [5] as a PAPR reduction method. Since SLM is a distortionless PAPR reduction method, it can reduce PAPR without inter-carrier interference. In conventional SLM schemes, phase rotation of subcarriers is employed as a mapping function [5, 6]. In [6], many sequence sets are evaluated as the set of sequences for phase rotation.

However, there are few researches about other kinds of mapping functions. In this study, we propose to use interleavers as a mapping function of SLM for MC-CDMA signals. Although SLM by interleaving is evaluated for OFDM systems [7], effects to MC-CDMA systems are not evaluated so far. In [7], subcarrier interleaving is weaker than phase rotation, however, the result of MC-CDMA systems is different as we will show. We show a pseudo random interleaver is the most effective compared with shift register and block interleaver and is comparable to the best phase rotation sequences of conventional SLM. Moreover, we propose a method to transmit the information of selected interleaver without sending explicit side information. The proposed method utilizes both scramble code and pilot symbol which are essential components of MC-CDMA system downlink. We show that the receiver can sufficiently accurately estimate the used interleaver and detect send data symbols.

2. System Model

In this section, we introduce a basic structures of transmitter and receiver of MC-CDMA system. The functions related to SLM are described in the next section.

2.1 Transmitter

Transmitter model of the system is shown in Fig. 1. Modulated data symbols of k-th user d_k are converted from serial to P parallel streams. When the number of effective subcarriers is N_c and the spreading factor is N, the number P can be shown as $P = N_c / N$. Each symbol is multiplied by the pre-assigned channelization code $c_{k,l}$, which is used for user identification, after copy to N symbols. Where $c_{k,l}$ is the l-th chip of k-th user's code. We assume the first code is assigned to PLCH (pilot channel). Although PLCH is generally used for channel estimation, our method utilizes the channel for interleaver estimation as described below. The number of users is shown by K, this number does not include PLCH.

Spread symbols of all active users are summed up at each subcarrier. Then a scramble code C_n whose length is N_c is multiplied by the symbol at the corresponding subcarrier. The scramble code is employed for base station identification aspect.

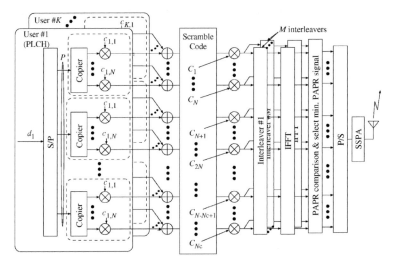

Figure 1. Transmitter Model of MC-CDMA System with SLM by Interleaving.

The subcarriers are interleaved for PAPR reduction purposes. The resultant sequence is input to N_{FFT}-point. Inverse Fast Fourier Transform (IFFT), and converted into a time-domain signal.

In this study, we define the PAPR of an MC-CDMA symbol as follows,

$$\mathrm{PAPR} = \frac{\max_{0 \leq t < T_s} |s(t)|^2}{E\left[|s(t)|^2\right]}. \qquad (1)$$

Where $s(t)$ is IFFT output and T_S is the MC-CDMA symbol duration. To obtain accurate PAPR value, we adopt oversampling by a factor of 4 according to the reference [8].

Before transmission, the signal passes SSPA (Solid State Power Amplifier). Rapp's model is used for the model of SSPA nonlinearity [9].

2.2 Receiver

The receiver structure of k-th user is shown in Fig. 2. Basically, the receiver performs the reverse operations of transmitter. The main difference from normal receivers is deinterleaving and its estimation. Since subcarrier components are interleaved at the transmitter, adequate deinterleaving is important for data detection.

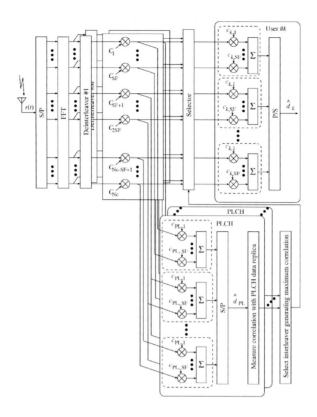

Figure 2. Receiver Model of MC-CDMA System with Interleaver Estimation.

3. SLM (Selected Mapping) with Interleavers

3.1 SLM method

In the transmitter (Fig. 1), SLM section is between scramble code multiplication and P/S converter. The mechanism of reducing PAPR by SLM is the following. A data sequence is applied to one of M mapping functions. IFFT generates a multicarrier signal from the sequence. Different signals are generated by applying different mapping functions and the PAPR values are compared, then the signal with minimum PAPR is selected and transmitted. The performance of PAPR reduction depend on the mapping functions. Therefore, preparation of adequate mapping functions is a crucial in SLM.

In the receiver (Fig. 2), deinterleavers perform deinterleaving after FFT. TIt is important to select the correct deinterleaver corresponding to the selected interleaver at the transmitter.

3.2 Interleavers as a Mapping Function of SLM

We propose to use interleavers as a mapping function of SLM. Interleavers perform that an IFFT input sequence is permuted without changing subcarrier components. Since even cyclic shift of IFFT input sequence enables to change the waveform of IFFT output, any interleaving may be applicable to mapping functions. Interleaving of subcarriers spent only unit time necessary to change the place, because the operation can be performed in parallel.

As denoted above, an SR (shift register) is possible as interleaver. By setting different cyclic shifts of an SR, different mapping functions can be provided.

BI (block interleaver) can be also employed as the mapping function. BI is defined as a normal rectangular shaped interleaver. As described above, large delay doesn't occur due to parallel operations. To provide different permutations by BI, different amount of cyclic shift is applied to IFFT input sequence before block interleaving.

Moreover, we consider PRI (pseudo-random interleaver). Note that the pseudo-random interleaver in [7] is different, although the name is same. An example of generating an interleaved sequence by PRI is shown in Fig. 3 where sequence length to be interleaved is $N = 8$. PRI utilizes a shift register which generates an M-sequence whose period satisfies the condition $N \leq 2^D - 1$, where D is the number of registers. Since $D = 4$ in this example, the sequence length must be less than 15. First, the state of registers shown in binary number is converted into decimal number. Then the decimal numbers which are larger than N is truncated. The remained sequence of decimal numbers becomes an order of the PRI. In this case, the input sequence $(X_1, X_2, ..., X_8)$ is interleaved into $(X_1, X_8, X_4, X_2, X_6, X_5, X_7, X_3)$. Different interleavers can be generated by changing the initial state of shift registers and/or connections between registers and adders denoted by primitive polynomials.

3.3 Interleaver Information Transmission Method

To deinterleave accurately, the receiver must select the correct interleaver. The information about correct interleaver should be transmitted with transmitted signals. Sending the information by a specific channel is the easiest way. However, the way decreases the number of channels available for users.

To minimize the additional costs for SLM, we propose the method exploiting essential components of MC-CDMA systems, i.e., PLCH and scramble code. In the proposed method, transmitters act in usual way

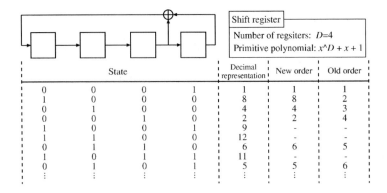

Figure 3. Example of generating a PR interleaved sequence.

except interleaving and selecting for SLM. On the other hand, receivers perform 1. deinterleaving, 2. decoding scramble code, 3. despreading PLCH, 4. correlating pilot symbols, 5. selecting deinterleaver, before channel decoding as shown in Fig. 2. In this study, since fading channel is not considered, channel estimation process is omitted.

Because scramble code tends to have some properties of random code, correlating the received signal with the pilot channel is expected to have sharp auto-correlation property.

4. Numerical Examples

The CCDF (Complementary Cumulative Distribution Function) of PAPR of MC-CDMA signals with SLM based on interleaver is shown in Fig. 4. The simulation parameters are shown in middle column of Table 1. As an interleaver, we employ SR, BI, and PRI. The PAPR distribution by conventional SLM whose mapping functions are phase rotations (labelled as CSSC) and that without PAPR reduction are also shown in the figure as references.

From the figure, we can find PAPR reduction ability depends on interleavers. All the interleavers can shift PAPR distribution to the lower direction compared with the system without PAPR reduction. PRI is the best amongst the three interleavers.

The performance of PRI is almost the same as that of CSSC (Cyclic-Shifted Scramble Code) method which is SLM based on phase rotation according to random sequence (actually, a scramble code is utilized as a phase rotation sequence) [10]. As stated in [6], random code or M-sequence can minimize PAPR distribution of MC-CDMA signals when SLM is

Table 1. Simulation parameters.

Parameter	Fig. 4	Fig. 5
Number of subcarriers N_c	768	192
FFT point N_{FFT}	1024	256
Oversampling factor	4	
Number of users K	5	
Modulation scheme	QPSK	
Channelization code	Hadamard code	
Spreading factor N	16	16, 32, 64
Scramble code	Random code	
Number of mapping functions M	16	

employed. Therefore PRI can be seen as the best mapping function set of SLM.

SLM can be said to be a diversity scheme. Therefore, independence between generated waveforms is important to effectively reduce PAPR. It can be easily estimated that generated waveforms by shift register possess high correlations. Therefore, it performs relatively worse. From a comparison to CSSC, the PRI will derive nearly independent waveforms.

The interleaver selecting error rate (ILER) for various spreading factors is shown in Fig. 5. The simulation parameters are shown in the third column of Table 1. The smoothness factor of Rapp model is set at 3, and IBO is set at 3dB. Note that the horizontal axis means the E_b/N_0 of data symbols. We assume the symbol energy of data and pilot be equivalent. When spreading factor is relatively low, the error performance seems to be improved. The improvement is caused from the number of pilot symbols. Since the number of total subcarriers is fixed, smaller spreading factor means use of more pilot symbols for interleaver estimation. Due to the same reason, BER performances of different spreading factors become almost equivalent.

5. Conclusions

In this study, we evaluated the effect of interleavers as mapping functions of SLM on PAPR reduction of MC-CDMA signals. Three kinds of interleavers, i.e., shift register, block interleaver, and pseudo-random interleaver, are used for comparison. Pseudo-random interleaver is the most effective to reduce PAPR and is comparable to SLM with random phase rotation. Therefore, we can conclude pseudo-random interleavers are suitable for SLM in MC-CDMA system.

We proposed interleaver information transmission method which is not required to send explicit side information. Through the evaluation of errors

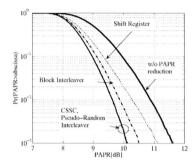 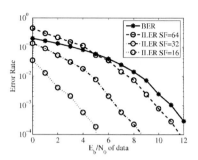

Figure 4. CCDF of PAPR of MC-CDMA signals with SLM based on interleavers.

Figure 5. BER and interleaver selection error rate of MC-CDMA systems with SLM by interleaving.

of decided interleaver and BER, we can conclude that the proposed method causes almost no degradation of BER performance,.

References

[1] S. Hara and R. Prasad. Overview of multicarrier CDMA. *IEEE Communications Magazine,* 35: 126–133, Dec. 1997.
[2] K. Fazel and S. Kaiser. *Multi-Carrier and Spread Spectrum Systems.* John Wiley and Sons, 2003.
[3] L. Hanzo, M. Münster, B. Choi, and T. Keller. *OFDM and MC-CDMA for Broadband Multi-user Communications, WLANs and Broadcasting.* John Wiley and Sons, 2003.
[4] S. Hara and R. Prasad. *Multicarrier Techniques for 4G Mobile Communications.* Artech House, 2003.
[5] R. W. Bäuml et al., R. F. H. Fischer, and J. B. Huber. Reducing the peak-to-average power ratio of multicarrier modulation by selected mapping. *Electronics Letters,* 32: 2056–2057, Oct. 1996.
[6] N. Ohkubo and T. Ohtsuki. Design criteria for phase sequences in selected mapping. *IEICE Transactions on Communications,* E86-B: 2628–2635, Sep. 2003.
[7] A. D. S. Jayalath and C. Tellambura. Use of data permutation to reduce the peak-to-average power ratio of an OFDM signal. *Wireless Communications and Mobile Computing,* 2: 187–203, Mar. 2002.
[8] C. Tellambura. Computation of the continuous-time PAR of an OFDM signal with BPSK subcarriers. *IEEE Communications Letters,* 5: 135–137, Apr. 2001.
[9] R. van Nee and R. Prasad. *OFDM for Wireless Multimedia Communications.* Artech House, 2000.
[10] M. Saito, A. Okuda, M. Okada, and H. Yamamoto. Peak-to-average power ratio reduction method suitable for MC-CDMA system downlink. *Proc. Of PIMRC 2005,* Sep. 2005.

ITERATIVE NONLINEAR CHANNEL COMPENSATION IN MC-CDMA SYSTEMS

Vincenzo Lottici and Filippo Giannetti
University of Pisa
Department of Information Engineering
Via G. Caruso, 16 – I-56122, Pisa, Italy
vincenzo.lottici@iet.unipi.it, filippo.giannetti@iet.unipi.it

Abstract   In this contribution we present an efficient multi-user data predistortion algorithm for the compensation of the nonlinear channel in the forward link of a wireless multiple access MC-CDMA system. A simplified method which eases the implementation of the iterative procedure by trading off performance and complexity is also described. Furthermore, we demonstrate that the performance of the proposed compensation scheme can be improved by combining it with a judicious allocation of the spreading signature codes. Computer simulations evidence a significantly better performance with respect to a conventional memoryless pre-distortion technique.

Keywords: MC-CDMA, nonlinear distortions, multiple access interference.

1.   Introduction

Multicarrier (MC) transmissions have been recently combined with Code Division Multiple Access (CDMA) in order to devise a technique offering robustness against both multipath fading channels and impulsive noise together with flexible and efficient multiple access capability [1]-[3]. Besides these appealing properties, however, MC-CDMA systems exhibit some critical features as well, such as a considerable vulnerability to nonlinear distortion induced by high power amplifiers (HPAs) operating at, or near, the saturation region. Indeed, in the presence of a nonlinear HPA device, the large fluctuations of the MC-CDMA signal amplitude (mainly due to its multicarrier structure [4]) give rise to intermodulation products affecting both the MC-CDMA signal itself and the adjacent channels [5]. This results in significant performance degradations [4], [6] that require adequate countermeasures. A simple, yet rather

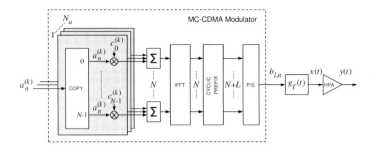

Figure 1. Block diagram of the MC-CDMA transmitter.

power expensive, method for the reduction of nonlinear distortion is to choose a large HPA output backoff (defined as the ratio between the saturation power of the HPA and the actual average transmitted power) so as to force the operating point of the amplifier in the linear region. Alternatively, a more efficient approach is to apply an appropriate technique at the transmitter side, such as the predistorter (PD) schemes recently proposed in [7]-[9].

In this paper we will pursue a different route towards the goal of compensating transmitter's nonlinearity. Our aim is to extend the PD method originally proposed in [9] for single-carrier single-user transmissions and to apply it to a forward link multiple access MC-CDMA system. The underlying idea is to jointly and iteratively optimize via a multi-user approach the signal constellation of *each* active user, such as to minimize the intersymbol interference component at *each* decision device, while keeping the computational complexity at a reasonable level. In order to further improve the overall performance, we also suggest a simple method based on a judicious allocation of the users' code signatures to reduce the peak-to-average power ratio (PAPR) of the transmitted signal.

2. System description

The block diagram of the transmitter for the forward link of a MC-CDMA system connecting the base station (BS) to N_u mobile stations (MSs) is depicted in Fig. 1. The n−th information-bearing symbol $a_n^{(k)}$ of the k−th user at rate $1/T$, belonging to a M−QAM constellation, is first copied into N branches (equal to the number of subcarries) each of which is then multiplied by the chip binary spreading sequence $c_m^{(k)} \in \{\pm 1\}$, $0 \leq m \leq N - 1$, belonging to the Walsh-Hadamard (WH) code set with length N which is the channel identification signature. The spread data of all N_u active users are summed together synchronously, and then mapped to the N available subcarriers using an IFFT (Inverse Fast Fourier Transform) unit. To avoid interference between successive blocks and preserve the orthogonality among subcarriers (at least under ideal propagation channel conditions), a cyclic prefix of L samples is inserted at the beginning of each IFFT output block to produce

$$b_{l,n} = \frac{1}{\sqrt{N}} \sum_{k=1}^{N_u} \sum_{m=0}^{N-1} a_n^{(k)} c_m^{(k)} e^{j2\pi ml/N}, \quad -L \leq l \leq N-1. \quad (1)$$

The complex envelope of the MC-CDMA signal so takes the form

$$x(t) = \sum_n \sum_{l=-L}^{N-1} b_{l,n} g(t - lT_c - nT), \quad (2)$$

where $T_c \triangleq T/N$ and T are the chip and the block interval, respectively, while $g(t)$ is a root-raised-cosine (RRC) pulse with rolloff α. The HPA is modeled as a nonlinear memoryless device, with the normalized AM/AM and AM/PM responses given by [3]-[4]

$$M(\rho) = \frac{\rho}{(1+\rho^q)^{1/q}}, \quad \Phi(\rho) = 0, \quad (3)$$

where q is an integer value which controls the smoothness of the transition between the linear region and the saturation one. At the user receiver (see Fig.2) the rate $1/T_c$ samples at the output of the matched filter (MF) are collected into blocks of size $N+L$. After removal of the cyclic prefix, they are transformed by a FFT unit of size N, despread by multiplying with the user's signature, and eventually, after channel equalization are fed to the decision device. Since, our interest addresses the performance degradation due to the HPA distortions, in the sequel carrier and timing (both code and chip) synchronization will be assumed ideal, whereas the only channel impairment is a stationary AWGN process added to the received signal.

3. Multiuser data predistortion

3.1 Rationale of the MUDP scheme

The two major effects produced by the nonlinear channel [8], [9] on the received signal constellation at the MF output are the *clustering*, i.e., each of the M points of the users constellations becomes a cluster reflecting the intersymbol interference (ISI) at the sampling instants, and the

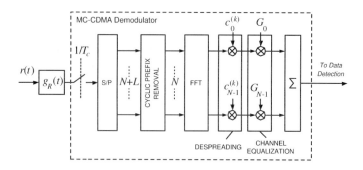

Figure 2. Block diagram of the MC-CDMA receiver.

warping, i.e., the center of mass of each cluster moves away from its nominal point. To tackle these detrimental effects (and ultimately, to reduce the overall nonlinear distortion and improve the system performance) a simple method, called Multi User Data Pedistortion (MUDP), which consists in modifying the transmitted symbols of each user's to obtain at the receiver's MF output a constellation that, in the average, looks like the one we would get in the absence of distortion. Following the above mentioned line and in view of the fact that we are concerned with the forward link of multiple access MC-CDMA system, we will resort, therefore, to a *multi-user extension* of the DP method proposed in [9] for single-carrier and single-user transmissions. The key idea herein is to *jointly and iteratively* optimize the constellations of all the active users taking into account the multiple access interference (MAI) arising due to the lack of orthogonality among the MC-CDMA signal subcarriers. This goal is achieved by considering the predistorted symbol transmitted by the k-th user as a nonlinear function of the data symbols sent out by all the N_u active users, i.e., the k-th user itself and the other $N_u - 1$ interfering users. In other words, the predistorted transmitted symbol has "memory" of the symbols currently transmitted by all other users. With the above plan in mind, the proposed iterative method can actually be summarized as follows (see the block diagram illustrated in Fig. 3):

1) let us focus on the M-order constellation of the k-th user ($1 \leq k \leq N_u$); the m-th clustered point of the constellation ($1 \leq m \leq M$) is then partitioned into the M^{N_u-1} (overlapping) sub-clusters $\Omega_{l,m}^{(k)}$ ($1 \leq l \leq M^{N_u-1}$) corresponding to a different combination of data symbols simultaneously transmitted by the $N_u - 1$ interferers;

2) the centers of mass of the sub-clusters $\Omega_{l,m}^{(k)}$ are estimated at the transmitter using a bank of local MC-CDMA receivers; we denote

with $z_{l,m}^{(k)}(n)$ the output of the local receiver, at instant nT, related to the l-th sub-cluster of the m-th point of the k-th user constellation; the estimate of the the centers of mass at the i-th iteration results then

$$q_{l,m}^{(k)}(i) = \frac{1}{N_{l,m}^{(k)}(i)} \sum_{n \in \Phi_{l,m}^{(k)}(i)} z_{l,m}^{(k)}(n), \qquad (4)$$

where $\Phi_{l,m}^{(k)}(i)$ contains $N_{l,m}^{(k)}(i)$ time indexes $n \in \left[(i-1)N_T^{(k)}(i), i N_T^{(k)}(i)\right)$, such that $z_{l,m}^{(k)}(n) \in \Omega_{l,m}^{(k)}$, and

$$N_T^{(k)}(i) = \sum_{m=1}^{M} \sum_{l=1}^{M^{N_u-1}} N_{l,m}^{(k)}(i) \qquad (5)$$

is the total number of symbols transmitted by the k-th user within the i-th iteration;

3) the complex-valued difference between the estimated center of mass given by (4) and the relevant nominal point, i.e.,

$$\varepsilon_{l,m}^{(k)}(i) \triangleq \left|\varepsilon_{l,m}^{(k)}(i)\right| \cdot e^{j \arg\left\{\varepsilon_{l,m}^{(k)}(i)\right\}} \triangleq q_{l,m}^{(k)}(i) - a_m^{(k)}, \qquad (6)$$

where $\left|\varepsilon_{l,m}^{(k)}(i)\right|$ and $\arg\left\{\varepsilon_{l,m}^{(k)}(i)\right\}$ are the amplitude and phase errors, respectively, is used as the error signal in a control loop to update the predistorted symbols $b_{l,m}^{(k)}$ corresponding to all the combinations of data symbols transmitted by the N_u users; designating with $\rho_{l,m}^{(k)}(i)$ and $\vartheta_{l,m}^{(k)}(i)$ the polar coordinates of $b_{l,m}^{(k)}$ at the i-th iteration, we employ the recursions

$$\begin{cases} \rho_{l,m}^{(k)}(i+1) = \rho_{l,m}^{(k)}(i) - \gamma_\rho \left|\varepsilon_{l,m}^{(k)}(i)\right| \\ \vartheta_{l,m}^{(k)}(i+1) = \vartheta_{l,m}^{(k)}(i) - \gamma_\vartheta \arg\left\{\varepsilon_{l,m}^{(k)}(i)\right\} \end{cases}, \qquad (7)$$

where γ_ρ and γ_ϑ are positive-valued step-sizes affecting the convergence speed of the algorithm;

4) applying (7), convergence is achieved when $\left|\varepsilon_{l,m}^{(k)}(i)\right| < \lambda_T$, $1 \leq l \leq M^{N_u-1}$, $1 \leq m \leq M$, $1 \leq k \leq N_u$, where λ_T is a given threshold.

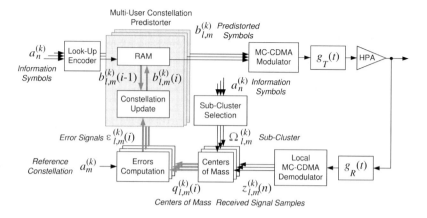

Figure 3. Block diagram of the MUDP scheme.

Some remarks are now in order:

- the n–th data symbol $a_n^{(k)}$ of the k–th user will be mapped, i.e., will be predistorted, to a value in accordance with the data symbols transmitted by the other $N_u - 1$ interfering users;

- the total number of recursions given by (7) is $2N_u M^{N_u}$ corresponding to $N_u M^{N_u}$ predistorted symbols;

- the MUPD scheme resides entirely at the base station transmitter, it does not require any feedback information in the reverse link, and the predistorted constellations of all the active users are updated all at once;

- as will be shown when discussing simulation results, after a given number of iterations the proposed procedure reaches the optimum solution, i.e., the centers of mass of all the sub-clusters are pushed toward their relevant nominal points, and consequently, both warping and the clustering effects are minimized;

- due to the lack of orthogonality among users (caused by the effects of the nonlinear channel), a considerable MAI level arises at each iteration which may hinder the iterative procedure; this means that the step-sizes γ_ρ and γ_ϑ have to be properly selected so as to minimize convergence time, while updating the predistorted constellations as much gradually as possible;

- the proposed MUDP scheme requires for each user only a modified signal constellation; therefore, it is well suited for a digital implementation that can easily adapt itself to possible variations, e.g., due to a different signature allocation of the active users, automatic control of transmitted power, thermal instability, etc..

3.2 Efficient implementation of MUDP

The iterative scheme illustrated so far efficiently compensates for nonlinear distortions in the forward link of a MC-CDMA multi-user scenario. However, it has the inherent drawback of high overall complexity, which is given by both the number of recursions (7) (which grows exponentially with the number N_u of active users) and the convergence time. Note that the latter is specified by the number of data symbols required to make at each iteration a reliable estimate of the center of mass of each constellation sub-clusters. To make a realistic example, let us choose 16-QAM as modulation format, i.e., $M = 16$, $N = 256$ as spreading factor, and $N_u = 128$ active users corresponding to a 50% cell loading. It is apparent that the proposed technique would be quite unfeasible in that the number of required recursions turns out to be $2N_u M^{N_u} = 2 \cdot 128 \cdot 16^{128} = 2^{520} \approx 2 \cdot 10^{156}$! To make the problem much more manageable three main approximations are pursued:

1) the application of the Bussgang theorem [10] allows for any given user to select the strongest interfering channels and to neglect all the others, as outlined in details in [11]; as a result, the number of recursions (7) and the convergence time is brought towards reasonable values;

2) the constellations of the interfering users can be considered as a whole, i.e., for 16-QAM as a set of $M = 16$ points, or can be partitioned into subsets of adjacent symbols belonging to the 4 quadrants, or not considered at all; in the sequel these options will be referred to as P_{16}, P_4 or P_0 partition, respectively.

3) the quadrant symmetry of the QAM constellation enables to reduce further the complexity by a factor 4.

3.3 Efficient allocation of spreading codes

It is well known that the MC-CDMA signal exhibits one major drawback, i.e., high PAPR, that makes it very sensitive to the nonlinear behavior of the transmitter HPA. Therefore, we can figure out that the task of any nonlinear channel compensation scheme, as for instance the MUPD scheme proposed above, would be made easier whether the PAPR of the transmitted signal could be reduced in some way. To adhere to that approach, let us first observe that the PAPR metric is strictly dependent on the features of the code signatures and on the way in which they are allocated to different users [12]. Indeed, it can be demonstrated that in a multi-user scenario the aperiodic cross-correlations and auto-correlations of the spreading codes play a significant role in determining the system sensitivity to nonlinear channel. That is, more specifically, the larger is the mean square value of the above functions, the higher is the PAPR of the transmitted signal [11]. Therefore, our goal can be summarized as follows. For a given number N_u of active users, let us consider the $N \times N_u$ array $\tilde{\mathbf{C}} = \left[\mathbf{c}^{(1)}, \mathbf{c}^{(2)}, \ldots, \mathbf{c}^{(N_u)}\right]$, where $\mathbf{c}^{(k)} = \left[c_0^{(k)}, c_1^{(k)}, \ldots, c_{N-1}^{(k)}\right]^T$ and T denoting transpose, made of N_u code signatures selected among the WH sequences with period N to be employed within the cell. Hence, choosing the optimal allocation of spreading codes (in the sense of PAPR reduction) means finding the array $\mathbf{C}_{opt}$ containing the set of codes that minimizes the metric

$$\Lambda(\tilde{\mathbf{C}}) = \frac{1}{N_u N} \left[\sum_{k=0}^{N_u-1} \sum_{n=0}^{N-1} \left[A_n^{(k)}\right]^2 + \sum_{k=0}^{N_u-1} \sum_{\substack{q=0 \\ q \neq k}}^{N_u-1} \sum_{n=0}^{N-1} \left[X_n^{(k,q)}\right]^2 \right], \quad (8)$$

i.e., the sum of the mean square value over the N_u users of the aperiodic auto-correlation $A_n^{(k)}$ and cross-correlation $X_n^{(k,q)}$ defined as

$$A_n^{(k)} \triangleq \sum_{j=0}^{N-1-n} c_j^{(k)} c_{j+n}^{(k)}, \quad X_n^{(k,q)} \triangleq \sum_{j=0}^{N-1-n} c_j^{(k)} c_{j+n}^{(q)}. \quad (9)$$

Note that (8) is a function of the matrix $\tilde{\mathbf{C}}$ containing N_u available spreading codes. In the sequel, the above procedure (8) will be referred to as "PAPR-Driven Allocation" (PDA) method, whereas the conventional allocation strategy wherein the spreading codes of $\tilde{\mathbf{C}}$ are selected randomly from the WH set, will be labeled as "Random Allocation" (RA) method. It is worth noting that for a given N_u and N the PDA procedure, if solved through an exhaustive search, would require a prohibitive number of combinations of signatures to be considered, thus revealing itself as quite unfeasible. Therefore, we will resort in the sequel to an alternative method with much lower complexity. Let us label univocally the N_u active users with an index in the range $[0, N_u - 1]$, and define

$$\Lambda_p(\tilde{\mathbf{C}}) = \frac{1}{N_u N} \left[\sum_{k=0}^{p} \sum_{n=0}^{N-1} \left[A_n^{(k)}\right]^2 + \sum_{k=0}^{p} \sum_{\substack{q=0 \\ q \neq k}}^{N_u-1} \sum_{n=0}^{N-1} \left[X_n^{(k,q)}\right]^2 \right], \quad (10)$$

as the metric (8) over the first p users, with $p \in [0, N_u - 1]$. So, according to (10) we can write

$$\Lambda_p(\tilde{\mathbf{C}}) = \Lambda_{p-1}(\tilde{\mathbf{C}}) + \delta\Lambda_p(\tilde{\mathbf{C}}), \qquad (11)$$

where the increment $\delta\Lambda_p(\tilde{\mathbf{C}})$ is given by

$$\delta\Lambda_p(\tilde{\mathbf{C}}) = \frac{1}{N_u N} \left[\sum_{n=0}^{N-1} \left[A_n^{(p)}\right]^2 + \sum_{\substack{q=0 \\ q \neq p}}^{N_u-1} \sum_{n=0}^{N-1} \left[X_n^{(p,q)}\right]^2 \right]. \qquad (12)$$

Now, making use of (10)-(12) the PDA method can be applied as outlined in the following. At the first step $p = 0$, the signature of the user #0 will be chosen by minimizing the metric $\Lambda_0(\tilde{\mathbf{C}}) = \delta\Lambda_0(\tilde{\mathbf{C}})$. At the next step $p = 1$, we obtain

$$\Lambda_1(\tilde{\mathbf{C}}) = \Lambda_0(\tilde{\mathbf{C}}) + \delta\Lambda_1(\tilde{\mathbf{C}}) \qquad (13)$$

and consequently, the signature of the user #1 will be chosen by minimizing only $\delta\Lambda_1(\tilde{\mathbf{C}})$ in that $\Lambda_0(\tilde{\mathbf{C}})$ has been already minimized in the previous step. The procedure goes on until $p = N_u - 1$, in a sort of tree-based search approach. Two remarks about the PDA procedure are just of interest: *i)* one of the main advantages of the proposed PDA procedure lies on the fact that in a typical mobile communications scenario, wherein the users continuously get in and out from the cell, the admission of a new user requires applying (11) for a further additional low-complexity step, i.,e., minimizing $\delta\Lambda_{N_u}(\tilde{\mathbf{C}})$ over all the not-yet-allocated signatures; *ii)* as the simulation results will point out, the joint application of the PDA method with the MUPD compensation scheme enables a considerable performance boost at the price of only a slight increase in complexity.

4.   Simulation results

In this section we will verify the effectiveness of both the MUDP scheme and PDA procedure. We will assume as performance metric the Total Degradation (TD), evaluated at a specified value of BER as the sum (in decibels) of the amplifier output BO and the loss due to the residual nonlinearity incurred in the energy-per-bit-to-noise-spectral-density ratio E_b/N_0 after the MUDP insertion [8]. Since BER performance depends on the different users, for each configuration we will consider the mean values of TD averaged over the active users, along with the TD $\pm\Delta$TD values, $\pm\Delta$TD being the relevant standard deviation. The numerical results obtained by computer simulations and presented in the following refer to the case of a solid state power amplifier (SSPA) at the transmitter complying with the definitions (3) with $q = 10$, 16-QAM as modulation alphabet, RRC pulse shaping with roll-off $\alpha = 0.125$, $N = 256$ subcarriers, WH codes with length $N = 256$ for synchronous orthogonal spreading allocated to the active users according to the RA or PDA methods, and the presence of AWGN transmission channel.

Figures 4 and 5 plot the averaged TD metric as a function of the output BO at the reference BER= 10^{-4}, for a system with $N_u = 32$ active users, and featuring the RA and PDA methods. The curves present the TD (solid marks) along with the TD $\pm \Delta$TD values (empty marks) in the uncompensated case (triangles), for the memoryless configuration P_0, P_0, P_0 (squares) and that with memory labeled as P_{16}, P_{16}, P_{16} (circles). The above results suggest the following remarks: i) the P_0, P_0, P_0 MUPD without memory outperforms the uncompensated system yielding a TD gain around 1 dB. However, whereas the RA method gets the minimum TD at BO = 8 dB for the uncompensated case and at BO = 6.8 dB for the P_0, P_0, P_0 configuration, for the PDA scheme performance gets better. Indeed, the uncompensated case achieves the minimum TD at BO = 6.2 dB and the P_0, P_0, P_0 MUPD at BO = 4.7 dB. ii) The configuration P_{16}, P_{16}, P_{16} enables a further TD improvement over P_0, P_0, P_0. Indeed, such a configuration attains the minimum TD at BO = 5.7 dB (RA) and at BO = 4.2 dB (PDA). iii) The dispersion curves TD $\pm \Delta$TD (dashed lines lines with empty marks) around TD for the RA case indicates that the gain of the MUDP depends on the user, and ultimately on its allocated signature, while the PDA scheme enables performance with reduced dispersion over the active users. iv) Considering that the higher complexity the better results, it can be shown that a reasonable tradeoff between implementation complexity and efficiency in the mitigation of nonlinear channel is given by the configurations (not shown here due to space limitations) P_{16}, P_{16}, P_0 and P_{16}, P_0, P_0 whose TD is 5.09 dB and 4.96 dB, respectively, both at BO = 4.4 dB (PDA).

5. Concluding remarks

In this paper we have illustrated an efficient MUDP procedure to be applied at the transmitter side in the forward link of a multiple access MC-CDMA system for the mitigation of the nonlinear distortions induced by the HPA. Several practical adjustments have been proposed to make the iterative scheme as more reliable as possible while keeping the implementation complexity at an affordable level. In addition, a simple method based on a proper allocation of the users' code signatures has been addressed to reduce the PAPR of the transmitted signal

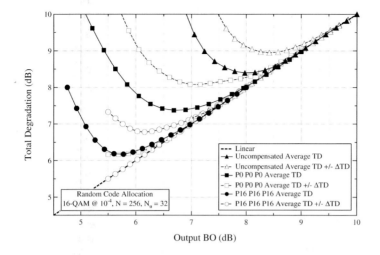

Figure 4. Total degradation for RA and $N_u = 32$ users.

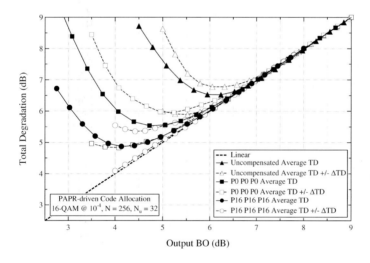

Figure 5. Total degradation for PDA and $N_u = 32$ users.

and, accordingly, to improve the MUDP performance. Simulation results demonstrated that the proposed schemes prove to be very efficient thus achieving a remarkable improvement over a memoryless conventional scheme.

Acknowledgments

This work was carried out in the framework of the project "PRIMO (Piattaforme Riconfigurabili per Interoperabilità in Mobilità)", RBNE01

8RFY, FIRB 2001, funded by MIUR (Italian Ministry for Education, University and Research).

References

[1] J. A. Bingham, "Multicarrier Modulation for Data Transmission: An Idea Whose Time Has Come," *IEEE Commun. Mag.*, vol. 28, n. 5, pp. 5-14, May 1990.

[2] H. Sari, G. Karam and I. Jeanclaude, "Transmission Techniques for Digital Terrestrial TV Broadcasting," *IEEE Commun. Mag.*, vol. 33, n. 2, pp. 100-109, Feb. 1995.

[3] N. Yee, J. P. Linnarz and G. Fettweis, "Multi-Carrier CDMA in Indoor Wireless Radio Systems," *IEEE PIMRC 1993*, Sept. 1993.

[4] K. Fazel and S. Kaiser, "Analysis of Nonlinear Distortions on MC-CDMA," *IEEE ICC 1998*, June 1998.

[5] V. Lottici, A. N. D'Andrea and R. Reggiannini, "Nonlinear Predistortion of OFDM Signals over Frequency-Selective Fading Channels," *IEEE Trans. Commun.*, vol. 49, no. 5, pp. 837-843, May 2001.

[6] M. Chrysochoos and Junghwan Kim, "Performance Analysis of an MC-CDMA Broadcasting System under High Power Amplifier Non-Linearities, Part I: System Proposal," *IEEE Trans. Broadcasting*, vol. 46, pp. 256-262, Dec. 2000.

[7] J. Jeon, Y. Shin and S. Im, "A Data Predistortion Technique for the Compensation of Nonlinear Distortion in MC-CDMA Systems," *IEEE SPAWC 1999*, May 1999.

[8] G. Karam and H. Sari, "Analysis of Predistortion, Equalization, and ISI Cancellation Techniques in Digi-tal Radio Systems with Nonlinear Transmit Amplifiers," *IEEE Trans. Commun.*, vol. 37, pp. 1245-1253, Dec. 1989.

[9] A. A. M. Saleh and J. Salz, "Adaptive Linearization of Power Amplifiers in Digital Radio Systems," *Bell Syst. Tech. Journal*, vol. 62, pp. 1019-1033, April 1983.

[10] D. Dardari, V. Tralli, A. Vaccari, "A theoretical Characterization of Nonlinear Distortion Effects in OFDM Systems" *IEEE Transactions On Communications*, Vol.48. No. 10, October 2000, pp. 1755-1764.

[11] F. Giannetti, V. Lottici and I. Stupia, "Iterative Multi-User Predistortion for MC-CDMA Communications," *FIRB Project Tech. Rep., Dept. Inf. Eng., University of Pisa*, June 2005.

[12] N. Hathi, M. Rodrigues, I. Darwazeh and J. OReilly, "Analysis of the Influence of Walsh-Hadamard Code Allocation Strategies on the Performance of Multi-Carrier CDMA Systems in the Presence of HPA Non-Linearities," *IEEE PIMRC 2002*, pp. 1305-1309, Sept. 2002.

SIDELOBE SUPPRESSION IN OFDM SYSTEMS

Ivan Cosovic
German Aerospace Center (DLR), Inst. of Communications and Navigation
Oberpfaffenhofen, 82234 Wessling, Germany
ivan.cosovic@dlr.de

Vijayasarathi Janardhanam
Munich University of Technology (TUM), 80333 Munich, Germany
vijayasarathi.j@mytum.de

Abstract In this contribution, we develop a method for reducing out-of-band emission caused by high sidelobes in OFDM systems. The method is termed multiple-choice sequences (MCS) and operates in the frequency domain of an OFDM system. The principle of MCS is to map the original transmission sequence onto a set of sequences and to choose, from this set, a sequence with the lowest power in sidelobes for the actual transmission. To enable successful signal detection, de-mapping of the received sequence onto the original sequence is required at the receiver. Hence, an index which uniquely identifies the selected sequence is signalled from the transmitter to receiver. From this generalized framework we derive several practical MCS algorithms. Simulation results show that the MCS method achieves a considerable sidelobe suppression which justifies the introduced signalling overhead.

1. Introduction

Orthogonal frequency-division multiplexing (OFDM) systems have gained a lot of popularity lately due to their high spectral efficiency and robustness to multi-path environments. OFDM has been chosen for many standards like ADSL, DAB, DVB, IEEE 802.11a [1]. One of the drawbacks of OFDM is the high out-of-band radiation caused by the high sidelobes of the OFDM transmission signal. The high sidelobes are particularly a critical issue in OFDM based overlay systems in which a broadband OFDM system is overlaid on top of existing narrowband systems [2]. As illustrated in Fig. 1, an overlay system exploits the unused parts of the spectrum assigned to the existing legacy systems, thus increasing the spectral efficiency. As this concept requires successful co-existence between the legacy system and the OFDM based overlay system, a crucial task in designing such an overlay system is the avoidance of interference towards the legacy system. Therefore, the reduction of out-of-band radiation becomes an essential topic, especially for the design of OFDM based overlay systems.

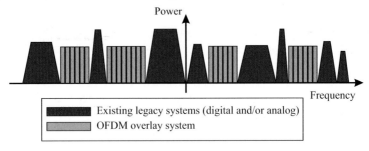

Figure 1. OFDM overlay concept - exploiting the frequency gaps in an existing frequency bandwidth.

The topic of sidelobe suppression in OFDM systems has not been extensively investigated so far. In [3], a multiplication of each OFDM symbol with a windowing function in time domain and insertion of empty guard bands are investigated. In [4] [5], insertion of a few dummy subcarriers at the edges of the used bands which are determined such that the sidelobes of the original OFDM signal are suppressed is presented. In [6], a technique in which the subcarriers are weighted so that the sidelobes of the transmission signal are minimized according to an optimization algorithm is proposed.

In this paper, a different method to significantly suppress the OFDM sidelobes is introduced. This technique, referred to as *multiple-choice sequences (MCS)*, performs mapping of the original transmission sequence onto a set of sequences. From this set, a sequence which offers maximum reduction of out-of-band radiation is chosen for the actual transmission. To enable successful signal detection, de-mapping of the received sequence onto the original sequence is required at the receiver. To this purpose, an index which uniquely identifies the selected sequence in the set of several MCS has to be signalled from the transmitter to receiver. This results in a slightly reduced data throughput. However, numerical results show that this moderate loss in throughput is justified by the significant sidelobe suppression achieved with this technique.

The paper is structured as follows. In Section 2 the signal model is introduced. The principle of MCS method is described and several MCS algorithms are proposed and analyzed in Section 3. The proposed MCS algorithms are compared by numerical simulations in Section 4. Finally, in Section 5 conclusions are drawn.

2. OFDM Signal Model

As illustrated in Fig.1, a real OFDM based overlay system might consist of several continuous transmission sub-bands in-between the legacy systems. The proposed algorithm can be applied to the OFDM transmission signal by considering all the sub-bands jointly or by considering each of the sub-bands separately. As we concentrate on the principle of MCS in this contribution, a simplified problem with a single continuous OFDM transmission band is considered in the following.

An OFDM system with a total number of N subcarriers is considered. The block diagram of the OFDM transmitter is illustrated in Fig. 2. The input bits are symbol-mapped and N complex-valued data symbols d_n, $n = 1, 2, \ldots, N$, are generated. These symbols are serial-to-parallel (S/P) converted resulting in an N-element data symbol array $\mathbf{d} = (d_1, d_2, \ldots, d_N)^{\mathrm{T}}$, where $(.)^{\mathrm{T}}$ denotes transposition. The array $\mathbf{d}$ is fed into the MCS sidelobe suppression unit which outputs the selected MCS, denoted with $\bar{\mathbf{d}} = (\bar{d}_1, \bar{d}_2, \ldots, \bar{d}_N)^{\mathrm{T}}$, and the index of the chosen MCS, denoted with Q. The MCS algorithms that determine $\bar{\mathbf{d}}$ and Q are described in the next section. Finally, the selected MCS sequence $\bar{\mathbf{d}}$ is modulated onto the N subcarriers using the inverse discrete Fourier transform (IDFT). After that, parallel-to-serial (P/S) conversion is performed and a guard interval that exceeds the delay spread of the multipath channel is added as cyclic prefix. In addition, the index of the selected MCS sequence Q is coded in bits and transmitted over the corresponding signaling channel. Note that in the following, for simplicity, we assume that the cyclic prefix is considerably shorter than the useful part of an OFDM symbol.

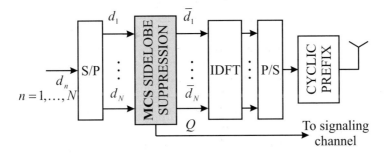

Figure 2. Block diagram of the OFDM transmitter with MCS sidelobe suppression.

3. Sidelobe Suppression by Multiple-Choice Sequences (MCS)

The Principle of MCS

The principle of MCS is illustrated in Fig. 3. A set of sequences $\mathbf{d}^{(p)} = (d_1^{(p)}, d_2^{(p)}, \ldots, d_N^{(p)})^{\mathrm{T}}$, $p = 1, 2, \ldots, P$, is produced from the sequence $\mathbf{d}$. For each sequence $\mathbf{d}^{(p)}$ the average sidelobe power, denoted with $A^{(p)}$, $p = 1, 2, \ldots, P$, is calculated. To determine $A^{(p)}$, a certain frequency range spanning several OFDM sidelobes, called optimization range, is considered using discrete frequency samples. Recalling that the spectrum of an individual subcarrier equals a si-function $\mathrm{si}(x) = \sin(x)/x$, $A^{(p)}$ is given by

$$A^{(p)} = \frac{1}{K} \sum_{k=1}^{K} \left| \sum_{n=1}^{N} d_n^{(p)} \mathrm{si}\left(\pi(y_k - x_n)\right) \right|^2, \quad p = 1, 2, \ldots, P, \qquad (1)$$

where x_n, $n = 1, 2, \ldots, N$, are the normalized subcarrier frequencies and y_k, $k = 1, 2, \ldots, K$, are normalized frequency samples within the optimization range. The index Q of the sequence with maximum sidelobe suppression is given by

$$Q = \arg\min_{p} A^{(p)}, \quad p = 1, 2, \ldots, P. \qquad (2)$$

Thus, the sequence $\bar{\mathbf{d}} = \mathbf{d}^{(Q)}$ is chosen for transmission and output from the MCS unit.

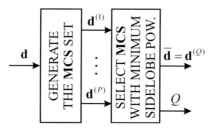

Figure 3. Block diagram of the MCS sidelobe suppression unit.

To enable successful data detection, the received sequence has to be de-mapped onto the original sequence at the receiver. The MCS set is constructed such that the knowledge about the index Q of the selected sequence is sufficient to perform this de-mapping. Thus, the index Q is coded in bits, passed from the MCS unit to the signalling channel, and sent to the receiver. For example, assuming an OFDM system with N subcarriers modulated with M-ary phase-shift-keying (M-PSK) or M-ary quadrature amplitude modulation (M-QAM) symbols, the overhead needed for the signalling information is

$$\lceil \log_2(P) \rceil / \bigl(\log_2(M) \cdot N + \lceil \log_2(P) \rceil \bigr), \tag{3}$$

which is negligible for large N and/or M. In (3), $\lceil x \rceil$ denotes the smallest integer greater than or equal to x.

At the receiver, an estimate $\tilde{\mathbf{d}}^{(Q)}$ of the transmitted sequence $\mathbf{d}^{(Q)}$ is obtained which is transformed into an estimate $\tilde{\mathbf{d}}$ of the original sequence $\mathbf{d}$ using the signalling information. Note that the signalling information is the index Q which indicates that the sequence $\mathbf{d}^{(Q)}$ out of the MCS set has been chosen for transmission.

In the following several computationally effective, but yet efficient algorithms to generate MCS sets are proposed and analyzed. The proposed methods do not degrade the bit-error rate performance at the receiver and require only a slightly increased signalling overhead.

Symbol Constellation Approach

This algorithm generates the set of MCS such that the elements $d_n^{(p)}$, $n = 1, 2, \ldots, N$, of $\mathbf{d}^{(p)}$ belong to the same symbol constellation as the elements of $\mathbf{d}$. With this approach the fact that different symbol sequences have sidelobes with different powers is exploited.

Assume that the symbol constellation consists of M points that are numbered as $0, 1, \ldots, M-1$. To each symbol d_n, $n = 1, 2, \ldots, N$, an index $i_n \in \{0, 1, \ldots, M-1\}$ is assigned which corresponds to the number of the respective constellation point. Then, the index $i_n^{(p)}$ that corresponds to the MCS symbol $d_n^{(p)}$, $n = 1, 2, \ldots, N$, $p = 1, 2, \ldots, P$, is given by

$$i_n^{(p)} = \bigl((i_n + r_n^{(p)}) \bmod M \bigr), \quad n = 1, 2, \ldots, N, \ p = 1, 2, \ldots, P. \tag{4}$$

In (4), $r_n^{(p)}$ is an integer randomly chosen from the set $r_n^{(p)} \in \{0, 1, \ldots, M-1\}$. After determining P index vectors $\mathbf{i}^{(p)} = (i_1^{(p)}, i_2^{(p)}, \ldots, i_N^{(p)})^\mathrm{T}$ the MCS vectors $\mathbf{d}^{(p)}$, $p = 1, 2, \ldots, P$, are obtained by taking the data symbols from the symbol constellation according to the vectors $\mathbf{i}^{(p)}$. We assume that the same random seed for generating $r_n^{(p)}$, $n = 1, 2, \ldots, N$, $p = 1, 2, \ldots, P$, is used at both transmitter and receiver. Hence, the transformation of the received sequence back to the original sequence can be easily performed by exploiting the transmitted signalling information.

Let p_α be the probability that a sequence at the input of the MCS unit has an average power in the optimization range above a certain threshold α. With the symbol constellation approach the P generated MCS sequences belong to the same symbol constellation as the sequence input to the MCS unit. Therefore, the corresponding probability for each of the P generated sequences is also p_α, whereas for the output MCS sequence this probability is

$$\bar{p}_\alpha = (p_\alpha)^P, \tag{5}$$

i.e., the probability is reduced from p_α to $(p_\alpha)^P$, proving the benefits of the proposed approach.

Interleaving Approach

The interleaving approach produces P MCS sequences by permutating the input sequence in a pseudorandom order. As a result, the resulting MCS symbols equal

$$d_n^{(p)} = d_{\Pi_n^{(p)}}, \quad n = 1, 2, \ldots, N, \quad p = 1, 2, \ldots, P, \tag{6}$$

where $\Pi_n^{(p)}$ are permutation indices stored at both transmitter and receiver. The permutation indices $\Pi_n^{(p)}$ take values from the set $\Pi_n^{(p)} \in \{0, 1, \ldots, N-1\}$ such that $\Pi_n^{(p)} \neq \Pi_m^{(p)}$ if $n \neq m$.

Similar to the symbol constellation approach, the MCS symbols $d_n^{(p)}$ produced by the interleaving approach stay in the same symbol constellation as the original symbols d_n. However, unlike the symbol constellation approach, the number of different MCS $\mathbf{d}^{(p)}$ possible with the interleaving approach decreases when the original sequence $\mathbf{d}$ contains reoccurring data symbols. For example, if $\mathbf{d} = (1, 1, \ldots, 1)^T$ the interleaving approach always produces $\mathbf{d}^{(p)} = (1, 1, \ldots, 1)^T$, $p = 1, 2, \ldots, P$, irrespective of the selected permutation indices. As a consequence, the probability $\tilde{p}_\alpha$ that an output MCS sequence has an average power in the optimization range above a certain threshold α satisfies the condition

$$(p_\alpha)^P \leq \tilde{p}_\alpha \leq p_\alpha. \tag{7}$$

Note that the equality in (7) is valid only if $P = 1$.

Phase Approach

In this approach the MCS symbols are obtained by applying random phase shifts to the original symbols. Hence, the resulting MCS symbols are formed as

$$d_n^{(p)} = d_n \exp(j\varphi_n^{(p)}), \quad n = 1, 2, \ldots, N, \quad p = 1, 2, \ldots, P, \tag{8}$$

where the phase shifts $\varphi_n^{(p)}$ lie in the interval $[0, 2\pi)$ and are generated as

$$\varphi_n^{(p)} = 2\pi \left(\frac{\bar{r}_n^{(p)}}{M}\right). \tag{9}$$

In (9), $\overline{M}$ is a constant integer and $\bar{r}_n^{(p)}$ is an integer randomly chosen from the set $\bar{r}_n^{(p)} \in \{0, 1, \ldots, \overline{M}-1\}$. Thus, $\varphi_n^{(p)}$ can take one of the $\overline{M}$ discrete phase values. Again, the same random seeds are used at the transmitter and receiver. Note that assuming a BPSK system and $\overline{M} = 2$, this approach becomes equivalent to the corresponding symbol constellation approach.

In the phase approach, the resulting MCS symbols do not necessarily belong to the same symbol constellation as the original symbols. Hence, a property similar to those described in (5) and (7) cannot be easily derived except for some special cases, e.g., $\overline{M} = 2$.

4. Simulation Results

In this section, several numerical results are given that illustrate the effectiveness of the proposed MCS methods.

BPSK modulation is applied and no channel coding is considered. The number of used subcarriers is set to $N = 12$. The optimization range consists of 16 sidelobes at each side of the spectrum and starts from the first sidelobe outside the OFDM transmission bandwidth. Different MCS methods are considered assuming different sizes of the MCS set P.

In Fig. 4, the normalized power spectrum of the OFDM signals averaged over all possible symbol vectors, i.e., 2^N symbol vectors, prior and after the MCS unit are compared. The symbol constellation approach is applied and the size of the MCS set is fixed to $P = 4$. The benefits of the MCS technique are clearly visible. In comparison to OFDM without MCS the sidelobes are suppressed by around 6.1 dB on average. In addition, from (3) it follows that these results are related to a reduction in system throughput of 14% for the chosen system parameters. This signalling overhead reduces if more subcarriers and/or higher modulation schemes are applied.

In Fig. 5, the sidelobe suppression averaged over all possible symbol vectors for different sizes P of the MCS set and different MCS methods is given. To calculate the average sidelobe suppression, standard OFDM without MCS block is taken as a reference. It can be seen that the symbol constellation approach outperforms the other techniques. In particular, the interleaving approach is outperformed as it offers less degrees of freedom in construction of the MCS set than the symbol constellation approach. The performance of the phase approach depends on the number of possible random phases $\overline{M}$. To obtain these simulation results $\overline{M}$ has been set to $\overline{M} = 64$. As already noted, setting $\overline{M} = 2$ would lead to the same sidelobe suppression results as obtainable by the symbol constellation approach. As expected, in all considered MCS approaches,

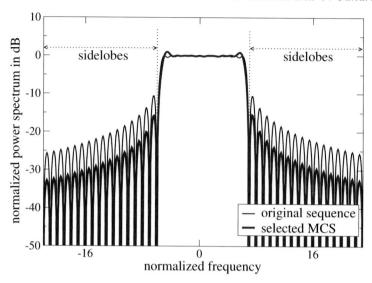

Figure 4. OFDM spectrum of the original transmission sequence and of the transmission sequence after the MCS unit averaged over all possible data sequences; symbol constellation approach; BPSK, $N = 12$, $P = 4$.

an increase in size of the MCS set improves sidelobe suppression, but simultaneously leads to a further increase in signalling overhead. As a consequence, there is a trade-off between the additional sidelobe suppression obtained by enlarging the set size P and the increased signalling overhead. Setting $P = 2$, 4, or 8 seems to be a good compromise. A further increase of P appears to be unjustified as it leads to a relatively high signalling overhead with only moderate further improvement in sidelobe suppression.

The probability that average power in the considered sidelobes of the chosen MCS exceeds the threshold α is presented in Fig. 6. Simulation results are given for $P = 4$ and $P = 16$ assuming different MCS algorithms. As reference, corresponding probability for standard OFDM without MCS block is given. As it can be seen, the symbol constellation approach with $P = 16$ performs better than other considered alternatives. Moreover, for $P = 4$, there is almost no difference in performance between the symbol constellation and interleaving approach, whereas the phase approach performs considerably worse. Again, for the phase approach $\overline{M}$ has been set to $\overline{M} = 64$. Finally, we note that the presented numerical results agree with the analytical results given in (5) and (7).

Note that the MCS technique can be easily combined with other sidelobe suppression methods, e.g., methods from [3]- [6]. However, due to the space limitation of this paper we skip details of such analysis.

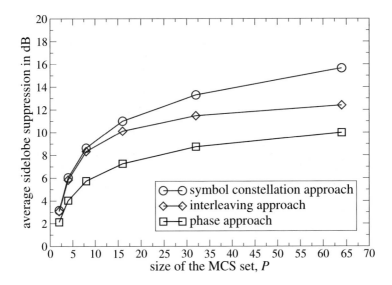

Figure 5. Average sidelobe suppression for different sizes of the MCS set P and for different MCS methods; BPSK, $N = 12$.

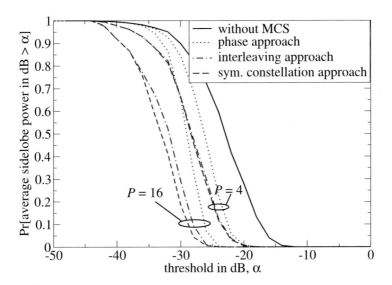

Figure 6. Probability that average power in the considered sidelobes of the chosen MCS exceeds the threshold α; BPSK, $N = 12$.

5. Conclusions

In this paper, we have introduced a new technique, termed multiple-choice sequences (MCS), to suppress sidelobes of OFDM transmission signals. The MCS technique can be used to improve the spectral efficiency of OFDM based transmission systems and/or to reduce interference of OFDM based overlay systems towards the legacy systems sharing the same frequency band. The proposed sidelobe suppression scheme is capable of easily reducing the sidelobes of OFDM transmission signals by several dB. The price to pay for this achievement is a moderate reduction in system throughput, since the transmission of additional signalling information is required.

Acknowledgment

This work was supported by the *Broadband VHF Aeronautical Communications System Based on MC-CDMA (B-VHF)* project [7] which is funded by the European Commission within the 6th Framework Programme.

References

[1] K. Fazel and S. Kaiser. *Multi-Carrier and Spread Spectrum Systems*. John Wiley & Sons, 2003.

[2] T. Weiss and F. Jondral. Spectrum pooling - an innovative strategy for the enhancement of spectrum efficiency. In *IEEE Communications Magazine, Radio Communications Supplement*, pages S8–S14, Mar. 2004.

[3] T. Weiss, J. Hillenbrand, A. Krohn, and F. Jondral. Mutual interference in OFDM-based spectrum pooling systems. In *Proceedings IEEE Vehicular Technology Conference (VTC'04, Spring)*, May 2004.

[4] J. Bingham. RFI suppression in multicarrier transmission systems. In *Proceedings IEEE Global Telecommunications Conference (GLOBECOM'96)*, Nov. 1996.

[5] S. Brandes, I. Cosovic, and M. Schnell. Sidelobe supression in OFDM systems by insertion of cancellation carriers. In *Proceedings IEEE Vehicular Technology Conference (VTC'05 Fall)*, Sept. 2005.

[6] I. Cosovic, S. Brandes, and M. Schnell. A technique for sidelobe suppression in OFDM systems. In *Proceedings IEEE Global Telecommunications Conference (GLOBECOM'05)*, Nov. 2005.

[7] http://www.b-vhf.org.

BLIND PHASE NOISE ESTIMATION IN OFDM SYSTEMS BY SEQUENTIAL MONTE CARLO METHOD

Erdal Panayırcı
Bilkent University
Department of Electrical and Electronics Engineering
Bilkent 06800, Ankara, Turkey

Hakan A. Çırpan
Istanbul University
Department of Electrical-Electronics Engineering
Avcılar 34850, Istanbul, Turkey

Marc Moeneclaey and Nele Noels
Ghent University
TELIN/DIGCOM Department
B9000 Gent, Belgium

Abstract   In this paper, based on a sequential Monte Carlo method, a computationally efficient algorithm is presented for estimating the residual phase noise, blindly, generated at the output the phase tracking loop employed in OFDM systems. The basic idea is to treat the transmitted symbols as "missing data" and draw samples sequentially of them based on the observed signal samples up to time t. This way, the Bayesian estimates of the phase noise is obtained through these samples, sequentially drawn, together with their importance weights. The proposed receiver structure is seen to be ideally suited for high-speed parallel implementation using VLSI technology.

1. Introduction

One of the main drawbacks of OFDM systems is the phase noise (PN) caused by the oscillator instabilities [1]. Unfortunately, due to the PN, the most valuable feature namely orthogonality between the carriers, is destroyed resulting in a significant degradation of the performance of OFDM systems [1]. Random PN causes two effects on OFDM systems, rotating each symbol by a random phase that is referred to as the common phase error (CPE) and producing intercarrier interference (ICI) term that adds to the channel noise due to the lost of orthogonality between subcarriers [2]. Several methods have been proposed in the literature for the estimation and compensation of the PN in OFDM systems [3, 4]. Most of the approaches however only addresses the estimation of CPE by assuming ICI terms is approximated by a Gaussian distribution and these techniques are implemented after the DFT process at the receiver [4]. The main drawback of these approaches is the data dependent ICI which introduces an additional random noise on top of the additive Gaussian channel noise causes a significant degradation in the CPE estimator performance. In contrast to these approaches, we try to solve PN estimation problem in the time domain before the DFT process at the OFDM receiver. As it will be seen next section this approach will not be faced with ICI effect during the estimation procedure resulting in more accurate random phase estimation. The method proposed is based on the sequential monte Carlo techniques. The basic idea is to treat the transmitted symbols as "missing data" and to sequentially draw samples of them based on the current observation and computing appropriate importance sampling weights. Based on sequentially drawn samples, the Kalman filter is used to estimate the unknown phase from a extended Kalman state-space model of the underlying system. Furthermore, the tracking the time-varying PN and the data detection are naturally integrated. The algorithm is self-adaptive and no training/pilot symbols or decision feedback are needed.

2. System Description

We consider an OFDM system with N subcarriers operating over a frequency selective Rayleigh fading channel. In this paper we assume that the multipath intensity profile has exponential distribution and the delay spread T_d is less than or equal to the guard interval L. With the aid of the discrete time channel model [6], the output of the frequency selective channel can be written as $y_t = \sum_{k=0}^{L} h_k s_{t-k}$ where the $h_k, k = 0, 1, \ldots, L$ denotes the kth tap gain and we assume to have ideal knowledge of these channel tap gains. Also, assuming perfect frequency and timing synchronization, the received signal, r_t, corrupted by the additive Gaussian noise n_t and distorted by the time-varying phase noise θ_t can be expressed as

$$r_t = y_t e^{j\theta_t} + n_t, \quad t = 1, \cdots T_0 \qquad (1)$$

where $s_t = \sum_{n=0}^{N-1} d_n e^{-j\frac{2\pi t n}{N}}$. Here $\{d_n\}$ denotes the independent data symbols transmitted on the nth subcarrier of an OFDM symbol. We assume that d_n's are M-PSK symbols taking values in the set $\{e^{-j\frac{2\pi r}{M}}, r = 0, 1, \cdots, M-1\}$. Hence, s_t is a linear combination of independent, identically distributed random variables. If the number of subcarriers is sufficiently large, s_t can be modelled a a complex Gaussian process whose real and imaginary parts are independent. It has zero mean and variance $\sigma_s^2 = E\{|s_t|^2\} = E_s$, where E_s is the symbol energy per subcarrier. n_t is the complex envelope of the additive white Gaussian noise with variance $\sigma_n^2 = E\{|n_t(k)|^2\}$. θ_t is the sample of the PN process at the output of the free-running local oscillator representing the phase noise. It can be shown that PN can be modelled as a Wiener process defined as

$$\theta_t = \theta_{t-1} + u_t \quad \text{where} \quad \theta_0 \sim \text{uniform}(-\pi, \pi) \qquad (2)$$

where u_t is zero-mean Gaussian random variable with variance $\sigma_u^2 = 2\pi BT_s$ where T_s is the sampling period of the OFDM receiver A/D converter and BT refers to the PN rate, where $T = T_s(N+L)$. It is assumed that u_t and n_t are independent of each other. Defining the vectors $\boldsymbol{R}_t = [r_0, r_1, \cdots, r_t]^T$, $\boldsymbol{S}_t = [s_0, s_1, \cdots s_t]^T$, $\boldsymbol{s}_t = [s_t, s_{t-1}, \cdots s_{t-L}]^T$, and $\boldsymbol{h}_t = [h_0, h_1, \cdots, h_L]^T$, combining (1), (2) and taking into account the structure of s_t, we obtain the following dynamic state-space representation of the communication system,

$$\theta_t = \theta_{t-1} + u_t, \qquad \boldsymbol{s}_t = \boldsymbol{F}\boldsymbol{s}_{t-1} + \boldsymbol{v}_t, \qquad r_t = \boldsymbol{h}^T \boldsymbol{s}_t e^{i\theta_t} + n_t \qquad (3)$$

where

$$\boldsymbol{F} = \begin{bmatrix} 0 & 0 & \cdots & 0 \\ 0 & 1 & \cdots & 0 \\ . & . & \cdots & . \\ 0 & 0 & \cdots & 1 \end{bmatrix} \qquad (4)$$

is a $(L+1) \times (L+1)$ shifting matrix, and $\boldsymbol{v}_t = [s_t, 0, \cdots, 0]$ is a $(L+1) \times 1$ perturbation vector that contains the new symbol s_t.

Since we are interested in estimating the the phase noise θ_t blindly at time t based on the observation $\boldsymbol{R}_t$, the Bayes solution requires the posterior distribution

$$p(\theta_t | \boldsymbol{R}_t) = \int p(\theta_t | \boldsymbol{R}_t, \boldsymbol{S}_t) p(\boldsymbol{S}_t | \boldsymbol{R}_t) d\boldsymbol{S}_t. \qquad (5)$$

Note that with a given $\boldsymbol{S}_t$, the nonlinear (Kalman filter) model (3) can be converted into a linear model by linearizing the observation equation (1) as follows [7]:

$$\theta_t = \theta_{t-1} + u_t, \quad \text{and} \quad r_t = \boldsymbol{h}^T \boldsymbol{s}_t (V_t \theta_t + Q_t) + n_t \qquad (6)$$

where $V_t = je^{j\hat{\theta}_{t|t-1}}$ and $Q_t = (1 - j\hat{\theta}_{t|t-1})e^{j\hat{\theta}_{t|t-1}}$. $\hat{\theta}_{t|t-1}$ denotes the estimator of θ_t based on the observations $\boldsymbol{R}_{t-1} = (r_0, r_1 \cdots, r_{t-1})$. Then the state-space model (3) becomes a linear Gaussian system. Hence, $p(\theta_t|\boldsymbol{S}_t, \boldsymbol{R}_t) \sim N(\mu_{\theta_t}(\boldsymbol{S}_t), \sigma^2_{\theta_t}(\boldsymbol{S}_t))$, where the mean $\mu_{\theta_t}(\boldsymbol{S}_t)$ and the variance $\sigma^2_{\theta_t}(\boldsymbol{S}_t)$ can be obtained as follows. Denoting $\mu_{\theta_t}(\boldsymbol{S}_t) \stackrel{\triangle}{=} \hat{\theta}_{t|t}$, and $\sigma^2_{\theta_t}(\boldsymbol{S}_t) \stackrel{\triangle}{=} M_{t|t}$.

$\hat{\theta}_{t|t}$ and $M_{t|t}$ can be calculated recursively by using the Extended Kalman Technique [[7], page 449-452] with the given $\boldsymbol{S}_t$ as:

$$\hat{\theta}_{t|t} = \hat{\theta}_{t|t-1} + K_t(r_t - \boldsymbol{h}^T \boldsymbol{s}_t e^{j\hat{\theta}_{t|t-1}}) \qquad (7)$$
$$M_{t|t} = (1 - K_t V_t M_{t|t-1})$$

where

$$K_t = \frac{M_{t|t-1} V_t^*}{(M_{t|t-1} + \sigma^2_n)}, \quad \hat{\theta}_{t|t-1} = \hat{\theta}_{t-1|t-1}, \quad M_{t|t-1} = M_{t-1|t-1} + \sigma^2_u. \qquad (8)$$

2.1 SMC for Blind Phase Noise Estimation

We can now make timely estimates of θ_t based on the currently available observation $\boldsymbol{R}_t$, up to time t, blindly, as follows. With the Bayes theorem, we realize that the optimal solution to this problem is

$$\hat{\theta}_t = E\{\theta_t|\boldsymbol{R}_t\} = \int_{\boldsymbol{S}_t} \underbrace{\left[\int_{\theta_t} \theta_t p(\theta_t|\boldsymbol{S}_t, \boldsymbol{R}_t) d\theta_t\right]}_{\mu_{\theta_t}(\boldsymbol{S}_t)} p(\boldsymbol{S}_t|\boldsymbol{R}_t), d\boldsymbol{S}_t. \qquad (9)$$

In most cases, an exact evaluations of the expectation (9) is analytically intractable. Sequential Monte Carlo technique can provide us an alternative way for the required computation. Specifically, following the notation adopted in [8], if we can draw m independent random samples $\{\boldsymbol{S}_t^{(j)}\}_{j=1}^m$ from the distribution $p(\boldsymbol{S}_t|\boldsymbol{R}_t)$, then we can approximate the quantity of interest $E\{\theta_t|\boldsymbol{R}_t\}$ in (9) by $E\{\theta t|\boldsymbol{R}_t\} \cong \frac{1}{m}\sum_{j=1}^m \mu_{\theta_t}(\boldsymbol{S}_t^{(j)})$. But, usually drawing samples from $p(\boldsymbol{S}_t|\boldsymbol{R}_t)$ directly is usually difficult. Instead, sample generation from some *trial distribution* may be easier. In this case, the idea of *importance sampling* can be used [8]. By associating the weight $w_t^{(j)} = \frac{p(\boldsymbol{S}_t^{(j)}|\boldsymbol{R}_t)}{q(\boldsymbol{S}_t^{(j)}|\boldsymbol{R}_t)}$ to the samples, the quantity of interest, $E\{\theta_t|\boldsymbol{S}_t\}$ can be approximated as follows:

$$E\{\theta_t|\boldsymbol{R}_t\} \cong \frac{1}{W_t}\sum_{j=1}^m \mu_t(\boldsymbol{S}_t^{(j)}) w_t^{(j)} \qquad (10)$$

with $W_t = \sum w_t^{(j)}$. The pair $(S_t^{(j)}, w_t^{(j)}), j = 1, 2, \ldots, m$ is called a properly weighted sample with respect to distribution $p(S_t|R_t)$.

Specifically, it was shown in [8] that a suitable choice for the trial distribution is of the form $q(s_t|R_t, S_{t-1}^{(j)}) = p(s_t|R_t, S_{t-1}^{(j)})$. For this trial sampling distribution, it is shown in [8] that the importance weight is updated according to

$$w_t^{(j)} = w_{t-1}^{(j)} p(r_t|R_{t-1}, S_{t-1}^{(j)}), \quad t = 1, 2, \cdots \quad (11)$$

The optimal trial distribution in (11) can be computed as follows:

$$p(s_t|R_t, S_{t-1}^{(j)}) = p(r_t|R_{t-1}, S_{t-1}^{(j)}, s_t) P(s_t|R_{t-1}, S_{t-1}^{(j)}) \quad (12)$$

Furthermore, it can be shown from the state and observation equations in (3) that $p(r_t|R_{t-1}, S_{t-1}^{(j)}, s_t) \sim \mathcal{N}(\mu_{r_t}^{(j)}, \sigma_{r_t}^{2(j)})$ with mean and variance given by

$$\mu_{r_t}^{(j)} = E\{r_t|R_{t-1}, S_{t-1}^{(j)}, s_t\} = h^T s_t (V_t \widehat{\theta}_{t|t-1}^{(j)} + Q_t) \quad (13)$$

$$\sigma_{r_t}^{2(j)} = \text{Var}\{r_t|R_{t-1}, S_{t-1}^{(j)}, s_t\} = |h^T s_t|^2 M_{t|t-1}^{(j)} + \sigma_n^2$$

where the quantities $\widehat{\theta}_{t|t-1}^{(j)}$ and $M_{t|t-1}^{(j)}$ in (13) can be computed recursively for the Extended Kalman equations given in (7), (8). Also since s_t is independent of S_{t-1} and R_{t-1}, the second term in (12) can be written as $p(s_t|R_{t-1}, S_{t-1}^{(j)}) = p(s_t)$ where it was pointed out earlier that $p(s_t) \sim \mathcal{N}(0, \sigma_s^2)$.

Note that dependency of the $\sigma_{r_t}^{2(j)}$ in (13) to s_t precludes combining the product of Gaussian densities in (12) into a single Gaussian, hence obtaining a tractable sampling distribution. This problem can be circumvented by approximating the $\sigma_{r_t}^{2(j)}$ as follows. From (3), we can use the approximation $s_t \approx F s_{t-1}$ in (13) to obtain

$$\sigma_{r_t}^{2(j)} \cong |h^T F s_{t-1}^{(j)}|^2 M_{t|t-1}^{(j)} + \sigma_n^2 . \quad (14)$$

Similarly using (9) in (13), the mean $\mu_{r_t}^{(j)}$ can be expressed as $\mu_{r_t}^{(j)} = (h^T F s_{t-1}^{(j)} + h_0 s_t) G_t^{(j)}$ where $G_t^{(j)} \triangleq V_t \widehat{\theta}_{t|t-1}^{(j)} + Q_t$. Then, the true trial sampling distribution $p(s_t|R_t, S_{t-1}^{(j)})$ in (12) can be expressed as follow:

$$p(s_t|R_t, S_{t-1}^{(j)}) \sim \mathcal{N}(\mu_{s_t}^{(j)}, \sigma_{s_t}^{2(j)}) \quad (15)$$

where

$$\mu_{s_t}^{(j)} = \frac{(r_t - \boldsymbol{h}^T \boldsymbol{F} \boldsymbol{s}_{t-1}^{(j)} G_t^{(j)})}{h_0 G_t^{(j)}} \left(\frac{\sigma_{r_t}^{2(j)}}{|h_0 G_t^{(j)}|^2 \sigma_s^2} + 1 \right)^{-1}$$

$$\sigma_{s_t}^{2(j)} = \frac{\sigma_{r_t}^{2(j)} \sigma_s^2}{\sigma_{r_t}^{2(j)} + |h_0 s_t G_t^{(j)}|^2 \sigma_s^2}$$

and $\sigma_{r_t}^{2(j)}$ is defined as (13).

In order to obtain the recursion for the weighting factor $w_t^{(j)}$, the predictive distribution $p(r_t|\boldsymbol{R}_{t-1}, \boldsymbol{S}_{t-1}^{(j)})$ in (12) should be evaluated. It is given by

$$\begin{aligned} p(r_t|\boldsymbol{R}_{t-1}, \boldsymbol{S}_{t-1}^{(j)}) &= \int_{s_t} p(r_t|\boldsymbol{R}_{t-1}, \boldsymbol{S}_{t-1}^{(j)}, s_t) p(s_t|\boldsymbol{R}_{t-1}, \boldsymbol{S}_{t-1}^{(j)}) ds_t \\ &= \int_{s_t} p(r_t|\boldsymbol{R}_{t-1}, \boldsymbol{S}_{t-1}^{(j)}, s_t) p(s_t) ds_t \end{aligned} \quad (16)$$

where (15) holds because s_t is independent of $\boldsymbol{S}_{t-1}$ and $\boldsymbol{R}_{t-1}$. Since the the both terms in the integrand of (15), are Gaussian densities, the product of the Gaussian densities are integrated with respect to s_t is also Gaussian. Therefore the predictive distribution is found to be

$$p(r_t|\boldsymbol{R}_{t-1}, \boldsymbol{S}_{t-1}^{(j)}) \sim \mathcal{N}(M_{r_t}^{(j)}, \Sigma_{r_t}^{2(j)}), \quad (17)$$

where $M_{r_t}^{(j)} = \boldsymbol{h}^T \boldsymbol{F} \boldsymbol{s}_{t-1}^{(j)} G_t^{(j)}$ and $\Sigma_{r_t}^{2(j)} = |h_0 G_t^{(j)}|^2 \sigma_{s_t}^2 + |\boldsymbol{h}^T \boldsymbol{F} \boldsymbol{s}_{t-1}^{(j)}|^2 M_{t|t-1}^{(j)} + \sigma_n^2$. We now summarize the SMC blind data phase noise estimation algorithm in Table I:

TABLE I
SMC ALGORITHM FOR BLIND PHASE NOISE ESTIMATION

Given $\{h_0, h_1, \cdots, h_L\}$

- Initialization:
    - *Initialize the extended Kalman filter:* Choose the initial mean and the variance of the estimated θ_t as

    $$\mu_{\theta_0}^{(j)} = \widehat{\theta}_{0|0}^{(j)} = 0, \quad \sigma_{\theta_0}^{2(j)} = M_{0|0}^{(j)} = \pi^2/12, \quad j = 1, 2, \cdots, m.$$

    - *Initialize the importance weights:* All importance weights are initialized as $w_0^{(j)} = 1, j = 1, 2, \cdots, m.$

For $j = 1, m$
For $t = 1, T_0$
Given $\widehat{\theta}_{t|t-1}, M_{t|t-1}^{(j)}, \boldsymbol{S}_{t-1}^{(j)}$

- Compute $\mu_{r_t}^{(j)}, \sigma_{r_t}^{2(j)}$ from equations (13).
- Compute sampling distribution mean/variance $\mu_{s_t}^{(j)}, \sigma_{s_t}^{2(j)}$ from the equation (15).
- Sample $s_t^{(j)} \sim N(\mu_{s_t}^{(j)}, \sigma_{s_t}^{(j)})$ and Append $s_t^{(j)}$ to $\boldsymbol{S}_{t-1}^{(j)}$ to obtain $\boldsymbol{S}_t^{(j)} = (s_t^{(j)}, \boldsymbol{S}_{t-1}^{(j)})$.
- Compute the importance weights:

$$w_t^{(j)} = w_{t-1}^{(j)} p(r_t | \boldsymbol{R}_{t-1}, \boldsymbol{S}_{t-1}^{(j)}),$$

where $p(r_t|\boldsymbol{R}_{t-1}, \boldsymbol{S}_{t-1}^{(j)})$ is computed from equation (17)
- Compute Kalman updates $\widehat{\theta}_{t|t-1}^{(j)}, M_{t|t-1}^{(j)}$ from equations (7).
- Update the a posteriori mean and variance of the phase noise:
  If the samples drawn up to time t is $\boldsymbol{S}_t$, set

$$\mu_{\theta_t}(\boldsymbol{S}_t^{(j)}) \stackrel{\Delta}{=} \mu_{\theta_t}^{(j)} = \widehat{\theta}_{t|t}^{(j)}$$
$$\sigma_{\theta_t}^{2(j)}(\boldsymbol{S}_t^{(j)}) \stackrel{\Delta}{=} \sigma_{\theta_t}^{2(j)} = M_{t|t}^{(j)} \quad j=1,2,\cdots,m.$$

  and update according to the Kalman equations (8).
- Do the re-sampling as described in [8].
  next j
- Estimate phase noise $\widehat{\theta}_t = \frac{1}{m}\sum_{j=1}^m \mu_{\theta_t}(\boldsymbol{S}_t^{(j)})$

  next t

3. Simulation Results

In this section, we provide some computer simulation examples to demonstrate the performance of the proposed SMC approach for blind phase noise estimation and data detection in OFDM systems. The phase process is modelled by AR process driven by a white Gaussian noise with $\sigma_u^2 = 0.001$. s_t is modelled as a complex Gaussian process which has zero mean and variance $\sigma_s^2 = 1$. In order to demonstrate the performance of the adaptive SMC approach, we present the tracking performance for both phase and symbols at $SNR = 20dB$ in Fig. 1. It is shown through simulations that the performance of the proposed SMC algorithm can track the phase as well as transmitted symbols close to the true values.

4. Conclusions

We have developed a new adaptive Bayesian approach for blind phase noise estimation and data detection for OFDM systems based on sequential monte carlo methodology. The optimal solutions to joint symbol detection and phase noise estimation problem is computationally prohibitive to implement by conventional methods. Thus the proposed sequential approach offers an novel and powerful approach to tackling this problem at a reasonable computational cost.

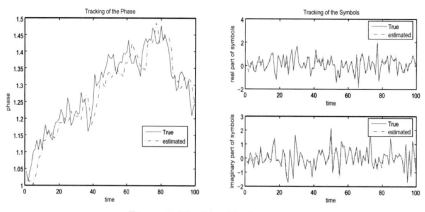

Figure 1. Tracking Performance

5. Acknowledgments

This paper has been produced as part of the NEWCOM Network of Excellence, a project funded by the European Commission's 6th Framework Programme. This work was also supported in part by the Research Fund of the University of Istanbul. Project number: UDP-599/28072005, 220/29042004.

References

[1] M.V.B.T. Pollet and M. Moeneclaey, "BER sensitivity of OFDM systems to carrier frequency offset and wiener phase noise," In *IEICE Transactions on Communications*, vol. 43, Feb/Mar/Apr 1995, pp. 192-193.

[2] H. Meyr, M. Moeneclaey and S.A. Fechtel, *Digital Communications Receivers*, John Wiley, New York, 1998.

[3] S. Wu and Y. Bar-Ness, "A Phase Noise Suppression Algorithm for OFDM-Based WLANs," *IEEE Communications Letters* vol. 44, no. 3, May 1998.

[4] D. Petrovic, W. Rave, and G. Fetweis, "Phase Noise Suppression in OFDM Including Intercarrier Interference" In Proc. Intl. OFDM Workshop, 2003.

[5] M. Pitt and N. Shephard, "Filtering via simulation: auxiliry particle filter", *J. Amer. Statist. Soc. B*, no. 63, 2001.

[6] J.J. van de Beek, O. Edfors, M. Sandell, S.K. Wilson, and P.O. Börjesson, "On channel estimation in OFDM systems," In *Proc. 45th IEEE Vehicular Technology Conf.(VTC'96)*, Atlanta, GA, April 1996.

[7] S.M. Kay, "Fundamentals of Statistical Signal Processing: Estimation Theory," *Prentice Hall* 1993.

[8] Z. Yang and X. Wang, "A sequential Monte Carlo blind receiver for OFDM systems in frequency-selective fading channels", *IEEE Trans. Signal Proc.*,vol. 50, no. 2, pp. 271-280, Feb. 2003.

SENSITIVITY COMPARISON OF MULTI-CARRIER AND SPREAD SPECTRUM SYSTEMS TO PHASE NOISE

Christelle Garnier, Matthieu Loosvelt and Yves Delignon
GET/INT/ENIC, IEMN-DHS UMR CNRS 8520 ; ENIC Télécom Lille 1, Cité Scientifique, rue Guglielmo Marconi, 59658 Villeneuve d'Ascq, France

Abstract: This paper compares the sensitivity to phase noise of different transmission and multiple access techniques : DS-CDMA, OFDM-TDMA, MC-CDMA. The comparison is based on the degradation expressed as the additional Eb/N0 required to achieve a BER of 10^{-3} in the presence of phase noise.

Key words: Phase noise, Wiener process, DS-CDMA, OFDM, MC-CDMA.

1. INTRODUCTION

High data rate transmission in indoor environment is an important challenge for the next generation of wireless communication systems. For high data rate links, the use of millimeter waves at 60 GHz is an attractive solution [1] because of the wide available bandwidth (about 5 GHz). However at very high frequencies, one issue is RF component mastery and especially the design of stable low cost local oscillators (LO). Phase noise is then a performance degradation factor whose impact must be taken into account when designing a system for this frequency range.

This paper deals with the sensitivity to phase noise of different candidate techniques for transmission and multiple access. We consider as candidates all the schemes capable of supporting high data rates and standing up to channel multipath effects, including the multi-carrier modulation (OFDM), the spectrum spreading (DS-CDMA) and the most promising combined technique (MC-CDMA) benefiting from the advantages of both techniques [2]. We consider here they are configured as in [3] to combat the 60 GHz channel frequency selectivity. The objective is to establish if one technique inherently offers a better robustness against phase noise.

The performance of these techniques in the presence of phase noise has been separately studied in many papers [4-16]. However, it is difficult to fairly compare their sensitivity because assumptions and performance indicators are quite different from one paper to another. In [4-9], the authors consider a free-running LO (only frequency-locked) and phase noise is modeled by a Wiener process. In other approaches, the LO is assumed to be phase-locked and phase noise is modeled either by a filtered white noise to obtain the power spectrum density (psd) of a typical LO [10-14] or by a random process with a Tychonov distribution [15-16]. Moreover small phase noise variance is often assumed to simplify analysis [4][10-14]. Concerning performance, results are given in various forms. In [4][6][12-14], the exact expression of the signal to noise (plus interference) ratio (SNR) is derived. BER performance is obtained from simulation for a few values of system parameters in [7][12-13][16] and from an analytical expression in [5][9] which assume that the additional interference due to phase noise is Gaussian. Furthermore in [8][10], the authors report the SNR penalty due to phase noise to achieve a specific BER value.

In this paper we use a unified approach to analyze phase noise effects on the three systems. Our approach is based on Tomba's work concerning OFDM [5] and MC-CDMA [9] (Wiener model, no small phase noise assumption) and we extend it to DS-CDMA. However our work differs in the way we do not use the Gaussian approximation to derive the BER because it is not always valid, as shown in [7] for OFDM. Technique comparison is based on the degradation expressed as the additional E_b/N_0 required to achieve a BER of 10^{-3} in the presence of phase noise.

2. TRANSMISSION SYSTEM MODEL

2.1 Phase noise model

In the presence of phase noise, the frequency conversion operation can be modeled, in its complex baseband equivalent form, by the following term :

$$e(t) = e^{j\theta(t)} \tag{1}$$

where $\theta(t)$ is the phase noise. In the case of a free-running LO, $\theta(t)$ is assumed to be a Wiener process with zero mean and variance of $2\pi\beta|t|$ [4]. The signal $e(t)$ delivered by the noisy LO then presents a Lorentzian shaped psd whose the two-sided -3dB linewidth is represented by the parameter β. The phase noise level is actually characterized by the normalized parameter $\beta.T_s$ defined as the ratio of the phase noise rate β to the symbol rate $f_s=1/T_s$. This model has the advantage of providing a statistical characterization of phase time-varying fluctuation and of producing a $1/f^2$ type noise power behavior that agrees with experimental measurements carried on RF oscillators [10-11].

2.2 DS-CDMA, OFDM, MC-CDMA system model

Figure 1 shows the block diagram of the transmission system in its complex baseband equivalent form. It corresponds to the most generic system : MC-CDMA which includes the OFDM modulation and the DS-CDMA spreading. In order to focus on phase noise effects, we consider a downlink multi-access environment in an AWGN channel and no other impairment factor (no selectivity, perfect synchronization). Data symbols are QPSK modulated, as described in [3], and are assumed to form an i.i.d. zero mean random process with unitary power.

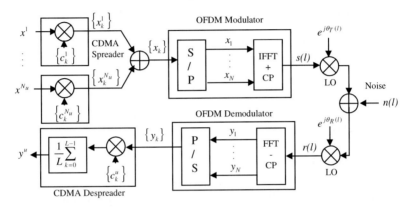

Figure 1. Block diagram of the transmission system

For each user u, the DS-CDMA spreader consists in multiplying each symbol x^u of duration T by the spreading code $\{c_k^u, k=0...L-1\}$ assigned to the user and made of L chips of duration $T_c=T/L$. The samples of the spread sequence are written as:

$$x_k^u = x^u \, c_k^u \, , \, k = 0...L-1 \qquad (2)$$

In the downlink, transmission towards the different users is synchronous, Walsh codes are then selected for their orthogonality property. At the reception, the DS-CDMA despreader of user u performs the correlation between the received sequence y_k and the user code. The signal at the output of the correlator is:

$$y^u = \frac{1}{L}\sum_{k=0}^{L-1} c_k^u \, y_k \qquad (3)$$

The OFDM modulator consists in transmitting symbols x_k of duration T_x in parallel on N orthogonal subcarriers using an IFFT. The resulting OFDM symbol of duration $T_s = NT_x$ is then extended with a cyclic prefix (CP) of N_{CP} samples, larger than the channel maximum excess delay, to prevent interference between adjacent OFDM symbols. The samples of the complete OFDM symbol are given by:

$$s(l) = \frac{1}{\sqrt{N}} \sum_{k=0}^{N-1} x_k \, e^{j2\pi \frac{kl}{N}}, \quad l = -N_{CP}...N-1 \tag{4}$$

Dual operations are implemented at the demodulator : CP removing, FFT and one tap equalization per carrier. In an AWGN channel, the last operation is useless and the signal obtained from the received sequence $r(l)$ on the kth subcarrier is :

$$y_k = \frac{1}{\sqrt{N}} \sum_{l=0}^{N-1} r(l) e^{-j2\pi \frac{lk}{N}} \tag{5}$$

In the case of MC-CDMA, L is chosen equal to N for simplicity. But in practice, it can be interesting to select different values to adjust the signal to the channel. The DS-CDMA system corresponds to $N = 1$ and the OFDM system to $L = 1$. Unlike DS-CDMA and MC-CDMA, OFDM is a transmission scheme which must be associated with a multiple access technique. TDMA is considered in this study, and since multiple access interference (MAI) can be avoided in such a system, we can consider $N_u=1$ and omit the index u in the OFDM case.

3. PHASE NOISE EFFECTS ON TRANSMISSION

The input signal $r(l)$ at the receiver of user u is the signal $s(l)$ transmitted towards the N_u active users affected by the phase noise $\theta_T(l)$ introduced at the radio access point, the additive noise $n(l)$ and the phase noise $\theta_R(l)$ at the user terminal :

$$r(l) = \left(s(l)e^{j\theta_T(l)} + n(l)\right) e^{j\theta_R(l)} = s(l)e^{j\theta(l)} + n(l)e^{j\theta_R(l)} \tag{6}$$

$\theta_T(l)$ and $\theta_R(l)$ are independent so we can define $\theta(l)$ a Wiener process accounting for phase noise generated at both transmitter and receiver. By replacing $s(l)$ by its expression and applying the operations (3) or/and (5), we can show that the decision variable for every transmission scheme, in the presence of phase noise, is written as[1] :

$$y_k^u = x_k^u \alpha_0 + x_{I,k} + \eta_k \tag{7}$$

This expression shows three distortion terms :
- α_0 is a complex multiplicative factor coming from an average of phase fluctuations over the symbol duration as shown by its expression[2]:

$$\alpha_0 = \frac{1}{N} \sum_{k=0}^{N-1} e^{j\theta(k)} \tag{8}$$

---

[1] The index u is omitted in the OFDM-TDMA case and the index k is omitted for DS-CDMA and MC-CDMA systems.

[2] The paramater N is replaced by L in the case of DS-CDMA.

Sensitivity Comparison of Multi-Carrier and Spread Spectrum Systems 495

This term causes both an attenuation and a phase rotation of the useful symbol. This effect is always independent of the index k. Because it identically affects the subcarriers of an OFDM symbol, it is called the common error phase (CEP). It can be removed by a pilot based CPE correction scheme [17].

- $x_{I,k}$ is an interference term coming from an alteration of code or/and subcarrier orthogonality by phase noise, as pointed out by expressions (9), (11) and (13).
- η_k is the additive thermal noise. According to [18], phase rotation does not modify the distribution of a complex Gaussian process, linear filtering either. The additive noise then remains Gaussian. Using independence between thermal and phase noises, calculations show that the variance is also unchanged.

To quantify the impact of phase noise, the statistical properties of α_0 and $x_{I,k}$ are investigated for every transmission technique.

3.1 Statistical properties of the multiplicative term α_0

The first and second moments of the real part $\alpha_{0\Re}$ and the imaginary part $\alpha_{0\Im}$ of α_0 can be calculated straight away using the statistical properties of a Wiener process. The expressions show that the resulting attenuation and phase rotation remain moderate for $\beta.T_s \leq 0.01$. Under the assumption of small phase noise as in [4] and [10-14], the following approximation can be used: $e^{j\theta(l)} \approx 1 + j\theta(l)$. Then $\alpha_{0\Re}$ is a constant and $\alpha_{0\Im}$ is a Gaussian random variable (rv). We have checked this approximation holds only for $\beta.T_s \leq 0.01$. To find the statistical characterization of α_0 in a more general case, we refer to several works in the context of optical communication systems [17-19]. They present different solutions to approximate the probability density functions (pdf) of $\alpha_{0\Re}$ and $\alpha_{0\Im}$: Monte-Carlo simulation in [17] and derivation of a recursive formula for the moments of $\alpha_{0\Re}$ and $\alpha_{0\Im}$ in [18] and [19]. The pdf is then be obtained from the moments using an orthogonal polynomial series expansion or a maximum entropy approach. The recursion derived in [19, expression (8)] has the advantage of being very simple.

3.2 Statistical properties of the interference term x_I

3.2.1 DS-CDMA system

The interference term corresponds to the contribution of the other active users and results in MAI as shown by its expression:

$$x_I = x_{MAI} = \sum_{m=0, m \neq u}^{N_u - 1} x^m \xi_{u,m} \quad \text{with} \quad \xi_{u,m} = \frac{1}{L} \sum_{l=0}^{L-1} c_l^u c_l^m e^{j\theta(l)} \quad (9)$$

The complex weighing coefficient $\xi_{u,m}$ can be considered as a "modified" (by phase noise) cross-correlation function between the codes of users u and m. Its value depends on code properties. For Walsh codes, the cross-correlation is zero without phase noise, but in the presence of phase noise code orthogonality is removed. Using the statistical properties of symbols x^m, the mean and the variance of MAI are:

$$E[x_{MAI}] = 0 \quad \text{and} \quad \sigma^2_{MAI} = E\left[|x_{MAI}|^2\right] = \sum_{m=0, m \neq u}^{N_u - 1} E\left[|\xi_{u,m}|^2\right] \tag{10}$$

In Appendix A, making use of the statistical characteristics of a Wiener process and Walsh codes, we obtain (A.5) expressing the power of $\xi_{u,m}$ versus the LO normalized linewidth $\beta.T_s$ and the code length L. The MAI power then is directly derived and represented in Figure 2 for Walsh codes of length $L=128$ and for different values of N_u. We can observe a linear relationship, in a logarithmic scale, between the MAI power and $\beta.T_s$. So the interference level, which is zero in the absence of phase noise, rapidly increases when $\beta.T_s$ gets higher because of the loss of code orthogonality. The curves also show that it is in direct proportion to the system load.

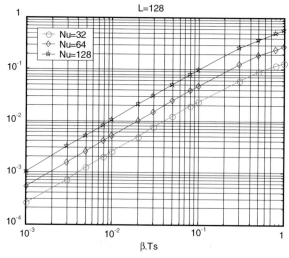

Figure 2. Power of the MAI term versus $\beta.T_s$

On the basis of the central limit theorem, it seems justified to model the distribution of the MAI by a Gaussian law. However by simulations and using a Kolmogorov-Smirnov test, we have shown that the Gaussian approximation is valid only for a full large enough load: $N_u = L \geq 32$. For a smaller load, the MAI can not be modeled by a complex Gaussian rv.

3.2.2 OFDM-TDMA system

The interference term corresponds to the contribution of all the subcarriers adjacent to the detected subcarrier k and results in ICI as shown by its expression:

$$x_{I,k} = x_{ICI,k} = \sum_{r=0, r \neq k}^{N-1} x_r \alpha_{r-k} \quad \text{with} \quad \alpha_{r-k} = \frac{1}{N} \sum_{l=0}^{N-1} e^{j\theta(l)} e^{j2\pi \frac{(r-k)l}{N}} \quad (11)$$

The complex weighing coefficients α_{r-k} are related to the phase noise affecting the LO during the OFDM symbol duration and the spacing $(r-k)$ between the detected subcarrier k and the other carriers of the OFDM system. Making use of the statistical properties of data symbols x_r, the ICI mean and variance are:

$$E\left[x_{ICI,k}\right] = 0 \quad \text{and} \quad \sigma_{ICI,k}^2 = E\left[\left|x_{ICI,k}\right|^2\right] = \sum_{r=0, r \neq k}^{N-1} E\left[\left|\alpha_{r-k}\right|^2\right] \quad (12)$$

In [5][9], the power of α_{r-k} is obtained from the Fokker-Planck equation verified by the joint pdf of real and imaginary parts and Laplace transformation. More simply, it can be derived from the statistical characteristics of a Wiener process and the obtained expression is similar to [6, expression (9)]. Figure 3 shows the power of the coefficients for the six nearest subcarriers of the detected subcarrier ($\alpha_1, \alpha_2 ... \alpha_6$) versus $\beta.T_s$. Because of the linear relationship, in a logarithmic scale, between the power of α_{r-k} and $\beta.T_s$, it rapidly increases when $\beta.T_s$ gets higher. The most significant contribution comes from the nearest carriers, especially from the first and second adjacent carriers. It decreases more gradually for more distant carriers.

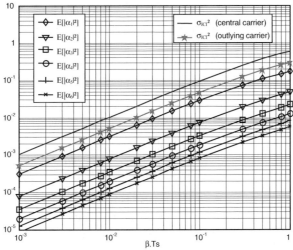

Figure 3. Power of α_{r-k} for the 6 closest subcarriers and ICI power for central and outlying subcarriers versus $\beta.T_s$

The figure also reports the ICI power for central and outlying carriers. The ICI level affecting a carrier results from the summation of the power of adjacent carrier weighing coefficients α_{r-k}. As expected, the central carrier corresponds to the worst case because of the greater number of neighbours.

We have also checked the validity of the Gaussian approximation to model ICI distribution. Simulations and the Kolmogorov-Smirnov test show this approximation does not hold, as observed in [7], even for a large value of the number of subcarriers N. It can be explained by the correlation between the weighing coefficients α_{r-k} involved in the ICI term. In the DS-CDMA system, the weighing coefficients $\zeta_{u,m}$ are less correlated because the correlation due to phase noise is altered by the random user codes.

3.2.3   MC-CDMA system

The interference term is caused by ICI which affects an OFDM system in the presence of phase noise and thus generates MAI. It is written as:

$$x_I = x_{ICI-MAI} = \sum_{m=0}^{N_u-1} x^m \zeta_{u,m} \quad \text{with} \quad \zeta_{u,m} = \frac{1}{N} \sum_{k=0}^{N-1} \sum_{r=0, r \neq k}^{N-1} c_k^u c_r^m \alpha_{r-k} \quad (13)$$

According to the statistical properties of data symbols x^m, the mean and the variance of the ICI-MAI term are expressed as:

$$E[x_{ICI-MAI}] = 0 \quad \text{and} \quad \sigma^2_{ICI-MAI} = E\left[|x_{ICI-MAI}|^2\right] = \sum_{m=0}^{N_u-1} E\left[|\zeta_{u,m}|^2\right] \quad (14)$$

From expression (A.2) concerning Walsh code statistical properties, the power of the complex weighing coefficients $\zeta_{u,m}$ can be written as:

$$E\left[|\zeta_{u,m}|^2\right] = \frac{1}{N^2} \sum_{k=0}^{N-1} \sum_{r=0, r \neq k}^{N-1} E\left[|\alpha_{r-k}|^2\right] = \frac{1}{N^2} \sum_{k=0}^{N-1} \sigma^2_{ICI,k} \quad (15)$$

where $\sigma^2_{ICI,k}$ is the power of the ICI affecting the kth subcarrier in an OFDM system. The power of the ICI-MAI then results from an average of the power of the ICI affecting each carrier. Figure 4 reports the power of the ICI-MAI versus $\beta.T_s$ for $N=L=128$ and for different values of N_u. The power is also a logarithmically linear function of $\beta.T_s$, the interference level then quickly increases when $\beta.T_s$ gets larger. As expected from (14) and (15), it is also in proportion to the system load.

$x_{ICI-MAI}$ is similar to a combination of the ICI and MAI terms which appear in the OFDM and DS-CDMA systems in the presence of phase noise. As for the MAI term in DS-CDMA, simulations and the Kolmogorov-Smirnov test show that the Gaussian approximation is valid for a full large enough load: $N_u=L \geq 32$. The random user codes alter the correlation between the weighing coefficients $\zeta_{u,m}$. For a smaller load, the ICI-MAI can not be modeled by a complex Gaussian rv anymore.

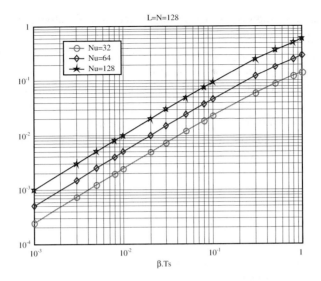

Figure 4. Power of the ICI-MAI term versus $\beta.T_s$

4. COMPARISON OF SENSITIVITY TO PHASE NOISE

The power of the different distortion terms involved in expression (7) is known. The exact expression of SINR in the presence of phase noise can then be derived as in [4][6][12-14] and SNR degradation, obtained by comparison to the SNR in an AWGN channel, can be used to characterize phase noise sensitivity. However, this indicator is not the most appropriate because the relationship with BER is not direct when the Gaussian approximation is not valid. For this reason, BER is more convenient to quantify the performance in the presence of phase noise and to compare the sensitivity of the three different techniques, an interesting indicator is the degradation due to phase noise in term of additional E_b/N_0 required to achieve a BER of 10^{-3}.

BER is obtained by simulation as a function of E_b/N_0 for a wide range of values of the normalized linewidth $\beta.T_s$ of the noisy LO. For every transmission scheme, BER degrades as soon as $\beta.T_s \geq 0.01$ with the apparition of an irreducible error floor. We can precise this error floor gets larger than 10^{-3} for $\beta.T_s \geq 0.03$. By comparison to the reference performance in an AWGN channel, we then derive the degradation due to phase noise in term of additional E_b/N_0 required to achieve a BER of 10^{-3}. Figure 5 shows, for the different techniques, this degradation versus $\beta.T_s$. For the DS-CDMA and MC-CDMA schemes with Walsh codes of length $L=128$, several load cases are considered. For the OFDM-TDMA system with a number of subcarriers $N=128$, the case corresponding to a central subcarrier has been selected because it is more representative.

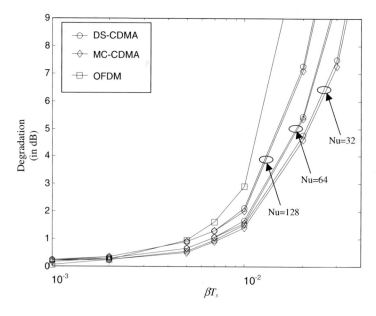

Figure 5. E_b/N_0 degradation at BER=10^{-3} versus $\beta.T_s$ for different values of N_u

First, it can be observed that, for all the systems, the normalized linewidth $\beta.T_s$ of the noisy oscillator must stay moderate. Otherwise the existence of a high ICI or MAI level combined with a phase rotation and an attenuation of the transmitted symbols results in a very strong degradation, which requires the use of phase correction algorithms at the receiver.

For acceptable values of $\beta.T_s$, the curves show that :
- For spectrum spreading based techniques, the performance degradation is more severe for a larger load of the system. It is due to the MAI whose power linearly increases with the number of active users. Moreover DS-CDMA and MC-CDMA sensitivity to phase noise is similar because both systems are approximately affected by the same amount of interference.
- By comparison, the OFDM-TDMA system suffers from a stronger performance degradation than DS-CDMA and MC-CDMA, even in the worst case of a full load. It can be surprising because the power of the ICI affecting most OFDM subcarriers is close to the MAI power in full loaded DS-CDMA and MC-CDMA systems. But it can be explained by the difference between ICI and MAI distribution. Unlike DS-CDMA and MC-CDMA with a full load, the interference in the OFDM-TDMA system is not Gaussian. The ICI distribution is actually Gaussian shaped but with broader tails, which makes BER performance worse than with the Gaussian distribution.

Sensitivity Comparison of Multi-Carrier and Spread Spectrum Systems

As a result, we can conclude that CDMA based systems using orthogonal Walsh codes inherently offer the greatest robustness to phase noise, whatever the system load. In further work, the sensitivity to phase noise will be studied in a more realistic context including the effects of the 60 GHz multipath channel and some phase noise correction methods.

APPENDIX A - Power of $\xi_{u,m}$

From expression (9) and independence of user codes and phase noise, the power of the weighing coefficient $\xi_{u,m}$ is written as :

$$E\left[|\xi_{u,m}|^2\right] = \frac{1}{L^2} \sum_{l=0}^{L-1} \sum_{l'=0}^{L-1} E\left[c_l^u c_{l'}^m c_l^{u*} c_{l'}^{m*}\right] E\left[e^{j(\theta(l)-\theta(l'))}\right] \quad (A.1)$$

In the case of Walsh codes, the first term of (A.1) is given by :

$$E\left[c_l^u c_{l'}^m c_k^{u*} c_{k'}^{m*}\right] = \delta_{l,k} \delta_{l',k'} - \frac{1}{L-1} \delta_{l,l'} \delta_{k,k'} \quad (A.2)$$

The second term of (A.1) corresponds to the characteristic function of $\theta(l)-\theta(l')$. $\theta(l)$ is a zero mean Wiener process with variance $2\pi\beta T_s |l|/L$. $\theta(l)-\theta(l')$ is then a zero mean Gaussian rv with variance $2\pi\beta T_s |l-l'|/L$ and its characteristic function is :

$$\psi_{\theta(l)-\theta(l')}(1) = E\left[e^{j(\theta(l)-\theta(l'))}\right] = e^{-\pi\beta T_s \frac{|l-l'|}{L}} \quad (A.3)$$

By respectively substituting the two terms by (A.2) and (A.3), (A.1) becomes :

$$E\left[|\xi_{u,m}|^2\right] = \frac{1}{L} - \frac{2}{L^2(L-1)} \sum_{l=0}^{L-1} \sum_{l'=l+1}^{L-1} e^{-\pi\beta T_s \frac{|l-l'|}{L}} \quad (A.4)$$

After several variable substitutions and identification of the obtained series, the power of $\xi_{u,m}$ is given by :

$$E\left[|\xi_{u,m}|^2\right] = \frac{1}{L} - \frac{2}{L^2(L-1)} \left(\frac{(L-1)e^{\frac{\pi\beta T_s}{L}} + e^{-\frac{\pi\beta T_s(L-1)}{L}} - L}{\left(1-e^{\frac{\pi\beta T_s}{L}}\right)^2} \right) \quad (A.5)$$

REFERENCES

[1] P. Smulders, "Exploiting the 60 GHz band for local wireless multimedia access : prospects and future direction", *IEEE Communications Magazine*, pp. 140-147, Jan. 2002.
[2] S. Hara, R. Prasad, "Overview of multi-carrier CDMA", *IEEE Communications Magazine*, pp. 126-133, Dec. 1997.
[3] L. Clavier, C. Garnier, "Multiple access techniques for the indoor 60 GHz channel", *COST 273, TD(02)042*, May 2002.
[4] T. Pollet, M. Van Bladel, M. Moeneclaey, "BER sensitivity of OFDM systems to carrier frequency offset and Wiener phase noise", *IEEE Trans. on Communications*, vol. 43, n°2/3/4, pp. 191-193, Feb./Mar./Avr. 1995.
[5] L. Tomba, "On the effect of Wiener phase noise in OFDM systems", *IEEE Trans. on Vehicular Technology*, vol. 46, n°5, pp. 580-583, May 1998.
[6] S. Wu, Y. Bar-Ness, "OFDM systems in the presence of phase noise : consequences and solutions", *IEEE Trans. on Communications*, vol. 52, n°11, pp. 1988-1996, Nov. 2004.
[7] D. Petrovic, W. Rave, G. Fettweis, "Intercarrier interference due to phase noise in OFDM - Estimation and suppression", *Proc. of VTC'04 Fall*, pp. 2191-2195, Sept. 2004.
[8] R. Corvaja, S. Pupolin, "Performance of CDMA with differential detection in the presence of phase noise and multiuser interference", *IEEE Trans. on Communications*, vol. 52, n°3, pp. 498-506, March 2004.
[9] L. Tomba, W.A. Krzymien, "Sensitivity of the MC-CDMA Access Scheme to Carrier Phase Noise and Frequency Offset", *IEEE Trans. on Vehicular Technology*, vol. 48, n°5, pp. 1657-1665, Sept. 1999.
[10] P. Robertson, S. Kaiser, "Analysis of the effect of phase noise in orthogonal frequency division multiplex (OFDM) systems", *Proc. of ICC'95*, pp. 1652-1657, Jun. 1995.
[11] C. Muschallik, "Influence of RF oscillators on an OFDM signal", *IEEE Trans. on Consumer Electronics*, vol. 41, n°3, pp. 592-603, Aug. 1995.
[12] A.G. Armada, "Understanding the effects of phase noise in orthogonal frequency division multiplexing (OFDM)", *IEEE Trans. on Broadcasting*, vol. 47, n°2, pp. 153-159, June 2001.
[13] Z. Jianhua, H. Rohling, Z. Ping, "Analysis of ICI cancellationscheme in OFDM systems with phase noise", *IEEE Trans. on Broadcasting*, vol. 50, n°2, pp. 97-106, June 2004.
[14] H. Steendam, M. Moeneclaey, "The effect of carrier phase jitter on MC-CDMA performance", *IEEE Trans. on Communications*, vol. 47, n°2, pp. 195-198, Feb. 1999.
[15] T. Eng, A. Chockalingam, L.B. Milstein, "Capacities of FDMA/CDMA systems in the presence of phase noise and multipath Rayleigh fading", *IEEE Trans. on Communications*, vol. 46, pp. 997-999, Aug. 1998.
[16] J.Y. Kim, "Performance of a CDMA-based satellite communication system with phase noise", *Proc. of MILCOM'99*, pp. 616-620, Nov. 1999.
[17] D. Petrovic, W. Rave, G. Fettweis, "Common phase error due to phase noise in OFDM - Estimation and suppression", *Proc. of PIMRC'04*, pp. 1901-1905, Sept. 2004.
[18] G.F. Foschini, G. Vannucci, "Characterizing filtered light waves corrupted by phase noise", *IEEE Trans. on Information Theory*, vol. 34, n°6, pp. 1437-1448, Nov. 1988.
[19] G. Pierobon, L. Tomba, "Moment characterization of phase noise in coherent optical systems", *Journal of Lightwave Technology*, vol. 9, n°8, pp. 996-1005, Aug. 1991.
[20] I.T. Monroy, G. Hooghiemstra, "On a recursive formula for the moments of phase noise", *IEEE Trans. on Communications*, vol. 48, n°6, pp. 917-920, June 2000.